VOLUME SEVENTY SEVEN

ADVANCES IN PHARMACOLOGY

VOLUME SEVENTY SEVEN

ADVANCES IN PHARMACOLOGY

Endothelium

Edited by

RAOUF A. KHALIL

Harvard Medical School, Brigham and Women's Hospital, Division of Vascular Surgery, Boston, Massachusetts, United States

Serial Editor

S.J. ENNA

University of Kansas Medical Center, Kansas City, Kansas, United States

Managing Editor

LYNN LECOUNT

University of Kansas Medical Center School of Medicine, Kansas City, Kansas, United States

ELSEVIER

AMSTERDAM • BOSTON • HEIDELBERG • LONDON
NEW YORK • OXFORD • PARIS • SAN DIEGO
SAN FRANCISCO • SINGAPORE • SYDNEY • TOKYO

Academic Press is an imprint of Elsevier

Academic Press is an imprint of Elsevier
50 Hampshire Street, 5th Floor, Cambridge, MA 02139, United States
525 B Street, Suite 1800, San Diego, CA 92101-4495, United States
The Boulevard, Langford Lane, Kidlington, Oxford OX5 1GB, United Kingdom
125 London Wall, London, EC2Y 5AS, United Kingdom

First edition 2016

ISBN: 978-0-12-804396-7
ISSN: 1054-3589

Publisher: Zoe Kruze
Acquisition Editor: Zoe Kruze
Editorial Project Manager: Shellie Bryant
Production Project Manager: Vignesh Tamil
Designer: Greg Harris

Typeset by SPi Global, India

CONTENTS

CONTRIBUTORS

O. Baretella
State Key Laboratory of Pharmaceutical Biotechnology, The University of Hong Kong, Hong Kong S.A.R., China

M. Barton
Molecular Internal Medicine, University of Zürich, Zürich, Switzerland

K. Bian
School of Medicine, George Washington University, Washington, DC, United States

S.L.K. Bowers
Dalton Cardiovascular Research Center, University of Missouri School of Medicine, Columbia, MO, United States

G.E. Davis
Dalton Cardiovascular Research Center, University of Missouri School of Medicine, Columbia, MO, United States

C. de Wit
Institut für Physiologie, Universität zu Lübeck; Deutsches Zentrum für Herz-Kreislauf-Forschung (DZHK) e.V. (German Center for Cardiovascular Research), Partner Site Hamburg/Kiel/Lübeck, Lübeck, Germany

L. Desbiens
Université de Sherbrooke, Sherbrooke, QC, Canada

P. D'Orléans-Juste
Université de Sherbrooke, Sherbrooke, QC, Canada

D.J.R. Fulton
Medical College of Georgia at Augusta University, Augusta, GA, United States

M.S. Goligorsky
New York Medical College at Touro University, Valhalla, NY; Yale University, New Haven, CT, United States

K. Hirschi
New York Medical College at Touro University, Valhalla, NY; Yale University, New Haven, CT, United States

M. Houde
Université de Sherbrooke, Sherbrooke, QC, Canada

J.D. Imig
Medical College of Wisconsin, Milwaukee, WI, United States

R.A. Khalil
Vascular Surgery Research Laboratories, Brigham and Women's Hospital, Harvard Medical School, Boston, MA, United States

R. Köhler
Aragon Institute of Health Sciences & IIS Aragon; Aragon Agency for Research and Development (ARAID), Zaragoza, Spain

M.R. Meyer
Triemli City Hospital, Zürich, Switzerland

F.Z. Mónica
School of Medicine, George Washington University, Washington, DC, United States; State University of Campinas (UNICAMP), Campinas, Brazil

F. Murad
School of Medicine, George Washington University, Washington, DC, United States

P.R. Norden
Dalton Cardiovascular Research Center, University of Missouri School of Medicine, Columbia, MO, United States

A. Oliván-Viguera
Aragon Institute of Health Sciences & IIS Aragon; Aragon Agency for Research and Development (ARAID), Zaragoza, Spain

J.S. Possomato-Vieira
Vascular Surgery Research Laboratories, Brigham and Women's Hospital, Harvard Medical School, Boston, MA, United States

K. Schmidt
Institut für Physiologie, Universität zu Lübeck; Deutsches Zentrum für Herz-Kreislauf-Forschung (DZHK) e.V. (German Center for Cardiovascular Research), Partner Site Hamburg/Kiel/Lübeck, Lübeck, Germany

P.M. Vanhoutte
State Key Laboratory of Pharmaceutical Biotechnology, The University of Hong Kong, Hong Kong S.A.R., China

R. Windler
Institut für Physiologie, Universität zu Lübeck; Deutsches Zentrum für Herz-Kreislauf-Forschung (DZHK) e.V. (German Center for Cardiovascular Research), Partner Site Hamburg/Kiel/Lübeck, Lübeck, Germany

H. Wulff
University of California, Davis, CA, United States

PREFACE

The vascular system plays an important role in the distribution of blood and nutrient supply to different tissues and organs. It comprises a vast and intricate network of large and small, superficial and deep arteries and veins. Structurally, a blood vessel is composed of three layers: the intima and endothelium, the media and vascular smooth muscle, and the adventitia and extracellular matrix. Each of these layers has a specific role in order to maintain the blood vessel diameter and to handle the changes in blood pressure. Interestingly, the blood vessel wall is lined with only one layer of endothelial cells. However, this single layer of cells is capable of releasing several relaxing factors including nitric oxide, prostacyclin, and hyperpolarizing factor that could promote vasodilation and decrease vascular resistance and blood pressure. Endothelial cells also release powerful contracting factors such as endothelin and thromboxane A2 with multiple targets in the blood vessels and other tissues, an important role in the control of the circulation, the hemodynamics, and cardiovascular and renal function. An imbalance between endothelium-derived vasodilators and vasoconstrictors could cause pathological changes that lead to major disorders such as hypertension, coronary heart disease, and diabetes.

This volume of *Advances in Pharmacology* focuses on the vascular endothelium and its role under physiological and pathological conditions. The volume will highlight the current knowledge regarding the various endothelium-derived mediators and how they may change with gender, during vascular development and senescence, and in hypertensive disorders. Renowned scientists and researchers have agreed to share their expertise and advanced knowledge on the endothelium. The first half of the volume will cover important topics regarding the basic function of the endothelium and the roles of nitric oxide, gap junctions, potassium channels, and endothelin. The second half will include topics that discuss the role of the endothelium in vascular development and the changes in endothelial and vascular function with gender and advanced age and in certain hypertensive states such as hypertension in pregnancy and preeclampsia. These important review articles were written by research investigators and clinician–scientists from different parts of the world, thus promoting different viewpoints and highlighting different approaches in the management of cardiovascular disease. Thanks to the good work of the contributing authors, and the careful review of our dedicated reviewers and Editors, the *Advances in Pharmacology* was able to put together these important topics, and present them to our readers in a clear, concise, and informative fashion. I encourage every researcher, clinician, medical, graduate, and undergraduate student with aspiration to work in the vascular field to read this state-of-the-art synopsis on the endothelium.

I would like to take this opportunity to express my deepest gratitude to Dr. S.J. Enna, the Series Editor of *Advances in Pharmacology*, who gave me the opportunity to be the Editor of this special and timely volume. I also would like to thank our outstanding Managing Editor, Ms. Lynn LeCount, and our hard-working Editorial Staff who spared no effort to ensure the highest quality of the articles. I also would like to acknowledge our contributing authors not only for their excellent articles but also for sharing some of the reviewers' duties, and for being very generous with their time and effort in providing other authors with timely review of the articles as well as helpful comments and constructive criticism. I particularly wish to thank our readers for their interest in vascular pharmacology and the endothelium. I encourage all of you to provide feedback and contact me directly if you have any questions, comments, suggestions, criticism, or ideas that could further enhance our knowledge and help us achieve our goals and meet the highest expectations of our readers.

Raouf A. Khalil
Harvard Medical School, Brigham and Women's Hospital,
Boston, MA, United States

CHAPTER ONE

The Endothelium-Dependent Nitric Oxide–cGMP Pathway

F.Z. Mónica*,†, K. Bian*,1, F. Murad*,1
*School of Medicine, George Washington University, Washington, DC, United States
†State University of Campinas (UNICAMP), Campinas, Brazil
[1]Corresponding authors: e-mail address: bcmkxb@gwu.edu; ferid.murad@gmail.com

Contents

Abstract

Nitric oxide (NO)–cyclic 3′-5′ guanosine monophosphate (cGMP) signaling plays a critical role on smooth muscle tone, platelet activity, cardiac contractility, renal function and fluid balance, and cell growth. Studies of the 1990s established endothelium dysfunction as one of the major causes of cardiovascular diseases. Therapeutic strategies that benefit NO bioavailability have been applied in clinical medicine

Advances in Pharmacology, Volume 77
ISSN 1054-3589
http://dx.doi.org/10.1016/bs.apha.2016.05.001

extensively. Basic and clinical studies of cGMP regulation through activation of soluble guanylyl cyclase (sGC) or inhibition of cyclic nucleotide phosphodiesterase type 5 (PDE5) have resulted in effective therapies for pulmonary hypertension, erectile dysfunction, and more recently benign prostatic hyperplasia. This section reviews (1) how endothelial dysfunction and NO deficiency lead to cardiovascular diseases, (2) how soluble cGMP regulation leads to beneficial effects on disorders of the circulation system, and (3) the epigenetic regulation of NO–sGC pathway components in the cardiovascular system. In conclusion, the discovery of the NO–cGMP pathway revolutionized the comprehension of pathophysiological mechanisms involved in cardiovascular and other diseases. However, considering the expression "from bench to bedside" the therapeutic alternatives targeting NO–cGMP did not immediately follow the marked biochemical and pathophysiological revolution. Some therapeutic options have been effective and released on the market for pulmonary hypertension and erectile dysfunction such as inhaled NO, PDE5 inhibitors, and recently sGC stimulators. The therapeutic armamentarium for many other disorders is expected in the near future. There are currently numerous active basic and clinical research programs in universities and industries attempting to develop novel therapies for many diseases and medical applications.

ABBREVIATIONS

ADMA asymmetric dimethylarginine
BH_4 tetrahydrobiopterin
cGMP cyclic 3′-5′ guanosine monophosphate
CHF congestive heart failure
COPD chronic obstructive pulmonary disease
CTEPH chronic thromboembolic pulmonary hypertension
EDCF endothelium-derived contracting factor
EDRF endothelium-derived relaxing factor
FAD flavin adenine nucleotide
FMN flavin mononucleotide
GTP guanosine triphosphate
HAT histone acetyltransferase
HDAC histone deacetylase
L-NMMA N^G-monomethtyl-L-arginine
NO nitric oxide
NOS nitric oxide synthase
$O_2^{\cdot-}$ superoxide anion
$ONOO^-$ peroxynitrite
PAH pulmonary arterial hypertension
PDE phosphodiesterases
PH pulmonary hypertension
sGC soluble guanylyl cyclase

1. INTRODUCTION

In Oct. 1998 the Nobel Prize in Physiology and Medicine was awarded to Robert Furchgott, Louis Ignarro, and Ferid Murad for their discoveries of nitric oxide (NO) as a key molecule in the cardiovascular system. In the 1970s, Ferid Murad's group showed that nitroglycerin and other nitrovasodilators acted by forming nitric oxide and increasing soluble guanylyl cyclase (sGC) activity and cyclic 3′-5′ guanosine monophosphate (cGMP) formation, thus leading to smooth muscle relaxation (Arnold, Mittal, Katsuki, & Murad, 1977; Katsuki, Arnold, & Murad, 1977; Murad, Mittal, Arnold, Katsuki, & Kimura, 1978). In 1980, Furchgott's group demonstrated that the presence of endothelial cells was essential for the relaxation induced by acetylcholine, bradykinin, adenosine triphosphate, and histamine, and these substances released a relaxing factor from endothelial cells, which they named at that time as endothelium-derived relaxing factor (EDRF) (Furchgott & Zawadzki, 1980). In 1986 Murad proposed that EDRF is an "endogenous nitrovasodilator" and that NO explained the effects of EDRF on cyclic GMP formation and vascular relaxation. Later in 1997 Ignarro, Furchgott, and Moncada concluded that EDRF is nitric oxide (Ignarro, Buga, Wood, Byrns, & Chaudhuri, 1987; Palmer, Ferrige, & Moncada, 1987).

Nitric oxide is an important gas and free radical that is ubiquitously synthesized in various cells by three different types of nitric oxide synthases (NOSs), namely endothelial (eNOS, NOS-3), neuronal (nNOS, NOS-1), and inducible (iNOS, NOS-2) (Nakane, Schmidt, & Murad, 1991). Nitric oxide stimulates its intracellular receptor sGC, which converts guanosine triphosphate (GTP) into cGMP. The physiological effects of cGMP are exerted through the activation of cGMP-dependent protein kinases, cyclic nucleotide-gated ion channels, and the activation and/or inhibition of PDEs. In the cardiovascular system, NO–sGC signaling plays an important role for antithrombotic and antiatherogenic agents by inhibiting platelet aggregation (Marcondes et al., 2006), leukocyte adhesion and chemotaxis (Conran et al., 2001), and vascular smooth muscle relaxation (Murad, 1986; Rapoport & Murad, 1983). Alterations on (1) formation and NO bioavailability mainly due to by lower production by NOS and/or greater inactivation and clearance and (2) cGMP levels are some of the major alterations seen in cardiovascular- and vascular-related diseases, such as hypertension, pulmonary hypertension (PH), congestive heart failure (CHF), diabetes, and erectile dysfunction.

This review will be divided into three sections: (1) how endothelial dysfunction and NO deficiency lead to cardiovascular diseases; (2) how sGC regulation leads to beneficial effects on disorders of the circulation system, and (3) the epigenetic regulation of NO–sGC pathway components in the cardiovascular system.

2. PART I: NO–sGC SIGNALING

2.1 Endothelial Dysfunction and NO Deficiency

The endothelium, the thin inner lining of vessels, can be activated by chemical and physical stimuli that lead to the formation and release of EDRF prostacyclin, bradykinin, and contracting (EDCF) factors such as endothelin-1 and angiotensin II (Versari, Daghini, Virdis, Ghiadoni, & Taddei, 2009). Endothelial cells produce a wide range of factors that control cellular adhesion, smooth muscle reactivity, proliferation, and vessel wall inflammation and atherogenesis. The imbalance between vasodilating and vasoconstricting substances leads to endothelium dysfunction, which contributes to several cardiovascular diseases including hypertension, heart failure (HF), PH, and vascular-related diseases such as diabetes and erectile dysfunction (Bian, Doursout, & Murad, 2008; Mónica, Murad, & Bian, 2014; Murad, 2006).

Nitric oxide plays a pivotal role in vascular hemostasis since a defect in NO production or activity has been proposed as a major mechanism that leads to endothelial dysfunction. Besides its vasorelaxing action that opposes the effects of EDCF, NO also inhibits platelet adherence and aggregation (Marcondes et al., 2006), leukocyte adhesion/infiltration (Benjamim et al., 2002), and proliferation of vascular smooth muscle cells (VSMCs) (Wolf, Rasmussen, Sherman, Bundens, & Hye, 1995) and prevents the oxidation of low-density lipoprotein cholesterol (Mehta, Bryant, & Mehta, 1995). Lower NO production and increased NO inactivation are major factors that contribute to the initiation, progression, and clinical phenotype of atherosclerosis (Fig. 1).

2.2 eNOS (NOS-3) Uncoupling

NOSs convert L-arginine into L-citrulline and NO in a reaction dependent on oxygen, heme, tetrahydrobiopterin (BH_4), nicotinamide adenine dinucleotide phosphate (NADPH), calmodulin, flavin adenine nucleotide

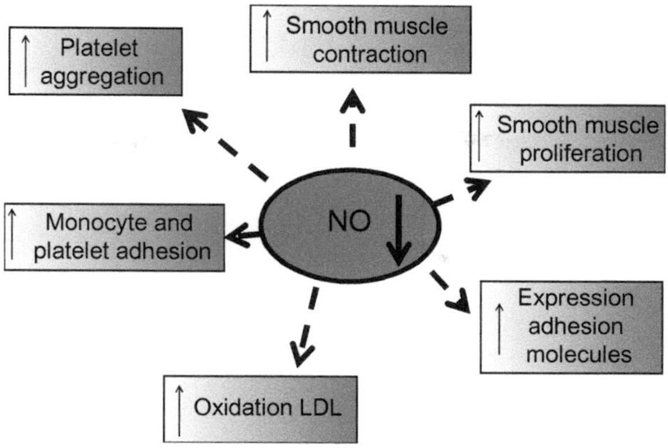

Fig. 1 Nitric oxide (NO) plays a key role in the initiation of endothelial dysfunction since the lower its level, the higher will be platelet and monocyte aggregation and adhesion, smooth muscle contractility, expression of adhesion molecules, and oxidation of low-density protein. These events contribute to the formation of atherosclerotic plaque.

(FAD), and flavin mononucleotide (FMN) (Schmidt, Smith, Nakane, & Murad, 1992). eNOS (NOS-3) is the major isoform expressed in the endothelial layer and is the critical enzyme in the maintenance of vascular homeostasis by NO production.

The functional eNOS (NOS-3) is a dimer that consists of a C-reductase domain (binds NADPH, FAD, and FMN), which is linked to the N-terminal oxygenase domain (binds the prosthetic group, BH_4, molecular oxygen, and the substrate L-arginine). Heme group is essential for the linkage between the reductase and oxygenase domains and allows the NOS dimerization. BH_4, an essential cofactor of eNOS (NOS-3), is necessary for optimal since it facilitates the NADPH electron transferring from the eNOS reductase to the oxygenase domain. There is evidence that eNOS uncoupling is the underlying cause of endothelial dysfunction seen in atherosclerotic lesions (Herranz et al., 2012; Margaritis et al., 2013), hypercholesterolemia (Warnholtz et al., 1999), hypertension (Li et al., 2006), and diabetes (Cassuto et al., 2014). In these situations, NAPDH oxidases are upregulated and the levels of reactive oxygen species as superoxide ($O_2^{\cdot-}$), which reacts with NO to produce peroxynitrite ($ONOO^-$), are high. The cofactor BH_4 is sensitive to oxidation by $ONOO^-$ and lower levels of BH_4 promote superoxide $O_2^{\cdot-}$ rather than NO production by eNOS in a process referred as eNOS uncoupling (Förstermann & Münzel, 2006). L-Arginine deficiency or the presence of NOS inhibitors may lead to eNOS uncoupling (Sydow & Münzel, 2003) (Fig. 2).

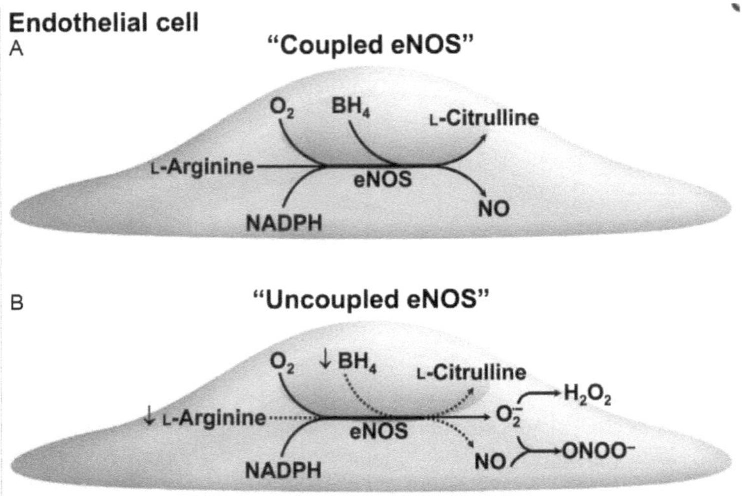

Fig. 2 (A) In the coupled state, eNOS (NOS-3) synthesizes NO in the presence of both L-arginine and tetrahydrobiopterin (BH_4). (B) In the uncoupled state, eNOS synthesizes anion superoxide (O_2^-) in conditions where BH_4 and/or L-arginine are low. *Katusic, Z. S., d'Uscio, L. V., & Nath, K. A. (2009). Vascular protection by tetrahydrobiopterin: Progress and therapeutic prospects.* Trends in Pharmacological Sciences, 30*(1), 48–54. Reproduction authorized with license number 3793391035392.*

Based on the aforementioned, pharmacological interventions to prevent eNOS uncoupling are of great interest. However, substances that increase eNOS levels are only beneficial when maintaining eNOS functionality. Several drugs in clinical use have been pleiotropic actions. Thus, statins, drugs commonly used to lower cholesterol levels, have improved endothelial dysfunction, stabilize atherosclerotic plaque, and inhibit oxidative stress and inflammation in part due to greater eNOS activity (Feron, Dessy, Desager, & Balligand, 2001) and higher BH_4 levels (Hattori, Nakanishi, Akimoto, Yoshida, & Kasai, 2003; Landmesser et al., 2003) in both animals and human. Drugs that act on the renin–angiotensin system as angiotensin II receptor antagonists, for example, are also alternatives since they improved endothelial dysfunction by increasing eNOS expression and prevent uncoupling mainly due lower expression and activity of NAPDH oxidase in an animal model of diabetes type I (Satoh et al., 2008). Other compounds such as eNOS transcription enhancers (AVE 9488 and AVE 3085) (Wohlfart et al., 2008) and *trans*-resveratrol, a polyphenolic phytoalexin found in red grapes, which increases eNOS activity (Mattagajasingh et al., 2007) and upregulates antioxidant enzymes are also options, but their therapeutic potential in cardiovascular diseases needs to be investigated in long-term in vivo studies (Förstermann & Li, 2011). A number of herbal extracts exert significant effects in reversing the inflammation process or are antioxidants restoring endothelium functions (Bian et al., 2008; Shi et al., 2014).

2.3 Endogenous Inhibitors of eNOS (NOS-3)

Asymmetric dimethylarginine (ADMA), N^G-monomethtyl-L-arginine (L-NMMA), and ϖ-N^G,N^G-symmetric dimethylarginine are three methylated arginine compounds identified in eukaryotes. The first two compounds are endogenous inhibitors of NOS, although the former is considered the main physiological inhibitor of NOS as circulating ADMA is present at higher levels than L-NMMA (Leiper & Nandi, 2011). ADMA is produced after posttranslational methylation of arginine residues within proteins by the action of methyltransferase enzymes and is eliminated mainly by renal excretion and metabolism by dimethyl-L-arginine dimethylaminohydrolase enzyme (Ogawa, Kimoto, Watanabe, & Sasaoka, 1987). Higher plasma ADMA concentrations have been found in patients with chronic renal failure (Vallance, Leone, Calver, Collier, & Moncada, 1992), with congenital heart disease with severe pulmonary arterial hypertension (PAH) (Fang et al., 2015), and with preeclampsia (Savvidou et al., 2003) and may contribute to the endothelial dysfunction seem in these dirsorders. Recently, it was shown that the levels of ADMA were greater in $COX2^{-/-}$ mice in comparison with the control mice and the relaxation response to acetylcholine, but not the NO-donor sodium nitroprusside was reduced in the aorta from these animals. The administration of nonsteroidal antiinflammatory drugs (NSAIDs) celecoxib (200 mg, bid) and naproxen (500 mg, bid) for 7 days in healthy volunteers increased the plasmatic levels of ADMA. This study suggests that the cardiovascular side effects caused by NSAIDs could be in part due to higher levels of ADMA that could be used as a biomarker for endothelial dysfunction (Ahmetaj-Shala et al., 2015).

2.4 NO Receptor: sGC

The activation of the sGC by NO, which leads to cGMP accumulation, requires the presence of a heme prosthetic group (Arnold et al., 1977 ; Katsuki et al., 1977 ; Murad et al., 1988). The heme of sGC is a group of five-membered rings containing nitrogen atoms surrounding an iron atom, which can be found in the reduced (Fe^{2+}) or in the oxidized (Fe^{3+}) form. The proposed mechanisms for cGMP-mediated relaxation include (1)

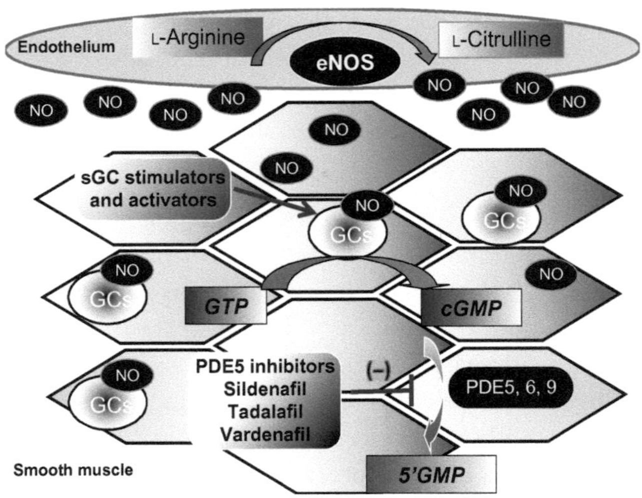

Fig. 3 In the vessels, endothelial nitric oxide synthase (eNOS) (NOS-3) expressed on the endothelium converts L-arginine into L-citrulline and nitric oxide (NO). NO diffuses to the smooth layer to activate sGC, which converts guanosine triphosphate (GTP) into cyclic 3′-5′ guanosine monophosphate (cGMP). The action of cGMP is terminated mainly by the actions of phosphodiesterases type 5, 6, and 9 (PDE5, 6, 9). The cGMP levels are also increased by sGC stimulators, activators, or PDE's inhibitors.

inhibition of inositol-1,4,5-triphosphate generation; (2) enhanced cytosolic Ca^{2+} extrusion; (3) dephosphorylation of myosin light chain kinase; (4) Ca^{2+} influx inhibition; (5) protein kinase activation; (6) stimulation of membrane Ca^{2+} ATPase; and (7) potassium channels opening. Removal of the heme group or its oxidation leads to an NO-insensitive form of the enzyme. Although the oxidized form of sGC becomes insensitive to NO, sGC can be still activated by sGC stimulators (BAY 41–2272, BAY 41–8543, BAY 63–2521) or activators (BAY 58–2667, BAY 60–2770, HMR 1766), which act independently of NO (Evgenov et al., 2006; Stasch et al., 2002). To date, compounds acting in the NO pathway by either by releasing NO or activating the enzyme sGC or inhibiting the enzyme phosphodiesterase-5 (PDE5) constitute the therapeutic options to treating cardiovascular and vascular-related diseases (Fig. 3).

2.5 Phosphodiesterases

The phosphodiesterases (PDEs) are a family of enzymes that hydrolyze 3′, 5′-cyclic nucleotide monophosphates into their respective 5′-monophosphates.

PDE5, PDE6, and PDE9 are selective for cGMP (Fig. 3). The PDE5, initially identified in the lung and subsequently purified and cloned (Francis, Lincoln, & Corbin, 1980), is present in platelet, vas deferens, testis, kidney, prostate, bladder, corpus cavernosum, gastrointestinal epithelial cells, and the Purkinje cells of the cerebellum (Shimizu-Albergine et al., 2003). PDE5 is a homodimer with a regulatory domain in the amino-terminal portion and a catalytic domain in the carboxy-terminal portion. PDE5 also contains two allosteric cGMP-binding sites and at least one phosphorylation site on each subunit. When cGMP binds to the allosteric site of PDE5, it stimulates the catalytic site for cGMP (Corbin, 2004).

3. PART II: NO–sGC PATHWAY AS A THERAPEUTIC TARGET FOR CARDIOVASCULAR DISEASES

3.1 Therapeutic Approach for Pulmonary Hypertension Therapy

PAH is a disorder in which mechanical obstruction leads to a rise in mean pulmonary arterial pressure. The pathogenesis of this disease involves sustained vasoconstriction, thrombosis, hypertrophy, and intimal proliferation. Endothelial dysfunction, inflammation and immune responses, and abnormal extracellular matrix function play a key role in this disease (Guignabert et al., 2015). PAH is a critical source of various substances involved in vascular remodeling, including growth factors (angiotensin II, serotonin, fibroblast growth factor), vasoactive substances (NO, prostacyclin), cytokines (IL-1, IL-6), chemokines, and adipokines (Archer & Rich, 2000; Guignabert et al., 2015; Peach & Loeb, 1987). With respect to endothelial dysfunction, strategies to increase prostacyclin and NO bioavailability or intracellular levels of cGMP are the therapeutic options available.

The World Health Organization divides PH into five groups: (1) PAH that includes idiopathic PAH, familial PAH, and PAH associated with other conditions such as connective tissue disease and congenital heart disease; (2) PH with left HF, where mitral valve disease or long-term high blood pressure can cause left HF and PH; (3) PH associated with lung diseases, such as chronic obstructive pulmonary disease (COPD) and interstitial (IN-ter-STISH-al) lung diseases; (4) PH due to pulmonary embolism in the lungs or blood clotting disorders, and (5) PH caused by other diseases such as blood disorders (polycythemia vera, essential thrombocythemia), systemic disorders (sarcoidosis, vasculitis), metabolic disorders (thyroid and glycogen storage diseases). In all groups, the increased pressure in the pulmonary arteries is 25 mmHg or higher. The pressure in normal pulmonary arteries is 8–20 mmHg at rest.

3.1.1 PAH in Newborn: Inhaled NO

Inhaled NO is used to treat newborns with PAH. Low doses of inhaled NO at 20 ppm for 4 h and then 6 ppm for 20 h resulted in improved oxygenation without systemic effects and decreased systolic blood pressure in nine newborn infants with persistent PH (Kinsella, Neish, Ivy, Shaffer, & Abman, 1993; Kinsella, Neish, Shaffer, & Abman, 1992). The ability of inhaled NO to increase systemic oxygen levels was confirmed in multicenter randomized controlled studies of term and near-term babies with hypoxemia and PAH (Roberts et al., 1997). Inhaled NO has only been formally approved by the Food and Drug Administration (FDA) and the European agency (EMEA) for clinical use in the treatment in term and near-term neonates with hypoxemia and PH.

3.1.2 Adult PAH: PDE5 Inhibitors

In adults with severe PAH, pulmonary vascular resistance was reduced significantly after inhaled NO and after prostacyclin (24 μg/h), whereas systemic vascular resistance was not affected by inhaled NO (Pepke-Zaba, Higenbottam, Dinh-Xuan, Stone, & Wallwork, 1991). In an open, prospective, randomized, controlled trial, the combination of inhaled NO and oxygen improved pulmonary hemodynamics better than inhaled NO alone (Barst et al., 2010). However, since a significant proportion of adults with PAH do not respond to inhaled NO, likely because of sGC impairment or irreversible pulmonary disease such as fibrosis, this therapy is only accepted for the usage of term and near-term infants. Organic nitrates are not effective for PH because their long-term use leads to tolerance and upon discontinuation the development of serious rebound PH. Furthermore, pulmonary smooth muscle is less sensitive to nitrovasodilator-induced relaxation than vascular smooth muscle (Boerrigter & Burnett, 2009).

It is worth noting that in patients with chronic thromboembolic pulmonary hypertension (CTEPH, WHO 4), pulmonary endarterectomy (a surgery to remove old blood clots from the pulmonary arteries) is a therapeutic choice to restore pulmonary hemodynamics. However, in the cases of CTEPH, it is deemed ineffective and those who develop persistent PH after surgery are eligible for using PAH-specific pharmacotherapy such as prostanoids, endothelin receptor antagonists, PDE5 inhibitors, and more recently sGC stimulators. Three completed randomized controlled trials in patients with inoperable CTEPH have used prostanoids (Olschewski et al., 2002), PDE5 inhibitors, sildenafil (Jaïs et al., 2008), and an endothelin-1 receptor antagonist, bosentan (Jaïs et al., 2008). Only bosentan demonstrated a positive therapeutic effect on hemodynamics, although no improvement was observed in exercise capacity. Further trials are needed to ascertain the role of pharmacotherapy on patients with CTEPH. In patients with PH associated with left heart disease (WHO group 2), the role of PAH-specific drugs is controversial and not well studied in large randomized controlled trials. However, recent studies suggest that acute (Alaeddini et al., 2004; Guazzi, Tumminello, Di Marco, & Guazzi, 2004) and chronic (Behling et al., 2008; Lewis et al., 2007) administration of oral sildenafil reduced pulmonary artery pressure and pulmonary vascular resistance without significant changes in systemic arterial pressure. It has been known for many years that PDE inhibitors can augment the accumulation of both cAMP and cGMP in most tissues including pulmonary tissues and blood vessels to increase their effects (Murad, 1986, 1998, 2006; Murad et al., 1978). As for patients with PH-associated lung diseases (WHO 3), there are little evidences showing the efficacy of PDE5 inhibitors. Sildenafil improved pulmonary hemodynamics at rest and during exercise in 20 patients with COPD-associated PH (Gimeno et al., 2010), whereas in another study sildenafil had no effect on stroke volume or exercise capacity in nine patients with COPD-associated PH (Rietema et al., 2008).

An experimental ovine fetal model for perinatal PH of the neonate was characterized by altered pulmonary vasoreactivity and structure. Infusion of PDE5 inhibitors, dipyridamole and zaprinast, lowered pulmonary resistance by 55% and 35%, respectively, in hypertensive animals. Moreover, PDE5 activity was 150% higher in hypertensive fetal lambs in comparison with healthy animals (Hanson et al., 1998). In chronically hypoxic mice with disruption of the gene encoding eNOS, sildenafil reduced pulmonary arterial pressure less than in wild-type controls and failed to decrease right ventricular hypertrophy (Zhao et al., 2001). In the same model where GC-A was blunted, PDE5 inhibition reduced pulmonary arterial pressure, right ventricular hypertrophy, and pulmonary vascular remodeling with less efficacy than wild-type animals (Zhao, Mason, Strange, Walker, & Wilkins, 2003), suggesting that NO or natriuretic peptides also influence the vasodilator response to sildenafil by enhancing cGMP levels. Patients with idiopathic PH ($n=9$, WHO 1), PAH ($n=2$, WHO group 1), or PH due to left heart diseases ($n=2$, WHO 2) were given sildenafil, inhaled NO, or sildenafil

+inhaled NO. The decrease in pulmonary vascular resistance was similar with inhaled NO ($-19 \pm 5\%$) and sildenafil ($-27 \pm 3\%$), whereas sildenafil+inhaled NO ($-32 \pm 5\%$, $P < 0.003$) was more effective than inhaled NO. In addition, sildenafil and sildenafil+inhaled NO increased cardiac index (Michelakis et al., 2002). Ten PAH patients with normal left ventricular function were included in a prospective randomized, placebo-controlled crossover study to evaluate the short-term effect of sildenafil. Patients in the sildenafil group had improvement of the 6-min walk distance from 163.9 to 266.7 m ($P < 0.0005$), a decrease in the Borg dyspnea index (from 5.2 to 3.6, $P < 0.01$), and pulmonary artery systolic pressure (from 80.8 to 55.3, $P < 0.05$) (Bharani, Mathew, Sahu, & Lunia, 2003). In another double-blind, placebo-controlled study, patients with PAH received placebo or sildenafil (20, 40, or 80 mg) for 12 weeks. The 6-min walk increased from baseline in all sildenafil groups. All sildenafil doses reduced the mean pulmonary artery pressure, showing the efficacy of sildenafil in improving exercise capacity and hemodynamics (Galiè et al., 2005). These observations provided the rationale to target PDE5 inhibition for the treatment of PH with or without agents that enhance cGMP formation such as natriuretic peptides and/or nitrovasodilators.

3.1.3 Adult PAH: sGC Modulators

Activating sGC has been approved as a new therapeutic method for PH. sGC activation can be reached by two mechanisms: the sGC stimulators (BAY 41–2272, BAY 63–2521, BAY 41–8543) are dependent on the presence of the reduced (ferrous-Fe^{2+}) prosthetic group of sGC heme and they work synergically with NO, whereas sGC activators (BAY 58–2667, BAY 60–2770, and HMR 1766) effectively activate sGC when it is oxidized (ferric-Fe^{3+}) or in a heme-free state (Fig. 2). PAH is associated with endothelial dysfunction and thus low bioavailability of NO. sGC stimulators or activators overcome such limitations, since these substances activate sGC in a NO-independent manner (Montani et al., 2014). More recently, the FDA approved riociguat (BAY 63–2521), an sGC stimulator for treating PAH and CTEPH. The rationale of using this class of drugs is that it produces vasodilation in conditions of lower NO bioavailability. Riociguat caused concentration-dependent increases of sGC activity in vitro, exhibited a higher degree of sGC specificity, and did not have off-target inhibitory effects on PDEs. Riociguat is the first sGC stimulator approved as a useful addition to the arsenal of pharmacological agents available to treat PAH. Riociguat appeared to be well tolerated, although its main adverse effect, systemic hypotension, may limit the doses used.

In a phase I study of 58 healthy male volunteers, oral riociguat (BAY 63–2521) was well tolerated. In a phase IIa study in 19 patients with diagnosis of PAH (WHO 1), distal chronic thromboembolic (WHO 4) or PH associated with mild to moderate interstitial lung disease (WHO 3) riociguat demonstrated hemodynamic efficacy and favorable tolerability to a greater extent than inhaled NO. After a 2.5 mg dose of riociguat, mean arterial pressure fell by 14% (Grimminger et al., 2009). In patients with CTEPH (WHO 4) and PAH (WHO 1), mean arterial blood pressure decreased 4.5 mmHg from baseline after riociguat treatment (1–2.5 mg three times daily for 3 weeks). Dyspnea and a 6-min walking distances showed a significant improvement (Ghofrani et al., 2010). In a phase III, multicenter, randomized, double-blind, placebo-controlled study, 261 patients with CPTEPH (WHO group 4) or persistent PH after surgery (WHO group 4) patients who received riociguat ($n=173$) had a mean increase of 39 m, whereas placebo ($n=88$) group a mean decrease of 6 m with the 6-min walk distance test. Pulmonary vascular resistance decreased and increased in riociguat and placebo group, respectively. Riociguat also significantly reduced NT-proBNP levels (Ghofrani et al., 2013). In October 2013, riociguat was the first-in-class sGC stimulator approved by FDA for PAH and CTEPH, excluding the prior use of many nitrovasodilators.

3.2 Therapeutic Approach for Heart Disease

3.2.1 HF: PDE5 Inhibitors

Endothelial dysfunction mainly due to a lack of NO-mediated vasodilation contributes to increased vasomotor tone in patients with HF. The 24-h urinary excretion of ^{15}N-nitrate (stable metabolite of NO) after infusion of L-[^{15}N]arginine at rest and during maximal exercise was significantly reduced in patients with CHF, suggesting a decrease on L-arginine-NO metabolic pathway in patients with CHF (Katz et al., 1999). Although the alteration of the NO pathway is believed to play an important role in the overloaded heart and cardiac remodeling in CHF, the use of conventional nitrates is limited, as discussed earlier. In patients with CHF, sildenafil (10 mg) decreased pulmonary artery systolic (−21.6%) and diastolic (−31.8%) pressure and arteriolar resistance (−36.9%), without affecting ejection fraction, cardiac index, wedge pulmonary pressure, and blood capillary volume, suggesting that sildenafil ameliorates pulmonary hemodynamics (Guazzi et al., 2004). In a fixed-dose double-blind, randomized, placebo-controlled, two-way crossover study, the effect of sildenafil (50 mg) was investigated in 23 men with CHF. Sildenafil reduced significantly heart rate, systolic and diastolic blood pressure before and after cardiopulmonary walking, and exercise test (Bocchi et al., 2002). However, a multicenter, double-blind, placebo-controlled, parallel-group, randomized clinical trial of 216 outpatients with HF (ejection fraction $\geq$50%, elevated BNP, and reduced exercise capacity), sildenafil (20 mg 8/8 h for 12 weeks followed by 60 mg 8/8 h for 12 weeks) did not result in significant improvement in exercise capacity or clinical status (Redfield et al., 2013). Chronic treatment with sildenafil (50 mg, 8/8 h) for 4 weeks improved maximal oxygen uptake, ventilator efficiency, and oxygen uptake kinetics in patients with CHF with an ejection fraction of $28 \pm 6\%$ in comparison to placebo (Behling et al., 2008).

3.2.2 HF: sGC Modulators

The sGC stimulator (riociguat, BAY 63–2521) and activator (cinaciguat, BAY 58–2667) are also being investigated in PH patients with left heart disease (WHO group 2). In patients with acute decompensated HF, infusion of cinaciguat (50–400 mg/h, 6 h) led to significant reductions in pulmonary capillary wedge pressure, mean right arterial pressure, mean pulmonary artery pressure, pulmonary vascular resistance, and systemic vascular resistance, while increasing heart rate and cardiac output (Lapp et al., 2009). In patients with HF with preserved left ventricular ejection fraction and PH, riociguat (2 mg) significantly increased stroke volume, decreased systolic blood pressure, and right ventricular end-diastolic area (Bonderman et al., 2014). Whether sGC stimulators or activators are effective in HF requires more clinical trials with a larger number of patients.

4. PART III: EPIGENETICS REGULATION OF NO–sGC PATHWAY COMPONENTS IN THE CARDIOVASCULAR SYSTEM

4.1 Outline of Epigenetic Regulation

Epigenetics is used to refer to heritable alterations in chromatin that are not due to changes in DNA sequence. DNA methylation and histone modifications represent the major epigenetic mechanisms involved on gene regulation in mammals. DNA methyltransferases (DNMTs) are the enzymes responsible for transferring a methyl group at carbon 5 of cytosine. In general, hypomethylation of the DNA surrounding the promoter region favors gene activation, while hypermethylation gene silencing. Histone modifications are under the effect of two classes of enzymes namely histone acetyltransferases (HATs) and histone deacetylases (HDACs), which opens and condenses the chromatin, respectively. There exist three families of HATs, namely GNAT, MYST, and CBP/p300, which acetylate histone-tail lysines. Acetylation enhances chromatin relaxation and facilitates transcriptional elongation by RNA polymerase II (Edmondson & Roth, 1996). On the other hand, removal of acetyl group is catalyzed by HDACs, which remove acetyl groups, restore a positive charge to lysine residues of the histone tails, compress chromatin, and promote transcriptional repression. There exist four families of HDACs: class I (HDAC1–3, HDAC8), class IIa (HDAC4–7), class IIb (HDCA9–10), class III (sirtuins (1–7)), and class IV (HDAC11) (Sengupta & Seto, 2004). Class I HDACs are mostly localized in the nucleus, whereas class II HDACs shuttle between nucleus and cytoplasm (Minucci & Pelicci, 2006). This section will be focused on how epigenetics regulate the NO-pathway components in health and disease of the cardiovascular system.

4.2 Epigenetic Regulation of eNOS (NOS-3)

Endothelial NOS is the main source of NO in the vascular endothelium. eNOS −/− mice are characterized by systemic and PH, impaired angiogenesis, and impaired mobilization of endothelial progenitor cells (Kubis et al., 2002). Due to the relevance of eNOS in the regulation of vascular endothelium, the mechanisms that control its expression have been studied in detail.

To date, it has been shown that DNA methylation and histone modifications are the major epigenetics mechanisms that regulate endothelial-driven eNOS expression. Whereas the eNOS proximal promoter was hypomethylated and hyperacetylated in human endothelial cells, the promoter was heavily methylated in VSMCs, leading to transcription repression. Treating human VSMCs and HeLa, cells that do not express appreciable amount of eNOS, with the nonselective inhibitor of HDAC, TSA (1 μM, 24 h) resulted in increased levels of eNOS mRNA expression and also histones H3 and H4 acetylation at eNOS promoter (Chan et al., 2004; Fish et al., 2005) (Fig. 4). Chromatin immunoprecipitation analysis revealed recruitment of Sp1, Sp3 Ets transcription factors, which favors transcription, and RNA polymerase II to eNOS promoter in endothelial cells, while in vascular smooth muscle methyl CPG-binding protein 2 (MeCP2) and HDAC1, which are involved on gene repression, were localized to eNOS promoter (Chan et al., 2004; Fish et al., 2005; Gan et al., 2005).

Under hypoxia state expression of eNOS gene, acetylation of lysine 4 (histone 3) and methylation of eNOS proximal promoter were decreased

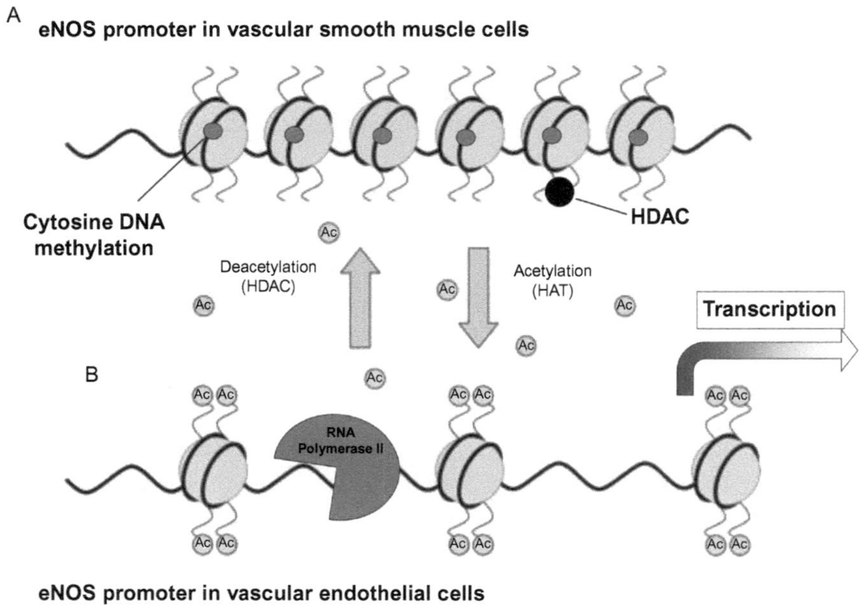

Fig. 4 (A) Endothelial nitric oxide synthase (eNOS) (NOS-3) promoter is found hypermethylated in the promoter region in the vascular smooth muscle cells and (B) hyperacetylated in endothelial cells, thus favoring its expression. *Modified from Pons, D., de Vries, F. R., van den Elsen, P. J., Heijmans, B. T., Quax, P. H., Jukema, J. W. (2009). Epigenetic histone acetylation modifiers in vascular remodelling: New targets for therapy in cardiovascular disease.* European Heart Journal, 30*(3), 266–277. Reproduction authorized with license number 3793360870849.*

(Fish et al., 2010), showing that hypoxia-inducing endothelial dysfunction can be in part due to lower eNOS expression. Children conceived by assisted reproductive technologies (ART) display vascular dysfunction similar to that seen in children with mothers with preeclampsia. Mice generated by ART showed lower potency values to acetylcholine in isolated mesenteric artery in comparison to the control. Besides, vascular stiffness was increased in ART mice. DNA methylation of the promoter of eNOS was increased in the aorta from ART mice, whereas the DNA methylation of the promoters for endothelin-1 and angiotensin-converting enzymes, pathways that favor vasoconstriction, did not differ between ART and control groups. Expression of eNOS was decreased in the carotid artery from ART mice in comparison with control animals, suggesting that DNA methylation of eNOS promoter contributes to endothelial dysfunction seen in ART mice (Rexhaj et al., 2013).

4.3 Epigenetic Regulation of iNOS

Inducible NOS (iNOS) is silenced in most tissues, but transcriptionally activated by inflammatory cytokines, which lead to increased levels of NO during an inflammatory response and contributing to free radical-mediated tissue injury in inflammatory disorders, including atherosclerotic plaque formation (Alderton, Cooper, & Knowles, 2001). Similar to eNOS, iNOS expression is also under the control of epigenetic mechanisms. In mesangial cells, specialized cells surrounding blood vessels in the kidney, treatment with DNA methylation inhibitor, 5-aza-CdR, resulted in higher nitrite production, iNOS protein levels, and iNOS promoter activity. Antisense knockdown of DNMT3b augmented IL-1β induction of NO and iNOS promoter activity (Yu & Kone, 2004). In cells lines that are resistant to iNOS induction as neonatal HMVEC, HUVEC, HMVEC-L, the iNOS promoter was heavily methylated at CpG dinucleotides. On the other hand, in human cell lines (A549 pulmonary adenocarcinoma, DLD-1 colon adenocarcinoma, and primary hepatocytes) where iNOS induction occurs, its promoter was hypomethylated and treatment with DNA methylation inhibitor induced iNOS promoter DNA hypomethylation and higher levels of iNOS mRNA after induction with cytokines (Chan et al., 2005). With respect to histone modification, in mouse macrophage cell line RAW 264.7, transient overexpression of HAT (p300) resulted in increased basal levels iNOS protein and the treatment with LPS/INFγ produced higher levels. Concentration-dependent augment of p300 transfection (0.2–1 μg) increased iNOS promoter activity under basal and LPS/INFγ-stimulated cells. Roscovitine, an indirect inhibitor of p300 HAT, inhibited basal and LPS/INFγ-stimulated iNOS promoter activity, suggesting that iNOS expression is regulated by p300 HAT (Deng & Wu, 2003).

4.4 Epigenetic Regulation of nNOS

Neuronal NOS (nNOS or NOS1) is the main source of NO in the myocardium. In the healthy heart, NO derived from nNOS reduces basal inotropism where it has a pivotal role in protecting myocardium from systolic/diastolic dysfunction, remodeling, and arrhythmias in the failing heart (Zhang, Jin, Jang, & Wang, 2014). nNOS −/− mice had increased basal contraction in isolated left ventricular cardiomyocytes as well as in vivo. The disruption of nNOS gene increased calcium current and prolonged significantly the inactivation of calcium current, thus leading to an increase in calcium influx and a greater calcium load in sarcoplasmic reticulum (Sears et al., 2003). In the diseased heart such as ischemia–reperfusion, injury (Aragón et al., 2011), infarct (Takimoto et al., 2002), hypertrophy, and HF (Damy et al., 2004), the nNOS expression and activity were increased in the myocardium, suggesting an NOS-dependent protection against stimulated ischemia–reperfusion prior to cardiomyocyte hypertrophy. In a study involving children with an average age of 11 years old the association between DNA methylation in arginase (ARG) and NOS gene and carotid artery intima-media thickness (CIMT) measurement was determined. The authors showed that ARG1, NOS1, NOS2A promoter, and NOS3 were highly methylated. Besides, after adjusting for sex, ancestry, body index, town, and age, they observed that DNA methylation in NOS1 was associated with CIMT and for every 1% increase in average DNA methylation of NOS1, CIMT increase by 1.2 μm ($P=0.02$), whereas no association was observed for methylation of NOS2A, NOS3, ARG1, and ARG2 (Breton et al., 2014).

4.5 Epigenetic Regulation of sGC

The mechanisms that control sGC subunits (α1 and β1) gene expression are much less studied than NOS genes in part because there exist few studies that focus on sGC transcriptional regulation (Marro et al., 2008; Sharina, Martin, Thomas, Uray, & Murad, 2003; Yamamoto & Suzuki, 2002). Only one study investigated sGC promoter activity in human vasculature (Marro et al., 2008). The decrease of sGC subunits has been shown in vascular- and nonvascular-related diseases; however, the mechanisms at the gene and protein levels that led to sGC subunits' downregulation were not explored in detail. Thus, in a model of late stage of atherosclerosis, the protein expression of sGC subunits was significantly reduced in rabbit aorta (Melichar et al., 2004). In animals with PH, sGC expression is reduced (Wedgwood et al., 2005) and cells isolated from pulmonary artery smooth muscle and exposed to hypoxia did not express sGC (Hassoun et al., 2004). Protein expression of sGCα1 and β1 was decreased in bladder tissues from obese mice and restored by sGC activator, BAY 60–2770 (Leiria et al., 2014). More recently, it was shown that the inoculation of mouse lung tissues in vitro with a lentivirus containing microRNA (miRNA) 34c-5p decreased both the expression of sGCβ1 and NO-induced sGC activity, suggesting that this miRNA regulates sGCβ1 expression (Xu et al., 2012).

Our previous study (Sharina et al., 2003) showed that site deletions of CCAAT- (sequence recognized by several transcriptional factors including

NF-Y) and growth independence 1 (GFl1)-binding sites decreased transcriptional activity, thus suggesting that these factors are important for sGCβ1 promoter function in a neuroblastoma cell line. Only one study explored the transcriptional regulation of sGC subunits in human vasculature. Human aortic smooth muscle cells (HASMCs) were transiently transfected with constructs covering 0.3–3.0 kb of human sGC α1 and 0.3–3.2 kb of human sGC β1. Functional analysis with luciferase reporters showed that c-Myb, GAGA, NFAT, and CCAAT-BF are important transcription factors that repressed sGC α1 promoter activity, while NF-Kb and NFY, which recognize CCAAT sequence, decreased and increased, respectively, sGC β1 activity in HASMC. With a proinflammatory mixture, a significant reduction of the sGC α1 and β1 promoter activity at 24 h and a time-dependent decrease of sGC β1 protein level in HASMC were also observed (Marro et al., 2008). This study suggested that in several cardiovascular diseases, specially atherosclerosis, where chronic inflammation occurs the downregulation of sGC subunits might contribute to neointima proliferation and vascular alterations.

With respect to sGC controlled by epigenetic mechanisms, only one study in breast cancer cell lines revealed that the promoter regions of sGC α1 and β1 were hypermethylated and the treatment with DNA methylation inhibitor increased their expression (Wen et al., 2015). Preliminary results from our lab showed that the HDAC inhibitors trichostatin, SAHA, LBH-589, and MS-275 increased sGC subunits at the gene and protein levels in breast cancer cell lines and the sGC β1 transcript was the most affected (unpublished data). DNA methylation and HDAC inhibitors could be used in cardiovascular diseases in situations where sGC subunits are downregulated or even lost.

4.6 Nitric Oxide Modulating Epigenetics Pathways

Nitric oxide can affect the transcription of several genes although the mechanisms are still unclear. For instance, in endothelial cells from human saphenous veins, NO inhibited vascular cell adhesion molecule-1 (VCAM-1) and E-selectin. This effect was unaffected by cGMP analogues (De Caterina et al., 1995). The idea of NO as an epigenetic regulator is novel. Laminar shear stress (SS) enhances NO production via eNOS activation, regulates genes expression in human endothelial cells (Illi et al., 2003), and directs embryonic stem cell differentiation (Illi et al., 2005). It was also observed that SS induced a transient increase on histone acetylation activity showing a peak between 1 and 2 h. Since SS also increases eNOS activity and NO production, the authors hypothesized that there was a link between NO production and histone modifications. Inhibitors of NOS, *S*-methylisothiourea (SMT), and 7-nitroindazole enhanced histone H3 acetylation under static and 4 h of SS. To understand the mechanisms involved on NO-regulating gene expression under SS conditions, it was observed that HDAC4 and 5 translocated to the nucleus of endothelial cells after 4 h of SS, being this effect reduced by NOS inhibitor. The treatment of endothelial cells with the NO-donor, DETA/NO, also increased the translocation of HDAC4 and 5 (Illi et al., 2008). This study suggests that one of the possible mechanisms by which NO inhibits apoptosis and cell adhesion and prevents atherosclerosis is by recruiting HDACs to silence genes.

5. CONCLUSION

Initially, the scientific community was reluctant to accept nitric oxide as a free radical and gas to be an important intracellular messenger, autacoid, paracrine substance, and hormone as we have predicted (Murad, 1999). The discovery of the NO–cGMP pathway by our lab in the 1970s revolutionized the comprehension of pathophysiological mechanisms involved on cardiovascular diseases. However, considering the expression "from bench to bedside," we concluded that therapeutic alternatives that target NO–cGMP did not immediately follow the biochemical and pathophysiological revolution. Therapeutic options coming from NO–cyclic GMP research have been proven effective and marketed recently for the treatment of cardiovascular disorders such as inhaled NO, PDE5 inhibitors, and more recently sGC stimulators. Several drugs in clinical use have pleiotropic effects by improving NO bioavailability. Today there are novel pharmacological options to prevent or reverse the endothelial dysfunction seen in the cardiovascular diseases.

CONFLICT OF INTEREST

The authors declare there are no conflicts of interest.

ACKNOWLEDGMENTS

F.Z.M. acknowledges FAPESP (2013/02246-7), CAPES (BEX 6718–14), and UNICAMP for research support.

REFERENCES

Ahmetaj-Shala, B., Kirkby, N. S., Knowles, R., Al'Yamani, M., Mazi, S., Wang, Z., et al. (2015). Evidence that links loss of cyclooxygenase-2 with increased asymmetric dimethylarginine: Novel explanation of cardiovascular side effects associated with anti-inflammatory drugs. *Circulation, 131*(7), 633–642.

Alaeddini, J., Uber, P. A., Park, M. H., Scott, R. L., Ventura, H. O., & Mehra, M. R. (2004). Efficacy and safety of sildenafil in the evaluation of pulmonary hypertension in severe heart failure. *American Journal of Cardiology, 94*(11), 1475–1477.

Alderton, W. K., Cooper, C. E., & Knowles, R. G. (2001). Nitric oxide synthases: Structure, function and inhibition. *Biochemical Journal, 357*(Pt. 3), 593–615.

Aragón, J. P., Condit, M. E., Bhushan, S., Predmore, B. L., Patel, S. S., Grinsfelder, D. B., et al. (2011). Beta3-adrenoreceptor stimulation ameliorates myocardial ischemia-reperfusion injury via endothelial nitric oxide synthase and neuronal nitric oxide synthase activation. *Journal of the American College of Cardiology, 58*(25), 2683–2691.

Archer, S., & Rich, S. (2000). Primary pulmonary hypertension: A vascular biology and translational research "Work in progress" *Circulation, 102*(22), 2781–2791.

Arnold, W. P., Mittal, C. K., Katsuki, S., & Murad, F. (1977). Nitric oxide activates guanylate cyclase and increases guanosine 3′:5′-cyclic monophosphate levels in various tissue preparations. *Proceedings of the National Academy of Sciences of the United States of America, 74*(8), 3203–3207.

Barst, R. J., Agnoletti, G., Fraisse, A., Baldassarre, J., Wessel, D. L., & NO Diagnostic Study Group. (2010). Vasodilator testing with nitric oxide and/or oxygen in pediatric pulmonary hypertension. *Pediatric Cardiology, 31*(5), 598–606.

Behling, A., Rohde, L. E., Colombo, F. C., Goldraich, L. A., Stein, R., & Clausell, N. (2008). Effects of 5′-phosphodiesterase four-week long inhibition with sildenafil in patients with chronic heart failure: A double-blind, placebo-controlled clinical trial. *Journal of Cardiac Failure, 14*(3), 189–197.

Benjamim, C. F., Silva, J. S., Fortes, Z. B., Oliveira, M. A., Ferreira, S. H., & Cunha, F. Q. (2002). Inhibition of leukocyte rolling by nitric oxide during sepsis leads to reduced migration of active microbicidal neutrophils. *Infection and Immunity, 70*(7), 3602–3610.

Bharani, A., Mathew, V., Sahu, A., & Lunia, B. (2003). The efficacy and tolerability of sildenafil in patients with moderate-to-severe pulmonary hypertension. *Indian Heart Journal, 55*(1), 55–59.

Bian, K., Doursout, M. F., & Murad, F. (2008). Vascular system: Role of nitric oxide in cardiovascular diseases. *Journal of Clinical Hypertension (Greenwich), 10*(4), 304–310.

Bocchi, E. A., Guimarães, G., Mocelin, A., Bacal, F., Bellotti, G., & Ramires, J. F. (2002). Sildenafil effects on exercise, neurohormonal activation, and erectile dysfunction in congestive heart failure: A double-blind, placebo-controlled, randomized study followed by a prospective treatment for erectile dysfunction. *Circulation, 106*(9), 1097–1103.

Boerrigter, G., & Burnett, J. C., Jr. (2009). Soluble guanylate cyclase: Not a dull enzyme. *Circulation, 119*(21), 2752–2754.

Bonderman, D., Pretsch, I., Steringer-Mascherbauer, R., Jansa, P., Rosenkranz, S., Tufaro, C., et al. (2014). Acute hemodynamic effects of riociguat in patients with pulmonary hypertension associated with diastolic heart failure (DILATE-1): A randomized, double-blind, placebo-controlled, single-dose study. *Chest, 146*(5), 1275–1285.

Breton, C. V., Park, C., Siegmund, K., Gauderman, W. J., Whitfield-Maxwell, L., Hodis, H. N., et al. (2014). NOS1 methylation and carotid artery intima-media thickness in children. *Circulation. Cardiovascular Genetics*, 7(2), 116–122.

Cassuto, J., Dou, H., Czikora, I., Szabo, A., Patel, V. S., Kamath, V., et al. (2014). Peroxynitrite disrupts endothelial caveolae leading to eNOS uncoupling and diminished

flow-mediated dilation in coronary arterioles of diabetic patients. *Diabetes*, *63*(4), 1381–1393.
Chan, Y., Fish, J. E., D'Abreo, C., Lin, S., Robb, G. B., Teichert, A. M., et al. (2004). The cell-specific expression of endothelial nitric-oxide synthase: A role for DNA methylation. *The Journal of Biological Chemistry*, *279*, 35087–35100.
Chan, G. C., Fish, J. E., Mawji, I. A., Leung, D. D., Rachlis, A. C., & Marsden, P. A. (2005). Epigenetic basis for the transcriptional hyporesponsiveness of the human inducible nitric oxide synthase gene in vascular endothelial cells. *Journal of Immunology*, *175*(6), 3846–3861.
Conran, N., Ferreira, H. H., Lorand-Metze, I., Thomazzi, S. M., Antunes, E., & De Nucci, G. (2001). Nitric oxide regulates human eosinophil adhesion mechanisms in vitro by changing integrin expression and activity on the eosinophil cell surface. *British Journal of Pharmacology*, *134*(3), 632–638.
Corbin, J. D. (2004). Mechanisms of action of PDE5 inhibition in erectile dysfunction. *International Journal of Impotence Research*, *16*(1), S4–S7.
Damy, T., Ratajczak, P., Shah, A. M., Camors, E., Marty, I., Hasenfuss, G., et al. (2004). Increased neuronal nitric oxide synthase-derived NO production in the failing human heart. *The Lancet*, *363*(9418), 1365–1367.
De Caterina, R., Libby, P., Peng, H. B., Thannickal, V. J., Rajavashisth, T. B., Gimbrone, M. A., Jr., et al. (1995). Nitric oxide decreases cytokine-induced endothelial activation. Nitric oxide selectively reduces endothelial expression of adhesion molecules and proinflammatory cytokines. *The Journal of Clinical Investigation*, *96*(1), 60–68.
Deng, W. G., & Wu, K. K. (2003). Regulation of inducible nitric oxide synthase expression by p300 and p50 acetylation. *The Journal of Immunology*, *171*(12), 6581–6588.
Edmondson, D. G., & Roth, S. Y. (1996). Chromatin and transcription. *The FASEB Journal*, *10*(10), 1173–1182.
Evgenov, O. V., Pacher, P., Schmidt, P. M., Haskó, G., Schmidt, H. H., & Stasch, J. P. (2006). NO-independent stimulators and activators of soluble guanylate cyclase: Discovery and therapeutic potential. *Nature Reviews. Drug Discovery*, *5*(9), 755–768.
Fang, Z. F., Huang, Y. Y., Tang, L., Hu, X. Q., Shen, X. Q., Tang, J. J., et al. (2015). Asymmetric dimethyl-L-arginine is a biomarker for disease stage and follow-up of pulmonary hypertension associated with congenital heart disease. *Pediatric Cardiology*, *36*(5), 1062–1069.
Feron, O., Dessy, C., Desager, J. P., & Balligand, J. L. (2001). Hydroxy-methylglutaryl-coenzyme A reductase inhibition promotes endothelial nitric oxide synthase activation through a decrease in caveolin abundance. *Circulation*, *103*(1), 113–118.
Fish, J. E., Matouk, C. C., Rachlis, A., Lin, S., Tai, S. C., D'Abreo, C., et al. (2005). The expression of endothelial nitric-oxide synthase is controlled by a cell-specific histone code. *The Journal of Biological Chemistry*, *280*, 24824–24838.
Fish, J. E., Yan, M. S., Matouk, C. C., St Bernard, R., Ho, J. J., Gavryushova, A., et al. (2010). Hypoxic repression of endothelial nitric-oxide synthase transcription is coupled with eviction of promoter histones. *The Journal of Biological Chemistry*, *285*(2), 810–826.
Förstermann, U., & Li, H. (2011). Therapeutic effect of enhancing endothelial nitric oxide synthase (eNOS) expression and preventing eNOS uncoupling. *British Journal of Pharmacology*, *164*(2), 213–223.
Förstermann, U., & Münzel, T. (2006). Endothelial nitric oxide synthase in vascular disease: From marvel to menace. *Circulation*, *113*(13), 1708–1714.
Francis, S. H., Lincoln, T. M., & Corbin, J. D. (1980). Characterization of a novel cGMP binding protein from rat lung. *The Journal of Biological Chemistry*, *255*(2), 620–626.
Furchgott, R. F., & Zawadzki, J. V. (1980). The obligatory role of endothelial cells in the relaxation of arterial smooth muscle by acetylcholine. *Nature*, *288*(5789), 373–376.

Galiè, N., Ghofrani, H. A., Torbicki, A., Barst, R. J., Rubin, L. J., Badesch, D., et al. (2005). Sildenafil citrate therapy for pulmonary arterial hypertension. *The New England Journal of Medicine, 353*(20), 2148–2157.

Gan, Y., Shen, Y. H., Wang, J., Wang, X., Utama, B., Wang, J., et al. (2005). Role of histone deacetylation in cell-specific expression of endothelial nitric-oxide synthase. *The Journal of Biological Chemistry, 280*, 16467–16475.

Ghofrani, H. A., Galiè, N., Grimminger, F., Grünig, E., Humbert, M., Jing, Z. C., et al. (2013). Riociguat for the treatment of pulmonary arterial hypertension. *New England Journal of Medicine, 369*(4), 330–340.

Ghofrani, H. A., Hoeper, M. M., Halank, M., Meyer, F. J., Staehler, G., Behr, J., et al. (2010). Riociguat for chronic thromboembolic pulmonary hypertension and pulmonary arterial hypertension: A phase II study. *European Respiratory Journal, 36*(4), 792–799.

Gimeno, E., Munoz, P. A., Pizarro, S., Gistau, C., Rodriguez-Roisin, R., Roca, J., et al. (2010). Hemodynamic and gas exchange effects of sildenafil in patients with chronic obstructive pulmonary disease and pulmonary hypertension. *American Journal of Respiratory and Critical Care Medicine, 181*(3), 270–278.

Grimminger, F., Weimann, G., Frey, R., Voswinckel, R., Thamm, M., Bölkow, D., et al. (2009). First acute haemodynamic study of soluble guanylate cyclase stimulator riociguat in pulmonary hypertension. *European Respiratory Journal, 33*(4), 785–792.

Guazzi, M., Tumminello, G., Di Marco, F., & Guazzi, M. D. (2004). Influences of sildenafil on lung function and hemodynamics in patients with chronic heart failure. *Clinical Pharmacology & Therapeutics, 76*(4), 371–378.

Guignabert, C., Tu, L., Girerd, B., Ricard, N., Huertas, A., Montani, D., et al. (2015). New molecular targets of pulmonary vascular remodeling in pulmonary arterial hypertension: Importance of endothelial communication. *Chest, 147*(2), 529–537.

Hanson, K. A., Ziegler, J. W., Rybalkin, S. D., Miller, J. W., Abman, S. H., & Clarke, W. R. (1998). Chronic pulmonary hypertension increases fetal lung cGMP phosphodiesterase activity. *American Journal of Physiology, 275*(5 Pt. 1), L931–L941.

Hassoun, P. M., Filippov, G., Fogel, M., Donaldson, C., Kayyali, U. S., Shimoda, L. A., et al. (2004). Hypoxia decreases expression of soluble guanylate cyclase in cultured rat pulmonary artery smooth muscle cells. *American Journal of Physiology. Lung Cellular and Molecular Physiology, 30*(6), 908–913.

Hattori, Y., Nakanishi, N., Akimoto, K., Yoshida, M., & Kasai, K. (2003). HMG-CoA reductase inhibitor increases GTP cyclohydrolase I mRNA and tetrahydrobiopterin in vascular endothelial cells. *Arteriosclerosis, Thrombosis, and Vascular Biology, 23*(2), 176–182.

Herranz, B., Marquez, S., Guijarro, B., Aracil, E., Aicart-Ramos, C., Rodriguez-Crespo, I., et al. (2012). Integrin-linked kinase regulates vasomotor function by preventing endothelial nitric oxide synthase uncoupling: Role in atherosclerosis. *Circulation Research, 110*(3), 439–449.

Ignarro, L. J., Buga, G. M., Wood, K. S., Byrns, R. E., & Chaudhuri, G. (1987). Endothelium-derived relaxing factor produced and released from artery and vein is nitric oxide. *Proceedings of the National Academy of Sciences of the United States of America, 84*(24), 9265–9269.

Illi, B., Dello Russo, C., Colussi, C., Rosati, J., Pallaoro, M., Spallotta, F., Rotili, D., et al. (2008). Nitric oxide modulates chromatin folding in human endothelial cells via protein phosphatase 2A activation and class II histone deacetylases nuclear shuttling. *Circulation Research, 102*(1), 51–58.

Illi, B., Nanni, S., Scopece, A., Farsetti, A., Biglioli, P., Capogrossi, M. C., et al. (2003). Shear stress-mediated chromatin remodeling provides molecular basis for flow-dependent regulation of gene expression. *Circulation Research, 93*(2), 155–161.

Illi, B., Scopece, A., Nanni, S., Farsetti, A., Morgante, L., Biglioli, P., et al. (2005). Epigenetic histone modification and cardiovascular lineage programming in mouse embryonic stem cells exposed to laminar shear stress. *Circulation Research, 96*(5), 501–508.

Jaïs, X., D'Armini, A. M., Jansa, P., Torbicki, A., Delcroix, M., Ghofrani, H. A., et al. (2008). Bosentan for treatment of inoperable chronic thromboembolic pulmonary hypertension: BENEFiT (Bosentan effects in iNopErable forms of chronic thromboembolic pulmonary hypertension), a randomized, placebo-controlled trial. *Journal of the American College of Cardiology, 52*(25), 2127–2134.

Katsuki, S., Arnold, W. P., & Murad, F. (1977). Effects of sodium nitroprusside, nitroglycerin, and sodium azide on levels of cyclic nucleotides and mechanical activity of various tissues. *Journal of Cyclic Nucleotide Research, 3*(4), 239–247.

Katz, S. D., Khan, T., Zeballos, G. A., Mathew, L., Potharlanka, P., Knecht, M., et al. (1999). Decreased activity of the L-arginine-nitric oxide metabolic pathway in patients with congestive heart failure. *Circulation, 99*(16), 2113–2117.

Kinsella, J. P., Neish, S. R., Ivy, D. D., Shaffer, E., & Abman, S. H. (1993). Clinical responses to prolonged treatment of persistent pulmonary hypertension of the newborn with low doses of inhaled nitric oxide. *Journal of Pediatrics, 123*(1), 103–108.

Kinsella, J. P., Neish, S. R., Shaffer, E., & Abman, S. H. (1992). Low-dose inhalation nitric oxide in persistent pulmonary hypertension of the newborn. *The Lancet, 340*(8823), 819–920.

Kubis, N., Besnard, S., Silvestre, J. S., Feletou, M., Huang, P. L., Lévy, B. I., et al. (2002). Decreased arteriolar density in endothelial nitric oxide synthase knockout mice is due to hypertension, not to the constitutive defect in endothelial nitric oxide synthase enzyme. *Journal of Hypertension, 20*(2), 273–280.

Landmesser, U., Dikalov, S., Price, S. R., McCann, L., Fukai, T., Holland, S. M., et al. (2003). Oxidation of tetrahydrobiopterin leads to uncoupling of endothelial cell nitric oxide synthase in hypertension. *The Journal of Clinical Investigation, 111*(8), 1201–1209.

Lapp, H., Mitrovic, V., Franz, N., Heuer, H., Buerke, M., Wolfertz, J., et al. (2009). Cinaciguat (BAY 58–2667) improves cardiopulmonary hemodynamics in patients with acute decompensated heart failure. *Circulation, 119*(21), 2781–2788.

Leiper, J., & Nandi, M. (2011). The therapeutic potential of targeting endogenous inhibitors of nitric oxide synthesis. *Nature Reviews. Drug Discovery, 10*(4), 277–291.

Leiria, L. O., Silva, F. H., Davel, A. P., Alexandre, E. C., Calixto, M. C., De Nucci, G., et al. (2014). The soluble guanylyl cyclase activator BAY 60–2770 ameliorates overactive bladder in obese mice. *Journal of Urology, 191*(2), 539–547.

Lewis, G. D., Lachmann, J., Camuso, J., Lepore, J. J., Shin, J., Martinovic, M. E., et al. (2007). Sildenafil improves exercise hemodynamics and oxygen uptake in patients with systolic heart failure. *Circulation, 115*(1), 59–66.

Li, H., Witte, K., August, M., Brausch, I., Gödtel-Armbrust, U., Habermeier, A., et al. (2006). Reversal of endothelial nitric oxide synthase uncoupling and up-regulation of endothelial nitric oxide synthase expression lowers blood pressure in hypertensive rats. *Journal of the American College of Cardiology, 47*(12), 2536–2544.

Marcondes, S., Cardoso, M. H., Morganti, R. P., Thomazzi, S. M., Lilla, S., Murad, F., et al. (2006). Cyclic GMP-independent mechanisms contribute to the inhibition of platelet adhesion by nitric oxide donor: A role for alpha-actinin nitration. *Proceedings of the National Academy of Sciences of the United States of America, 103*(9), 3434–3439.

Margaritis, M., Antonopoulos, A. S., Digby, J., Lee, R., Reilly, S., Coutinho, P., et al. (2013). Interactions between vascular wall and perivascular adipose tissue reveal novel roles for adiponectin in the regulation of endothelial nitric oxide synthase function in human vessels. *Circulation, 127*(22), 2209–2221.

Marro, M. L., Peiró, C., Panayiotou, C. M., Baliga, R. S., Meurer, S., Schmidt, H. H., et al. (2008). Characterization of the human alpha1 beta1 soluble guanylyl cyclase promoter: Key role for NF-kappaB(p50) and CCAAT-binding factors in regulating expression of the nitric oxide receptor. *The Journal of Biological Chemistry, 283*(29), 20027–20036.
Mattagajasingh, I., Kim, C. S., Naqvi, A., Yamamori, T., Hoffman, T. A., Jung, S. B., et al. (2007). SIRT1 promotes endothelium-dependent vascular relaxation by activating endothelial nitric oxide synthase. *Proceedings of the National Academy of Sciences of the United States of America, 104*(37), 14855–14860.
Mehta, J. L., Bryant, J. L., Jr., & Mehta, P. (1995). Reduction of nitric oxide synthase activity in human neutrophils by oxidized low-density lipoproteins. Reversal of the effect of oxidized low-density lipoproteins by high-density lipoproteins and L-arginine. *Biochemical Pharmacology, 50*(8), 1181–1185.
Melichar, V. O., Behr-Roussel, D., Zabel, U., Uttenthal, L. O., Rodrigo, J., Rupin, A., et al. (2004). Reduced cGMP signaling associated with neointimal proliferation and vascular dysfunction in late-stage atherosclerosis. *Proceedings of the National Academy of Sciences of the United States of America, 101*(47), 16671–16676.
Michelakis, E., Tymchak, W., Lien, D., Webster, L., Hashimoto, K., & Archer, S. (2002). Oral sildenafil is an effective and specific pulmonary vasodilator in patients with pulmonary arterial hypertension: Comparison with inhaled nitric oxide. *Circulation, 105*(20), 2398–2403.
Minucci, S., & Pelicci, P. G. (2006). Histone deacetylase inhibitors and the promise of epigenetic (and more) treatments for cancer. *Nature Reviews. Cancer, 6*(1), 38–51.
Mónica, F. Z., Murad, F., & Bian, K. (2014). Modulating cGMP levels as therapeutics drug targets in cardiovascular and non-cardiovascular diseases. *OA Biochemistry, 2*(1), 3–12.
Montani, D., Chaumais, M. C., Guignabert, C., Günther, S., Girerd, B., Jaïs, X., et al. (2014). Targeted therapies in pulmonary arterial hypertension. *Pharmacology & Therapeutics, 141*(2), 172–191.
Murad, F. (1986). Cyclic guanosine monophosphate as a mediator of vasodilation. *Journal of Clinical Investigation, 78*(1), 1–5.
Murad, F. (1998). Nitric oxide signaling: Would you believe that a simple free radical could be a second messenger, autacoid, paracrine substance, neurotransmitter, and hormone? *Recent Progress in Hormone Research, 53*, 43–59.
Murad, F. (1999). Cellular signaling with nitric oxide. In *Association for academic surgeons meeting, Philadelphia, Pennsylvania*. November.
Murad, F. (2006). Shattuck Lecture. Nitric oxide and cyclic GMP in cell signaling and drug development. *New England Journal of Medicine, 355*(19), 2003–2011.
Murad, F., Leitman, D., Waldman, S., Chang, C. H., Hirata, M., & Kohse, K. (1988). Effects of nitrovasodilators, endothelium-dependent vasodilators, and atrial peptides on cGMP. *Cold Spring Harbor Symposia on Quantitative Biology, 53*(2), 1005–1009.
Murad, F., Mittal, C. K., Arnold, W. P., Katsuki, S., & Kimura, H. (1978). Guanylate cyclase: Activation by azide, nitro compounds, nitric oxide, and hydroxyl radical and inhibition by hemoglobin and myoglobin. *Advances in Cyclic Nucleotide Research, 9*, 145–158.
Nakane, M., Schmidt, H. H. H. W., & Murad, F. (1991). Purification and molecular cloning of guanylyl cyclase activating factor synthase. In *Neuroscience meeting*. November.
Ogawa, T., Kimoto, M., Watanabe, H., & Sasaoka, K. (1987). Metabolism of NG, NG-and NG, N'G-dimethylarginine in rats. *Archives of Biochemistry and Biophysics, 252*(2), 526–537.
Olschewski, H., Simonneau, G., Galiè, N., Higenbottam, T., Naeije, R., Rubin, L. J., et al. (2002). Inhaled iloprost for severe pulmonary hypertension. *New England Journal of Medicine, 347*(5), 322–329.
Palmer, R. M., Ferrige, A. G., & Moncada, S. (1987). Nitric oxide release accounts for the ; biological activity of endothelium-derived relaxing factor. *Nature, 327*(6122), 524–526.

Peach, M. J., & Loeb, A. L. (1987). Changes in vascular endothelium and its function in systemic arterial hypertension. *American Journal of Cardiology*, *60*(17), 110I–115I.

Pepke-Zaba, J., Higenbottam, T. W., Dinh-Xuan, A. T., Stone, D., & Wallwork, J. (1991). Inhaled nitric oxide as a cause of selective pulmonary vasodilatation in pulmonary hypertension. *The Lancet*, *338*(8776), 1173–1174.

Rapoport, R. M., & Murad, F. (1983). Endothelium-dependent and nitrovasodilator-induced relaxation of vascular smooth muscle: Role of cyclic GMP. *Journal of Cyclic Nucleotide and Protein Phosphorylation Research*, *9*(4–5), 281–296.

Redfield, M. M., Chen, H. H., Borlaug, B. A., Semigran, M. J., Lee, K. L., Lewis, G., et al. (2013). Effect of phosphodiesterase-5 inhibition on exercise capacity and clinical status in heart failure with preserved ejection fraction: A randomized clinical trial. *JAMA*, *309*(12), 1268–1277.

Rexhaj, E., Paoloni-Giacobino, A., Rimoldi, S. F., Fuster, D. G., Anderegg, M., Somm, E., et al. (2013). Mice generated by in vitro fertilization exhibit vascular dysfunction and shortened life span. *The Journal of Clinical Investigation*, *123*(12), 5052–5060.

Rietema, H., Holverda, S., Bogaard, H. J., Marcus, J. T., Smit, H. J., Westerhof, N., et al. (2008). Sildenafil treatment in COPD does not affect stroke volume or exercise capacity. *European Respiratory Journal*, *31*(4), 759–764.

Roberts, J. D., Jr., Fineman, J. R., Morin, F. Cr, Shaul, P. W., Rimar, S., Schreiber, M. D., et al. (1997). Inhaled nitric oxide and persistent pulmonary hypertension of the newborn. *New England Journal of Medicine*, *336*(9), 605–610.

Satoh, M., Fujimoto, S., Arakawa, S., Yada, T., Namikoshi, T., Haruna, Y., et al. (2008). Angiotensin II type 1 receptor blocker ameliorates uncoupled endothelial nitric oxide synthase in rats with experimental diabetic nephropathy. *Nephrology, Dialysis, Transplant*, *23*(12), 3806–3813.

Savvidou, M. D., Hingorani, A. D., Tsikas, D., Frölich, J. C., Vallance, P., & Nicolaides, K. H. (2003). Endothelial dysfunction and raised plasma concentrations of asymmetric dimethylarginine in pregnant women who subsequently develop pre-eclampsia. *The Lancet*, *361*(9368), 1511–1517.

Schmidt, H. H., Smith, R. M., Nakane, M., & Murad, F. (1992). Ca2+/calmodulin-independent diaphorase and reductase activities. *Biochemistry*, *31*(12), 3243–3249.

Sears, C. E., Bryant, S. M., Ashley, E. A., Lygate, C. A., Rakovic, S., Wallis, H. L., et al. (2003). Cardiac neuronal nitric oxide synthase isoform regulates myocardial contraction and calcium handling. *Circulation Research*, *92*(5), e52–e59.

Sengupta, N., & Seto, E. (2004). Regulation of histone deacetylase activities. *Journal of Cellular Biochemistry*, *93*(1), 57–67.

Sharina, I. G., Martin, E., Thomas, A., Uray, K. L., & Murad, F. (2003). CCAAT-binding factor regulates expression of the beta1 subunit of soluble guanylyl cyclase gene in the BE2 human neuroblastoma cell line. *Proceedings of the National Academy of Sciences of the United States of America*, *100*(20), 11523–11528.

Shi, H. X., Yang, J., Yang, T., Xue, Y. L., Liu, J., Li, Y. J., et al. (2014). Alpha-asarone protects endothelial cells from injury by angiotensin II. *Evidence-Based Complementary and Alternative Medicine*, *2014*, 682041.

Shimizu-Albergine, M., Rybalkin, S. D., Rybalkina, I. G., Feil, R., Wolfsgruber, W., Hofmann, F., et al. (2003). Individual cerebellar Purkinje cells express different cGMP phosphodiesterases (PDEs): In vivo phosphorylation of cGMP-specific PDE (PDE5) as an indicator of cGMP-dependent protein kinase (PKG) activation. *The Journal of Neuroscience*, *23*(16), 6452–6459.

Stasch, J. P., Schmidt, P., Alonso-Alija, C., Apeler, H., Dembowsky, K., Haerter, M., et al. (2002). NO- and haem-independent activation of soluble guanylyl cyclase: Molecular basis and cardiovascular implications of a new pharmacological principle. *British Journal of Pharmacology*, *136*(5), 773–783.

Sydow, K., & Münzel, T. (2003). ADMA and oxidative stress. *Atherosclerosis Supplements*, *4*(4), 41–51.

Takimoto, Y., Aoyama, T., Tanaka, K., Keyamura, R., Yui, Y., & Sasayama, S. (2002). Augmented expression of neuronal nitric oxide synthase in the atria parasympathetically decreases heart rate during acute myocardial infarction in rats. *Circulation*, *105*(4), 490–496.

Vallance, P., Leone, A., Calver, A., Collier, J., & Moncada, S. (1992). Accumulation of an endogenous inhibitor of nitric oxide synthesis in chronic renal failure. *The Lancet*, *339*(8793), 572–575.

Versari, D., Daghini, E., Virdis, A., Ghiadoni, L., & Taddei, S. (2009). Endothelium-dependent contractions and endothelial dysfunction in human hypertension. *British Journal of Pharmacology*, *157*(4), 527–536.

Warnholtz, A., Nickenig, G., Schulz, E., Macharzina, R., Bräsen, J. H., Skatchkov, M., et al. (1999). Increased NADH-oxidase-mediated superoxide production in the early stages of atherosclerosis: Evidence for involvement of the renin-angiotensin system. *Circulation*, *99*(15), 2027–2033.

Wedgwood, S., Steinhorn, R. H., Bunderson, M., Wilham, J., Lakshminrusimha, S., Brennan, L. A., et al. (2005). Increased hydrogen peroxide downregulates soluble guanylate cyclase in the lungs of lambs with persistent pulmonary hypertension of the newborn. *American Journal of Physiology. Lung Cellular and Molecular Physiology*, *289*(4), L660–L666.

Wen, H. C., Chuu, C. P., Chen, C. Y., Shiah, S. G., Kung, H. J., King, K. L., et al. (2015). Elevation of soluble guanylate cyclase suppresses proliferation and survival of human breast cancer cells. *PLoS One*, *10*(4), e0125518.

Wohlfart, P., Xu, H., Endlich, A., Habermeier, A., Closs, E. I., Hubschle, T., et al. (2008). Antiatherosclerotic effects of small-molecular-weight compounds enhancing endothelial nitric-oxide synthase (eNOS) expression and preventing eNOS uncoupling. *Journal of Pharmacology and Experimental Therapeutics*, *325*, 370–379.

Wolf, Y. G., Rasmussen, L. M., Sherman, Y., Bundens, W. P., & Hye, R. J. (1995). Nitroglycerin decreases medial smooth muscle cell proliferation after arterial balloon injury. *Journal of Vascular Surgery*, *21*(3), 499–504.

Xu, X., Wang, S., Liu, J., Dou, D., Liu, L., Chen, Z., et al. (2012). Hypoxia induces downregulation of soluble guanylyl cyclase β1 by miR-34c-5p. *Journal of Cell Science*, *125*(Pt. 24), 6117–6126.

Yamamoto, T., & Suzuki, N. (2002). Promoter activity of the 5′-flanking regions of medaka fish soluble guanylate cyclase alpha1 and beta1 subunit genes. *Biochemical Journal*, *361*(Pt. 2), 337–345.

Yu, Z., & Kone, B. C. (2004). Hypermethylation of the inducible nitric-oxide synthase gene promoter inhibits its transcription. *The Journal of Biological Chemistry*, *279*(45), 46954–46961.

Zhang, Y. H., Jin, C. Z., Jang, J. H., & Wang, Y. (2014). Molecular mechanisms of neuronal nitric oxide synthase in cardiac function and pathophysiology. *The Journal of Physiology*, *592*(Pt. 15), 3189–3200.

Zhao, L., Mason, N. A., Morrell, N. W., Kojonazarov, B., Sadykov, A., Maripov, A., et al. (2001). Sildenafil inhibits hypoxia-induced pulmonary hypertension. *Circulation*, *104*(4), 424–428.

Zhao, L., Mason, N. A., Strange, J. W., Walker, H., & Wilkins, M. R. (2003). Beneficial effects of phosphodiesterase 5 inhibition in pulmonary hypertension are influenced by natriuretic peptide activity. *Circulation*, *107*(2), 234–237.

CHAPTER TWO

Transcriptional and Posttranslational Regulation of eNOS in the Endothelium

D.J.R. Fulton[1]
Medical College of Georgia at Augusta University, Augusta, GA, United States
[1]Corresponding author: e-mail address: dfulton@gru.edu

Contents

Abstract

Nitric oxide (NO) is a highly reactive free radical gas and these unique properties have been adapted for a surprising number of biological roles. In neurons, NO functions as a neurotransmitter; in immune cells, NO contributes to host defense; and in endothelial cells, NO is a major regulator of blood vessel homeostasis. In the vasculature, NO is

Advances in Pharmacology, Volume 77
ISSN 1054-3589
http://dx.doi.org/10.1016/bs.apha.2016.04.001

synthesized on demand by a specific enzyme, endothelial nitric oxide synthase (eNOS) that is uniquely expressed in the endothelial cells that form the interface between the circulating blood and the various tissues of the body. NO regulates endothelial and blood vessel function via two distinct pathways, the activation of soluble guanylate cyclase and cGMP-dependent signaling and the S-nitrosylation of proteins with reactive thiols (S-nitrosylation). The chemical properties of NO also serve to reduce oxidation and regulate mitochondrial function. Reduced synthesis and/or compromised biological activity of NO precede the development of cardiovascular disease and this has generated a high level of interest in the mechanisms controlling the synthesis and fate of NO in the endothelium. The amount of NO produced results from the expression level of eNOS, which is regulated at the transcriptional and posttranscriptional levels as well as the acute posttranslational regulation of eNOS. The goal of this chapter is to highlight and integrate past and current knowledge of the mechanisms regulating eNOS expression in the endothelium and the posttranslational mechanisms regulating eNOS activity in both health and disease.

ABBREVIATIONS

aa amino acid
CSD caveolin-scaffolding domain
EDRF endothelium-derived relaxing factor
eNOS endothelial nitric oxide synthase (NOSIII)
Hsp90 heat-shock protein 90
iNOS inducible nitric oxide synthase (NOSII)
nNOS neuronal nitric oxide synthase (NOSI)
NO nitric oxide
NOS nitric oxide synthase

1. INTRODUCTION: DISCOVERY OF ENDOTHELIAL NITRIC OXIDE SYNTHASE

The unmasking of the important role of the endothelium in controlling the function of isolated blood vessels was a breakthrough in cardiovascular physiology (Furchgott & Zawadzki, 1980) that provided new and critical insights into the mechanisms by which vascular function might be compromised in cardiovascular disease states. The endothelium, a single layer of cells lining the lumen of all blood vessels, was previously thought to be an inert barrier to circulating blood. A functional role of the diminutive endothelial cell in regulating the contractile force of the much larger smooth muscle was not anticipated and generated a high-level enthusiasm. From c.1980 onward the race to discover the entity responsible for endothelium-derived relaxing factor (EDRF) was underway. The culmination of the contributions of multiple laboratories was the discovery that nitric oxide (NO) was the chemical mediator of EDRF (Arnold, Mittal, Katsuki, & Murad, 1977; Ignarro, Buga, Wood, Byrns, & Chaudhuri, 1987; Palmer, Ferrige, & Moncada, 1987; Rapoport & Murad, 1983). NO is a lipophilic and rapidly diffusible free radical gas (Dudzinski, Igarashi, Greif, & Michel, 2006) with a half-life in biological fluids that is in the order of seconds (Thomas, Liu, Kantrow, & Lancaster, 2001). NO produced in the vascular endothelium is an autacoid that acts locally on target cells, in particular, vascular smooth muscle cells, via the direct binding to its cognate receptor soluble guanylate cyclase and the subsequent elevation of cGMP. NO is also produced to varying degrees in the nervous, digestive, and immunological systems where it exerts a variety of biological effects under both physiological and pathological conditions (Nathan, 1992). Following the discovery of novel enzymes isolated from neurons and immune cells that synthesize NO from L-arginine, another race was underway to identify the enzyme responsible for endothelium-derived NO (Bredt & Snyder, 1990; Xie et al., 1992). This culminated in the discovery of endothelial nitric oxide synthase (eNOS) (Janssens, Shimouchi, Quertermous, Bloch, & Bloch, 1992; Lamas, Marsden, Li, Tempst, & Michel, 1992; Nishida et al., 1992; Sessa et al., 1992) a calcium-dependent enzyme that is preferentially expressed in the vascular endothelium and utilizes L-arginine to synthesize NO. The genetic deletion of eNOS in mice results in the loss of endothelium-derived relaxation (particularly in large conduit vessels where there are no alternative pathways of endothelium-dependent vasorelaxation, ie, EDHF), impaired vascular remodeling, and hypertension (Huang et al., 1995; Rudic et al., 1998; Shesely et al., 1996) which together confirmed the importance of endothelium-derived NO in cardiovascular physiology and pathophysiology. Many studies have since shown that the reduced expression and dysregulation of eNOS, manifest by the decreased bioavailability of NO and increased production of its endogenous antagonist, superoxide, increase the severity of cardiovascular disease (Forstermann & Munzel, 2006). Thus, the crucial role of eNOS-derived NO in the cardiovascular system has stimulated a high level of interest in defining the mechanisms that control eNOS enzyme activity.

2. EXPRESSION OF eNOS IN VASCULAR ENDOTHELIUM

eNOS (also designated NOSIII as it was the third nitric oxide synthase (NOS) isoform identified) belongs to a family of three distinct NOS genes that include neuronal (nNOS, also designated NOSI) and inducible (iNOS or NOSII) NOS. In the cardiovascular system, nNOS and iNOS isoforms can also be detected and particularly evident in the heart (Balligand & Cannon, 1997; Balligand et al., 1994; Xu, Huso, Dawson, Bredt, & Becker, 1999) and smooth muscle (Ichihara, Inscho, Imig, & Navar, 1998) with the relative expression levels of nNOS and iNOS increasing in disease states (Beasley, Schwartz, & Brenner, 1991; Wilcox et al., 1997). In blood vessels, nNOS can be detected in vascular smooth muscle cells of certain types of blood vessels, in particular the microcirculation and kidney (Ichihara et al., 1998; Segal, Brett, & Sessa, 1999). The iNOS is not normally present in blood vessels but can be rapidly expressed in response to infection, injury, and/or cytokine stimulation which results in high levels of expression in vascular smooth muscle cells as well as in infiltrating immune cells (Beasley et al., 1991; Gross, Jaffe, Levi, & Kilbourn, 1991; Kibbe, Billiar, & Tzeng, 1999). However, the predominant NOS isoform that is expressed in cardiovascular tissues is eNOS (Huang et al., 1995; Segal et al., 1999; Shesely et al., 1996) and this is best evidenced by the loss of endothelium-dependent relaxation in the aorta (where EDRF is 100% NO dependent) in eNOS knockout mice. The molecular mechanisms responsible for the restricted expression of eNOS in endothelial cells are complex. There are numerous other endothelial-specific genes and some such as VE-cadherin and TIE2 are preferentially expressed in the endothelium largely due to properties inherent in their proximal promoters. In contrast, the endothelial-specific expression of eNOS appears to be more complex and arises from multiple mechanisms. The human gene for eNOS resides on chromosome 7. Cloning and identification of the 5′ region of the eNOS gene revealed a TATA-less promoter and the presence of binding sites for a number of transcription factors including Sp1, Sp3, Ets1, GATA2, AP-1, and KLF2 among others (Fish & Marsden, 2006). Unlike genes that are specific to smooth muscle cells, the expression of endothelial-specific genes is not governed by a master regulator such as SRF and there are no equivalent *cis* binding elements such as the CArG box, which is present in genes preferentially expressed in smooth muscle, to guide the endothelial-specific expression of genes such as eNOS. In transgenic mice, truncated eNOS proximal promoters (1.6 kb) can drive endothelial expression of a transgene (β-galactosidase) in most, but not all, tissues (Guillot et al., 1999). Longer eNOS promoter regions (5 kb) can more faithfully recapitulate the expression pattern of natively expressed eNOS (Teichert et al., 2000) and these data are in agreement with the discovery, using DNAse mapping, of an enhancer region on the eNOS gene that is 5 kb upstream of the eNOS transcriptional start site. This site was shown to contribute to the endothelial-specific expression of eNOS and further suggested that chromatin structure may guide the endothelial-specific expression of eNOS. Epigenetic regulation of eNOS expression has been shown through DNA methylation of the proximal promoter (Chan et al., 2004) which prevents eNOS expression in nonendothelial cells and posttranslation acetylation and methylation of local histones (Fish et al., 2005; Gan et al., 2005).

The expression level of eNOS in the endothelium was originally thought to be constitutive and invariant (in contrast to the inducible NOS isoform, iNOS), but numerous studies have shown that shear stress (Nishida et al., 1992), reactive oxygen species (H_2O_2) (Drummond, Cai, Davis, Ramasamy, & Harrison, 2000), hormones (estrogen) (Tan, Gurjar, Sharma, & Bhalla, 1999), statins (Laufs, La Fata, Plutzky, & Liao, 1998), and growth factors (VEGF) (Bouloumie, Schini-Kerth, & Busse, 1999) can increase the abundance of eNOS, whereas other stimuli such as cytokines (TNFα) (Nishida et al., 1992) and hypoxia (McQuillan, Leung, Marsden, Kostyk, & Kourembanas, 1994) can decrease the expression level of eNOS. The expected outcome of increased expression of eNOS should be a proportional change in the amount of NO produced. However, it was soon discovered that in cells, the stoichiometric relationship between eNOS expression and the amount of NO produced did not always correlate well. Indeed, one of the first observations of the cloned enzyme was the ability to acutely increase NO synthesis in response to increased calcium levels (Sessa et al., 1992). Numerous studies have since revealed that the dynamic production of NO results from posttranslational regulation of eNOS enzyme activity. However, the posttranslational regulation of eNOS is much more complex than just calcium and this is detailed later.

3. ENZYMOLOGY OF eNOS

The structural organization of eNOS is that of a bidomain enzyme comprised of an N-terminal oxidase domain which binds to heme, zinc, tetrahydrobiopterin (BH4), and calcium–calmodulin and a C-terminal reductase domain which binds NADPH as well as the flavins mononucleotide (FMN) and flavin adenine dinucleotide (FAD). The N-terminal region also coordinates the formation of the eNOS homodimer which is essential for enzyme activity (Baek, Thiel, Lucas, & Stuehr, 1993). Electrons derived from NADPH are shuttled via FAD and FMN on the C-terminal reductase domain of one NOS monomer to the N-terminal oxygenase domain of the other NOS monomer (Siddhanta et al., 1998). Two distinct monooxygenase reactions are then carried out. In the first step of NO synthesis, an electron reduces ferric to ferrous heme which binds molecular oxygen and then reacts with L-arginine in the substrate-binding pocket to form the intermediate molecule N(ω)-hydroxy-L-arginine (NOHA). In the second reaction, NOHA reacts with reduced oxygen to form NO with the byproduct L-citrulline. To synthesize NO, eNOS consumes 1 mol of L-arginine, 1.5 mol of NADPH, and 2 mol of molecular oxygen (Dudzinski et al., 2006). The catalytic activity of eNOS requires the complex coordination and interplay between numerous domains, substrates, and cofactors. Disruption of this highly coordinated catalysis can occur in the presence of subthreshold concentrations of substrate and cofactors resulting in the reduced synthesis of NO. In particular, decreased levels of the cofactor BH4 or increased levels of its oxidized form (BH2) can result in enzyme uncoupling in which superoxide is generated at the expense of NO (Vasquez-Vivar et al., 1998; Wever, van Dam, van Rijn, de Groot, & Rabelink, 1997). In comparison to iNOS, the rate of heme reduction and thus NO synthesis is lower in eNOS and elegant domain-swapping experiments have shown that the rate-limiting step resides within the reductase domain and more specifically within autoinhibitory elements that constrain eNOS activity (Chen & Wu, 2003; Nishida & Ortiz de Montellano, 1998).

4. REGULATION OF eNOS ACTIVITY BY PROTEIN: PROTEIN INTERACTIONS

4.1 Calcium–Calmodulin

Early pharmacological studies of endothelium-dependent relaxation of isolated blood vessels and measurements of NO release by endothelial cells in culture revealed that removal of extracellular calcium prevented agonist-stimulated release of NO and demonstrated an essential role for calcium in the synthesis of NO (Forstermann, Pollock, Schmidt, Heller, & Murad, 1991; Singer & Peach, 1982). Further studies showed that removal or antagonism of calmodulin could also prevent the generation of NO and suggested that eNOS was a calcium–calmodulin-dependent enzyme (Busse & Mulsch, 1990). It was subsequently shown that calcium–calmodulin binds directly to purified eNOS and is necessary for its activity (Pollock et al., 1991) and this observation marked the first described protein: protein interaction that regulates the activity of eNOS. These findings were consistent with earlier observations of the neuronal isoform (nNOS) which was also shown to be a calmodulin-binding, calcium-dependent enzyme (Bredt & Snyder, 1990). The molecular cloning of eNOS revealed a canonical calmodulin-binding site that lies between the N-terminal oxygenase and C-terminal reductase domains of eNOS (amino acid (aa) 493–512, type 1-8-14). Later studies revealed an additional site in the reductase domain (aa 594–613) of eNOS that may also influence enzyme activity. The binding of calcium-activated calmodulin to eNOS displaces adjacent autoinhibitory loops and "unlocks" NADPH-dependent electron flux from the C- to the N-terminal and ultimately to the heme moiety for production of NO (Nishida & Ortiz de Montellano, 1999; Salerno et al., 1997). eNOS activity is proportional to the level of intracellular calcium and the binding of calcium-activated calmodulin and this property accounts for the stepwise increase in endothelium-dependent relaxation in response to increasing concentrations of muscarinic agonist in aortic rings as observed by Furchgott prior to 1980 (Furchgott & Zawadzki, 1980). In contrast to the calcium-dependent activity of eNOS and nNOS, the activity of iNOS was found to be independent of calcium (Xie et al., 1992). Despite its calcium independence, iNOS also contains a calmodulin-binding site between its N- and C-terminal domains. However, this site is similar to the eNOS calmodulin-binding site and alone is not sufficient to confer calcium independence. Instead, the calcium-independent activity of iNOS derives from the absence of C-terminal autoinhibitory domains that are present in eNOS and restrict its activity in the absence of calcium (Chen & Wu, 2003). Isolated eNOS enzyme activity assays have demonstrated an essential role for calcium–calmodulin in the synthesis of NO. Consistent with this concept, in intact endothelial cells, certain agonists that elicit a robust increase in intracellular calcium (ie, acetylcholine and bradykinin) also stimulate NO synthesis. However, there are numerous other stimuli that can increase NO synthesis in cells without appreciably changing the level of intracellular calcium (insulin, estrogen, ceramide, shear stress, and phosphatase inhibitors) (Caulin-Glaser, Garcia-Cardena, Sarrel, Sessa, & Bender, 1997; Fleming, Bauersachs, & Busse, 1997; Fleming, Bauersachs, Fisslthaler, & Busse, 1998; Igarashi, Thatte, Prabhakar, Golan, & Michel, 1999; Tsukahara, Gordienko, Tonshoff, Gelato, & Goligorsky, 1994). The explanation for this discrepancy is the existence of additional posttranslational mechanisms (beyond calcium–calmodulin) that will be discussed later including additional protein: protein interactions, the subcellular localization of eNOS, phosphorylation, and other modifications (summarized in Table 1).

Table 1 eNOS-Interacting Proteins

Interacting Protein	Binding Site	Effect on eNOS Activity
AKT	ND	↑
ASL	ND	↑
β-actin	326–333	↓
Calmodulin	493–512, 594–613	↑
CAT-1	ND	↑
Caveolin-1	348–356	↓
CHIP	ND	↓
Endoglin	ND	↑
GPCR (B2)	ND	↓
GRX-1	ND	↑
HDAC3	ND	↓
Hsp90	310–323	↑
NOSIP	ND	↓
NOSTRIN	ND	↓
PIN1	114	↓
PP2A	ND	↓
PP1	ND	↓
SIRT1	ND	↑
VDAC1/2	ND	↑

ND, not determined.

4.2 Caveolins

In the mid-1990s, a second eNOS-interacting protein was identified by three separate groups as the integral membrane protein, caveolin-1 (Feron et al., 1996; Garcia-Cardena, Fan, Stern, Liu, & Sessa, 1996; Ju, Zou, Venema, & Venema, 1997). Caveolae are flask-shaped plasma membrane organelles that are enriched in endothelial cells and are formed via the actions and interactions of caveolar proteins that include caveolin-1. Caveolae participate in a variety of biological processes, including (but not limited to) cellular signaling and endocytosis (Patel, Murray, & Insel, 2008). Like calmodulin, caveolin-1 binds to eNOS in a stimulus-dependent manner. However in contrast to calmodulin, calcium-mobilizing agonists such as carbachol (analogous to acetylcholine) and calcium ionophore stimulate the disassociation of eNOS and caveolin-1 (Feron, Saldana, Michel, & Michel, 1998). The primary region of caveolin-1 that binds to eNOS lies between aa 60–101 and, to a lesser extent, aa 135–178 (Garcia-Cardena et al., 1997; Ju et al., 1997). This region is referred to as the caveolin-scaffolding domain (CSD) and peptides flanking this region (aa 82 and 101, DGIWKASFTTFTVTKYWFYR) can mimic the effect of caveolin-1 binding and disrupt the eNOS–caveolin-1 interaction (Garcia-Cardena et al., 1997; Michel, Feron, Sase, Prabhakar, & Michel, 1997). A further distinction with calmodulin is that caveolin-1 is a negative regulator of eNOS activity and NO production. This is seen with full-length caveolin-1 as well as CSD peptides and a CSD-fusion protein called cavtratin which consists of the cell-penetrating sequence from antennapedia linked to CSD (RQIKIWFQNRRMKWKK-CSD) (Garcia-Cardena et al., 1997; Ju et al., 1997; Michel et al., 1997). In addition to caveolin-1, caveolin-2 and caveolin-3 are present in caveolae isolated from endothelial cells and cardiac myocytes, respectively. The scaffolding domain of caveolin-3 is also capable of binding to eNOS and inhibiting its activity, but the equivalent region from caveolin-2 fails to inhibit eNOS (Feron et al., 1996; Garcia-Cardena et al., 1997). Caveolin-3 is predominantly expressed in muscle cells but is believed to interact with eNOS that is expressed in cardiac myocytes. This mechanism may have implications for the regulation of NO synthesis in the heart. Caveolin-1 has been shown to bind to a consensus caveolin-binding motif on eNOS that is located within its N-terminal domain (aa 348–356) and disruption of this site on eNOS via site-directed mutagenesis impairs the ability of caveolin-1 to antagonize eNOS (Garcia-Cardena et al., 1997). The ability of caveolin and CSD peptides to suppress eNOS activity is reversed by calcium–calmodulin and in cells by agonists that promote calcium–calmodulin binding to eNOS. This observation revealed a reciprocal relationship between allosteric binding partners with calmodulin stimulating and caveolin-1 inhibiting eNOS activity (Feron et al., 1998). In vivo, the loss of caveolin-1 in knockout mice results in altered cardiovascular and pulmonary physiology. Caveolin-1 knockout mice display enhanced endothelium-dependent relaxation and reduced blood pressure, findings that are consistent with the removal of a negative regulator of eNOS activity (Drab et al., 2001; Murata et al., 2007; Razani et al., 2001; Zhao et al., 2002). Caveolin-1 knockout mice also develop pulmonary hypertension and fibrosis as well as dilated cardiomyopathy. As caveolin-1 is expressed in other cell types (ie, fibroblasts), the phenotype of caveolin-1 knockout mice may result from the loss of caveolin-1 in other cell types where eNOS is not expressed. However, the endothelial-specific reexpression of caveolin-1 in caveolin-1 knockout mice corrects most of these phenotypes including the exaggerated blood vessel relaxation, the pulmonary hypertension, and cardiomyopathy and reverses the reduced systemic blood pressure. These data reflect the important role that endothelial caveolin-1 has in repressing eNOS activity and NO production in the endothelium (Murata et al., 2007). While some of these findings (pulmonary hypertension and fibrosis) may appear at odds with the well-accepted beneficial role of a NO-synthesizing eNOS, there are conditions where the fidelity of eNOS synthesis (reduced NO, increased superoxide production) can be compromised or where excessive NO signaling is inappropriate. In these settings, suppressing aberrant eNOS activity would be expected to have a protective effect and the loss of caveolin-1 pathologic (Zhao et al., 2009). Along these lines, caveolin-1 expression is decreased significantly in cancer, fibrosis, and pulmonary hypertension (Chen, Barman, et al., 2014; Koleske, Baltimore, & Lisanti, 1995; Wang et al., 2006) and cavtratin, which mimics some of the actions of caveolin-1, has beneficial actions in reducing endothelial hyperpermeability, tumor progression (Gratton et al., 2003) as well as the amelioration of pulmonary hypertension (Jasmin, Mercier, Dupuis, Tanowitz, & Lisanti, 2006) and fibrosis (Tourkina et al., 2008). A complicating factor is that the loss of eNOS-derived NO increases atherosclerosis (Fernandez-Hernando, Yu, Davalos, Prendergast, & Sessa, 2010; Kuhlencordt et al., 2001) and strategies to bolster endothelial NO, such as caveolin-1 deletion, are highly protective against atherosclerosis (Frank et al., 2004). Statins are widely used in the prevention of cardiovascular disease and also promote eNOS-derived NO, at least in part, via the decreased expression of caveolin-1 (Feron, Dessy, Desager, & Balligand, 2001). Additional mapping of the caveolin-1:eNOS interface revealed that aa 89–95, and particularly F92, of caveolin-1 mediates binding to eNOS (Bernatchez et al., 2005). Cell permeable CSD mutant peptides that incorporate the F92A mutation result in increased NO synthesis and vasodilation and reduced blood pressure when administered to mice (Bernatchez, Sharma, Bauer, Marin, & Sessa, 2011). These data are in agreement with experiments in genetically modified mice where endothelial expression of the F92A-modified caveolin-1 results in increased NO and decreased blood pressure without affecting the stability of caveolin-2 or the formation of caveolae (Kraehling et al., 2015). This is advantageous as the loss of caveolin-2 alone can promote fibrosis, endothelial cell proliferation, and pulmonary hypertension. Collectively, these results position caveolin-1 as an endogenous repressor of eNOS activity and the development of cell permeable peptides to mimic or inhibit the actions of caveolin-1 as useful tools in developing a greater understanding of the role of NO in disease states where caveolin-1 is implicated such as pulmonary hypertension and atherosclerosis.

4.3 G-Protein-Coupled Receptors

Shortly after the discovery of caveolin-1 as an eNOS regulatory protein, it was shown that the intracellular domain 4 (ID4) of the bradykinin 2 (B2) receptor and later that the angiotensin II R1 (AT1) and endothelin B (ETB) receptors can negatively regulate eNOS activity in vitro (Ju, Venema, Marrero, & Venema, 1998; Marrero et al., 1999). eNOS was shown to coprecipitate with the bradykinin B2 receptor and bind to GST fusions containing the ID4. Synthetic peptides derived from the B2 receptor ID4 (aa 310–334) potently inhibit the activity of eNOS (Ju et al., 1998). The binding of eNOS to the intracellular domains of G-protein-coupled receptors (GPCR) is decreased in response to agonists in a manner analogous to caveolin-1. Agonist-stimulated phosphorylation of the ID4 region is thought to mediate dissociation of the GPCR and eNOS and facilitate activation of eNOS (Marrero et al., 1999). Mechanistically, the ID4 peptide reduces the catalytic activity of eNOS by interfering with flavin to heme electron transfer (Golser et al., 2000). Other GPCR-regulated proteins have also been shown to regulate eNOS activity. The GPCR kinase interactor-1 (GIT1) binds to eNOS and increases NO production in a calcium-independent manner via the increased phosphorylation of eNOS (Liu, Premont, & Rockey, 2012).

4.4 Molecular Chaperones—Hsp90

Heat-shock protein 90 (Hsp90) belongs to a family of molecular chaperones that work in concert with other chaperones and proteins to enable the correct folding of substrates or client proteins. There are multiple forms of Hsp90 including cytosolic Hsp90s comprised of Hsp90α (inducible, HSP90AA) and Hsp90β (constitutive, HSP90AB), the endoplasmic reticulum restricted GRP94 (HSP90B1) and the mitochondrial HSP90L (TRAP1). In 1998, Hsp90 was found to be associated with eNOS and was the first example of the ability of molecular chaperones to regulate

NO synthesis. Endothelial agonists such as VEGF, histamine, estrogen, and shear stress promote the increased association of Hsp90 and eNOS. Increased amounts of Hsp90 bound to eNOS stimulate NO production, whereas pharmacological inhibitors of Hsp90 decrease NO synthesis (Garcia-Cardena et al., 1998; Russell et al., 2000; Venema, Zou, Ju, Marrero, & Venema, 1997). Domain-mapping studies have shown that eNOS binds to the M (substrate-binding region) of Hsp90 (aa 400–600) (Fontana et al., 2002; Xu et al., 2007) and that Hsp90 binds eNOS on the N-terminal oxygenase domain (aa 310–323). The isoforms of Hsp90 that bind to eNOS have also been investigated. Both Hsp90α (HSP90AA) and Hsp90β (HSP90AB) have been shown to bind to eNOS (Fontana et al., 2002). While there are reported differences in the ability of these isoforms to influence NO production, with Hsp90α stimulating and Hsp90β inhibiting eNOS activity (Cortes-Gonzalez et al., 2010), others have found that Hsp90β positively regulates eNOS activity (Fontana et al., 2002; Garcia-Cardena et al., 1998). Hsp90 also coordinates the close association of eNOS and soluble guanylate cyclase which has been proposed as a novel mechanism reducing the diffusion distance of NO that facilitates the efficient production of cGMP (Papapetropoulos et al., 2005). CHIP (carboxyl terminus of Hsp70-interacting protein) is an Hsp90 cochaperone that has also been shown to interact with eNOS. Rather than being an allosteric modulator of enzyme activity, CHIP instead promotes the translocation of eNOS from the Golgi complex into an insoluble and inactive intracellular compartment. Although CHIP is involved in the ubiquitination and degradation of Hsp90 client proteins, it does not promote eNOS degradation. The ability of CHIP to regulate eNOS activity has also been shown in vivo and CHIP knockout mice have greater expression of eNOS in the Golgi and plasma membrane and increased eNOS activity (Jiang, Cyr, Babbitt, Sessa, & Patterson, 2003).

4.5 NOSIP/NOSTRIN

Nitric oxide synthase-interacting protein (NOSIP) and nitric oxide synthase traffic inducer (NOSTRIN) have been identified as eNOS-associated proteins (Dedio et al., 2001; Zimmermann et al., 2002). Both NOSIP and NOSTRIN promote the translocation of eNOS from plasma membrane caveolae to intracellular compartments, such as the Golgi and cytoskeleton (Schleicher, Brundin, Gross, Muller-Esterl, & Oess, 2005). NOSIP and NOSTRIN decrease eNOS activity and NO release by displacing eNOS from the optimal signaling milieu at the plasma membrane in a manner analogous to CHIP which also binds to eNOS and promotes its intracellular redistribution.

4.6 Pin1

Pin1 (peptidylprolyl *cis/trans* isomerase, NIMA-interacting 1) binds to substrates that are phosphorylated on serine/threonine–proline motifs by proline-directed kinases and alters the function of substrates via *cis/trans* isomerization of the peptide bonds that link phosphoserine or phosphothreonine to proline. Pin1 associates with eNOS that is phosphorylated on S114 (discussed in more detail later). Pin1 is a negative regulator of eNOS activity and in endothelial cells the silencing of Pin1; expression of a dominant negative Pin1 or a pharmacological inhibitor of Pin1 increases NO release whereas the increased expression of Pin1 decreases NO levels (Ruan et al., 2011).

4.7 Cytoskeletal Proteins. Beta-Actin

In pulmonary endothelial cells, eNOS has been shown to colocalize and bind to components of the cytoskeleton such as F- and G-actin. Exposure of eNOS to purified actin increases NO production. The polymerization state of actin regulates the extent of eNOS activation and phalloidin, which selectively binds to F-actin filaments and prevents depolymerization, inhibits the eNOS–actin interaction, and reduces NO release (Su, Edwards-Bennett, Bubb, & Block, 2003). Using synthetic peptides and site-directed mutagenesis, the β-actin-binding site on eNOS was identified on the N-terminus (aa 326–333) (Kondrikov et al., 2010). eNOS can also be localized to microtubules in pulmonary endothelial cells and the ability of microtubule-disrupting agents such as nocodazole to reduce eNOS activity suggests that this interaction is functionally relevant (Su, Zharikov, & Block, 2002).

4.8 Endoglin

Endoglin is a homodimeric membrane glycoprotein that is highly expressed in proliferating endothelium and contributes to the process of angiogenesis (Nassiri et al., 2011). Endoglin associates with eNOS in plasma membrane caveolae where it stabilizes eNOS expression, facilitates the interaction of eNOS and Hsp90, and promotes NO release (Toporsian et al., 2005).

Genetic deficiency of endoglin reduces NO release in vitro and, in mice, reduces agonist-induced vasodilation (Jerkic et al., 2004).

4.9 Voltage-Dependent Anion Channels

Using an MS screening approach, voltage-dependent anion channel (VDAC) 1 (Porin) was identified as an eNOS-associated protein. Endothelial agonists such as bradykinin promote the binding of porin to eNOS and overexpression of porin augments eNOS activity (Sun & Liao, 2002). Other studies have shown that both VDAC1 and VDAC2 can bind to eNOS, but only VDAC2 regulates eNOS activity (Alvira et al., 2012). The mechanism by which VDACs regulate eNOS activity is not yet established.

4.10 Arginine Regulatory Enzymes

The substrate for eNOS is the amino acid L-arginine. Low concentrations of L-arginine have been proposed to account for the inability of eNOS to synthesize NO in disease states and a cause of endothelial dysfunction. However, this concept remains controversial and poorly understood. The high concentration of L-arginine in endothelial cells (>100 μM) and the high affinity of eNOS for L-arginine (k_m ~2–3 μM) coupled with the ability of high (mM) concentrations of extracellular L-arginine to stimulate NO synthesis have prompted the term "arginine paradox" to explain the apparent ability of L-arginine to regulate eNOS activity when it should not be possible. One explanation for the arginine paradox is the close association and compartmentalization of eNOS with arginine transporters and recycling enzymes such as CAT-1 and argininosuccinate lyase (ASL) (Erez et al., 2011; McDonald, Zharikov, Block, & Kilberg, 1997). However, while eNOS has been shown to interact directly with the arginine transporter, CAT-1, its ability to regulate eNOS activity is actually independent of L-arginine transport (Li, Huang, Harris, Goolsby, & Venema, 2005). The compartmentalization of L-arginine is unlikely to impact eNOS activity significantly as the activity of eNOS in different intracellular locations under conditions of L-arginine deprivation and excess is consistent with that of a freely diffusible substrate (Elms et al., 2013).

5. SUBCELLULAR LOCALIZATION OF eNOS

In the endothelial cell, eNOS can be detected at the plasma membrane as well as the Golgi complex (Sessa et al., 1995). The subcellular targeting of eNOS to cellular membranes and intracellular compartmentalization is mediated by protein fatty acid acylation. This involves two N-terminal lipid modifications: the cotranslational attachment of myristic acid to glycine-2 (G2) and posttranslational palmitoylation of cysteines-15 and 26 (C15, C26). Myristoylation is the first step and is necessary for the membrane association of eNOS and a requirement for subsequent palmitoylation (Liu, Garcia-Cardena, & Sessa, 1995; Prabhakar, Cheng, & Michel, 2000). Palmitoylation of eNOS occurs in the Golgi by protein acyl transferases such as ZDHHC21 which then targets eNOS to plasmalemmal caveolae (Fernandez-Hernando et al., 2006; Garcia-Cardena, Oh, Liu, Schnitzer, & Sessa, 1996; Shaul et al., 1996). The palmitoylation of eNOS is a reversible modification and depalmitoylation has been proposed as a mechanism that regulates the intracellular traffic and distribution of eNOS in endothelial cells (Yeh, Duncan, Yamashita, & Michel, 1999). The attachment of fatty acids is crucial for eNOS activity in intact cells and this is demonstrated by the dramatically lower amounts of NO released from cells expressing G2A mutants of eNOS which lack both myristoylation and palmitoylation (Sessa, Barber, & Lynch, 1993). In isolated enzyme activity assays, the G2A eNOS has kinetic properties that are equal to that of the WT enzyme which suggests that its reduced activity in intact cells results from the inability to target optimal cellular subdomains (Sessa et al., 1993). In addition to acylation, polybasic regions on eNOS have been shown to facilitate membrane association (Gao et al., 2004; Venema, Sayegh, Arnal, & Harrison, 1995) but in the absence of acylation are not sufficient to promote membrane association. The trafficking of eNOS to plasma membrane caveolae is consistent with the concept that this location is optimal for the synthesis of NO. At the plasma membrane, eNOS is in close proximity to ligand-accessible GPCRs, scavenging receptors, ion channels regulating calcium entry, kinases and arginine transporters, and recycling enzymes (Shaul, 2002). Numerous factors (oxLDL, hypochlorite, NOSIP, NOSTRIN, and CHIP) have been shown to promote the redistribution of eNOS away from the plasma membrane to intracellular sites which results in a concomitant decrease in eNOS activity (Dedio et al., 2001; Nuszkowski et al., 2001; Uittenbogaard, Shaul, Yuhanna, Blair, & Smart, 2000; Zimmermann et al., 2002). While these data support the plasma membrane and caveolae as premier destinations and optimal platforms for NO synthesis, in mature blood vessels only a fraction of the eNOS pool can be detected at the plasma membrane. The bulk of the eNOS pool resides in perinuclear organelles such as the Golgi (Andries, Brutsaert, & Sys, 1998; O'Brien, Young, Povey, & Furness, 1995) and these results are consistent with observations in cultured cells (Govers, van der Sluijs, van Donselaar, Slot, & Rabelink, 2002). To more precisely determine the impact of location on eNOS activity at the Golgi or plasma membranes, nonacylated (G2A) eNOS was targeted to either the cytoplasmic face of the Golgi or the plasma membrane via fusion to protein domains that target these sites. Plasma membrane-targeted eNOS was constitutively active and highly sensitive to transmembrane calcium fluxes. In contrast, *cis*-Golgi-targeted eNOS was less responsive to calcium and maximally activated by the protein kinase Akt. Targeting of eNOS to the *trans*-Golgi or to the cytoplasm was less optimal and resulted in both reduced calcium and Akt-dependent activation (Fulton et al., 2004). The mechanism underlying the ability of intracellular location to influence eNOS activity was identified using calcium-insensitive NOS constructs (iNOS and eNOS) which were targeted to the cytosol, Golgi, and plasma membrane. These studies demonstrate that calcium, but not other eNOS cofactors or substrates (including L-arginine), is the primary variable constraining NO production in different parts of the cell (Church & Fulton, 2006; Jagnandan, Sessa, & Fulton, 2005). As discussed later in more detail, the phosphorylation pattern of eNOS also changes with intracellular location and is a major mechanism underlying the calcium sensitization and calcium-dependent regulation of eNOS (Fulton et al., 2002; McCabe, Fulton, Roman, & Sessa, 2000). Studies have also shown that eNOS and caveolin-1 can translocate into the nucleus and that eNOS is present on mitochondrial membranes (Brodsky, Gao, Li, & Goligorsky, 2002; Feng, Venema, Venema, Tsai, & Caldwell, 1999). However, the targeting of eNOS to the nucleus or the mitochondria results in an enzyme that produces very little NO. In contrast, targeting calcium-insensitive (Jagnandan et al., 2005) NOS to the same locations does not affect NO synthesis.

While intracellular location is clearly an important factor regulating the degree of eNOS activity and NO synthesis, much less is known about how the intracellular compartmentalization of eNOS contributes to downstream signal transduction. It has been shown that Golgi-localized eNOS can influence the S-nitrosylation of nearby proteins in the Golgi (Iwakiri et al., 2006; Sangwung et al., 2012). Given that localization can profoundly impact eNOS activity, an important factor regulating S-nitrosylation is also the amount of NO synthesized. When expressed in endothelial cells, PM-targeted eNOS, which synthesizes higher levels of NO, promotes greater cGMP-dependent signaling in adjacent smooth muscle cells, enhanced endothelium-dependent relaxation, and greater S-nitrosylation of NSF than a Golgi-targeted enzyme (Zhang et al., 2006). However, using calcium-insensitive forms of eNOS, which produce equal amounts of NO regardless of location, it was shown that the amount of NO generated and not the location of synthesis is the most important variable influencing cGMP production, S-nitrosylation of NSF, and suppression of vWF secretion (Qian et al., 2010). While the plasma membrane can be an optimal location for the highest production of NO from eNOS, it is not without negative consequences. At the plasma membrane, eNOS is more susceptible to extracellular influences such as oxidized LDL which selectively reduces NO release from PM-targeted eNOS (Blair, Shaul, Yuhanna, Conrad, & Smart, 1999), whereas a Golgi-targeted eNOS is more resilient (Zhang et al., 2006).

6. eNOS PHOSPHORYLATION

Early studies showed that in response to agonists and shear stress, eNOS becomes hyperphosphorylated on serine residues and to a lesser extent on tyrosine and threonine residues (Corson et al., 1996; Garcia-Cardena, Fan, et al., 1996; Michel, Li, & Busconi, 1993). To date, seven primary sites of eNOS phosphorylation have been identified which include Y81, S114, T495, S615, S633, Y657, and S1177 (equivalent to Y83, S116, T497, S617, S635, Y659, and S1179 of the bovine eNOS isoform due to an additional 2 aa in the bovine sequence, summarized in Table 2). The phosphorylation of human eNOS S1177 on the C-terminal reductase domain was one of the first eNOS phosphorylation sites identified and its phosphorylation by the protein kinase Akt has been shown to be important in the

Table 2 eNOS Phosphorylation Sites and Kinases

Phosphorylation Site	Effect on eNOS Activity	Kinase
S114	↓	ERK
Y83	↑	SRC
T495	↓	PKC, ROCK
S615	↑	AMPK
S633	↑	AMPK, PKA
Y659	↓	PYK2
S1179	↑	Akt, PKA, AMPK

activation of eNOS in response to VEGF and shear stress (Dimmeler et al., 1999; Fulton et al., 1999; Gallis et al., 1999). Other protein kinases have also been shown to phosphorylate eNOS at S1177, including adenosine monophosphate-activated kinase (AMPK) (Chen et al., 1999), CaM kinase II (Fleming, Fisslthaler, Dimmeler, Kemp, & Busse, 2001), protein kinase A (PKA) (Boo et al., 2002; Michell et al., 2001), checkpoint kinase 1 (Chk1) (Park et al., 2011), and protein kinase G (PKG) (Butt et al., 2000). The increased phosphorylation of S1177 directly impacts enzyme activity and mimicking the S1177 phosphorylated form of eNOS using a genetically encoded mutant that mimics the negative charge of phosphorylation (S1177D eNOS) reveals that there is both increased electron flow in the reductase domain and increased sensitivity to calcium–calmodulin. These properties enable eNOS to remain active at basal or low levels of calcium and explain, at least in part, reports of calcium-independent eNOS activity in response to shear stress and endothelial agonists that do not mobilize intracellular calcium (Butt et al., 2000; McCabe et al., 2000). Other sites of phosphorylation on eNOS can also influence enzyme activity. Increased phosphorylation of both S615 and S633 has also been shown to contribute to eNOS-derived NO release (Boo et al., 2003; Mount, Kemp, & Power, 2007). The kinases responsible for S615/633 phosphorylation include PKA, AMPK, Akt, and Pim1 (Bauer et al., 2003; Boo et al., 2003; Chen et al., 2016; Michell et al., 2002). Phosphomimetics of S615 (eNOS S615D) display greater sensitivity to calcium–calmodulin without an overall change in eNOS enzyme activity (Michell et al., 2002). Phosphomimetics of S633 (eNOS S633D) have enhanced overall activity compared to WT as well as increased sensitivity to calcium–calmodulin. It is noteworthy that all three of these phosphorylation sites impact the calcium sensitivity of eNOS, and S615/S633 and S1179 are located within two key autoinhibitory regions in the reductase domain of eNOS. The calcium-independent iNOS lacks these autoinhibitory regions and deletion of both of these regions in eNOS results in an enzyme that, like iNOS, is insensitive to changes in calcium. These results strongly support the concept that the C-terminal phosphorylation of eNOS at S615/S633 and S1179 increases the calcium sensitivity and overall activity of eNOS via displacement of autoinhibitory domains.

Not all of the eNOS phosphorylation sites enhance enzyme activity. The phosphorylation of T497 has been proposed to inhibit eNOS activity and phosphomimetic mutants (eNOS T495D) display substantially reduced ability to synthesize NO. Mechanistically, the phosphorylation of T495 is thought to interfere with the binding of calcium-activated calmodulin to eNOS as well as alter eNOS fidelity in favor of increased superoxide production and reduced nitric oxide (Chen, Kumar, et al., 2014; Fleming et al., 2001; Lin et al., 2003). The phosphorylation of T497 occurs primarily by members of the protein kinase C family (PKC) (Chen, Kumar, et al., 2014; Fleming et al., 2001; Michell et al., 2001; Oubaha & Gratton, 2009) as well as the Rho kinase, ROCK (Sugimoto et al., 2007), and AMPK (Chen et al., 1999). Agonists such as bradykinin, which stimulate NO release and S1177 phosphorylation, promote the reciprocal dephosphorylation of T495 (Fleming et al., 2001; Harris et al., 2001), which enables calmodulin binding and eNOS activation.

S114 was first discovered as a site of phosphorylation resulting from increased shear stress on endothelial cells (Gallis et al., 1999). Although initially controversial, the consensus on the impact of S114 phosphorylation on eNOS activity is that it is inhibitory (Mount et al., 2007). Evidence to support this results from the ability of VEGF and bradykinin to induce the dephosphorylation of S114, the increased activity of a phospho-null mutation (eNOS S114A) (Bauer et al., 2003; Kou, Greif, & Michel, 2002; Li et al., 2007) and the reduced activity of a phosphomimetic (eNOS S114D) which decreases basal NO release from endothelial cells and impairs endothelium-dependent relaxation in aortic rings (Li et al., 2007). The mechanism by which S114 phosphorylation impacts eNOS activity is complex and isolated enzyme activity of S114A and S114D mutants of eNOS is not different from WT (Ruan et al., 2013). However, in cells, the increased phosphorylation promotes greater association with negative regulators of eNOS activity, notably caveolin-1 and Pin1 (described earlier) (Li et al., 2007; Ruan et al., 2011). Moreover, dephosphorylation of S114 increases c-Src binding and eNOS activation via the phosphorylation of Y81 (Ruan et al., 2013).

Y81 is a recently identified eNOS phosphorylation site (Fulton et al., 2005, 2008) which is phosphorylated by Src kinase in response to VEGF and other eNOS-activating agonists. Phosphorylation of Y81 increases eNOS activity and NO production in both cotransfected COS-7 cells and endothelial cells. Phospho-null mutants (eNOS Y81F) produce less NO and exhibit impaired endothelium-dependent relaxation when reconstituted in the endothelium of aortae from eNOS knockout mice (Fulton et al., 2008). Tyrosine phosphorylation has also been reported on Y657 by the proline-rich tyrosine kinase 2 (PYK2). Phosphorylation of this site is thought to impair eNOS activity (Loot, Schreiber, Fisslthaler, & Fleming, 2009).

Our understanding of the mechanisms by which eNOS becomes dephosphorylated, while equally as important as phosphorylation, remains poorly understood by comparison. Pharmacological inhibition of both serine/threonine phosphatases and tyrosine phosphatases results in robust increases in eNOS phosphorylation and activity (Church & Fulton, 2006; Papapetropoulos et al., 2004), suggesting that a dominant effect of phosphatase inhibition is the increased phosphorylation of stimulatory eNOS phosphorylation sites. PP2A is a negative regulator of eNOS activity and this is likely due to multiple effects including the direct dephosphorylation of eNOS as well as the dephosphorylation and inhibition of kinases such as Akt that phosphorylate eNOS directly (Loot et al., 2009). PP2A binds to eNOS and promotes the dephosphorylation of multiple sites including S1177 and T495 (Bharath et al., 2015; Greif, Kou, & Michel, 2002; Michell et al., 2001; Park et al., 2013). PP1 also binds eNOS, but is more specific for T495 (Michell et al., 2001). The dephosphorylation of T495 and S114 is also blocked by calcineurin inhibitors (Harris et al., 2001; Kou et al., 2002; Ruan et al., 2013). A greater appreciation of the roles of individual phosphatases regulating eNOS phosphorylation is limited by imperfect tools of discovery and the challenges inherent in studying dephosphorylation. Serine/threonine phosphatases function as multi-subunit complexes and catalyze the dephosphorylation of sites on both upstream kinases (which may be activated or inhibited) and substrates leading to complex interpretations of changes in substrate phosphorylation. Evidence also exists for cross talk between the substrate phosphorylation sites and the phosphorylation of one site may mask another.

7. NITROSATIVE MODIFICATION OF eNOS

S-Nitrosylation is the covalent modification of protein cysteine thiols by NO to form S-nitrosothiols (Lipton et al., 1993). S-Nitrosylation has been proposed as an alternative signaling pathway for NO that mediates cGMP-independent effects (Lima, Forrester, Hess, & Stamler, 2010). Given that eNOS is the primary source of NO in the endothelium, it is no surprise that eNOS itself is S-nitrosylated. More surprising is the paradoxical ability of endothelial agonists, which increase NO synthesis, to promote the rapid denitrosylation (loss of NO) of eNOS (Erwin, Lin, Golan, & Michel, 2005). These data suggest that the mechanisms mediating the denitrosylation of proteins predominate following agonist stimulation despite the increased levels of NO. S-Nitrosylated eNOS exhibits reduced catalytic activity and this can be reversed with the release of bound NO (Erwin et al., 2005). The primary sites of eNOS S-nitrosylation are C94 and C98 which ordinarily participate in the formation of a zinc tetrathiolate cluster, a structure important in regulating the dimerization of eNOS. Like the phosphorylation of S1177, the localization of eNOS also influences the degree of S-nitrosylation. eNOS targeted to the cytosol is less nitrosylated than a membrane-targeted form of eNOS (Erwin et al., 2005; Erwin, Mitchell, Sartoretto, Marletta, & Michel, 2006). The degree of S-nitrosylation is proportional to the level of NO generated at rest from these constructs (as described earlier). The tyrosine nitration of eNOS has also been reported on multiple sites (Zickus, Fonseca, Tummala, Black, & Ryzhov, 2008) and may be caused by nitrating specifies such a peroxynitrite (ONOO–) which is formed by the diffusion-limited reaction of NO and superoxide. Tyrosine nitration is considered a stable modification and indices of the reduced biological activity of NO. Despite the stability of this modification, mechanisms of denitration have been reported including enzymatic denitrases (Deeb et al., 2013). The primary sites of tyrosine nitration are surface accessible tyrosine residues such as Y81, Y163, Y210, Y331, Y357, Y373, and Y410 (Zickus et al., 2008). The specific impact of tyrosine nitration on eNOS activity is not well described, but peroxynitrite has been shown by numerous groups to inactivate eNOS (Zou, Shi, & Cohen, 2002).

8. PROTEIN ACETYLATION

Aspirin stimulates the production of NO from the endothelium and promotes the acetylation of eNOS on K609 as well as S765 and S771. The ability of aspirin to stimulate eNOS activity is reduced in the eNOS mutant, K609R and to a lesser degree by mutation of S765/771R (Jung et al., 2010; O'Kane et al., 2009; Taubert et al., 2004). The acetylation of K609 and resultant increased activity of eNOS are reversed by the histone deacetylase, HDAC3 (Jung et al., 2010). SIRT1 also deacetylates eNOS, but in contrast to the inhibitory effects of HDAC3, it promotes an increase in eNOS activity (Mattagajasingh et al., 2007). An explanation for this discrepancy is that SIRT1 deacetylates different residues on eNOS, namely K494 and K504 which lie within the calmodulin-binding domain of eNOS. Deacetylation of these residues is proposed to enable more efficient binding of calcium–calmodulin and thus eNOS activation.

9. GLUTATHIONYLATION

Glutathionylation involves the reversible attachment of the tripeptide glutathione to cysteine residues in target proteins. The levels of protein glutathionylation are increased by conditions of elevated oxidative stress. Oxidized glutathione has been shown to promote the glutathionylation of eNOS which results in the decreased ability of eNOS to make NO and an increase in superoxide generation. Glutathionylation occurs in the C-terminal reductase domain of eNOS at C689 and C908. The glutathionylation of these residues results in the disruption of flavin-dependent electron transport and reduced NO synthesis. Mutation of C689/908S prevents the glutathionylation of eNOS, maintains NO generation, and reverses the increase in superoxide (Chen et al., 2010). The glutathionylation of eNOS is reversed by the enzyme glutaredoxin (grx1) which forms a close association with eNOS and the loss of grx1 via genetic silencing increases the glutathionylation of eNOS and decreases NO synthesis (Chen, De Pascali, Basye, Hemann, & Zweier, 2013).

10. INTEGRATION OF eNOS REGULATORY MECHANISMS

The myriad pathways of eNOS regulation can seem bewildering. The simplest view of eNOS regulation is that calcium–calmodulin is a dynamic and robust mechanism of enzyme activation. However, this mechanism alone is insufficient to connect eNOS activity to the diverse stimuli presented to endothelial cells in different types of blood vessels with different shear rates, and nascent and growing blood vessels as well as in conditions of inflammation and injury. Greater clarity in how the multiplicity of mechanisms operates can be achieved when viewed from an integrated perspective. For example, while there are multiple positive and negative regulatory proteins that interact with eNOS, the degree of association can change in a coordination fashion. Endothelial agonists promote increased association of calmodulin and Hsp90 but decreased binding of inhibitory interactors such as caveolin-1 and GPCRs with the net effect being an increase in eNOS activity. Similarly, the increased binding of Hsp90 can facilitate the recruitment of kinases resulting in increased phosphorylation of eNOS. Phosphorylation occurs on specific sites that promote enzyme activity (Y83, S615, S633, and S1177) and there is reciprocal dephosphorylation of sites that limit enzyme activity (S114, Y652, and T495). S-Nitrosylation of eNOS negatively regulates activity, yet endothelial agonists promote denitrosylation of eNOS in the face of increased NO. The proper intracellular targeting of eNOS also involves increased binding of Hsp90 and changes in eNOS phosphorylation that work together to enable more efficient production of NO at lower levels of calcium. Over the longer term, pathways such as shear stress which stimulate numerous posttranslational mechanisms of eNOS regulation can transition to changes in transcriptional mechanisms that drive increased expression of eNOS. Disease may bias the pathways of eNOS activation or inactivation and disrupt or supersede these coordinated pathways of eNOS activation.

11. CONCLUSION

Given the importance of eNOS to cardiovascular physiology and pathophysiology the mechanisms regulating its expression level in endothelial cells and posttranslational control of NO synthesis have been intensely studied for the past two and a half decades. The fruits of this labor are that we have acquired a deep understanding of the mechanisms coordinating eNOS function as well as those contributing to dysfunction. This knowledge can be readily applied to a greater understanding of the mechanisms of disease as well as the development of selective tools such as the caveolin mimetic peptides that can selectively increase or decrease eNOS function as necessary to treat disease.

CONFLICT OF INTEREST

The author has no financial relationships with any organizations that might have an interest in the submitted work.

REFERENCES

Alvira, C. M., Umesh, A., Husted, C., Ying, L., Hou, Y., Lyu, S. C., … Cornfield, D. N. (2012). Voltage-dependent anion channel-2 interaction with nitric oxide synthase enhances pulmonary artery endothelial cell nitric oxide production. *American Journal of Respiratory Cell and Molecular Biology*, *47*(5), 669–678. http://dx.doi.org/10.1165/rcmb.2011-0436OC.

Andries, L. J., Brutsaert, D. L., & Sys, S. U. (1998). Nonuniformity of endothelial constitutive nitric oxide synthase distribution in cardiac endothelium. *Circulation Research*, *82*(2), 195–203.

Arnold, W. P., Mittal, C. K., Katsuki, S., & Murad, F. (1977). Nitric oxide activates guanylate cyclase and increases guanosine 3':5'-cyclic monophosphate levels in various tissue preparations. *Proceedings of the National Academy of Sciences of the United States of America*, *74*(8), 3203–3207.

Baek, K. J., Thiel, B. A., Lucas, S., & Stuehr, D. J. (1993). Macrophage nitric oxide synthase subunits. Purification, characterization, and role of prosthetic groups and substrate in regulating their association into a dimeric enzyme. *The Journal of Biological Chemistry, 268*(28), 21120–21129.

Balligand, J. L., & Cannon, P. J. (1997). Nitric oxide synthases and cardiac muscle. Autocrine and paracrine influences. *Arteriosclerosis, Thrombosis, and Vascular Biology, 17*(10), 1846–1858.

Balligand, J. L., Ungureanu-Longrois, D., Simmons, W. W., Pimental, D., Malinski, T. A., Kapturczak, M., … Kelly, R. A. (1994). Cytokine-inducible nitric oxide synthase (iNOS) expression in cardiac myocytes. Characterization and regulation of iNOS expression and detection of iNOS activity in single cardiac myocytes in vitro. *The Journal of Biological Chemistry, 269*(44), 27580–27588.

Bauer, P. M., Fulton, D., Boo, Y. C., Sorescu, G. P., Kemp, B. E., Jo, H., & Sessa, W. C. (2003). Compensatory phosphorylation and protein-protein interactions revealed by loss of function and gain of function mutants of multiple serine phosphorylation sites in endothelial nitric-oxide synthase. *The Journal of Biological Chemistry, 278*(17), 14841–14849. http://dx.doi.org/10.1074/jbc.M211926200.

Beasley, D., Schwartz, J. H., & Brenner, B. M. (1991). Interleukin 1 induces prolonged L-arginine-dependent cyclic guanosine monophosphate and nitrite production in rat vascular smooth muscle cells. *The Journal of Clinical Investigation, 87*(2), 602–608. http://dx.doi.org/10.1172/JCI115036.

Bernatchez, P. N., Bauer, P. M., Yu, J., Prendergast, J. S., He, P., & Sessa, W. C. (2005). Dissecting the molecular control of endothelial NO synthase by caveolin-1 using cell-permeable peptides. *Proceedings of the National Academy of Sciences of the United States of America, 102*(3), 761–766. http://dx.doi.org/10.1073/pnas.0407224102.

Bernatchez, P., Sharma, A., Bauer, P. M., Marin, E., & Sessa, W. C. (2011). A noninhibitory mutant of the caveolin-1 scaffolding domain enhances eNOS-derived NO synthesis and vasodilation in mice. *The Journal of Clinical Investigation, 121*(9), 3747–3755. http://dx.doi.org/10.1172/JCI44778.

Bharath, L. P., Ruan, T., Li, Y., Ravindran, A., Wan, X., Nhan, J. K., … Symons, J. D. (2015). Ceramide-initiated protein phosphatase 2A activation contributes to arterial dysfunction in vivo. *Diabetes, 64*(11), 3914–3926. http://dx.doi.org/10.2337/db15-0244.

Blair, A., Shaul, P. W., Yuhanna, I. S., Conrad, P. A., & Smart, E. J. (1999). Oxidized low density lipoprotein displaces endothelial nitric-oxide synthase (eNOS) from plasmalemmal caveolae and impairs eNOS activation. *The Journal of Biological Chemistry, 274*(45), 32512–32519.

Boo, Y. C., Sorescu, G. P., Bauer, P. M., Fulton, D., Kemp, B. E., Harrison, D. G., … Jo, H. (2003). Endothelial NO synthase phosphorylated at SER635 produces NO without requiring intracellular calcium increase. *Free Radical Biology & Medicine, 35*(7), 729–741.

Boo, Y. C., Sorescu, G., Boyd, N., Shiojima, I., Walsh, K., Du, J., & Jo, H. (2002). Shear stress stimulates phosphorylation of endothelial nitric-oxide synthase at Ser1179 by Akt-independent mechanisms: Role of protein kinase A. *The Journal of Biological Chemistry, 277*(5), 3388–3396. http://dx.doi.org/10.1074/jbc.M108789200.

Bouloumie, A., Schini-Kerth, V. B., & Busse, R. (1999). Vascular endothelial growth factor up-regulates nitric oxide synthase expression in endothelial cells. *Cardiovascular Research, 41*(3), 773–780.

Bredt, D. S., & Snyder, S. H. (1990). Isolation of nitric oxide synthetase, a calmodulin-requiring enzyme. *Proceedings of the National Academy of Sciences of the United States of America, 87*(2), 682–685.

Brodsky, S. V., Gao, S., Li, H., & Goligorsky, M. S. (2002). Hyperglycemic switch from mitochondrial nitric oxide to superoxide production in endothelial cells. *American Journal of Physiology. Heart and Circulatory Physiology, 283*(5), H2130–H2139. http://dx.doi.org/10.1152/ajpheart.00196.2002.

Busse, R., & Mulsch, A. (1990). Calcium-dependent nitric oxide synthesis in endothelial cytosol is mediated by calmodulin. *FEBS Letters, 265*(1–2), 133–136.

Butt, E., Bernhardt, M., Smolenski, A., Kotsonis, P., Frohlich, L. G., Sickmann, A., ... Schmidt, H. H. (2000). Endothelial nitric-oxide synthase (type III) is activated and becomes calcium independent upon phosphorylation by cyclic nucleotide-dependent protein kinases. *The Journal of Biological Chemistry*, *275*(7), 5179–5187.

Caulin-Glaser, T., Garcia-Cardena, G., Sarrel, P., Sessa, W. C., & Bender, J. R. (1997). 17 beta-estradiol regulation of human endothelial cell basal nitric oxide release, independent of cytosolic Ca2+ mobilization. *Circulation Research*, *81*(5), 885–892.

Chan, Y., Fish, J. E., D'Abreo, C., Lin, S., Robb, G. B., Teichert, A. M., ... Marsden, P. A. (2004). The cell-specific expression of endothelial nitric-oxide synthase: A role for DNA methylation. *The Journal of Biological Chemistry*, *279*(33), 35087–35100. http://dx.doi.org/10.1074/jbc.M405063200.

Chen, F., Barman, S., Yu, Y., Haigh, S., Wang, Y., Black, S. M., ... Fulton, D. J. (2014). Caveolin-1 is a negative regulator of NADPH oxidase-derived reactive oxygen species. *Free Radical Biology & Medicine*, *73*, 201–213. http://dx.doi.org/10.1016/j.freeradbiomed.2014.04.029.

Chen, C. A., De Pascali, F., Basye, A., Hemann, C., & Zweier, J. L. (2013). Redox modulation of endothelial nitric oxide synthase by glutaredoxin-1 through reversible oxidative post-translational modification. *Biochemistry*, *52*(38), 6712–6723. http://dx.doi.org/10.1021/bi400404s.

Chen, F., Kumar, S., Yu, Y., Aggarwal, S., Gross, C., Wang, Y., ... Fulton, D. J. (2014). PKC-dependent phosphorylation of eNOS at T495 regulates eNOS coupling and endothelial barrier function in response to G+ -toxins. *PLoS One*, *9*(7), e99823. http://dx.doi.org/10.1371/journal.pone.0099823.

Chen, Z. P., Mitchelhill, K. I., Michell, B. J., Stapleton, D., Rodriguez-Crespo, I., Witters, L. A., ... Kemp, B. E. (1999). AMP-activated protein kinase phosphorylation of endothelial NO synthase. *FEBS Letters*, *443*(3), 285–289.

Chen, C. A., Wang, T. Y., Varadharaj, S., Reyes, L. A., Hemann, C., Talukder, M. A., ... Zweier, J. L. (2010). S-glutathionylation uncouples eNOS and regulates its cellular and vascular function. *Nature*, *468*(7327), 1115–1118. http://dx.doi.org/10.1038/nature09599.

Chen, P. F., & Wu, K. K. (2003). Structural elements contribute to the calcium/calmodulin dependence on enzyme activation in human endothelial nitric-oxide synthase. *The Journal of Biological Chemistry*, *278*(52), 52392–52400. http://dx.doi.org/10.1074/jbc.M305469200.

Chen, M., Yi, B., Zhu, N., Wei, X., Zhang, G. X., Huang, S., & Sun, J. (2016). Pim1 kinase promotes angiogenesis through phosphorylation of endothelial nitric oxide synthase at Ser-633. *Cardiovascular Research*, *109*(1), 141–150. http://dx.doi.org/10.1093/cvr/cvv250.

Church, J. E., & Fulton, D. (2006). Differences in eNOS activity because of subcellular localization are dictated by phosphorylation state rather than the local calcium environment. *The Journal of Biological Chemistry*, *281*(3), 1477–1488. http://dx.doi.org/10.1074/jbc.M505968200.

Corson, M. A., James, N. L., Latta, S. E., Nerem, R. M., Berk, B. C., & Harrison, D. G. (1996). Phosphorylation of endothelial nitric oxide synthase in response to fluid shear stress. *Circulation Research*, *79*(5), 984–991.

Cortes-Gonzalez, C., Barrera-Chimal, J., Ibarra-Sanchez, M., Gilbert, M., Gamba, G., Zentella, A., ... Bobadilla, N. A. (2010). Opposite effect of Hsp90alpha and Hsp90beta on eNOS ability to produce nitric oxide or superoxide anion in human embryonic kidney cells. *Cellular Physiology and Biochemistry*, *26*(4-5), 657–668. http://dx.doi.org/10.1159/000322333.

Dedio, J., Konig, P., Wohlfart, P., Schroeder, C., Kummer, W., & Muller-Esterl, W. (2001). NOSIP, a novel modulator of endothelial nitric oxide synthase activity. *The FASEB Journal*, *15*(1), 79–89. http://dx.doi.org/10.1096/fj.00-0078com.

Deeb, R. S., Nuriel, T., Cheung, C., Summers, B., Lamon, B. D., Gross, S. S., & Hajjar, D. P. (2013). Characterization of a cellular denitrase activity that reverses nitration of cyclooxygenase. *American Journal of Physiology. Heart and Circulatory Physiology, 305*(5), H687–H698. http://dx.doi.org/10.1152/ajpheart.00876.2012.

Dimmeler, S., Fleming, I., Fisslthaler, B., Hermann, C., Busse, R., & Zeiher, A. M. (1999). Activation of nitric oxide synthase in endothelial cells by Akt-dependent phosphorylation. *Nature, 399*(6736), 601–605. http://dx.doi.org/10.1038/21224.

Drab, M., Verkade, P., Elger, M., Kasper, M., Lohn, M., Lauterbach, B., ... Kurzchalia, T. V. (2001). Loss of caveolae, vascular dysfunction, and pulmonary defects in caveolin-1 gene-disrupted mice. *Science, 293*(5539), 2449–2452. http://dx.doi.org/10.1126/science.1062688.

Drummond, G. R., Cai, H., Davis, M. E., Ramasamy, S., & Harrison, D. G. (2000). Transcriptional and posttranscriptional regulation of endothelial nitric oxide synthase expression by hydrogen peroxide. *Circulation Research, 86*(3), 347–354.

Dudzinski, D. M., Igarashi, J., Greif, D., & Michel, T. (2006). The regulation and pharmacology of endothelial nitric oxide synthase. *Annual Review of Pharmacology and Toxicology, 46*, 235–276. http://dx.doi.org/10.1146/annurev.pharmtox.44.101802.121844.

Elms, S., Chen, F., Wang, Y., Qian, J., Askari, B., Yu, Y., ... Fulton, D. J. (2013). Insights into the arginine paradox: Evidence against the importance of subcellular location of arginase and eNOS. *American Journal of Physiology. Heart and Circulatory Physiology, 305*(5), H651–H666. http://dx.doi.org/10.1152/ajpheart.00755.2012.

Erez, A., Nagamani, S. C., Shchelochkov, O. A., Premkumar, M. H., Campeau, P. M., Chen, Y., ... Lee, B. (2011). Requirement of argininosuccinate lyase for systemic nitric oxide production. *Nature Medicine, 17*(12), 1619–1626. http://dx.doi.org/10.1038/nm.2544.

Erwin, P. A., Lin, A. J., Golan, D. E., & Michel, T. (2005). Receptor-regulated dynamic S-nitrosylation of endothelial nitric-oxide synthase in vascular endothelial cells. *The Journal of Biological Chemistry, 280*(20), 19888–19894. http://dx.doi.org/10.1074/jbc.M413058200.

Erwin, P. A., Mitchell, D. A., Sartoretto, J., Marletta, M. A., & Michel, T. (2006). Subcellular targeting and differential S-nitrosylation of endothelial nitric-oxide synthase. *The Journal of Biological Chemistry, 281*(1), 151–157. http://dx.doi.org/10.1074/jbc.M510421200.

Feng, Y., Venema, V. J., Venema, R. C., Tsai, N., & Caldwell, R. B. (1999). VEGF induces nuclear translocation of Flk-1/KDR, endothelial nitric oxide synthase, and caveolin-1 in vascular endothelial cells. *Biochemical and Biophysical Research Communications, 256*(1), 192–197. http://dx.doi.org/10.1006/bbrc.1998.9790.

Fernandez-Hernando, C., Fukata, M., Bernatchez, P. N., Fukata, Y., Lin, M. I., Bredt, D. S., & Sessa, W. C. (2006). Identification of Golgi-localized acyl transferases that palmitoylate and regulate endothelial nitric oxide synthase. *The Journal of Cell Biology, 174*(3), 369–377. http://dx.doi.org/10.1083/jcb.200601051.

Fernandez-Hernando, C., Yu, J., Davalos, A., Prendergast, J., & Sessa, W. C. (2010). Endothelial-specific overexpression of caveolin-1 accelerates atherosclerosis in apolipoprotein E-deficient mice. *The American Journal of Pathology, 177*(2), 998–1003. http://dx.doi.org/10.2353/ajpath.2010.091287.

Feron, O., Belhassen, L., Kobzik, L., Smith, T. W., Kelly, R. A., & Michel, T. (1996). Endothelial nitric oxide synthase targeting to caveolae. Specific interactions with caveolin isoforms in cardiac myocytes and endothelial cells. *The Journal of Biological Chemistry, 271*(37), 22810–22814.

Feron, O., Dessy, C., Desager, J. P., & Balligand, J. L. (2001). Hydroxy-methylglutaryl-coenzyme A reductase inhibition promotes endothelial nitric oxide synthase activation through a decrease in caveolin abundance. *Circulation, 103*(1), 113–118.

Feron, O., Saldana, F., Michel, J. B., & Michel, T. (1998). The endothelial nitric-oxide synthase-caveolin regulatory cycle. *The Journal of Biological Chemistry*, *273*(6), 3125–3128.

Fish, J. E., & Marsden, P. A. (2006). Endothelial nitric oxide synthase: Insight into cell-specific gene regulation in the vascular endothelium. *Cellular and Molecular Life Sciences*, *63*(2), 144–162. http://dx.doi.org/10.1007/s00018-005-5421-8.

Fish, J. E., Matouk, C. C., Rachlis, A., Lin, S., Tai, S. C., D'Abreo, C., & Marsden, P. A. (2005). The expression of endothelial nitric-oxide synthase is controlled by a cell-specific histone code. *The Journal of Biological Chemistry*, *280*(26), 24824–24838. http://dx.doi.org/10.1074/jbc.M502115200.

Fleming, I., Bauersachs, J., & Busse, R. (1997). Calcium-dependent and calcium-independent activation of the endothelial NO synthase. *Journal of Vascular Research*, *34*(3), 165–174.

Fleming, I., Bauersachs, J., Fisslthaler, B., & Busse, R. (1998). Ca2+-independent activation of the endothelial nitric oxide synthase in response to tyrosine phosphatase inhibitors and fluid shear stress. *Circulation Research*, *82*(6), 686–695.

Fleming, I., Fisslthaler, B., Dimmeler, S., Kemp, B. E., & Busse, R. (2001). Phosphorylation of Thr(495) regulates Ca(2+)/calmodulin-dependent endothelial nitric oxide synthase activity. *Circulation Research*, *88*(11), E68–E75.

Fontana, J., Fulton, D., Chen, Y., Fairchild, T. A., McCabe, T. J., Fujita, N., ... Sessa, W. C. (2002). Domain mapping studies reveal that the M domain of hsp90 serves as a molecular scaffold to regulate Akt-dependent phosphorylation of endothelial nitric oxide synthase and NO release. *Circulation Research*, *90*(8), 866–873.

Forstermann, U., & Munzel, T. (2006). Endothelial nitric oxide synthase in vascular disease: From marvel to menace. *Circulation*, *113*(13), 1708–1714. http://dx.doi.org/10.1161/CIRCULATIONAHA.105.602532.

Forstermann, U., Pollock, J. S., Schmidt, H. H., Heller, M., & Murad, F. (1991). Calmodulin-dependent endothelium-derived relaxing factor/nitric oxide synthase activity is present in the particulate and cytosolic fractions of bovine aortic endothelial cells. *Proceedings of the National Academy of Sciences of the United States of America*, *88*(5), 1788–1792.

Frank, P. G., Lee, H., Park, D. S., Tandon, N. N., Scherer, P. E., & Lisanti, M. P. (2004). Genetic ablation of caveolin-1 confers protection against atherosclerosis. *Arteriosclerosis, Thrombosis, and Vascular Biology*, *24*(1), 98–105. http://dx.doi.org/10.1161/01.ATV.0000101182.89118.E5.

Fulton, D., Babbitt, R., Zoellner, S., Fontana, J., Acevedo, L., McCabe, T. J., ... Sessa, W. C. (2004). Targeting of endothelial nitric-oxide synthase to the cytoplasmic face of the Golgi complex or plasma membrane regulates Akt- versus calcium-dependent mechanisms for nitric oxide release. *The Journal of Biological Chemistry*, *279*(29), 30349–30357. http://dx.doi.org/10.1074/jbc.M402155200.

Fulton, D., Church, J. E., Ruan, L., Li, C., Sood, S. G., Kemp, B. E., ... Venema, R. C. (2005). Src kinase activates endothelial nitric-oxide synthase by phosphorylating Tyr-83. *The Journal of Biological Chemistry*, *280*(43), 35943–35952. http://dx.doi.org/10.1074/jbc.M504606200.

Fulton, D., Fontana, J., Sowa, G., Gratton, J. P., Lin, M., Li, K. X., ... Sessa, W. C. (2002). Localization of endothelial nitric-oxide synthase phosphorylated on serine 1179 and nitric oxide in Golgi and plasma membrane defines the existence of two pools of active enzyme. *The Journal of Biological Chemistry*, *277*(6), 4277–4284. http://dx.doi.org/10.1074/jbc.M106302200.

Fulton, D., Gratton, J. P., McCabe, T. J., Fontana, J., Fujio, Y., Walsh, K., ... Sessa, W. C. (1999). Regulation of endothelium-derived nitric oxide production by the protein kinase Akt. *Nature*, *399*(6736), 597–601. http://dx.doi.org/10.1038/21218.

Fulton, D., Ruan, L., Sood, S. G., Li, C., Zhang, Q., & Venema, R. C. (2008). Agonist-stimulated endothelial nitric oxide synthase activation and vascular relaxation. Role of eNOS phosphorylation at Tyr83. *Circulation Research*, *102*(4), 497–504. http://dx.doi.org/10.1161/CIRCRESAHA.107.162933.

Furchgott, R. F., & Zawadzki, J. V. (1980). The obligatory role of endothelial cells in the relaxation of arterial smooth muscle by acetylcholine. *Nature*, *288*(5789), 373–376.

Gallis, B., Corthals, G. L., Goodlett, D. R., Ueba, H., Kim, F., Presnell, S. R., … Corson, M. A. (1999). Identification of flow-dependent endothelial nitric-oxide synthase phosphorylation sites by mass spectrometry and regulation of phosphorylation and nitric oxide production by the phosphatidylinositol 3-kinase inhibitor LY294002. *The Journal of Biological Chemistry*, *274*(42), 30101–30108.

Gan, Y., Shen, Y. H., Wang, J., Wang, X., Utama, B., Wang, J., & Wang, X. L. (2005). Role of histone deacetylation in cell-specific expression of endothelial nitric-oxide synthase. *The Journal of Biological Chemistry*, *280*(16), 16467–16475. http://dx.doi.org/10.1074/jbc.M412960200.

Gao, S., Chen, J., Brodsky, S. V., Huang, H., Adler, S., Lee, J. H., … Goligorsky, M. S. (2004). Docking of endothelial nitric oxide synthase (eNOS) to the mitochondrial outer membrane: A pentabasic amino acid sequence in the autoinhibitory domain of eNOS targets a proteinase K-cleavable peptide on the cytoplasmic face of mitochondria. *The Journal of Biological Chemistry*, *279*(16), 15968–15974. http://dx.doi.org/10.1074/jbc.M308504200.

Garcia-Cardena, G., Fan, R., Shah, V., Sorrentino, R., Cirino, G., Papapetropoulos, A., & Sessa, W. C. (1998). Dynamic activation of endothelial nitric oxide synthase by Hsp90. *Nature*, *392*(6678), 821–824. http://dx.doi.org/10.1038/33934.

Garcia-Cardena, G., Fan, R., Stern, D. F., Liu, J., & Sessa, W. C. (1996). Endothelial nitric oxide synthase is regulated by tyrosine phosphorylation and interacts with caveolin-1. *The Journal of Biological Chemistry*, *271*(44), 27237–27240.

Garcia-Cardena, G., Martasek, P., Masters, B. S., Skidd, P. M., Couet, J., Li, S., … Sessa, W. C. (1997). Dissecting the interaction between nitric oxide synthase (NOS) and caveolin. Functional significance of the nos caveolin binding domain in vivo. *The Journal of Biological Chemistry*, *272*(41), 25437–25440.

Garcia-Cardena, G., Oh, P., Liu, J., Schnitzer, J. E., & Sessa, W. C. (1996). Targeting of nitric oxide synthase to endothelial cell caveolae via palmitoylation: Implications for nitric oxide signaling. *Proceedings of the National Academy of Sciences of the United States of America*, *93*(13), 6448–6453.

Golser, R., Gorren, A. C., Leber, A., Andrew, P., Habisch, H. J., Werner, E. R., … Mayer, B. (2000). Interaction of endothelial and neuronal nitric-oxide synthases with the bradykinin B2 receptor. Binding of an inhibitory peptide to the oxygenase domain blocks uncoupled NADPH oxidation. *The Journal of Biological Chemistry*, *275*(8), 5291–5296.

Govers, R., van der Sluijs, P., van Donselaar, E., Slot, J. W., & Rabelink, T. J. (2002). Endothelial nitric oxide synthase and its negative regulator caveolin-1 localize to distinct perinuclear organelles. *The Journal of Histochemistry and Cytochemistry*, *50*(6), 779–788.

Gratton, J. P., Lin, M. I., Yu, J., Weiss, E. D., Jiang, Z. L., Fairchild, T. A., … Sessa, W. C. (2003). Selective inhibition of tumor microvascular permeability by cavtratin blocks tumor progression in mice. *Cancer Cell*, *4*(1), 31–39.

Greif, D. M., Kou, R., & Michel, T. (2002). Site-specific dephosphorylation of endothelial nitric oxide synthase by protein phosphatase 2A: Evidence for crosstalk between phosphorylation sites. *Biochemistry*, *41*(52), 15845–15853.

Gross, S. S., Jaffe, E. A., Levi, R., & Kilbourn, R. G. (1991). Cytokine-activated endothelial cells express an isotype of nitric oxide synthase which is tetrahydrobiopterin-dependent, calmodulin-independent and inhibited by arginine analogs with a rank-order of potency characteristic of activated macrophages. *Biochemical and Biophysical Research Communications, 178*(3), 823–829.

Guillot, P. V., Guan, J., Liu, L., Kuivenhoven, J. A., Rosenberg, R. D., Sessa, W. C., & Aird, W. C. (1999). A vascular bed-specific pathway. *The Journal of Clinical Investigation, 103*(6), 799–805. http://dx.doi.org/10.1172/JCI6017.

Harris, M. B., Ju, H., Venema, V. J., Liang, H., Zou, R., Michell, B. J., … Venema, R. C. (2001). Reciprocal phosphorylation and regulation of endothelial nitric-oxide synthase in response to bradykinin stimulation. *The Journal of Biological Chemistry, 276*(19), 16587–16591. http://dx.doi.org/10.1074/jbc.M100229200.

Huang, P. L., Huang, Z., Mashimo, H., Bloch, K. D., Moskowitz, M. A., Bevan, J. A., & Fishman, M. C. (1995). Hypertension in mice lacking the gene for endothelial nitric oxide synthase. *Nature, 377*(6546), 239–242. http://dx.doi.org/10.1038/377239a0.

Ichihara, A., Inscho, E. W., Imig, J. D., & Navar, L. G. (1998). Neuronal nitric oxide synthase modulates rat renal microvascular function. *The American Journal of Physiology, 274*(3 Pt 2), F516–F524.

Igarashi, J., Thatte, H. S., Prabhakar, P., Golan, D. E., & Michel, T. (1999). Calcium-independent activation of endothelial nitric oxide synthase by ceramide. *Proceedings of the National Academy of Sciences of the United States of America, 96*(22), 12583–12588.

Ignarro, L. J., Buga, G. M., Wood, K. S., Byrns, R. E., & Chaudhuri, G. (1987). Endothelium-derived relaxing factor produced and released from artery and vein is nitric oxide. *Proceedings of the National Academy of Sciences of the United States of America, 84*(24), 9265–9269.

Iwakiri, Y., Satoh, A., Chatterjee, S., Toomre, D. K., Chalouni, C. M., Fulton, D., … Sessa, W. C. (2006). Nitric oxide synthase generates nitric oxide locally to regulate compartmentalized protein S-nitrosylation and protein trafficking. *Proceedings of the National Academy of Sciences of the United States of America, 103*(52), 19777–19782. http://dx.doi.org/10.1073/pnas.0605907103.

Jagnandan, D., Sessa, W. C., & Fulton, D. (2005). Intracellular location regulates calcium-calmodulin-dependent activation of organelle-restricted eNOS. *American Journal of Physiology. Cell Physiology, 289*(4), C1024–C1033. http://dx.doi.org/10.1152/ajpcell.00162.2005.

Janssens, S. P., Shimouchi, A., Quertermous, T., Bloch, D. B., & Bloch, K. D. (1992). Cloning and expression of a cDNA encoding human endothelium-derived relaxing factor/nitric oxide synthase. *The Journal of Biological Chemistry, 267*(21), 14519–14522.

Jasmin, J. F., Mercier, I., Dupuis, J., Tanowitz, H. B., & Lisanti, M. P. (2006). Short-term administration of a cell-permeable caveolin-1 peptide prevents the development of monocrotaline-induced pulmonary hypertension and right ventricular hypertrophy. *Circulation, 114*(9), 912–920. http://dx.doi.org/10.1161/CIRCULATIONAHA.106.634709.

Jerkic, M., Rivas-Elena, J. V., Prieto, M., Carron, R., Sanz-Rodriguez, F., Perez-Barriocanal, F., … Lopez-Novoa, J. M. (2004). Endoglin regulates nitric oxide-dependent vasodilatation. *The FASEB Journal, 18*(3), 609–611. http://dx.doi.org/10.1096/fj.03-0197fje.

Jiang, J., Cyr, D., Babbitt, R. W., Sessa, W. C., & Patterson, C. (2003). Chaperone-dependent regulation of endothelial nitric-oxide synthase intracellular trafficking by the co-chaperone/ubiquitin ligase CHIP. *The Journal of Biological Chemistry, 278*(49), 49332–49341. http://dx.doi.org/10.1074/jbc.M304738200.

Ju, H., Venema, V. J., Marrero, M. B., & Venema, R. C. (1998). Inhibitory interactions of the bradykinin B2 receptor with endothelial nitric-oxide synthase. *The Journal of Biological Chemistry, 273*(37), 24025–24029.

Ju, H., Zou, R., Venema, V. J., & Venema, R. C. (1997). Direct interaction of endothelial nitric-oxide synthase and caveolin-1 inhibits synthase activity. *The Journal of Biological Chemistry*, *272*(30), 18522–18525.

Jung, S. B., Kim, C. S., Naqvi, A., Yamamori, T., Mattagajasingh, I., Hoffman, T. A., ... Irani, K. (2010). Histone deacetylase 3 antagonizes aspirin-stimulated endothelial nitric oxide production by reversing aspirin-induced lysine acetylation of endothelial nitric oxide synthase. *Circulation Research*, *107*(7), 877–887. http://dx.doi.org/10.1161/CIRCRESAHA.110.222968.

Kibbe, M., Billiar, T., & Tzeng, E. (1999). Inducible nitric oxide synthase and vascular injury. *Cardiovascular Research*, *43*(3), 650–657.

Koleske, A. J., Baltimore, D., & Lisanti, M. P. (1995). Reduction of caveolin and caveolae in oncogenically transformed cells. *Proceedings of the National Academy of Sciences of the United States of America*, *92*(5), 1381–1385.

Kondrikov, D., Fonseca, F. V., Elms, S., Fulton, D., Black, S. M., Block, E. R., & Su, Y. (2010). Beta-actin association with endothelial nitric-oxide synthase modulates nitric oxide and superoxide generation from the enzyme. *The Journal of Biological Chemistry*, *285*(7), 4319–4327. http://dx.doi.org/10.1074/jbc.M109.063172.

Kou, R., Greif, D., & Michel, T. (2002). Dephosphorylation of endothelial nitric-oxide synthase by vascular endothelial growth factor. Implications for the vascular responses to cyclosporin A. *The Journal of Biological Chemistry*, *277*(33), 29669–29673. http://dx.doi.org/10.1074/jbc.M204519200.

Kraehling, J. R., Hao, Z., Lee, M. Y., Vinyard, D., Velazquez, H., Liu, X., ... Sessa, W. C. (2015). Uncoupling caveolae from intracellular signaling in vivo. *Circulation Research*, *118*(1), 48–55. http://dx.doi.org/10.1161/CIRCRESAHA.115.307767.

Kuhlencordt, P. J., Gyurko, R., Han, F., Scherrer-Crosbie, M., Aretz, T. H., Hajjar, R., ... Huang, P. L. (2001). Accelerated atherosclerosis, aortic aneurysm formation, and ischemic heart disease in apolipoprotein E/endothelial nitric oxide synthase double-knockout mice. *Circulation*, *104*(4), 448–454.

Lamas, S., Marsden, P. A., Li, G. K., Tempst, P., & Michel, T. (1992). Endothelial nitric oxide synthase: Molecular cloning and characterization of a distinct constitutive enzyme isoform. *Proceedings of the National Academy of Sciences of the United States of America*, *89*(14), 6348–6352.

Laufs, U., La Fata, V., Plutzky, J., & Liao, J. K. (1998). Upregulation of endothelial nitric oxide synthase by HMG CoA reductase inhibitors. *Circulation*, *97*(12), 1129–1135.

Li, C., Huang, W., Harris, M. B., Goolsby, J. M., & Venema, R. C. (2005). Interaction of the endothelial nitric oxide synthase with the CAT-1 arginine transporter enhances NO release by a mechanism not involving arginine transport. *The Biochemical Journal*, *386*(Pt 3), 567–574. http://dx.doi.org/10.1042/BJ20041005.

Li, C., Ruan, L., Sood, S. G., Papapetropoulos, A., Fulton, D., & Venema, R. C. (2007). Role of eNOS phosphorylation at Ser-116 in regulation of eNOS activity in endothelial cells. *Vascular Pharmacology*, *47*(5–6), 257–264. http://dx.doi.org/10.1016/j.vph.2007.07.001.

Lima, B., Forrester, M. T., Hess, D. T., & Stamler, J. S. (2010). S-nitrosylation in cardiovascular signaling. *Circulation Research*, *106*(4), 633–646. http://dx.doi.org/10.1161/CIRCRESAHA.109.207381.

Lin, M. I., Fulton, D., Babbitt, R., Fleming, I., Busse, R., Pritchard, K. A., Jr., & Sessa, W. C. (2003). Phosphorylation of threonine 497 in endothelial nitric-oxide synthase coordinates the coupling of L-arginine metabolism to efficient nitric oxide production. *The Journal of Biological Chemistry*, *278*(45), 44719–44726. http://dx.doi.org/10.1074/jbc.M302836200.

Lipton, S. A., Choi, Y. B., Pan, Z. H., Lei, S. Z., Chen, H. S., Sucher, N. J., ... Stamler, J. S. (1993). A redox-based mechanism for the neuroprotective and neurodestructive effects of nitric oxide and related nitroso-compounds. *Nature*, *364*(6438), 626–632. http://dx.doi.org/10.1038/364626a0.

Liu, J., Garcia-Cardena, G., & Sessa, W. C. (1995). Biosynthesis and palmitoylation of endothelial nitric oxide synthase: Mutagenesis of palmitoylation sites, cysteines-15 and/or -26, argues against depalmitoylation-induced translocation of the enzyme. *Biochemistry, 34*(38), 12333–12340.

Liu, S., Premont, R. T., & Rockey, D. C. (2012). G-protein-coupled receptor kinase interactor-1 (GIT1) is a new endothelial nitric-oxide synthase (eNOS) interactor with functional effects on vascular homeostasis. *The Journal of Biological Chemistry, 287*(15), 12309–12320. http://dx.doi.org/10.1074/jbc.M111.320465.

Loot, A. E., Schreiber, J. G., Fisslthaler, B., & Fleming, I. (2009). Angiotensin II impairs endothelial function via tyrosine phosphorylation of the endothelial nitric oxide synthase. *The Journal of Experimental Medicine, 206*(13), 2889–2896. http://dx.doi.org/10.1084/jem.20090449.

Marrero, M. B., Venema, V. J., Ju, H., He, H., Liang, H., Caldwell, R. B., & Venema, R. C. (1999). Endothelial nitric oxide synthase interactions with G-protein-coupled receptors. *The Biochemical Journal, 343*(Pt 2), 335–340.

Mattagajasingh, I., Kim, C. S., Naqvi, A., Yamamori, T., Hoffman, T. A., Jung, S. B., … Irani, K. (2007). SIRT1 promotes endothelium-dependent vascular relaxation by activating endothelial nitric oxide synthase. *Proceedings of the National Academy of Sciences of the United States of America, 104*(37), 14855–14860. http://dx.doi.org/10.1073/pnas.0704329104.

McCabe, T. J., Fulton, D., Roman, L. J., & Sessa, W. C. (2000). Enhanced electron flux and reduced calmodulin dissociation may explain "calcium-independent" eNOS activation by phosphorylation. *The Journal of Biological Chemistry, 275*(9), 6123–6128.

McDonald, K. K., Zharikov, S., Block, E. R., & Kilberg, M. S. (1997). A caveolar complex between the cationic amino acid transporter 1 and endothelial nitric-oxide synthase may explain the "arginine paradox" *The Journal of Biological Chemistry, 272*(50), 31213–31216.

McQuillan, L. P., Leung, G. K., Marsden, P. A., Kostyk, S. K., & Kourembanas, S. (1994). Hypoxia inhibits expression of eNOS via transcriptional and posttranscriptional mechanisms. *The American Journal of Physiology, 267*(5 Pt 2), H1921–H1927.

Michel, J. B., Feron, O., Sase, K., Prabhakar, P., & Michel, T. (1997). Caveolin versus calmodulin. Counterbalancing allosteric modulators of endothelial nitric oxide synthase. *The Journal of Biological Chemistry, 272*(41), 25907–25912.

Michel, T., Li, G. K., & Busconi, L. (1993). Phosphorylation and subcellular translocation of endothelial nitric oxide synthase. *Proceedings of the National Academy of Sciences of the United States of America, 90*(13), 6252–6256.

Michell, B. J., Chen, Z., Tiganis, T., Stapleton, D., Katsis, F., Power, D. A., … Kemp, B. E. (2001). Coordinated control of endothelial nitric-oxide synthase phosphorylation by protein kinase C and the cAMP-dependent protein kinase. *The Journal of Biological Chemistry, 276*(21), 17625–17628. http://dx.doi.org/10.1074/jbc.C100122200.

Michell, B. J., Harris, M. B., Chen, Z. P., Ju, H., Venema, V. J., Blackstone, M. A., … Kemp, B. E. (2002). Identification of regulatory sites of phosphorylation of the bovine endothelial nitric-oxide synthase at serine 617 and serine 635. *The Journal of Biological Chemistry, 277*(44), 42344–42351. http://dx.doi.org/10.1074/jbc.M205144200.

Mount, P. F., Kemp, B. E., & Power, D. A. (2007). Regulation of endothelial and myocardial NO synthesis by multi-site eNOS phosphorylation. *Journal of Molecular and Cellular Cardiology, 42*(2), 271–279. http://dx.doi.org/10.1016/j.yjmcc.2006.05.023.

Murata, T., Lin, M. I., Huang, Y., Yu, J., Bauer, P. M., Giordano, F. J., & Sessa, W. C. (2007). Reexpression of caveolin-1 in endothelium rescues the vascular, cardiac, and pulmonary defects in global caveolin-1 knockout mice. *The Journal of Experimental Medicine, 204*(10), 2373–2382. http://dx.doi.org/10.1084/jem.20062340.

Nassiri, F., Cusimano, M. D., Scheithauer, B. W., Rotondo, F., Fazio, A., Yousef, G. M., … Lloyd, R. V. (2011). Endoglin (CD105): A review of its role in angiogenesis and tumor diagnosis, progression and therapy. *Anticancer Research, 31*(6), 2283–2290.

Nathan, C. (1992). Nitric oxide as a secretory product of mammalian cells. *The FASEB Journal, 6*(12), 3051–3064.
Nishida, K., Harrison, D. G., Navas, J. P., Fisher, A. A., Dockery, S. P., Uematsu, M., ... Murphy, T. J. (1992). Molecular cloning and characterization of the constitutive bovine aortic endothelial cell nitric oxide synthase. *The Journal of Clinical Investigation, 90*(5), 2092–2096. http://dx.doi.org/10.1172/JCI116092.
Nishida, C. R., & Ortiz de Montellano, P. R. (1998). Electron transfer and catalytic activity of nitric oxide synthases. Chimeric constructs of the neuronal, inducible, and endothelial isoforms. *The Journal of Biological Chemistry, 273*(10), 5566–5571.
Nishida, C. R., & Ortiz de Montellano, P. R. (1999). Autoinhibition of endothelial nitric-oxide synthase. Identification of an electron transfer control element. *The Journal of Biological Chemistry, 274*(21), 14692–14698.
Nuszkowski, A., Grabner, R., Marsche, G., Unbehaun, A., Malle, E., & Heller, R. (2001). Hypochlorite-modified low density lipoprotein inhibits nitric oxide synthesis in endothelial cells via an intracellular dislocalization of endothelial nitric-oxide synthase. *The Journal of Biological Chemistry, 276*(17), 14212–14221. http://dx.doi.org/10.1074/jbc.M007659200.
O'Brien, A. J., Young, H. M., Povey, J. M., & Furness, J. B. (1995). Nitric oxide synthase is localized predominantly in the Golgi apparatus and cytoplasmic vesicles of vascular endothelial cells. *Histochemistry and Cell Biology, 103*(3), 221–225.
O'Kane, P., Xie, L., Liu, Z., Queen, L., Jackson, G., Ji, Y., & Ferro, A. (2009). Aspirin acetylates nitric oxide synthase type 3 in platelets thereby increasing its activity. *Cardiovascular Research, 83*(1), 123–130. http://dx.doi.org/10.1093/cvr/cvp120.
Oubaha, M., & Gratton, J. P. (2009). Phosphorylation of endothelial nitric oxide synthase by atypical PKC zeta contributes to angiopoietin-1-dependent inhibition of VEGF-induced endothelial permeability in vitro. *Blood, 114*(15), 3343–3351. http://dx.doi.org/10.1182/blood-2008-12-196584.
Palmer, R. M., Ferrige, A. G., & Moncada, S. (1987). Nitric oxide release accounts for the biological activity of endothelium-derived relaxing factor. *Nature, 327*(6122), 524–526. http://dx.doi.org/10.1038/327524a0.
Papapetropoulos, A., Fulton, D., Lin, M. I., Fontana, J., McCabe, T. J., Zoellner, S., ... Sessa, W. C. (2004). Vanadate is a potent activator of endothelial nitric-oxide synthase: Evidence for the role of the serine/threonine kinase Akt and the 90-kDa heat shock protein. *Molecular Pharmacology, 65*(2), 407–415. http://dx.doi.org/10.1124/mol.65.2.407.
Papapetropoulos, A., Zhou, Z., Gerassimou, C., Yetik, G., Venema, R. C., Roussos, C., ... Catravas, J. D. (2005). Interaction between the 90-kDa heat shock protein and soluble guanylyl cyclase: Physiological significance and mapping of the domains mediating binding. *Molecular Pharmacology, 68*(4), 1133–1141. http://dx.doi.org/10.1124/mol.105.012682.
Park, J. H., Kim, W. S., Kim, J. Y., Park, M. H., Nam, J. H., Yun, C. W., ... Jo, I. (2011). Chk1 and Hsp90 cooperatively regulate phosphorylation of endothelial nitric oxide synthase at serine 1179. *Free Radical Biology & Medicine, 51*(12), 2217–2226. http://dx.doi.org/10.1016/j.freeradbiomed.2011.09.021.
Park, J. H., Sung, H. Y., Lee, J. Y., Kim, H. J., Ahn, J. H., & Jo, I. (2013). B56alpha subunit of protein phosphatase 2A mediates retinoic acid-induced decreases in phosphorylation of endothelial nitric oxide synthase at serine 1179 and nitric oxide production in bovine aortic endothelial cells. *Biochemical and Biophysical Research Communications, 430*(2), 476–481. http://dx.doi.org/10.1016/j.bbrc.2012.12.011.
Patel, H. H., Murray, F., & Insel, P. A. (2008). Caveolae as organizers of pharmacologically relevant signal transduction molecules. *Annual Review of Pharmacology and Toxicology, 48*, 359–391. http://dx.doi.org/10.1146/annurev.pharmtox.48.121506.124841.

Pollock, J. S., Forstermann, U., Mitchell, J. A., Warner, T. D., Schmidt, H. H., Nakane, M., & Murad, F. (1991). Purification and characterization of particulate endothelium-derived relaxing factor synthase from cultured and native bovine aortic endothelial cells. *Proceedings of the National Academy of Sciences of the United States of America, 88*(23), 10480–10484.

Prabhakar, P., Cheng, V., & Michel, T. (2000). A chimeric transmembrane domain directs endothelial nitric-oxide synthase palmitoylation and targeting to plasmalemmal caveolae. *The Journal of Biological Chemistry, 275*(25), 19416–19421. http://dx.doi.org/10.1074/jbc.M001952200.

Qian, J., Zhang, Q., Church, J. E., Stepp, D. W., Rudic, R. D., & Fulton, D. J. (2010). Role of local production of endothelium-derived nitric oxide on cGMP signaling and S-nitrosylation. *American Journal of Physiology. Heart and Circulatory Physiology, 298*(1), H112–H118. http://dx.doi.org/10.1152/ajpheart.00614.2009.

Rapoport, R. M., & Murad, F. (1983). Agonist-induced endothelium-dependent relaxation in rat thoracic aorta may be mediated through cGMP. *Circulation Research, 52*(3), 352–357.

Razani, B., Engelman, J. A., Wang, X. B., Schubert, W., Zhang, X. L., Marks, C. B., ... Lisanti, M. P. (2001). Caveolin-1 null mice are viable but show evidence of hyperproliferative and vascular abnormalities. *The Journal of Biological Chemistry, 276*(41), 38121–38138. http://dx.doi.org/10.1074/jbc.M105408200.

Ruan, L., Torres, C. M., Buffett, R. J., Kennard, S., Fulton, D., & Venema, R. C. (2013). Calcineurin-mediated dephosphorylation of eNOS at serine 116 affects eNOS enzymatic activity indirectly by facilitating c-Src binding and tyrosine 83 phosphorylation. *Vascular Pharmacology, 59*(1-2), 27–35. http://dx.doi.org/10.1016/j.vph.2013.05.004.

Ruan, L., Torres, C. M., Qian, J., Chen, F., Mintz, J. D., Stepp, D. W., ... Venema, R. C. (2011). Pin1 prolyl isomerase regulates endothelial nitric oxide synthase. *Arteriosclerosis, Thrombosis, and Vascular Biology, 31*(2), 392–398. http://dx.doi.org/10.1161/ATVBAHA.110.213181.

Rudic, R. D., Shesely, E. G., Maeda, N., Smithies, O., Segal, S. S., & Sessa, W. C. (1998). Direct evidence for the importance of endothelium-derived nitric oxide in vascular remodeling. *The Journal of Clinical Investigation, 101*(4), 731–736. http://dx.doi.org/10.1172/JCI1699.

Russell, K. S., Haynes, M. P., Caulin-Glaser, T., Rosneck, J., Sessa, W. C., & Bender, J. R. (2000). Estrogen stimulates heat shock protein 90 binding to endothelial nitric oxide synthase in human vascular endothelial cells. Effects on calcium sensitivity and NO release. *The Journal of Biological Chemistry, 275*(7), 5026–5030.

Salerno, J. C., Harris, D. E., Irizarry, K., Patel, B., Morales, A. J., Smith, S. M., ... Gross, S. S. (1997). An autoinhibitory control element defines calcium-regulated isoforms of nitric oxide synthase. *The Journal of Biological Chemistry, 272*(47), 29769–29777.

Sangwung, P., Greco, T. M., Wang, Y., Ischiropoulos, H., Sessa, W. C., & Iwakiri, Y. (2012). Proteomic identification of S-nitrosylated Golgi proteins: New insights into endothelial cell regulation by eNOS-derived NO. *PLoS One*, 7(2), e31564. http://dx.doi.org/10.1371/journal.pone.0031564.

Schleicher, M., Brundin, F., Gross, S., Muller-Esterl, W., & Oess, S. (2005). Cell cycle-regulated inactivation of endothelial NO synthase through NOSIP-dependent targeting to the cytoskeleton. *Molecular and Cellular Biology, 25*(18), 8251–8258. http://dx.doi.org/10.1128/MCB.25.18.8251-8258.2005.

Segal, S. S., Brett, S. E., & Sessa, W. C. (1999). Codistribution of NOS and caveolin throughout peripheral vasculature and skeletal muscle of hamsters. *The American Journal of Physiology, 277*(3 Pt 2), H1167–H1177.

Sessa, W. C., Barber, C. M., & Lynch, K. R. (1993). Mutation of N-myristoylation site converts endothelial cell nitric oxide synthase from a membrane to a cytosolic protein. *Circulation Research*, *72*(4), 921–924.

Sessa, W. C., Garcia-Cardena, G., Liu, J., Keh, A., Pollock, J. S., Bradley, J., … Desai, K. M. (1995). The Golgi association of endothelial nitric oxide synthase is necessary for the efficient synthesis of nitric oxide. *The Journal of Biological Chemistry*, *270*(30), 17641–17644.

Sessa, W. C., Harrison, J. K., Barber, C. M., Zeng, D., Durieux, M. E., D'Angelo, D. D., … Peach, M. J. (1992). Molecular cloning and expression of a cDNA encoding endothelial cell nitric oxide synthase. *The Journal of Biological Chemistry*, *267*(22), 15274–15276.

Shaul, P. W. (2002). Regulation of endothelial nitric oxide synthase: Location, location, location. *Annual Review of Physiology*, *64*, 749–774. http://dx.doi.org/10.1146/annurev.physiol.64.081501.155952.

Shaul, P. W., Smart, E. J., Robinson, L. J., German, Z., Yuhanna, I. S., Ying, Y., … Michel, T. (1996). Acylation targets endothelial nitric-oxide synthase to plasmalemmal caveolae. *The Journal of Biological Chemistry*, *271*(11), 6518–6522.

Shesely, E. G., Maeda, N., Kim, H. S., Desai, K. M., Krege, J. H., Laubach, V. E., … Smithies, O. (1996). Elevated blood pressures in mice lacking endothelial nitric oxide synthase. *Proceedings of the National Academy of Sciences of the United States of America*, *93*(23), 13176–13181.

Siddhanta, U., Presta, A., Fan, B., Wolan, D., Rousseau, D. L., & Stuehr, D. J. (1998). Domain swapping in inducible nitric-oxide synthase. Electron transfer occurs between flavin and heme groups located on adjacent subunits in the dimer. *The Journal of Biological Chemistry*, *273*(30), 18950–18958.

Singer, H. A., & Peach, M. J. (1982). Calcium- and endothelial-mediated vascular smooth muscle relaxation in rabbit aorta. *Hypertension*, *4*(3 Pt 2), 19–25.

Su, Y., Edwards-Bennett, S., Bubb, M. R., & Block, E. R. (2003). Regulation of endothelial nitric oxide synthase by the actin cytoskeleton. *American Journal of Physiology. Cell Physiology*, *284*(6), C1542–C1549. http://dx.doi.org/10.1152/ajpcell.00248.2002.

Su, Y., Zharikov, S. I., & Block, E. R. (2002). Microtubule-active agents modify nitric oxide production in pulmonary artery endothelial cells. *American Journal of Physiology. Lung Cellular and Molecular Physiology*, *282*(6), L1183–L1189. http://dx.doi.org/10.1152/ajplung.00388.2001.

Sugimoto, M., Nakayama, M., Goto, T. M., Amano, M., Komori, K., & Kaibuchi, K. (2007). Rho-kinase phosphorylates eNOS at threonine 495 in endothelial cells. *Biochemical and Biophysical Research Communications*, *361*(2), 462–467. http://dx.doi.org/10.1016/j.bbrc.2007.07.030.

Sun, J., & Liao, J. K. (2002). Functional interaction of endothelial nitric oxide synthase with a voltage-dependent anion channel. *Proceedings of the National Academy of Sciences of the United States of America*, *99*(20), 13108–13113. http://dx.doi.org/10.1073/pnas.202260999.

Tan, E., Gurjar, M. V., Sharma, R. V., & Bhalla, R. C. (1999). Estrogen receptor-alpha gene transfer into bovine aortic endothelial cells induces eNOS gene expression and inhibits cell migration. *Cardiovascular Research*, *43*(3), 788–797.

Taubert, D., Berkels, R., Grosser, N., Schroder, H., Grundemann, D., & Schomig, E. (2004). Aspirin induces nitric oxide release from vascular endothelium: A novel mechanism of action. *British Journal of Pharmacology*, *143*(1), 159–165. http://dx.doi.org/10.1038/sj.bjp.0705907.

Teichert, A. M., Miller, T. L., Tai, S. C., Wang, Y., Bei, X., Robb, G. B., … Marsden, P. A. (2000). In vivo expression profile of an endothelial nitric oxide synthase promoter-reporter transgene. *American Journal of Physiology. Heart and Circulatory Physiology*, *278*(4), H1352–H1361.

Thomas, D. D., Liu, X., Kantrow, S. P., & Lancaster, J. R., Jr. (2001). The biological lifetime of nitric oxide: Implications for the perivascular dynamics of NO and O2. *Proceedings of the National Academy of Sciences of the United States of America, 98*(1), 355–360. http://dx.doi.org/10.1073/pnas.011379598.

Toporsian, M., Gros, R., Kabir, M. G., Vera, S., Govindaraju, K., Eidelman, D. H., … Letarte, M. (2005). A role for endoglin in coupling eNOS activity and regulating vascular tone revealed in hereditary hemorrhagic telangiectasia. *Circulation Research, 96*(6), 684–692. http://dx.doi.org/10.1161/01.RES.0000159936.38601.22.

Tourkina, E., Richard, M., Gooz, P., Bonner, M., Pannu, J., Harley, R., … Hoffman, S. (2008). Antifibrotic properties of caveolin-1 scaffolding domain in vitro and in vivo. *American Journal of Physiology Lung Cellular and Molecular Physiology, 294*(5), L843–L861. http://dx.doi.org/10.1152/ajplung.00295.2007.

Tsukahara, H., Gordienko, D. V., Tonshoff, B., Gelato, M. C., & Goligorsky, M. S. (1994). Direct demonstration of insulin-like growth factor-I-induced nitric oxide production by endothelial cells. *Kidney International, 45*(2), 598–604.

Uittenbogaard, A., Shaul, P. W., Yuhanna, I. S., Blair, A., & Smart, E. J. (2000). High density lipoprotein prevents oxidized low density lipoprotein-induced inhibition of endothelial nitric-oxide synthase localization and activation in caveolae. *The Journal of Biological Chemistry, 275*(15), 11278–11283.

Vasquez-Vivar, J., Kalyanaraman, B., Martasek, P., Hogg, N., Masters, B. S., Karoui, H., … Pritchard, K. A., Jr. (1998). Superoxide generation by endothelial nitric oxide synthase: The influence of cofactors. *Proceedings of the National Academy of Sciences of the United States of America, 95*(16), 9220–9225.

Venema, R. C., Sayegh, H. S., Arnal, J. F., & Harrison, D. G. (1995). Role of the enzyme calmodulin-binding domain in membrane association and phospholipid inhibition of endothelial nitric oxide synthase. *The Journal of Biological Chemistry, 270*(24), 14705–14711.

Venema, V. J., Zou, R., Ju, H., Marrero, M. B., & Venema, R. C. (1997). Caveolin-1 detergent solubility and association with endothelial nitric oxide synthase is modulated by tyrosine phosphorylation. *Biochemical and Biophysical Research Communications, 236*(1), 155–161. http://dx.doi.org/10.1006/bbrc.1997.6921.

Wang, X. M., Zhang, Y., Kim, H. P., Zhou, Z., Feghali-Bostwick, C. A., Liu, F., … Choi, A. M. (2006). Caveolin-1: A critical regulator of lung fibrosis in idiopathic pulmonary fibrosis. *The Journal of Experimental Medicine, 203*(13), 2895–2906. http://dx.doi.org/10.1084/jem.20061536.

Wever, R. M., van Dam, T., van Rijn, H. J., de Groot, F., & Rabelink, T. J. (1997). Tetrahydrobiopterin regulates superoxide and nitric oxide generation by recombinant endothelial nitric oxide synthase. *Biochemical and Biophysical Research Communications, 237*(2), 340–344. http://dx.doi.org/10.1006/bbrc.1997.7069.

Wilcox, J. N., Subramanian, R. R., Sundell, C. L., Tracey, W. R., Pollock, J. S., Harrison, D. G., & Marsden, P. A. (1997). Expression of multiple isoforms of nitric oxide synthase in normal and atherosclerotic vessels. *Arteriosclerosis, Thrombosis, and Vascular Biology, 17*(11), 2479–2488.

Xie, Q. W., Cho, H. J., Calaycay, J., Mumford, R. A., Swiderek, K. M., Lee, T. D., … Nathan, C. (1992). Cloning and characterization of inducible nitric oxide synthase from mouse macrophages. *Science, 256*(5054), 225–228.

Xu, K. Y., Huso, D. L., Dawson, T. M., Bredt, D. S., & Becker, L. C. (1999). Nitric oxide synthase in cardiac sarcoplasmic reticulum. *Proceedings of the National Academy of Sciences of the United States of America, 96*(2), 657–662.

Xu, H., Shi, Y., Wang, J., Jones, D., Weilrauch, D., Ying, R., … Pritchard, K. A., Jr. (2007). A heat shock protein 90 binding domain in endothelial nitric-oxide synthase influences enzyme function. *The Journal of Biological Chemistry, 282*(52), 37567–37574. http://dx.doi.org/10.1074/jbc.M706464200.

Yeh, D. C., Duncan, J. A., Yamashita, S., & Michel, T. (1999). Depalmitoylation of endothelial nitric-oxide synthase by acyl-protein thioesterase 1 is potentiated by Ca(2+)-calmodulin. *The Journal of Biological Chemistry*, *274*(46), 33148–33154.

Zhang, Q., Church, J. E., Jagnandan, D., Catravas, J. D., Sessa, W. C., & Fulton, D. (2006). Functional relevance of Golgi- and plasma membrane-localized endothelial NO synthase in reconstituted endothelial cells. *Arteriosclerosis, Thrombosis, and Vascular Biology*, *26*(5), 1015–1021. http://dx.doi.org/10.1161/01.ATV.0000216044.49494.c4.

Zhao, Y. Y., Liu, Y., Stan, R. V., Fan, L., Gu, Y., Dalton, N., ... Chien, K. R. (2002). Defects in caveolin-1 cause dilated cardiomyopathy and pulmonary hypertension in knockout mice. *Proceedings of the National Academy of Sciences of the United States of America*, *99*(17), 11375–11380. http://dx.doi.org/10.1073/pnas.172360799.

Zhao, Y. Y., Zhao, Y. D., Mirza, M. K., Huang, J. H., Potula, H. H., Vogel, S. M., ... Malik, A. B. (2009). Persistent eNOS activation secondary to caveolin-1 deficiency induces pulmonary hypertension in mice and humans through PKG nitration. *The Journal of Clinical Investigation*, *119*(7), 2009–2018. http://dx.doi.org/10.1172/JCI33338.

Zickus, M. A., Fonseca, F. V., Tummala, M., Black, S. M., & Ryzhov, V. (2008). Identification of the tyrosine nitration sites in human endothelial nitric oxide synthase by liquid chromatography-mass spectrometry. *European Journal of Mass Spectrometry (Chichester, England)*, *14*(4), 239–247. http://dx.doi.org/10.1255/ejms.927.

Zimmermann, K., Opitz, N., Dedio, J., Renne, C., Muller-Esterl, W., & Oess, S. (2002). NOSTRIN: A protein modulating nitric oxide release and subcellular distribution of endothelial nitric oxide synthase. *Proceedings of the National Academy of Sciences of the United States of America*, *99*(26), 17167–17172. http://dx.doi.org/10.1073/pnas.252345399.

Zou, M. H., Shi, C., & Cohen, R. A. (2002). Oxidation of the zinc-thiolate complex and uncoupling of endothelial nitric oxide synthase by peroxynitrite. *The Journal of Clinical Investigation*, *109*(6), 817–826. http://dx.doi.org/10.1172/JCI14442.

CHAPTER THREE

Endothelial Small- and Intermediate-Conductance K Channels and Endothelium-Dependent Hyperpolarization as Drug Targets in Cardiovascular Disease

R. Köhler[*,†,1], A. Oliván-Viguera[*,†], H. Wulff[‡]
[*]Aragon Institute of Health Sciences & IIS Aragon, Zaragoza, Spain
[†]Aragon Agency for Research and Development (ARAID), Zaragoza, Spain
[‡]University of California, Davis, CA, United States
[1]Corresponding author: e-mail address: rkohler.iacs@aragon.es

Contents

Advances in Pharmacology, Volume 77
ISSN 1054-3589
http://dx.doi.org/10.1016/bs.apha.2016.04.002

Abstract

Endothelial calcium/calmodulin-gated K channels of small (KCa2.3) and intermediate conductance (KCa3.1) produce membrane hyperpolarization and endothelium-dependent hyperpolarization (EDH)-mediated vasodilation. Dysfunctions of the two channels and ensuing EDH impairments are found in several cardiovascular pathologies such as diabetes, atherosclerosis, postangioplastic neointima formation, but also inflammatory disease, cancer, and organ fibrosis. Moreover, KCa3.1 plays an important role in endothelial barrier dysfunction, edema formation in cardiac and pulmonary disease, and in ischemic stroke. Concerning KCa2.3, genome-wide association studies revealed an association of KCa2.3 channels with atrial fibrillation in humans. Accordingly, both channels are considered potential drug targets for cardio- and cerebrovascular disease states. In this chapter, we briefly review the function of the two channels in EDH-type vasodilation and systemic circulatory regulation and then highlight their pathophysiological roles in ischemic stroke as well as in pulmonary and brain edema. Finally, the authors summarize recent advances in the pharmacology of the channels and explore potential therapeutic utilities of novel channel modulators.

1. INTRODUCTION

In the vasculature, potassium channels provide an important electrochemical negative feedback on depolarizing calcium and sodium currents in smooth muscle by producing a K^+ countercurrent resulting in repolarization and/or hyperpolarization (Brayden, Quayle, Standen, & Nelson, 1991). This is an important mechanism to prevent vasospasm and to promote vasodilation (Fig. 1). Today we know that one or more members of nearly all K channel families (Gutman et al., 2003; Köhler & Ruth, 2010) are expressed in the vascular wall and—according to their unique biophysical properties and activation mechanisms—provide specific control of tone in order to meet hemodynamic and metabolic demands. In this chapter, we focus on small- and intermediate-conductance calcium/calmodulin-gated K channels, KCa2.3 and KCa3.1 (Wei et al., 2005), and their contribution to endothelium-dependent mechanisms of arterial dilations and specifically to the endothelium-dependent hyperpolarization (EDH) phenomenon of endothelium-dependent dilation (Edwards, Feletou, & Weston, 2010; Feletou, 2015; Garland & Plane, 1996). EDH-mediated vasodilation is not directly related to other transferable EDH factors, like EETs (Fisslthaler et al., 1999; Larsen et al., 2006), H_2S (Wang, 2002), or H_2O_2 (Shimokawa & Morikawa, 2005) that use smooth muscle KCa1.1 (BK_{Ca} channels), and is independent of the classical endothelium-derived relaxing

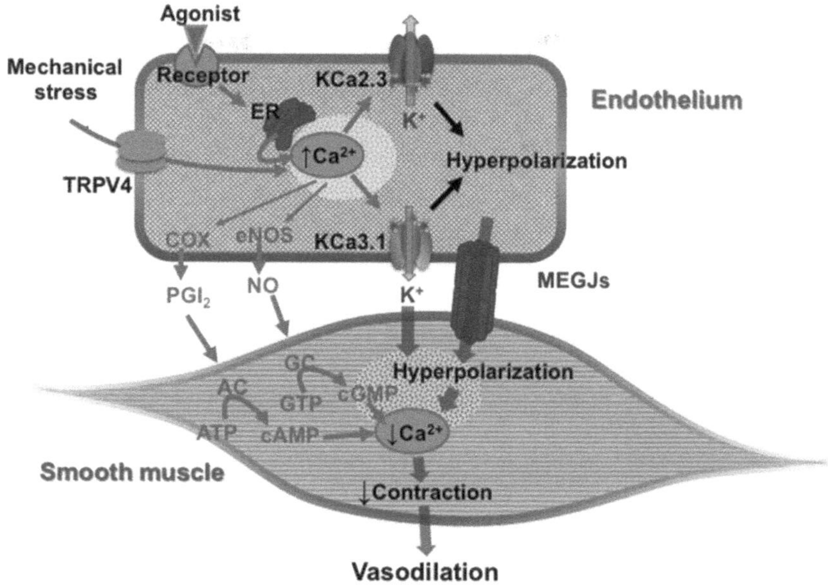

Fig. 1 Mechanisms of endothelium-dependent hyperpolarization (EDH)-type vasodilation. *COX*, cyclooxygenase; *eNOS*, endothelial nitric oxide synthase; *ER*, endoplasmic reticulum; *KCa2.3*, small-conductance calcium/calmodulin-gated potassium channel subtype 3; *KCa3.1*, intermediate-conductance calcium/calmodulin-gated potassium channel; *MEGJ*, myoendothelial gap junction; *TRPV4*, transient receptor potential channel of the vanilloid type 4.

factors, nitric oxide (Arnold, Mittal, Katsuki, & Murad, 1977; Furchgott & Zawadzki, 1980; Ignarro, Buga, Wood, Byrns, & Chaudhuri, 1987; Radomski, Palmer, & Moncada, 1987), prostacyclin (Moncada, Gryglewski, Bunting, & Vane, 1976), and other smooth muscle-relaxing prostaglandins (Feletou, Kohler, & Vanhoutte, 2010, 2012).

For other vascular K channels than KCa3.1 and KCa2 we would like to direct the reader to several excellent reviews that deal with, eg, K_{ATP} (Foster & Coetzee, 2016), K_V channels (Stott, Jepps, & Greenwood, 2014), and KCa1.1 channels (a.k.a. BK channels; Bentzen, Olesen, Ronn, & Grunnet, 2014; Fernandez-Fernandez et al., 2004; Köhler & Ruth, 2010).

Here, we discuss the importance of KCa3.1 and KCa2.3 channels for systemic cardiovascular regulation and highlight current insights mostly from the last 2 years that derived from genetic models of channel deficiency or from pharmacological experimentation, in which rodents and larger mammals were challenged systemically with specific small-molecule modulators. In keeping with the channels as mediators of K^+ release and concomitant Cl^- and water movements across epithelia and endothelia, we also explore their roles as drug targets in edema formation related to pulmonary and cardiac disease as well as in ischemic stroke and inflammatory disease. But before touching on these matters we will start with a brief introduction into molecular and biophysical features of the channels. We then will finish by reviewing recent advances in channel pharmacology and discuss the possible utility of channel modulation in human disease.

2. GENETIC RELATIONSHIPS AND BIOPHYSICAL PROPERTIES OF SMALL- AND INTERMEDIATE-CONDUCTANCE CALCIUM/CALMODULIN-GATED K CHANNELS

The mammalian gene family of calcium/calmodulin-gated potassium channels (Wei et al., 2005) consists of four members, of which three members (encoded by the KCNN1–KCNN3 genes, KCa2.1–3) have a small conductance of 5–10 pS, while the fourth member (gene name KCNN4, KCa3.1) has a larger conductance of 30–40 pS (Fig. 2). The respective genes encode for one subunit of the channel and four subunits assemble into a functional homotetrameric channel. An important feature of the

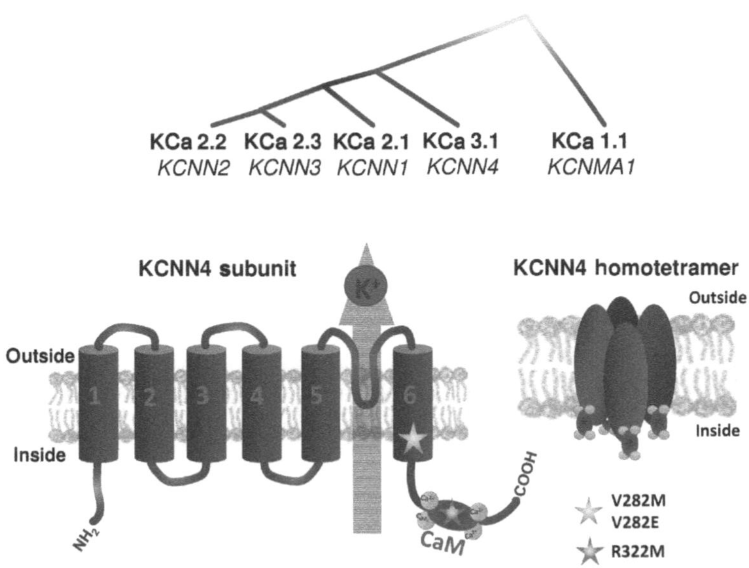

Fig. 2 *Upper panel*: Phylogenetic relationships of calcium/calmodulin-gated potassium channels. *Lower panel*: Schematic illustration of the structure of KCa3.1. *Stars* indicate locations of gene mutations related to human disease states.

channels is that they do not contain an arginine-rich S4-voltage-sensor as found in K_V and KCa1.1 channels preventing KCa2.1–3 and KCa3.1 channels from closing at negative membrane potentials as it occurs in K_V and KCa1.1 channels. So KCa2.1–3 and KCa3.1 are capable of producing solid and lasting hyperpolarization to values near the K equilibrium potential (~ -89 mV) in response to increases in intracellular calcium.

2.1 Mechanisms of Channel Activation

Channel gating is initiated upon binding of calcium to calmodulin, which is constitutively bound at the intracellular C-terminus (Fig. 2, lower panel) and serves as the calcium-sensing subunit. Activation occurs at calcium concentrations of >100 nM and the channels open with an EC_{50} of 250–900 nM and an extremely steep Hill coefficient in response to increases in intracellular calcium.

While KCa2.2 channels have been shown to be modulated by the casein kinase 2 and protein phosphatase 2A, which change calcium sensitivity by phosphorylation of T79 in the channel-associated CAM (Adelman, Maylie, & Sah, 2012), KCa3.1 activation has been suggested to require activity of nucleoside diphosphate kinase B in T cells (Di et al., 2010) and proliferating smooth muscle (Zhou et al., 2015). In contrast, protein histidine phosphatase-1 activity has been shown to cause KCa3.1 inhibition (Srivastava et al., 2008). Moreover, phosphorylation of KCa3.1 by cAMP-regulated protein kinase A seems to result in channel inhibition, while cGMP-regulated protein kinase G and subsequent activation of CaMKII have been reported to increase channel activity in microglia (Ferreira, Wong, & Schlichter, 2015). ATP has also been shown to directly and specifically stimulate human cloned KCa3.1 by interacting with residues in the C-terminus (Gerlach, Syme, Giltinan, Adelman, & Devors, 2001) and to prevent rundown of channel activity over time.

Polyunsaturated fatty acids (PUFA) of the omega-3 and the omega-6 type (eg, α-linolenic acid and arachidonic acid, respectively) as well as some CYP450-generated eicosanoid metabolites (EET-14-15 and 20-HETE) have been proposed to act as negative endogenous regulators of endothelial KCa3.1 (Kacik, Olivan-Viguera, & Kohler, 2014). Concerning the potential physiological meaning of PUFA regulation of KCa3.1, one may speculate that these PUFA–KCa3.1 interactions could provide a negative feedback on EDH dilations. However, in keeping with KCa3.1 promoting inflammatory responses and cell proliferation in the vascular wall (Köhler et al., 2003; Toyama et al., 2008) on the one hand and with

PUFAs having antiinflammatory cardiovascular protective effects (Endo & Arita, 2016) on the other hand, it might well be possible that some of the cardiovascular protective actions of PUFAs are mediated by KCa3.1 inhibition.

Important structural requirements in KCa3.1 conferring PUFA sensitivity (as well as to some small-molecule blockers, see later) are the residues T250/V275 in the inner pore. These residues are absent in KCa2.1–3 rendering these channels PUFA insensitive (Hamilton, Syme, & Devor, 2003; Kacik et al., 2014). In contrast, these residues are conserved in many vascular K_V channels making them sensitive to PUFA inhibition (Moreno et al., 2012). However, it is worth mentioning that other vascular K channels such as KCa1.1 (BK) (Wang, Chai, Lu, & Lee, 2011) and $K_V7.1$ (Liin et al., 2015) and some two-pore-domain K channels (K2P) such as K2P2.1 and K2P6.1 channels expressed in smooth muscle and/or endothelium are activated by PUFA and this activation results in vasorelaxation in vitro (Nielsen et al., 2013). Intriguingly, K2P6.1$^{-/-}$ mice develop pulmonary hypertension (Pandit et al., 2014), suggesting crucial roles of this PUFA-activated channel in maintaining normal pulmonary arterial resistance. Moreover, upregulation of K2P6.1 expression has been observed in hypoxic pulmonary hypertension in mice, suggesting potential compensatory roles for this channel in this condition (Nielsen et al., 2013).

However, taken together, the physiological meaning of the highly complex PUFA regulation of vascular K channels, particularly in the lung circuit, remains unclear at present. Yet, the inhibitory actions of PUFA on KCa3.1 and $K_V1.3$ channels in neointimal tissue and activated immune cells (Cidad et al., 2015; Köhler et al., 2010) may add to the cardiovascular protective actions of PUFAs.

Regarding the regulation of KCa3.1 protein trafficking, we would like to direct the reader to an excellent recent review summarizing the mechanisms of channel trafficking to the cell membrane in polarized epithelia and ubiquitination and degradation of KCa3.1 (Bertuccio et al., 2014).

3. KCa3.1 AND KCa2.3 IN THE VASCULAR WALL

Both KCa3.1 and KCa2.3 are widely expressed in the endothelium of many, if not all, vascular beds of many species including humans, pigs, cows, dogs, and rodents (Burnham et al., 2002; Damkjaer et al., 2012; Grgic, Kaistha, Hoyer, & Kohler, 2009; Köhler et al., 2001; Longden et al., 2011; Wulff &

Köhler, 2013; Fig. 1). It is worth mentioning that some nonendothelial cells like Kupffer cells in the liver and perivascular macrophages and presumably astrocyte endfeet in murine cerebral arterioles express KCa3.1 (Longden et al., 2011). The latter has been shown to mechanistically contribute to neurovascular coupling and cerebral blood flow regulation (Longden et al., 2011). In the lung, KCa3.1 as well as KCa2.3 is expressed in bronchial epithelium (Kroigaard et al., 2012), where particularly KCa2.3 may enable local bronchovascular coupling and vasorelaxation in tissues.

Subtype-specific amounts of functional channels in the cell membrane and protein as well as mRNA expression patterns are similar (Brähler et al., 2009; Eichler et al., 2003). For instance, single-cell RT-PCR on intact human endothelium from mesenteric arteries from adult and aged patients (18–80 years old) showed that ≈30% of the endothelial cells contained transcripts of KCa3.1 and KCa2.3 genes (Köhler & Ruth, 2010). For comparison, expression of eNOS was found in 75% of the cell samples. Current amplitudes produced by either channel are also alike (Brähler et al., 2009; Köhler & Ruth, 2010; Olivan-Viguera et al., 2013), suggesting similar amounts of functional proteins in the cell membrane, with perhaps larger quantities of KCa2.3 because of its smaller unitary conductance.

Whether or not the KCa2.1 and KCa2.2 types are also expressed is less clear. But mRNA expression was found in canine and porcine endothelium (Damkjaer et al., 2012). However, these channel subtypes are considered neuronal channels (Köhler et al., 1996).

Concerning subcellular channel compartmentalization, intracellular localization of KCa3.1 and KCa2.3 has been reported to differ substantially (for review, see Grgic, Kaistha, Hoyer, et al., 2009). Briefly, KCa3.1 protein has been found in areas of the cell membrane that are in proximity to ER calcium stores and at sites where smooth muscle cells and endothelial cells form myoendothelial gap-junctional contacts through holes in the internal elastic lamina (Dora, Gallagher, McNeish, & Garland, 2008; Ledoux et al., 2008; Sandow, Neylon, Chen, & Garland, 2006). In contrast, KCa2.3 protein was found at intercellular contacts within the endothelial monolayer as well as in caveolin-1-rich membrane fractions as measured by a biochemical approach using membrane fractioning (Absi et al., 2007; Weston et al., 2005).

This differential intracellular compartmentalization of the two channels may explain their different contributions to EDH dilations, with KCa3.1 sensing mainly ER calcium release following stimulation of

G-protein-coupled receptors (eg, of acetylcholine or bradykinin). In contrast, KCa2.3 at EC–EC junctions may sense calcium influx through colocalized mechanosensitive or receptor-operated channels of the TRP type (Andrade et al., 2005; Brähler et al., 2009; Du et al., 2014; Watanabe et al., 2002). However, this pattern of subcellular distribution of the channels has not been verified by proper immune-electron microscopy as it has been done for hippocampal KCa2.3 channels (Ballesteros-Merino et al., 2014).

Contractile, nonproliferating vascular smooth muscle cells normally do not express KCa3.1 or KCa2.3 channels (for review and discussion, see Feletou, 2009; Grgic, Kaistha, Hoyer, et al., 2009). However, this situation can change when vascular smooth muscle cells switch from a contractile phenotype toward a proliferative phenotype. In fact, expression of KCa3.1 in neointimal cells has been reported to occur in response to vascular damage caused by balloon catheter intervention in rodents and pigs or in atherosclerosis in humans (Köhler et al., 2010; Tharp et al., 2008; Wulff & Köhler, 2013) and, accordingly, the channel has been suggested as a pharmacological target for inhibiting neointima formation (see the following sections).

4. THE HUMAN TRANSCRIPTOME OF KCa3.1 AND KCa2.3

For a better understanding of KCa3.1 and KCa2.3 in human disease as well as their utility as potential drug targets, it is critical to know that the channels are expressed in several other tissues but with very distinct patterns. Human KCa3.1 is expressed in white and red blood cell lineages such as human T helper and cytotoxic T cells, B cells, monocyte/macrophages, and erythrocyte precursors (see http://biogps.org/#goto=genereport&id=3783). In these tissues, the channel has been linked to calcium-dependent immune responses (by providing a positive feedback loop on calcium entry) and erythrocyte volume regulation, respectively. Moreover, expression is solid in secretory epithelia of the intestine, lung, or salivary glands where calcium activation of KCa3.1 causes K efflux and water movements, while the resulting membrane hyperpolarization provides the electrical driving force to push Cl^- efflux through cAMP-regulated CFTR channel or concomitantly active Ca^{2+}-activated Cl^- channels (anoctamin-1, TMEM16A). This is considered an important epithelial mechanism to enable or improve, eg, mucus production in bronchia and saliva production in, eg, the parotid gland.

KCa2.3 transcriptomes (see http://www.ncbi.nlm.nih.gov/geo/tools/profileGraph.cgi?ID=GDS596:205903_s_at) also show a relative widespread expression pattern, but are predominantly found in the CNS, particularly in dopaminergic systems, where the channel mediates the apamin-sensitive medium afterhyperpolarization and thereby controls refractory times and firing frequency. Moreover, the channel is expressed in skeletal muscle, uterus, spinal cord, peripheral afferent nerves and dorsal root ganglia of rats, atrioventricular node, heart, and, similar to KCa3.1, in human endothelium and bronchial epithelium.

5. MOLECULAR MECHANISMS OF GENE TRANSCRIPTION

Expression of KCa3.1 depends essentially on intact MEK/ERK signaling following stimulation of tyrosine kinase receptors and translocation of the transcription factor, AP1 (c-jun/c-fos) as positive regulator to the nucleus, and removal of the repressor element 1-silencing transcription factor (REST) as suppressor of gene expression out of the nucleus (Cheong et al., 2005; Si, Grgic, et al., 2006). Augmentation of gene expression of KCa3.1 in human umbilical vein endothelial cells has been reported after laminar shear stress stimulation and involved activation of the MEK/ERK pathway (Brakemeier et al., 2003). In human coronary endothelial cells, shear stress causes upregulation of both, KCa2.3 and KCa3.1 (but none of the other subtypes) and mechanistically involved activation of the CaMKK/Akt/p300 cascade (Takai et al., 2013). These data suggest that transcriptional modification of KCa3.1/KCa2.3 channels is connected to vascular homeostasis on the one hand and to protection of the vascular wall against mechanical stress caused by stretch or altered hemodynamics on the other.

Notably, endothelial KCa3.1 gene expression can be induced by proangiogenic vascular endothelial growth factor and fibroblast growth factor, both using the MEK/ERK pathway (Grgic et al., 2005). This suggests some mitogenic potency of KCa3.1 during angiogenesis or perhaps endothelial repair.

Proproliferative roles of KCa3.1 have also been demonstrated during pathological organ remodeling such as of trachea transplants (Hua et al., 2013), transplant vasculopathy (Chen, Lam, Gregory, Schrepfer, & Wulff, 2013), diabetic nephropathy (Huang, Pollock, & Chen, 2014), liver and kidney fibrosis (Freise et al., 2014; Grgic et al., 2005), human liver cirrhosis

(Sevelsted-Moeller, 2015), idiopathic lung fibrosis (Roach, Wulff, Feghali-Bostwick, Amrani, & Bradding, 2014), and in activated T cells, B cells, and monocytes (Feske, Wulff, & Skolnik, 2015). Moreover, KCa3.1 has been proposed to promote pathological left ventricular remodeling after myocardial infarction (Ju et al., 2015) and cardiac fibrosis caused by pressure overload (Zhao et al., 2015).

Regulation of KCa2.3 gene expression differs substantially from that of KCa3.1. A major factor in the control of gene transcription is reportedly 17β-estradiol and its interaction between estrogen receptor alpha and the transcription factor, specificity protein 1 (Sp1) (Jacobson, Pribnow, Herson, Maylie, & Adelman, 2003) that acts as an enhancer. Moreover, 17β-estradiol has been reported to augment KCa2.3 in the uterus of rats (Rahbek et al., 2014) and of mice (Pierce & England, 2010). This process can be antagonized by SP3 revealing complex channel regulation by estrogen, which is presumably important for enabling appropriate uterus contractile function as concluded from the parturition defect in genetically modified KCa2.3-overexpressing mice (Bond et al., 2000).

6. EPIGENETIC ALTERATIONS OF KCa3.1 AND KCa2.3 CHANNEL FUNCTION IN CARDIOVASCULAR DISEASE

There are several reports on epigenetic alterations of channel function in cardiovascular disease (CVD) and other disease. For instance, a recent study on coronary arterioles reported reduced functions of endothelial KCa3.1/KCa2.3 channels in diabetic patients with no apparent differences in the amounts of channel proteins and transcripts (Liu et al., 2015). Moreover, the same group showed that microvascular dysfunction seen after cardioplegic arrest during cardiopulmonary bypass surgery and reperfusion is partially explained by a loss of KCa3.1 and KCa2.3 functions (Liu et al., 2008) rather than by a reduced gene expression, suggesting alterations of signaling pathways upstream or downstream of channel functions in this situation. In another study, improved EDH-type relaxation has been shown to compensate for a loss of nitric oxide-dependent relaxation of subcutaneous arteries from diabetics (Mokhtar et al., 2016). This suggests also compensatory roles of this system in human diabetes.

Epigenetic alterations of KCa2.3 channel expression at the arterial level have also been shown in a small Danish cohort of diabetic, uremic, and/or hypertensive patients (Odense Arteria Biobank:

http://open.rsyd.dk/OpenProjects/da/openProject.jsp?openNo=6). In this cohort, arterial KCa2.3 mRNA levels correlated negatively with blood pressure but not with diabetes or uremia. A very simple explanation for this association could be that low KCa2.3 expression is an epigenetic sign of endothelial dysfunction in hypertensive patients. In contrast, KCa3.1 mRNA expression levels were not associated positively or negatively with any other of these risk factors in this cohort. Another recent study showed that expression levels of KCa2.3 together with the inward-rectifying K channel $K_{IR}2.1$ correlated negatively with dilated cardiomyopathy and sustained monomorphic ventricular tachycardia (Ortega et al., 2015), suggesting pathogenic roles of high KCa2.3 expression in this condition.

Moreover, there are several other studies from the last two decades that found variable levels of dysfunctions of the channels and associated downstream pathways of EDH in experimental animal models of hypertension and diabetes (for reviews, see Ding & Triggle, 2005, 2010; Feletou, 2015; Young, Hill, Wiehler, Triggle, & Reid, 2008). In particular, alterations of channel function and EDH dilations have been reported in rodent models of type-2 diabetes: For instance, upregulation of KCa3.1/KCa2.3 functions leading to preserved EDH-type relaxation has been reported in coronary arteries of insulin-resistant obese Zucker rats as a model of type-2 diabetes (Climent et al., 2014). Another group found an elevated contribution of KCa3.1 to relaxation of small mesenteric artery in obese Zucker rats (Schach, Resch, Schmid, Riegger, & Endemann, 2014). This finding correlated with increased KCa3.1 mRNA and protein expression in this vessel. However, overall EDH-type relaxation was reduced in the diabetic animals.

In stroke-prone hypertensive rats, reduced expression of KCa2.3 and increased expression of KCa3.1 due to reduced expression of the suppressor of gene expression, REST-1, have been described (Giachini et al., 2009), suggesting also compensatory roles of KCa3.1 at reduced KCa2.3 functions. In a model of severe combined hypertension and renal failure, KCa3.1 and KCa2.3 expression was found largely preserved, showing at least availability of the channels as potential drug targets (Waeckel et al., 2014). Interestingly, loss of KCa3.1 function in endothelium has been reported to occur in Fabry disease, a hereditary lysosomal storage disorder that ensues disabling CVD. Here, accumulation of globotriaosylceramide has been shown to accelerate lysosomal degradation of the KCa3.1. This can explain at least in part the severe endothelial dysfunction in this condition (Choi et al., 2014).

Abnormal inflammatory activity within the vessel wall has been suggested to be of crucial pathogenic importance in the development of hypertension and cardiovascular remodeling in animal models and in humans (Luft, Dechend, & Muller, 2012; Suematsu, Suzuki, Delano, & Schmid-Schonbein, 2002). So, alternatively and in analogy to proliferating immune cells and neointimal smooth muscle cells, upregulation of endothelial KCa3.1 as found in experimental genetic hypertension and in type-2 diabetes may be interpreted as a sign of cell activation, endothelial repair, and vascular inflammation. This might also be extended to COPD as bronchoepithelial KCa3.1 upregulation was found in asthmatic patients and may indicate epithelial activation and inflammatory activity (Arthur et al., 2015).

In sum, these findings suggest that altered functions of KCa3.1 and KCa2.3 can be a sign of endothelial dysfunction or endothelial activation in the presence of major cardiovascular risk factors.

7. KCa3.1 AND CANCER

Although it is not the prime focus of this chapter, it is worth mentioning that high mRNA expression and function of KCa3.1 are a feature of many cancer cell lines and malignant cancers such as, for instance, glioblastoma, chronic lymphocytic leukemia (Grossinger et al., 2014), pancreatic cancer, adenocarcinoma, lung cancer, and clear cell renal carcinoma (Rabjerg et al., 2015; for recent review, see: Arcangeli & Becchetti, 2015). In the latter two cancers, the higher levels of KCa3.1 expression serve as a biomarker of poor prognosis. More recently, KCa3.1 has been mechanistically involved in anticancer immunity (Panyi, Beeton, & Felipe, 2014).

Also KCa2.3 appears to be linked to cancer since KCa2.3 expression is associated with tumor outgrowth and bone metastasis in breast and prostate cancer (Chantome et al., 2013).

At present it is unclear, at least in our hands, whether these changes are a mere epiphenomenon or whether this is of mechanistic importance for cancer growth, cancer vascularization, and progression. However, it seems questionable that KCa3.1 and KCa2.3 channels themselves are prime targets for cancer treatment. But pharmacological manipulation of the channels in the tumors, tumor vessels, and possible stromal cells may sensitize to conventional treatment with cytostatic drugs and tyrosine kinase inhibitors.

8. MUTATIONS AND POLYMORPHISMS IN KCa3.1 AND KCa2.3 GENES

Several missense mutations have been reported in the KCNN4 gene on Chr. 19q13.2 (Fig. 2, lower panel). One of these mutations is located within the region of the CAM-binding domain and has been reported to cause gain of function (Glogowska, Lezon-Geyda, Maksimova, Schulz, & Gallagher, 2015; Rapetti-Mauss et al., 2015). A second missense mutation has been found in the region encoding the ion channel domain and causes dehydrated hereditary stomatocytosis (xerocytosis) and mild hemolytic anemia in humans (Andolfo et al., 2015). This pathology is in line with a major role of KCa3.1 in erythrocyte volume regulation that possibly involves a functional interplay with mechanosensitive PIEZO channels (Cahalan et al., 2015). However, so far there are no reports available on eventual CVD in these patients or other disease in keeping with expression of the channel in various tissues. Still, a single-nucleotide polymorphism (SNP) has been associated with myocardial infarction in a Japanese cohort (Yamaguchi et al., 2009).

Concerning KCa2.3, recent genome-wide association studies (GWAS) found an association between an intraintronic SNP (rs13376333) in the KCNN3 gene on Chr. 1p21 and susceptibility to lone atrial fibrillation (AF) (Ellinor et al., 2010). Moreover, next-generation sequencing of AF candidate genes (including KCNN3) in patients with extreme trait of AF identified a de novo missense mutations in the KCNN3 gene that could increase AF susceptibility (Tsai et al., 2015).

Another very recent study showed that a de novo mutation in KCNN3 is associated with autosomal dominant idiopathic noncirrhotic portal hypertension (Koot, Alders, Verheij, Beuers, & Cobben, 2016). However, whether or not endothelial dysfunction in the portal vein is the reason here and whether or not there is also arterial endothelial dysfunction in these patients remains unclear.

Taken together, these data suggest that mutations in the KCNN3 gene are potentially pathogenic at the level of the human cardiovascular system. However, data available so far are still scarce making definite conclusions difficult.

9. KCa3.1 AND KCa2.3 AND VASCULAR REGULATION

Endothelial KCa3.1 and KCa2.3 as calcium-activated channels require an increase of intracellular calcium that can be produced by calcium release from stores and/or calcium entry through calcium-permeable cation channels (Fig. 1). This occurs for instance after $G_{q/11}$-receptor signaling to acetylcholine, ATP, or bradykinin stimulation. As a result the endothelial cell membrane lastingly hyperpolarizes from a "resting" membrane potential of the endothelium (−20 to −40 mV) toward more negative values (−50 to −89 mV). This hyperpolarization is then transmitted to the underlying smooth muscle where the hyperpolarization reduces the open probability of voltage-gated (depolarization-gated) calcium channels leading to a drop in $[Ca^{2+}]_i$ and finally relaxation. There are two major hypotheses about the mechanisms of transmission: 1. Myoendothelial gap junctions formed by Cx37/Cx40 proteins allow direct charge transfer from endothelium to smooth muscle (for review and recent insight, see de Wit & Griffith, 2010; Morton et al., 2015; Radtke, Schmidt, Wulff, Kohler, & de Wit, 2013). 2. K release by the KCa channels increases K concentrations at basal sites, which then stimulates K_{IR} channel functions in smooth muscle producing hyperpolarization (Edwards, Dora, Gardener, Garland, & Weston, 1998). Evidence for either hypothesis mostly derives from in vitro experimentation and, in fact, there are pros and cons for either hypothesis that have been addressed in several recent excellent reviews (Edwards et al., 2010; Feletou, 2015; Garland & Dora, 2016).

Therefore, in the following sections we wish to focus on the role of KCa3.1 and KCa2.3 channels in systemic circulatory regulation in vivo, specifically the KCa3.1/KCa2.3-EDH system, and review recent insights from cardiovascular monitoring in genetically modified mice as well as from pharmacological manipulation of the channels by selective channel modulators and blockers in mice, pigs, and dogs.

9.1 Cardiovascular Phenotypes in KCa3.1- and/or KCa2.3-Deficient Mice

Genetically encoded deficiency of KCa3.1 in mice (Si, Heyken, et al., 2006) causes an overall mild phenotype (Fig. 3): The most perceptible anatomical alteration is mild splenomegaly, which first manifests in adult mice and is progressive (Grgic, Kaistha, Paschen, et al., 2009). This alteration is reportedly linked to faster splenic erythrocyte degradation as indicated by higher bilirubin levels and iron deposition in the spleens of $KCa3.1^{-/-}$ mice (Sevelsted-Moeller, 2015). The primary reason for this accelerated erythrocyte turnover is presumably a defect of erythrocyte volume regulation. K efflux through KCa3.1 and concomitantly of water and Cl^- allows the volume adjustments necessary for squeezing through small arterioles and capillaries, the splenic filter, and kidneys, where erythrocytes need to handle

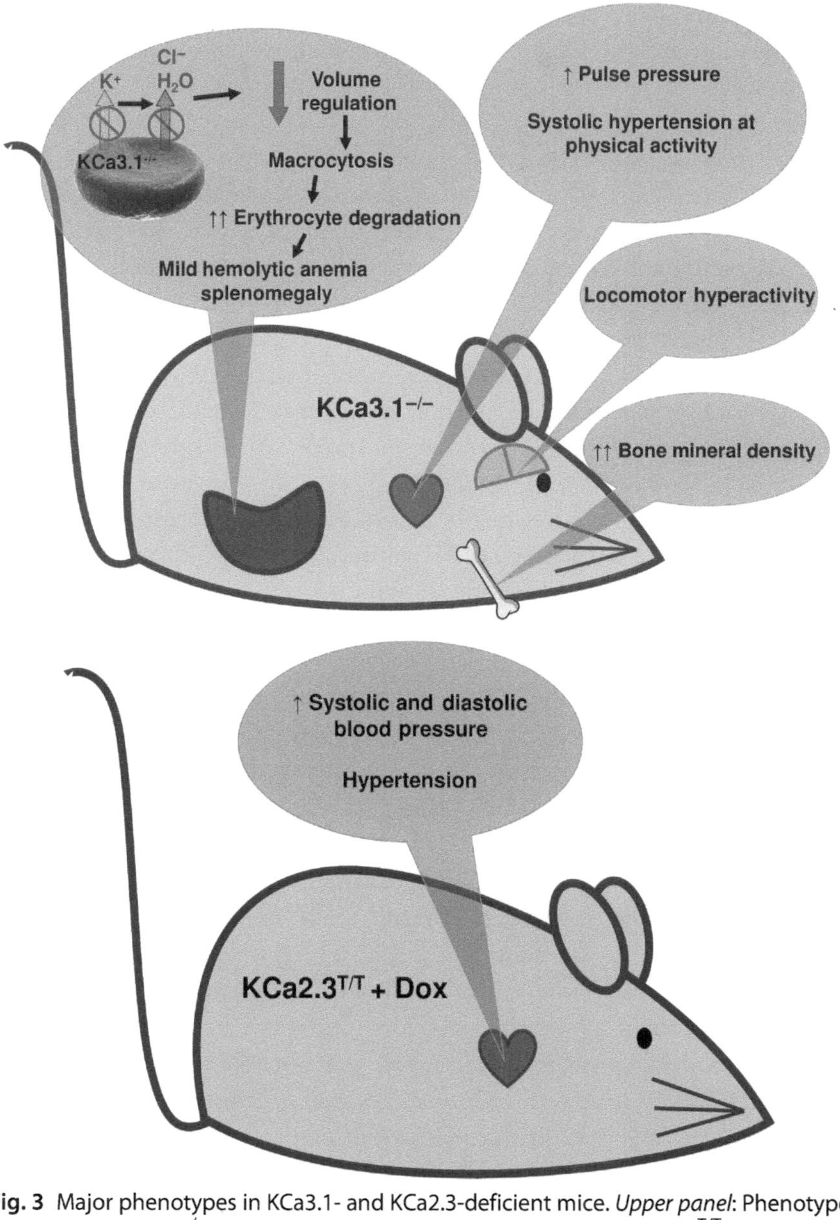

Fig. 3 Major phenotypes in KCa3.1- and KCa2.3-deficient mice. *Upper panel*: Phenotype observed in KCa3.1$^{-/-}$ mice. *Lower panel*: Phenotype observed in KCa2.3$^{T/T}$ mice treated with doxycycline (Dox).

high osmotic gradients. In fact, erythrocytes from KCa3.1$^{-/-}$ mice show mild overhydration (macrocytosis) and reduced filterability but no other erythrocyte abnormalities like altered shape or hemoglobin contents. In this regard, the erythrocyte phenotype in KCa3.1$^{-/-}$ mice is opposite to the dehydrated hereditary stomatocytosis found in patients with a gain-of-function mutation in the KCNN4 gene. However, mild hemolytic anemia and splenomegaly may be the common final path of either loss of function or gain of function.

Despite KCa3.1's relevance for T-cell activation and proliferation (Feske et al., 2015), KCa3.1$^{-/-}$ mice are not immunosuppressed and can handle *Leishmania* infection as well as wt mice (unpublished data by our group). Here, K_V1.3 channels coexpressed in T cells and macrophages are likely capable of compensating for the loss of KCa3.1.

Other phenotypes are an unexpected increased bone mineral density (hind limb BMD: ≈123% of wt), locomotor hyperactivity (≈+140% of wt), and significant alterations of central monoamine levels, with increased noradrenaline turnover (≈200% of wt) in frontal cortex as the most remarkable change (Lambertsen et al., 2012). The latter alterations were unexpected because KCa3.1 has not been linked to bone homeostasis and is believed to be absent from central neurons. However, these views have been challenged by recent reports showing an involvement of KCa3.1 in macrophage multinucleation and osteoclastogenesis (Kang et al., 2014) as well as a role for KCa3.1 in the slow afterhyperpolarization (sAHP) and thus in firing frequency in CA1 hippocampal neurons (King et al., 2015). However, the latter finding has been questioned by a recent report showing no influence of either KCa3.1 blockade or genetically encoded deficiency of the channel on sAHP in CA1 neurons (Wang et al., 2016). Still, there is another report showing KCa3.1 functions at nodes of Ranvier, which has been suggested to secure axonal spike propagation in spontaneously firing cerebellar Purkinje cells (Grundemann & Clark, 2015). Therefore, at present it cannot be completely excluded that KCa3.1 is expressed in neurons of selected brain regions of rodents.

The question as to whether KCa3.1 protein is expressed in human neurons remains unanswered. In fact, we were not able to detect significant neuronal KCa3.1 expression in human *postmortem* brain slices (Lambertsen et al., 2012), while we did see KCa3.1 protein expression in the BBB and endothelium of meningeal arteries.

In conclusion, potentially additional roles of KCa3.1 in CNS functions and possibly in sympathetic input on blood vessels should be taken into account when interpreting data from cardiovascular monitoring.

So do KCa3.1$^{-/-}$ mice have a cardiovascular phenotype? The answer is yes. But the phenotype is mild with a small increase in systolic blood pressure seen at physical activity but not at rest (Fig. 3). This may suggest that

KCa3.1-initiated EDH is not of overall importance for controlling basal arterial tone but important for counterbalancing the high arterial tone during sympathetically driven exercise. Diastolic pressure was found statistically unchanged. Heart rate at activity or rest was unchanged (Köhler & Ruth, 2010). However, pulse pressure (≈+5 mmHg) was increased, which fits with the finding of a higher systolic pressure and may reflect increased arterial stiffness or possibly higher end-systolic pressures as signs of cardiovascular remodeling as a consequence of lifelong KCa3.1 deficiency (Si, Heyken, et al., 2006).

Another recent study using another KCa3.1$^{-/-}$ strain showed defective depressor responses to pharmacological activation of purinergic receptors and to acetylcholine (Dominguez Rieg, Burt, Ruth, & Rieg, 2015). Interestingly, genetically encoded loss of connexin-37 (Cx37), which is a presumed component of interendothelial and myoendothelial gap junctions and key electrical pathway for transmission of hyperpolarization to smooth muscle downstream of KCa3.1-mediated hyperpolarization, showed similar impairments of depressor responses following purinergic receptor activation (Dominguez Rieg et al., 2015).

At the arterial level, KCa3.1$^{-/-}$ exhibited a solid defect in acetylcholine-induced EDH dilation in large carotid arteries as well as in small arterioles of skeletal muscle (Hasenau et al., 2011; Wolfle, Schmidt, Hoyer, Kohler, & de Wit, 2009). NO-mediated dilations were intact but total endothelial vasodilator capacity was still reduced in carotid arteries and skeletal muscle arterioles. Mechanistically, these deficits can be explained by reduced EDH-type dilations but also by a defective potential-driven calcium entry through, eg, endothelial TRPV4 channels following acetylcholine receptor activation causing impaired positive feedback on calcium-dependent hyperpolarization (Qian et al., 2014) but likely also an impaired calcium-dependent NO synthesis (Fig. 1).

Insights from KCa2.3-deficient mice: KCa2.3$^{T/T}$ mice contain a tetracycline-sensitive genetic switch and dietary doxycycline (Dox) shuts down KCa2.3 mRNA expression from the overexpression mode to levels substantially below wild-type levels (Bond et al., 2000). In contrast to the changes seen in KCa3.1$^{-/-}$ mice, the Dox-treated KCa2.3$^{T/T}$+Dox mice exhibited elevated systolic and diastolic blood pressure with no alterations in pulse pressure and heart rate (Köhler & Ruth, 2010; Taylor et al., 2003; Fig. 3). A very recent study on a endothelium-specific KCa2.3 knockout mice showed a similarly increased blood pressure, which further supports the idea that specifically the loss of endothelial KCa2.3 and resulting impairments of endothelium dependent and likely of EDH-type vasodilation but not loss of neuronal KCa2.3 or KCa2.3 in other tissue were responsible for this higher systemic pressure (Yap et al., 2016).

Considering that overexpression of KCa2.3 and ensuing hyperpolarization caused a strong tonic dilatory EDH impact on tone, it was somewhat surprising that the mice overexpressing the channel (in the absence of Dox) do not have lower blood pressure than normal wt mice (Köhler & Ruth, 2010; Taylor et al., 2003). Most likely there are compensatory mechanisms preventing this such as, eg, an increased sympathetic input on arteries as indicated by higher levels of plasma noradrenaline in untreated KCa2.3$^{T/T}$ (144% of wt). This was not seen in Dox-treated KCa2.3$^{T/T}$ (104% of wt) (Lambertsen et al., 2012). We did not observe changes in adrenaline and dopamine. Moreover, in the KCa3.1$^{-/-}$/KCa2.3$^{T/T}$ strain and independent of subchronic Dox treatment, there are several visible anatomic changes such as larger arterial diameters and a more generalized larger size of hollow organs related to the lifelong overexpression of KCa2.3 in KCa2.3$^{T/T}$ (Taylor et al., 2003; Wandall-Frostholm et al., 2014). This demonstrates that KCa2.3 channels exert control in cardiovascular development. Intriguingly, KCa2.3$^{T/T}$ mice are susceptible to inducible AF and sudden death (Mahida et al., 2014), which seems to be related to shortening of action potential duration with KCa2.3 overexpression. This is of interest considering that a GWAS (see earlier) found an association between susceptibility to AF and the KCa2.3-encoding KCNN3 gene (Ellinor et al., 2010). Together, these data further support particularly cardiac pathogenicity of KCa2.3 dysregulation.

At the arterial level, neither reduced expression nor overexpression caused notable impairments of acetylcholine-stimulated EDH- or NO-dependent dilation (Brähler et al., 2009). However, noradrenaline-induced tone was reduced because of a tonic hyperpolarization of the endothelium in untreated KCa2.3-overexpressing KCa2.3$^{T/T}$ mice (Taylor et al., 2003).

In endothelium-specific KCa2.3$^{-/-}$ (Yap et al., 2016), acetylcholine-induced vasorelaxation in mesenteric artery was reduced although there was no change of half-maximal responses to acetylcholine as it was seen in KCa3.1$^{-/-}$ mice (Hasenau et al., 2011). Moreover, KCa2.3 deficiency in KCa2.3$^{T/T}$Dox/KCa3.1$^{-/-}$ mice abrogated dilations of arterioles of the cremaster muscle during muscle twitching (Milkau, Kohler, & de Wit, 2010), suggesting a specific role of KCa2.3 in ensuring higher blood and by this oxygen and nutrition supply to the working skeletal muscle. Another example for the need of KCa2.3 in adequate muscle work is that overexpression of KCa2.3 channels in the myometrium depressed phasic uterine contraction (Brown et al., 2007), which could be interpreted as a way to prevent preterm labor (Pierce, Kresowik, Lamping, & England, 2008).

Double deficiency of KCa3.1 and KCa2.3 in KCa3.1$^{-/-}$/KCa2.3$^{T/T}$Dox mice resulted in a combined cardiovascular phenotype, ie, hypertension (mainly because of KCa2.3 deficiency) and pulse pressure increase and locomotor hyperactivity (because of KCa3.1 deficiency). However, blood pressure alterations seen in either strain were not cumulative (for review, see Köhler & Ruth, 2010), suggesting counterbalancing by other mechanisms. Moreover, KCa3.1$^{-/-}$/KCa2.3$^{T/T}$ + Dox mice showed significant anatomical alteration in the heart such as increased right wall thickness (Wandall-Frostholm et al., 2014), while KCa3.1$^{-/-}$ mice developed sizeable left ventricular hypertrophy (Si, Heyken, et al., 2006), again suggesting that deficiency of the channels leads to structural alterations as perhaps signs of adaptation to increased total peripheral resistance.

These findings in mice together with earlier pharmacological clues suggest that KCa3.1/KCa2.3 channels are integral parts of the EDH-signaling pathway in vivo and demonstrate an influence of KCa3.1/KCa2.3-dependent EDH on systemic blood pressure regulation.

10. PHARMACOLOGY OF KCa3.1 AND KCa2.3 CHANNELS

The pharmacology of KCa3.1 and KCa2.3 channels has advanced substantially in recent years (for recent reviews, see Christophersen & Wulff, 2015; Wulff & Köhler, 2013). We currently have several potent and selective small molecules available, particularly to modulate KCa3.1 functions in vivo. The pharmacology for KCa2 channels is likewise developed. However, because of their high similarity, achieving selectivity between the KCa2 channel subtypes remains a challenge.

Classical small molecules to study KCa3.1 channels in vivo are pore blockers, TRAM-34 (Wulff et al., 2000) and Senicapoc (ICA-17043; Ataga et al., 2008), both of which were developed on the basis of the rather unselective antifungal drug, clotrimazole (Fig. 4). Senicapoc entered clinical trials for sickle cell disease and was found to be safe, in particular with respect to blood pressure and cardiac functions. However, the treatment failed to reduce painful vaso-occlusive crisis as the primary endpoint in phase III clinical trials (Ataga et al., 2011). Pfizer subsequently deposited the compound in the NIH NCATs library and made it potentially accessible to

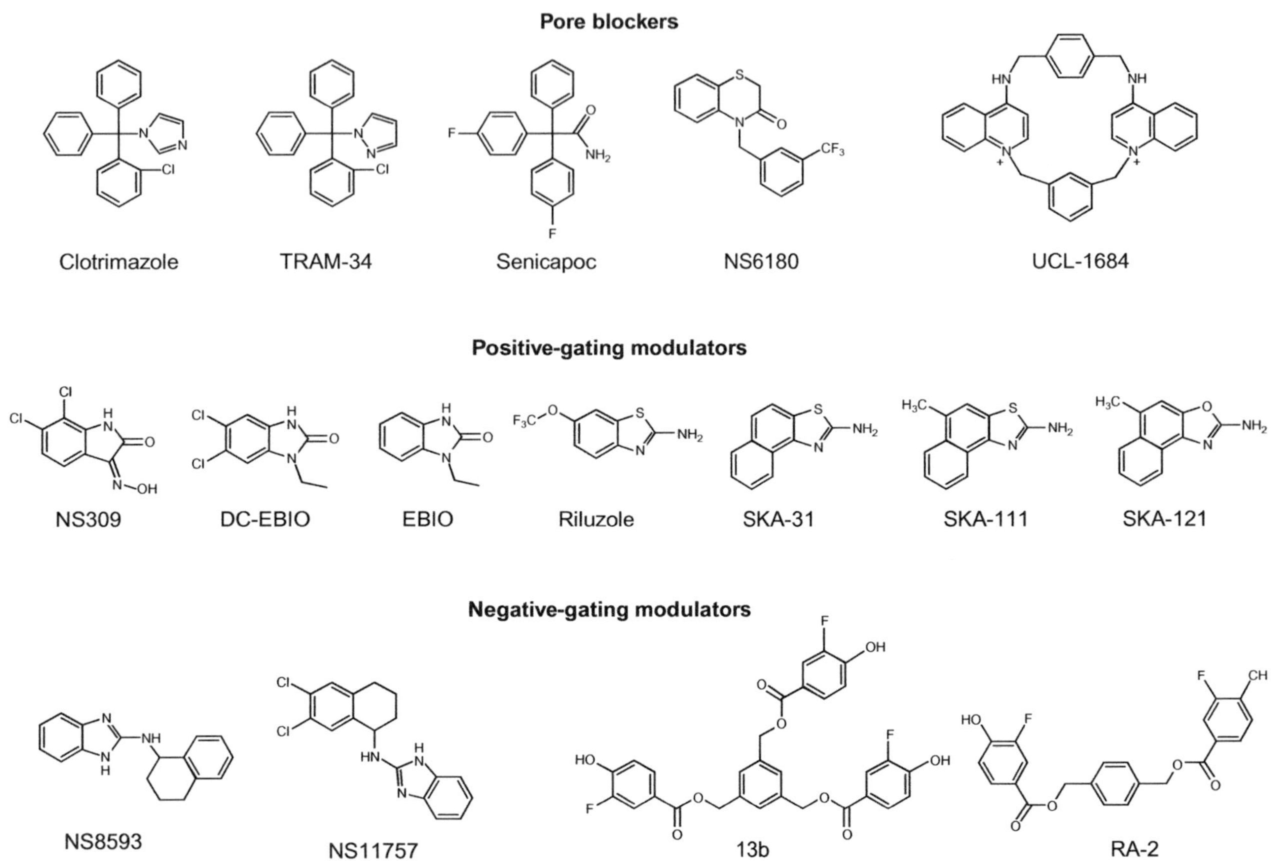

Fig. 4 Structures of KCa2 and KCa3.1 blockers and gating modulators.

clinical investigators that are interested in repurposing this KCa3.1 blocker for other disease such as ulcerative colitis in which KCa3.1 may play pathogenic roles (Koch Hansen et al., 2014).

TRAM-34 has never been advanced into humans but has become a widely used animal tool compound for experimental studies investigating the role of KCa3.1 in inflammatory diseases, fibrotic organ remodeling, or cancer (for review, see Wulff & Castle, 2010). For instance, TRAM-34 was found to reduce neointima formation and atherosclerosis as well as fibrosis of kidneys and heart in experimental studies on rodents and pigs. Moreover, TRAM-34 has been shown to reduce portal perfusion pressure in bile duct-ligated rats developing liver fibrosis (Freise et al., 2014). However, neither TRAM-34 nor Senicapoc has so far been evaluated for cardiovascular or lung disease in clinical trials.

11. RECENT ADVANCES AND CURRENT DEVELOPMENTS

The venom peptides charybdotoxin (blocking KCa3.1) and apamin (blocking KCa2) were classically used to identify and characterize KCa3.1 and KCa2.3 functions in vascular tissues in vitro. However, these peptides were of limited use for in vivo studies because of their high price and need for parenteral application and most of the more recent in vivo and in vitro studies have therefore been performed with the small-molecule blockers, TRAM-34 and Senicapoc (for KCa3.1) or the apamin-mimetic UCL-1684 (for all KCa2) (Rosa, Galanakis, Ganellin, Dunn, & Jenkinson, 1998; Fig. 4). However, when using these compounds it should be kept in mind that TRAM-34, Senicapoc, and the newer NS6180, which inhibit KCa3.1 with IC_{50}s of 10–20 nM, lose their selectivity at micromolar concentrations and affect other ion channels when used at concentrations of 5 or 10 μM (Strobaek et al., 2013). They also exhibit solubility issues at such high concentrations.

From the pharmaceutical viewpoint, the patents claiming TRAM-34 and Senicapoc are close to expiring, which prevents any attempts at further clinical development of the compounds. In recent years, however, the pharmacology of these channels has made substantial progress (see also Christophersen & Wulff, 2015; Wulff & Köhler, 2013) based on the discovery of new types of molecules that modulate channel activity not by blocking the pore and ion flow like the classical blockers but by producing gating modulation (Fig. 4). An obvious advantage of gating modulators of KCa3.1 and KCa2 channels is that they interact with the active channel (showing calcium-dependent gating) and at a given calcium concentration either potentiate their activity as in the case of a positive-gating modulator (leftward shift of the concentration–response curve for Ca^{2+}) or reduce gating as in the case of a negative-gating modulator (rightward-shift).

Established positive-gating modulators of KCa3.1 and KCa2 are in the order of potency, NS309, 1-EBIO, DC-EBIO, and the neuroprotective agent riluzole, which all show a roughly fivefold preference for KCa3.1 over KCa2 channels (Christophersen & Wulff, 2015; Wulff & Köhler, 2013; Fig. 4). Regarding "off"-target effects, these compounds have been reported to interfere with a series of other channels, such as voltage-gated sodium channels (riluzole), calcium channels (DC-EBIO, NS309) (Morimura, Yamamura, Ohya, & Imaizumi, 2006), or cardiac hERG channels (K_V11.1) (NS309). Recently, 9-phenanthrol, a presumed TRPM4 blocker, has been reported to activate rat KCa3.1 channels (Garland et al., 2015).

The more recently developed riluzole derivative, SKA-31, showed improved selectivity over voltage-gated sodium channels, calcium channels, and hERG (Sankaranarayanan et al., 2009). However, it still did not discriminate well between KCa3.1 and KCa2 channels (fivefold more potent on KCa3.1). The subsequently developed SKA-111 and SKA-121 (Fig. 4) showed an improved selectivity profile for KCa3.1 over KCa2 channels (120-fold and 40-fold, respectively) making these compounds particularly useful for specifically studying the effect of positive-gating modulation of KCa3.1 in vivo (Coleman et al., 2014).

Negative-gating modulation of KCa3.1 and KCa2 channels has been achieved more recently. The first reported negative-gating modulators are NS8593 and NS11757 (Fig. 4), which act selectively on KCa2.2 and KCa2.3 channels with potencies in the lower nanomolar range (Jenkins et al., 2011; Christophersen & Wulff, 2015). A recent small-scale screen of polyphenols and polyphenols esters as potential novel KCa3.1/KCa2 inhibitors identified the trifluorobenzoic ester, 13b (Lamoral-Theys et al., 2010; Olivan-Viguera et al., 2013), as a low nanomolar negative-gating modulator of both KCa2 and KCa3.1 channels, making it the first negative-gating modulators of KCa3.1. The smaller and thus more drug-like difluorobenzoic ester, RA-2, is similarly potent in inactivating cloned and endothelial KCa2 and KCa3.1 channels (Olivan-Viguera et al., 2015) and has been shown to specifically inhibit EDH-type relaxation in small-caliber porcine coronary arteries elicited by bradykinin and the larger "SKA-111-potentiated" response to bradykinin (Olivan-Viguera et al., 2016).

Importantly, negative-gating modulation elicited by these compounds could be concentration-dependently reversed by the positive-gating modulator of KCa3.1 and KCa2.3 channels, SKA-31, suggesting direct competition at the level of calmodulin and its binding domain (CAMBD) at the C-terminus. In this regard, negative-gating modulators differ clearly from blockers like TRAM-34 and the KCa2-selective negative-gating modulators, which bind to a site in the inner channel cavity below the K-selectivity filter (Wulff, Gutman, Cahalan, & Chandy, 2001) and do not compete with activators. Unfortunately, structural information from crystals showing the compounds' binding site similar to that reported for EBIO and NS309 complexed with the KCa2.2 CAM-binding domain and CAM (Zhang, Pascal, Schumann, Armen, & Zhang, 2012) is not yet available.

12. SYSTEMIC EFFECTS OF POSITIVE-GATING MODULATORS

Pharmacological activation of endothelial KCa2.3 and KCa3.1 channels at the systemic levels is expected to produce an endothelium-selective EDH-type depressor response and may theoretically be a strategy for endothelium-selective therapy for blood flow and pressure regulation. At present, systemically administered positive-gating modulators have been tested in vivo in mice, pigs, and dogs using SKA-31 (Wulff & Köhler, 2013). The aim here was to elucidate the consequences of KCa3.1 and KCa2 activation and EDH-type vasodilation for overall cardiovascular regulation. A main outcome of a study in conscious and trained dogs was that intravenous bolus injection of SKA-31 and of another analog, SKA-20, at a single dose of 2 mg/kg produced relatively short-lived depressor responses lasting for <1 min and significant and longer-lasting reflex tachycardia (Damkjaer et al., 2012). Moreover, the compound enhanced a likewise short-lived depressor response elicited by low-dose acetylcholine (20 ng/kg). We did not observe sedation, locomotor impairment, or any other overt side effects during the experiment or hours thereafter, suggesting that positive-gating modulation might be safe in dogs. In anesthetized pigs infusion of SKA-31 at a dose of 3–5 mg/mL/kg basically had the same effect and transiently reduced mean aortic pressure and increased systemic conductance (Mishra et al., 2016). Hence it is worth mentioning that in the anesthetized pig SKA-31 had no direct impact on cardiac function.

Data from telemetry in freely moving mice showed that intraperitoneal injection of a depot of 10–100 mg/kg SKA-31 in peanut oil caused long-lasting reductions of normal systemic pressures as well as reversed hypertension caused by angiotensin-II-infusion (Sankaranarayanan et al., 2009) or excessive renin release in Cx40-deficient mice (Radtke et al., 2013). These blood pressure lowering effects were not seen in KCa3.1-deficient mice. However, wt and KCa3.1$^{-/-}$ mice rapidly developed severe bradycardia and locomotor impairment, suggesting additional effects presumably mediated by activation of cardiac or neuronal KCa2 channels in rodents. Taken together, these data suggest that KCa3.1 channels and ensuing EDH dilation can be activated pharmacologically in vivo and that this activation results in an immediate depressor response. Whether NO signaling can also be potentiated by KCa3.1-mediated hyperpolarization is an intriguing question based on a recent study that showed that activation of KCa3.1 mediates NO-mediated myoendothelial feedback on tone (Kerr et al., 2015).

Interestingly, SKA-31 also binds to 5-HT_{2B} receptors with an IC_{50} of about 200 nM and to 5-HT_{1A} and 5-HT_{2C} with a higher IC_{50} of 5–6 μM that may also explain some of the neurological effects (Waeckel et al., 2014).

Therefore, SKA-31 is maybe not an ideal tool to study the subtype-specific role of endothelial KCa3.1 and KCa2 channels and associated EDH-type dilations for systemic circulatory control. The recently developed SKA-111 and SKA-121 with an improved selectivity for KCa3.1 over KCa2 (Coleman et al., 2014) showed similar efficacy in terms of blood pressure lowering, although the response in the case of SKA-121 was shorter (2–3 h) than in the case of SKA-111 (6–8 h). Bradycardia was still evident in wt- and KCa3.1-deficient mice treated with the more brain-penetrant SKA-111, while bradycardia was much less pronounced in the SKA-121-treated animals. This suggests that SKA-121 is the first more selective tool to study the consequences of positive-gating modulation of KCa3.1 and probably EDH-type dilation in vivo. However, we need to be careful when interpreting data from mouse telemetry as these animals have a much higher heart frequency and sympathetic drive than larger mammals.

13. SYSTEMIC EFFECTS OF NEGATIVE-GATING MODULATORS

Telemetry experiments in mice have also been recently performed using RA-2, the pan-negative-gating modulator of KCa3.1 and all KCa2 channels (Olivan-Viguera et al., 2015). A main finding was that i.p. injection

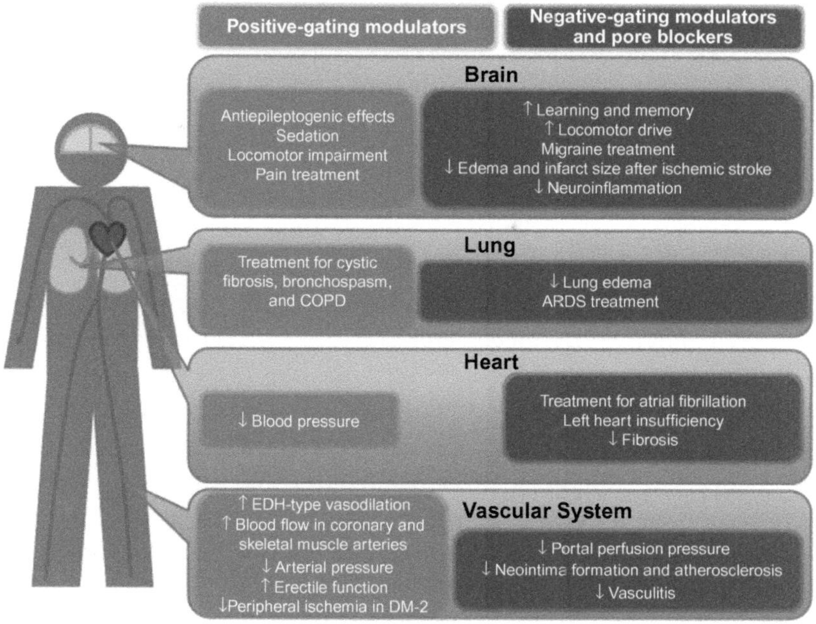

Fig. 5 Hypothetical therapeutic utilities of positive-/negative-gating modulators and blockers of KCa3.1 and KCa2.3.

of amounts of RA-2, which yielded plasma levels 10 times above the IC_{50}s for the channels within the first hour after injection, had no appreciable effects on blood pressure (Olivan-Viguera et al., 2015). However, we observed mild bradycardia in the animals that correlated well with plasma levels within the first 2 h after injection (Fig. 5). However, bradycardia was still evident at later time points where virtually all RA-2 was metabolized by ester hydrolysis and removed from the plasma. This could possibly reflect the continuing presence of RA-2 in cardiac or neuronal tissues or, alternatively, an active metabolite.

When treating the animals with the nitric oxide synthase blocker, L-NAME, to increase blood pressure by inhibiting NO synthesis, RA-2 did not worsen either blood pressure changes or bradycardia (Olivan-Viguera et al., 2015).

Taken together, these data provide the first evidence that pharmacological negative-gating modulation of KCa2 and KCa3.1 channels is relatively safe and suggest that, at least in mice, the channels and EDH are not essential for systemic circulatory regulation (even with compromised NO production).

It is worth mentioning that also in sickle cell patients and control subjects, treatment with Senicapoc (giving plasma concentrations that are theoretically sufficient for KCa3.1 blockade) was not found to alter cardiovascular parameters (Ataga et al., 2008).

Thus, given the marginal impact of channel blockers or negative-gating modulators on systemic circulatory control, a main question still remains unanswered: What is the real physiological role of these endothelial channels and for what is agonist-induced EDH-type dilation really good?

Perhaps we need to understand endothelial KCa channel activation and related EDH not only as a mechanism to achieve vasodilation in response to circulating or locally released factors activating $G_{q/11}$-coupled receptors on the endothelium but rather as a mechanism to regulate blood flow in situations of vascular stress (ischemia/reperfusion—mechanical stress or calcium overload), or nutrient demand during skeletal muscle work at sympathetic drive. Besides this, it might also be prudent to consider additional roles of KCa3.1 in fluid homeostasis and salt/water movements across endothelia and epithelia because K efflux through the channel and the resulting more negative membrane potential drives Cl^- and water movement out of the cells. This mechanism has been recognized as relevant in secretory epithelia of the salivary glands, intestine, and airways in which KCa3.1 is highly expressed. However, such a role of KCa3.1 in endothelia has not been fully established and the question arises as to whether KCa3.1 activation after receptor stimulation is meant to regulate K^+ and Cl^- secretion and fluid extravasation from arterioles and the microcirculation into the tissue. In contrast, in muscularized feeder arteries, KCa3.1-induced EDH dilation could be a mechanism for further increasing local blood supply to the tissue. Pathomechanistically, activation of the channel in the microcirculation could therefore contribute to edema formation in the tissue.

There are a series of very recent papers that support this possibility: For instance, KCa3.1 has been identified in rat blood–brain barrier endothelial cells and its blockade by TRAM-34 has been shown to reduce Na^+ uptake and cytotoxic edema after experimental ischemic stroke (middle cerebral artery occlusion) in rats (Chen, Raman, Bodendiek, O'Donnell, & Wulff, 2011; Chen, Wallace, et al., 2015). More recently, the same group showed that TRAM-34 treatment starting 12 h after MCAO in mice decreased infarct size, reduced neuroinflammatory activity by reducing microglial cytokine production, of IL-1β, IFNγ, and TGF-β 1 and neurological deficits following 8 days of reperfusion after MCAO (Chen, Nguyen, et al., 2015). Genetic KCa3.1 deficiency had similar beneficial effects. Another study showed that other KCa3.1 blockers (a triazole similar to clotrimazole and a cyclohexadiene developed at Bayer; Mauler et al., 2004) effectively reduced edema formation within 24 h and decreased infarct volume after 7 days in a rat acute subdural hematoma model, suggesting that inhibition of KCa3.1 in the BBB has substantial neuroprotective efficacy also in this model of traumatic brain injury.

Together, these studies demonstrate important novel pathomechanistic roles of KCa3.1 in the BBB, in edema formation, and in activated microglia after ischemic stroke and suggest therapeutic utilities of KCa3.1 blockade to lessen neurological damage in such conditions.

In a recent study by our collaborating groups investigating the role of KCa3.1 in experimental lung injury and edema caused by deleterious endothelial calcium overload after pharmacological TRPV4 channel activation (Wandall-Frostholm et al., 2015), we could show that KCa3.1 deficiency not only reduced endothelium-dependent pulmonary artery dilations to pharmacological opening of TRPV4 activator but also reduced perivascular fluid cuffs and hemorrhage and, intriguingly, prevented fatal lung circulatory collapse as an important primary endpoint.

Therefore, blockers of KCa3.1 channels could be of use to treat pressure overload-induced lung edema and alveolar and interstitial flooding caused by left-heart insufficiency, noncardiogenic acute lung injury, and adult respiratory distress syndrome.

In line with the notion that KCa2.3 mutations have been associated with the occurrence of AF and a role of KCa2.3 in particularly shaping atrial action potentials, KCa2 channels, especially the KCa2.3 subtype, have emerged as novel atrium-selective target for the treatment of AF (Diness, Bentzen, Sorensen, & Grunnet, 2015). Accordingly, negative-gating modulators of KCa2.3, like NS8593, are currently being developed by the Danish spin-off Acesion Pharma for the treatment of AF. Indeed, this negative-gating modulator has been shown to produce cardioversion in acute-pacing-induced AF and to reduce overall AF duration in anesthetized horses without having any effect on the ventricular conduction system and action potential (Haugaard et al., 2015). However, it remains to be seen whether this strategy holds true for the treatment of human AF. In keeping with the roles of KCa2.3 in endothelial function and thus systemic blood pressure regulation, a general concern could be that long-term treatment with such negative-gating modulators aggravates endothelial dysfunction and thereby exacerbates CVD in patients.

Interestingly, KCa3.1 expression in human embryonic stem cell-derived cardiomyocytes has been shown to facilitate pacemaker activity as concluded from the blocking action of the classical KCa3.1 blocker, clotrimazole (Weisbrod, Khun, Bueno, Peretz, & Attali, 2016).

14. CONCLUSION

KCa3.1 and KCa2.3 channels are without doubt the principal KCa channel types responsible for membrane hyperpolarization in vascular endothelium and EDH-mediated vasodilation. The function of the channels is compromised in animal models of disease, and channel expression is upregulated in a disease-related fashion in patients with CVD and other diseases characterized by chronic inflammation and abnormal cell proliferation. So from the translational perspective, the channels are indeed markers of disease activity and, in terms of CVD, of endothelial dysfunction or pathological vascular remodeling. Whether dysfunctional channels are pathogenic themselves and are capable of producing endothelial arterial (and perhaps venous) dysfunction and CVD in patients is unclear, but alterations in mice deficient of KCa3.1 and/or KCa2.3 or overexpressing of KCa2.3 channels cannot rule this out.

KCa3.1 and KCa2.3 channels are expressed in vascular and nonvascular tissues and have been suggested for the treatment of multiple pathophysiological conditions (Fig. 5). But do we consider KCa3.1 and KCa2.3 channels targets for antihypertensive therapy? The data from animal experimentation testing selective activators in rodents, pigs, and dogs support this idea in principle. However, the duration of the depressor responses is extremely short (<1 min), which can be due to rapid desensitization of the channels in vivo or immediate counteractions as indicated by the reflex bradycardia seen in conscious dogs. So it is likely that targeting KCa3.1/KCa2.3 channels and the EDH relaxation is less advantageous than directly targeting voltage-gated calcium channels on the vascular smooth muscle or employing other standard treatments (ACE inhibitors, AngII receptor blockers or beta blockers) that are known to produce long-lasting pressure reduction in patients when used alone or in combination. However, there is a large group of treatment-resistant hypertensive patients that may benefit from an alternative to classical calcium antagonists. Furthermore, KCa3.1/KCa2.3 activation may serve in situations of acute hypertension, eg, intrasurgical hypertension and vasospasm in patients with known insensitivity to NO donors or COPD patients in which prophylactic treatment with antianginal beta blockers is contraindicated because they can promote bronchospasm. KCa2.3 has been shown to regulate human detrusor smooth muscle function. Therefore, positive-gating modulation—for instance with SKA-31—may be an attractive way to decrease detrusor muscle activity and overactive bladder (Soder, Parajuli, Hristov, Rovner, & Petkov, 2013).

Neurological side effects (sedation), cardiac side effects (bradycardia and AV node blockade), migraine, and epithelial and gastroenterological perturbations (eg, diarrhea) are probably safety issues. Moreover, activators are suspected to produce rapid desensitization of their targets that may further limit the utility of the compounds and the EDH pathway for long-term treatments.

However, it is still worth speculating that direct delivery of activators to the lung using inhalers instead of systemic treatment may be useful for lung disorders like cystic fibrosis (characterized by abnormal fluid secretion and mucus production) or COPD and bronchospasm. However, edema formation could be an issue considering the role of KCa3.1 in fluid extravasation in the lung.

Topical treatments for peripheral ischemia and pain in diabetic patients may be other venues for development. Considering the presence of KCa2.3 (and possibly KCa3.1) in corpus cavernosum and penile artery, topical treatments with activators may serve in erectile dysfunction.

In contrast, channel blockers and negative-gating modulators of KCa3.1 and KCa2.3 may be of broader utility in CVD (Fig. 5). Inhibitors could be especially suitable treatments of CVD that is characterized by abnormal remodeling of the vessel wall (atherosclerosis) and signs of vascular inflammation that further aggravates pathological vessel remodeling. Moreover, there is recent evidence highlighting KCa3.1 channels as targets to prevent cerebral edema, lung edema, as well as neuroinflammation and neuronal death after ischemic stroke. Cardiac KCa2.3 channels emerged as target for AF. A general concern might still be that blocking vascular KCa3.1 and KCa2.3 may evoke endothelial dysfunction and increased tone by loss of EDH. However, the rather mild hypertension in KCa3.1/KCa2.3 deficiency as well as the minimal impact of negative-gating modulation of both KCa2 and KCa3.1 channels in mice—even during pharmacologically produced NO deficiency—does not substantiate this concern. Moreover, the fact that the KCa3.1 blocker, Senicapoc, did not produce hypertension or other cardiovascular side effects in clinical trials for sickle cell disease further suggests that strategies targeting KCa3.1 and KCa2.3 are presumably safe. However, this needs to be thoroughly studied and confirmed in human subjects. Repurposing of Senicapoc as a safe KCa3.1 blocker in humans may show whether KCa3.1 inhibitors have clinical value for the treatment of pathological vascular remodeling, chronic inflammation, and organ fibrosis. Chronic inflammatory skin diseases like atopical dermatitis and psoriasis, scar formation, and actinic keratosis could be other indications for the use of KCa3.1 blockers or negative-gating modulators. Regarding KCa2.3, it remains to be seen whether KCa2.3 negative-gating modulators fulfill the promise they are holding as a treatment of AF in humans.

CONFLICT OF INTEREST

The authors have no conflicts of interest to declare. The authors' institutions filled patents claiming SKA-121 and related compounds for the treatment of hypertension (PCTUS1527636) and RA-2 (PCTES2015070662) as a negative modulator for the study and manipulation of KCa2/3 channels in health and disease.

REFERENCES

Absi, M., Burnham, M. P., Weston, A. H., Harno, E., Rogers, M., & Edwards, G. (2007). Effects of methyl beta-cyclodextrin on EDHF responses in pig and rat arteries; association between SK(Ca) channels and caveolin-rich domains. *British Journal of Pharmacology*, *151*(3), 332–340.

Adelman, J. P., Maylie, J., & Sah, P. (2012). Small-conductance Ca^{2+}-activated K^{+} channels: Form and function. *Annual Review of Physiology*, *74*, 245–269.

Andolfo, I., Russo, R., Manna, F., Shmukler, B. E., Gambale, A., Vitiello, G., et al. (2015). Novel Gardos channel mutations linked to dehydrated hereditary stomatocytosis (xerocytosis). *American Journal of Hematology*, *90*(10), 921–926.

Andrade, Y. N., Fernandes, J., Vazquez, E., Fernandez-Fernandez, J. M., Arniges, M., Sanchez, T. M., et al. (2005). TRPV4 channel is involved in the coupling of fluid viscosity changes to epithelial ciliary activity. *Journal of Cell Biology*, *168*(6), 869–874.

Arcangeli, A., & Becchetti, A. (2015). Novel perspectives in cancer therapy: Targeting ion channels. *Drug Resistance Updates*, *21–22*, 11–19.

Arnold, W. P., Mittal, C. K., Katsuki, S., & Murad, F. (1977). Nitric oxide activates guanylate cyclase and increases guanosine 3′:5′-cyclic monophosphate levels in various tissue preparations. *Proceedings of the National Academy of Sciences of the United States of America*, *74*(8), 3203–3207.

Arthur, G. K., Duffy, S. M., Roach, K. M., Hirst, R. A., Shikotra, A., Gaillard, E. A., et al. (2015). KCa3.1 K^{+} channel expression and function in human bronchial epithelial cells. *PLoS One*, *10*(12), e0145259.

Ataga, K. I., Reid, M., Ballas, S. K., Yasin, Z., Bigelow, C., James, L. S., et al. (2011). Improvements in haemolysis and indicators of erythrocyte survival do not correlate with acute vaso-occlusive crises in patients with sickle cell disease: A phase III randomized, placebo-controlled, double-blind study of the Gardos channel blocker senicapoc (ICA-17043). *British Journal of Haematology*, *153*(1), 92–104.

Ataga, K. I., Smith, W. R., De Castro, L. M., Swerdlow, P., Saunthararajah, Y., Castro, O., et al. (2008). Efficacy and safety of the Gardos channel blocker, senicapoc (ICA-17043), in patients with sickle cell anemia. *Blood, 111*(8), 3991–3997.
Ballesteros-Merino, C., Watanabe, M., Shigemoto, R., Fukazawa, Y., Adelman, J. P., & Lujan, R. (2014). Differential subcellular localization of SK3-containing channels in the hippocampus. *European Journal of Neuroscience, 39*(6), 883–892.
Bentzen, B. H., Olesen, S. P., Ronn, L. C., & Grunnet, M. (2014). BK channel activators and their therapeutic perspectives. *Frontiers in Physiology, 5*, 389.
Bertuccio, C. A., Lee, S. L., Wu, G., Butterworth, M. B., Hamilton, K. L., & Devor, D. C. (2014). Anterograde trafficking of KCa3.1 in polarized epithelia is Rab1- and Rab8-dependent and recycling endosome-independent. *PLoS One, 9*(3), e92013.
Bond, C. T., Sprengel, R., Bissonnette, J. M., Kaufmann, W. A., Pribnow, D., Neelands, T., et al. (2000). Respiration and parturition affected by conditional overexpression of the Ca^{2+}-activated K^+ channel subunit, SK3. *Science, 289*(5486), 1942–1946.
Brähler, S., Kaistha, A., Schmidt, V. J., Wolfle, S. E., Busch, C., Kaistha, B. P., et al. (2009). Genetic deficit of SK3 and IK1 channels disrupts the endothelium-derived hyperpolarizing factor vasodilator pathway and causes hypertension. *Circulation, 119*(17), 2323–2332.
Brakemeier, S., Kersten, A., Eichler, I., Grgic, I., Zakrzewicz, A., Hopp, H., et al. (2003). Shear stress-induced up-regulation of the intermediate-conductance Ca(2+)-activated K (+) channel in human endothelium. *Cardiovascular Research, 60*(3), 488–496.
Brayden, J. E., Quayle, J. M., Standen, N. B., & Nelson, M. T. (1991). Role of potassium channels in the vascular response to endogenous and pharmacological vasodilators. *Blood Vessels, 28*(1–3), 147–153.
Brown, A., Cornwell, T., Korniyenko, I., Solodushko, V., Bond, C. T., Adelman, J. P., et al. (2007). Myometrial expression of small conductance Ca^{2+}-activated K^+ channels depresses phasic uterine contraction. *American Journal of Physiology. Cell Physiology, 292*(2), C832–C840.
Burnham, M. P., Bychkov, R., Feletou, M., Richards, G. R., Vanhoutte, P. M., Weston, A. H., et al. (2002). Characterization of an apamin-sensitive small-conductance Ca(2+)-activated K(+) channel in porcine coronary artery endothelium: Relevance to EDHF. *British Journal of Pharmacology, 135*(5), 1133–1143.
Cahalan, S. M., Lukacs, V., Ranade, S. S., Chien, S., Bandell, M., & Patapoutian, A. (2015). Piezo1 links mechanical forces to red blood cell volume. *eLife, 4*, e07370.
Chantome, A., Potier-Cartereau, M., Clarysse, L., Fromont, G., Marionneau-Lambot, S., Gueguinou, M., et al. (2013). Pivotal role of the lipid Raft SK3-Orai1 complex in human cancer cell migration and bone metastases. *Cancer Research, 73*(15), 4852–4861.
Chen, Y. J., Lam, J., Gregory, C. R., Schrepfer, S., & Wulff, H. (2013). The Ca(2)(+)-activated K(+) channel KCa3.1 as a potential new target for the prevention of allograft vasculopathy. *PLoS One, 8*(11), e81006.
Chen, Y. J., Nguyen, H. M., Maezawa, I., Grossinger, E. M., Garing, A. L., Kohler, R., et al. (2015). The potassium channel KCa3.1 constitutes a pharmacological target for neuroinflammation associated with ischemia/reperfusion stroke. *Journal of Cerebral Blood Flow & Metabolism*. pii: 0271678X15611434. [Epub ahead of print].
Chen, Y. J., Raman, G., Bodendiek, S., O'Donnell, M. E., & Wulff, H. (2011). The KCa3.1 blocker TRAM-34 reduces infarction and neurological deficit in a rat model of ischemia/reperfusion stroke. *Journal of Cerebral Blood Flow & Metabolism, 31*(12), 2363–2374.
Chen, Y. J., Wallace, B. K., Yuen, N., Jenkins, D. P., Wulff, H., & O'Donnell, M. E. (2015). Blood-brain barrier KCa3.1 channels: Evidence for a role in brain Na uptake and edema in ischemic stroke. *Stroke, 46*(1), 237–244.
Cheong, A., Bingham, A. J., Li, J., Kumar, B., Sukumar, P., Munsch, C., et al. (2005). Downregulated REST transcription factor is a switch enabling critical potassium channel expression and cell proliferation. *Molecular Cell, 20*(1), 45–52.

Choi, S., Kim, J. A., Na, H. Y., Cho, S. E., Park, S., Jung, S. C., et al. (2014). Globotriaosylceramide induces lysosomal degradation of endothelial KCa3.1 in fabry disease. *Arteriosclerosis, Thrombosis, and Vascular Biology*, *34*(1), 81–89.

Christophersen, P., & Wulff, H. (2015). Pharmacological gating modulation of small- and intermediate-conductance Ca(2+)-activated K(+) channels (KCa2.x and KCa3.1). *Channels (Austin, Tex.)*, *9*(6), 336–343.

Cidad, P., Miguel-Velado, E., Ruiz-McDavitt, C., Alonso, E., Jimenez-Perez, L., Asuaje, A., et al. (2015). Kv1.3 channels modulate human vascular smooth muscle cells proliferation independently of mTOR signaling pathway. *Pflügers Archiv/European Journal of Physiology*, *467*(8), 1711–1722.

Climent, B., Moreno, L., Martinez, P., Contreras, C., Sanchez, A., Perez-Vizcaino, F., et al. (2014). Upregulation of SK3 and IK1 channels contributes to the enhanced endothelial calcium signaling and the preserved coronary relaxation in obese Zucker rats. *PLoS One*, *9*(10), e109432.

Coleman, N., Brown, B. M., Olivan-Viguera, A., Singh, V., Olmstead, M. M., Valero, M. S., et al. (2014). New positive Ca^{2+}-activated K^+ channel gating modulators with selectivity for KCa3.1. *Molecular Pharmacology*, *86*(3), 342–357.

Damkjaer, M., Nielsen, G., Bodendiek, S., Staehr, M., Gramsbergen, J. B., de Wit, C., et al. (2012). Pharmacological activation of KCa3.1/KCa2.3 channels produces endothelial hyperpolarization and lowers blood pressure in conscious dogs. *British Journal of Pharmacology*, *165*(1), 223–234.

de Wit, C., & Griffith, T. M. (2010). Connexins and gap junctions in the EDHF phenomenon and conducted vasomotor responses. *Pflügers Archiv: European Journal of Physiology*, *459*(6), 897–914.

Di, L., Srivastava, S., Zhdanova, O., Sun, Y., Li, Z., & Skolnik, E. Y. (2010). Nucleoside diphosphate kinase B knock-out mice have impaired activation of the K^+ channel KCa3.1, resulting in defective T cell activation. *The Journal of Biological Chemistry*, *285*(50), 38765–38771.

Diness, J. G., Bentzen, B. H., Sorensen, U. S., & Grunnet, M. (2015). Role of calcium-activated potassium channels in atrial fibrillation pathophysiology and therapy. *Journal of Cardiovascular Pharmacology*, *66*(5), 441–448.

Ding, H., & Triggle, C. R. (2005). Endothelial cell dysfunction and the vascular complications associated with type 2 diabetes: Assessing the health of the endothelium. *Vascular Health and Risk Management*, *1*(1), 55–71.

Ding, H., & Triggle, C. R. (2010). Endothelial dysfunction in diabetes: Multiple targets for treatment. *Pflügers Archiv/European Journal of Physiology*, *459*(6), 977–994.

Dominguez Rieg, J. A., Burt, J. M., Ruth, P., & Rieg, T. (2015). P2Y(2) receptor activation decreases blood pressure via intermediate conductance potassium channels and connexin 37. *Acta Physiologica (Oxford)*, *213*(3), 628–641.

Dora, K. A., Gallagher, N. T., McNeish, A., & Garland, C. J. (2008). Modulation of endothelial cell KCa3.1 channels during endothelium-derived hyperpolarizing factor signaling in mesenteric resistance arteries. *Circulation Research*, *102*(10), 1247–1255.

Du, J., Ma, X., Shen, B., Huang, Y., Birnbaumer, L., & Yao, X. (2014). TRPV4, TRPC1, and TRPP2 assemble to form a flow-sensitive heteromeric channel. *The FASEB Journal*, *28*(11), 4677–4685.

Edwards, G., Dora, K. A., Gardener, M. J., Garland, C. J., & Weston, A. H. (1998). K^+ is an endothelium-derived hyperpolarizing factor in rat arteries. *Nature*, *396*(6708), 269–272.

Edwards, G., Feletou, M., & Weston, A. H. (2010). Endothelium-derived hyperpolarising factors and associated pathways: A synopsis. *Pflügers Archiv/European Journal of Physiology*, *459*(6), 863–879.

Eichler, I., Wibawa, J., Grgic, I., Knorr, A., Brakemeier, S., Pries, A. R., et al. (2003). Selective blockade of endothelial Ca^{2+}-activated small- and intermediate-conductance K^+-channels suppresses EDHF-mediated vasodilation. *British Journal of Pharmacology*, *138*(4), 594–601.

Ellinor, P. T., Lunetta, K. L., Glazer, N. L., Pfeufer, A., Alonso, A., Chung, M. K., et al. (2010). Common variants in KCNN3 are associated with lone atrial fibrillation. *Nature Genetics*, *42*(3), 240–244.
Endo, J., & Arita, M. (2016). Cardioprotective mechanism of omega-3 polyunsaturated fatty acids. *Journal of Cardiology*, *67*(1), 22–27.
Feletou, M. (2009). Calcium-activated potassium channels and endothelial dysfunction: Therapeutic options? *British Journal of Pharmacology*, *156*(4), 545–562.
Feletou, M. (2015). Endothelium-dependent hyperpolarization and endothelial dysfunction. *Journal of Cardiovascular Pharmacology*, (Epub ahead of print).
Feletou, M., Kohler, R., & Vanhoutte, P. M. (2010). Endothelium-derived vasoactive factors and hypertension: Possible roles in pathogenesis and as treatment targets. *Current Hypertension Reports*, *12*(4), 267–275.
Feletou, M., Kohler, R., & Vanhoutte, P. M. (2012). Nitric oxide: Orchestrator of endothelium-dependent responses. *Annals of Medicine*, *44*(7), 694–716.
Fernandez-Fernandez, J. M., Tomas, M., Vazquez, E., Orio, P., Latorre, R., Senti, M., et al. (2004). Gain-of-function mutation in the KCNMB1 potassium channel subunit is associated with low prevalence of diastolic hypertension. *The Journal of Clinical Investigation*, *113*(7), 1032–1039.
Ferreira, R., Wong, R., & Schlichter, L. C. (2015). KCa3.1/IK1 channel regulation by cGMP-dependent protein kinase (PKG) via reactive oxygen species and CaMKII in microglia: An immune modulating feedback system? *Frontiers in Immunology*, *6*, 153.
Feske, S., Wulff, H., & Skolnik, E. Y. (2015). Ion channels in innate and adaptive immunity. *Annual Review of Immunology*, *33*, 291–353.
Fisslthaler, B., Popp, R., Kiss, L., Potente, M., Harder, D. R., Fleming, I., et al. (1999). Cytochrome P450 2C is an EDHF synthase in coronary arteries. *Nature*, *401*(6752), 493–497.
Foster, M. N., & Coetzee, W. A. (2016). KATP channels in the cardiovascular system. *Physiological Reviews*, *96*(1), 177–252.
Freise, C., Heldwein, S., Erben, U., Hoyer, J., Kohler, R., Johrens, K., et al. (2014). K^+-channel inhibition reduces portal perfusion pressure in fibrotic rats and fibrosis associated characteristics of hepatic stellate cells. *Liver International*, *35*(4), 1244–1252.
Furchgott, R. F., & Zawadzki, J. V. (1980). The obligatory role of endothelial cells in the relaxation of arterial smooth muscle by acetylcholine. *Nature*, *288*(5789), 373–376.
Garland, C. J., & Dora, K. A. (2016). EDH: Endothelium-dependent hyperpolarization and microvascular signaling. *Acta Physiologica (Oxford, England)*, http://dx.doi.org/10.1111/apha.12649 (Epub ahead of print).
Garland, C. J., & Plane, F. (1996). Relative importance of endothelium-derived hyperpolarizing factor for the relaxation of vascular smooth muscle in different arterial beds. In P. M. Vanhoutte (Ed.), *Endothelium-derived hyperpolarizing factor: Vol. 1* (pp. 173–179). Amsterdam: Harwood Academic Publishers.
Garland, C. J., Smirnov, S. V., Bagher, P., Lim, C. S., Huang, C. Y., Mitchell, R., et al. (2015). TRPM4 inhibitor 9-phenanthrol activates endothelial cell intermediate conductance calcium-activated potassium channels in rat isolated mesenteric artery. *British Journal of Pharmacology*, *172*(4), 1114–1123.
Gerlach, A. C., Syme, C. A., Giltinan, L., Adelman, J. P., & Devors, D. C. (2001). ATP-dependent activation of the intermediate conductance, Ca^{2+}-activated K^+ channel, hIK1, is conferred by a C-terminal domain. *The Journal of Biological Chemistry*, *276*(24), 10963–10970.
Giachini, F. R., Carneiro, F. S., Lima, V. V., Carneiro, Z. N., Dorrance, A., Webb, R. C., et al. (2009). Upregulation of intermediate calcium-activated potassium channels counterbalance the impaired endothelium-dependent vasodilation in stroke-prone spontaneously hypertensive rats. *Translational Research*, *154*(4), 183–193.

Glogowska, E., Lezon-Geyda, K., Maksimova, Y., Schulz, V. P., & Gallagher, P. G. (2015). Mutations in the Gardos channel (KCNN4) are associated with hereditary xerocytosis. *Blood*, *126*(11), 1281–1284.

Grgic, I., Eichler, I., Heinau, P., Si, H., Brakemeier, S., Hoyer, J., et al. (2005). Selective blockade of the intermediate-conductance Ca^{2+}-activated K^{+} channel suppresses proliferation of microvascular and macrovascular endothelial cells and angiogenesis in vivo. *Arteriosclerosis, Thrombosis, and Vascular Biology*, *25*(4), 704–709.

Grgic, I., Kaistha, B. P., Hoyer, J., & Kohler, R. (2009a). Endothelial Ca^{+}-activated K^{+} channels in normal and impaired EDHF-dilator responses—Relevance to cardiovascular pathologies and drug discovery. *British Journal of Pharmacology*, *157*(4), 509–526.

Grgic, I., Kaistha, B. P., Paschen, S., Kaistha, A., Busch, C., Si, H., et al. (2009b). Disruption of the Gardos channel (KCa3.1) in mice causes subtle erythrocyte macrocytosis and progressive splenomegaly. *Pflügers Archiv/European Journal of Physiology*, *458*(2), 291–302.

Grossinger, E. M., Weiss, L., Zierler, S., Rebhandl, S., Krenn, P. W., Hinterseer, E., et al. (2014). Targeting proliferation of chronic lymphocytic leukemia (CLL) cells through KCa3.1 blockade. *Leukemia*, *28*(4), 954–958.

Grundemann, J., & Clark, B. A. (2015). Calcium-activated potassium channels at nodes of Ranvier secure axonal spike propagation. *Cell Reports*, *12*(11), 1715–1722.

Gutman, G. A., Chandy, K. G., Adelman, J. P., Aiyar, J., Bayliss, D. A., Clapham, D. E., et al. (2003). International Union of Pharmacology. XLI. Compendium of voltage-gated ion channels: Potassium channels. *Pharmacological Reviews*, *55*(4), 583–586.

Hamilton, K. L., Syme, C. A., & Devor, D. C. (2003). Molecular localization of the inhibitory arachidonic acid binding site to the pore of hIK1. *The Journal of Biological Chemistry*, *278*(19), 16690–16697.

Hasenau, A. L., Nielsen, G., Morisseau, C., Hammock, B. D., Wulff, H., & Kohler, R. (2011). Improvement of endothelium-dependent vasodilations by SKA-31 and SKA-20, activators of small- and intermediate-conductance Ca^{2+}-activated K^{+}-channels. *Acta Physiologica (Oxford)*, *203*(1), 117–126.

Haugaard, M. M., Hesselkilde, E. Z., Pehrson, S., Carstensen, H., Flethoj, M., Praestegaard, K. F., et al. (2015). Pharmacologic inhibition of small-conductance calcium-activated potassium (SK) channels by NS8593 reveals atrial antiarrhythmic potential in horses. *Heart Rhythm*, *12*(4), 825–835.

Hua, X., Deuse, T., Chen, Y. J., Wulff, H., Stubbendorff, M., Kohler, R., et al. (2013). The potassium channel KCa3.1 as new therapeutic target for the prevention of obliterative airway disease. *Transplantation*, *95*(2), 285–292.

Huang, C., Pollock, C. A., & Chen, X. M. (2014). Role of the potassium channel KCa3.1 in diabetic nephropathy. *Clinical Science (London, England: 1979)*, *127*(7), 423–433.

Ignarro, L. J., Buga, G. M., Wood, K. S., Byrns, R. E., & Chaudhuri, G. (1987). Endothelium-derived relaxing factor produced and released from artery and vein is nitric oxide. *Proceedings of the National Academy of Sciences of the United States of America*, *84*(24), 9265–9269.

Jacobson, D., Pribnow, D., Herson, P. S., Maylie, J., & Adelman, J. P. (2003). Determinants contributing to estrogen-regulated expression of SK3. *Biochemical and Biophysical Research Communications*, *303*(2), 660–668.

Jenkins, D. P., Strobaek, D., Hougaard, C., Jensen, M. L., Hummel, R., Sorensen, U. S., et al. (2011). Negative gating modulation by (R)-N-(benzimidazol-2-yl)-1,2,3,4-tetrahydro-1-naphthylamine (NS8593) depends on residues in the inner pore vestibule: Pharmacological evidence of deep-pore gating of K(Ca)2 channels. *Molecular Pharmacology*, *79*(6), 899–909.

Ju, C. H., Wang, X. P., Gao, C. Y., Zhang, S. X., Ma, X. H., & Liu, C. (2015). Blockade of KCa3.1 attenuates left ventricular remodeling after experimental myocardial infarction. *Cellular Physiology and Biochemistry*, *36*(4), 1305–1315.

Kacik, M., Olivan-Viguera, A., & Kohler, R. (2014). Modulation of KCa3.1 channels by eicosanoids, omega-3 fatty acids, and molecular determinants. *PLoS One*, *9*(11), e112081.

Kang, H., Kerloc'h, A., Rotival, M., Xu, X., Zhang, Q., D'Souza, Z., et al. (2014). Kcnn4 is a regulator of macrophage multinucleation in bone homeostasis and inflammatory disease. *Cell Reports*, *8*(4), 1210–1224.

Kerr, P. M., Wei, R., Tam, R., Sandow, S. L., Murphy, T. V., Ondrusova, K., et al. (2015). Activation of endothelial IKCa channels underlies NO-dependent myoendothelial feedback. *Vascular Pharmacology*, *74*, 130–138.

King, B., Rizwan, A. P., Asmara, H., Heath, N. C., Engbers, J. D., Dykstra, S., et al. (2015). IKCa channels are a critical determinant of the slow AHP in CA1 pyramidal neurons. *Cell Reports*, *11*(2), 175–182.

Koch Hansen, L., Sevelsted-Moller, L., Rabjerg, M., Larsen, D., Hansen, T. P., Klinge, L., et al. (2014). Expression of T-cell KV1.3 potassium channel correlates with pro-inflammatory cytokines and disease activity in ulcerative colitis. *Journal of Crohn's and Colitis*, *8*(11), 1378–1391.

Köhler, R., Brakemeier, S., Kuhn, M., Behrens, C., Real, R., Degenhardt, C., et al. (2001). Impaired hyperpolarization in regenerated endothelium after balloon catheter injury. *Circulation Research*, *89*(2), 174–179.

Köhler, M., Hirschberg, B., Bond, C. T., Kinzie, J. M., Marrion, N. V., Maylie, J., et al. (1996). Small-conductance, calcium-activated potassium channels from mammalian brain. *Science*, *273*(5282), 1709–1714.

Köhler, R., Kaistha, B. P., & Wulff, H. (2010). Vascular KCa-channels as therapeutic targets in hypertension and restenosis disease. *Expert Opinion on Therapeutic Targets*, *14*(2), 143–155.

Köhler, R., & Ruth, P. (2010). Endothelial dysfunction and blood pressure alterations in K^+-channel transgenic mice. *Pflügers Archiv/European Journal of Physiology*, *459*(6), 969–976.

Köhler, R., Wulff, H., Eichler, I., Kneifel, M., Neumann, D., Knorr, A., et al. (2003). Blockade of the intermediate-conductance calcium-activated potassium channel as a new therapeutic strategy for restenosis. *Circulation*, *108*(9), 1119–1125.

Koot, B. G., Alders, M., Verheij, J., Beuers, U. H., & Cobben, J. M. (2016). A de novo mutation in KCNN3 associated with autosomal dominant idiopathic non-cirrhotic portal hypertension. *Journal of Hepatology*, *64*(4), 974–977. http://dx.doi.org/10.1016/j.jhep.2015.11.027 (Epub ahead of print).

Kroigaard, C., Dalsgaard, T., Nielsen, G., Laursen, B. E., Pilegaard, H., Kohler, R., et al. (2012). Activation of endothelial and epithelial K(Ca) 2.3 calcium-activated potassium channels by NS309 relaxes human small pulmonary arteries and bronchioles. *British Journal of Pharmacology*, *167*(1), 37–47.

Lambertsen, K. L., Gramsbergen, J. B., Sivasaravanaparan, M., Ditzel, N., Sevelsted-Moller, L. M., Olivan-Viguera, A., et al. (2012). Genetic KCa3.1-deficiency produces locomotor hyperactivity and alterations in cerebral monoamine levels. *PLoS One*, 7(10), e47744.

Lamoral-Theys, D., Pottier, L., Kerff, F., Dufrasne, F., Proutiere, F., Wauthoz, N., et al. (2010). Simple di- and trivanillates exhibit cytostatic properties toward cancer cells resistant to pro-apoptotic stimuli. *Bioorganic & Medicinal Chemistry*, *18*(11), 3823–3833.

Larsen, B. T., Miura, H., Hatoum, O. A., Campbell, W. B., Hammock, B. D., Zeldin, D. C., et al. (2006). Epoxyeicosatrienoic and dihydroxyeicosatrienoic acids dilate human coronary arterioles via BK(Ca) channels: Implications for soluble epoxide hydrolase inhibition. *American Journal of Physiology. Heart and Circulatory Physiology*, *290*(2), H491–H499.

Ledoux, J., Taylor, M. S., Bonev, A. D., Hannah, R. M., Solodushko, V., Shui, B., et al. (2008). Functional architecture of inositol 1,4,5-trisphosphate signaling in restricted spaces of myoendothelial projections. *Proceedings of the National Academy of Sciences of the United States of America*, *105*(28), 9627–9632.

Liin, S. I., Silvera Ejneby, M., Barro-Soria, R., Skarsfeldt, M. A., Larsson, J. E., Starck Harlin, F., et al. (2015). Polyunsaturated fatty acid analogs act antiarrhythmically on the cardiac IKs channel. *Proceedings of the National Academy of Sciences of the United States of America*, *112*(18), 5714–5719.

Liu, Y., Sellke, E. W., Feng, J., Clements, R. T., Sodha, N. R., Khabbaz, K. R., et al. (2008). Calcium-activated potassium channels contribute to human skeletal muscle microvascular endothelial dysfunction related to cardiopulmonary bypass. *Surgery*, *144*(2), 239–244.

Liu, Y., Xie, A., Singh, A. K., Ehsan, A., Choudhary, G., Dudley, S., et al. (2015). Inactivation of endothelial small/intermediate conductance of calcium-activated potassium channels contributes to coronary arteriolar dysfunction in diabetic patients. *Journal of the American Heart Association*, *4*(8), e002062.

Longden, T. A., Dunn, K. M., Draheim, H. J., Nelson, M. T., Weston, A. H., & Edwards, G. (2011). Intermediate-conductance calcium-activated potassium channels participate in neurovascular coupling. *British Journal of Pharmacology*, *164*(3), 922–933.

Luft, F. C., Dechend, R., & Muller, D. N. (2012). Immune mechanisms in angiotensin II-induced target-organ damage. *Annals of Medicine*, *44*(Suppl. 1), S49–S54.

Mahida, S., Mills, R. W., Tucker, N. R., Simonson, B., Macri, V., Lemoine, M. D., et al. (2014). Overexpression of KCNN3 results in sudden cardiac death. *Cardiovascular Research*, *101*(2), 326–334.

Mauler, F., Hinz, V., Horvath, E., Schuhmacher, J., Hofmann, H. A., Wirtz, S., et al. (2004). Selective intermediate-/small-conductance calcium-activated potassium channel (KCNN4) blockers are potent and effective therapeutics in experimental brain oedema and traumatic brain injury caused by acute subdural haematoma. *European Journal of Neuroscience*, *20*(7), 1761–1768.

Milkau, M., Kohler, R., & de Wit, C. (2010). Crucial importance of the endothelial K^+ channel SK3 and connexin40 in arteriolar dilations during skeletal muscle contraction. *The FASEB Journal*, *24*(9), 3572–3579.

Mishra, R. C., Mitchell, J. R., Gibbons-Kroeker, C., Wulff, H., Belenkie, I., Tyberg, J. V., et al. (2016). A pharmacologic activator of endothelial KCa channels increases systemic conductance and reduces arterial pressure in an anesthetized pig model. *Vascular Pharmacology*, *79*, 24–31.

Mokhtar, S. S., Vanhoutte, P. M., Leung, S. W., Yusof, M. I., Wan Sulaiman, W. A., Mat Saad, A. Z., et al. (2016). Endothelium dependent hyperpolarization-type relaxation compensates for attenuated nitric oxide-mediated responses in subcutaneous arteries of diabetic patients. *Nitric Oxide*, *53*, 35–44.

Moncada, S., Gryglewski, R., Bunting, S., & Vane, J. R. (1976). An enzyme isolated from arteries transforms prostaglandin endoperoxides to an unstable substance that inhibits platelet aggregation. *Nature*, *263*(5579), 663–665.

Moreno, C., Macias, A., Prieto, A., de la Cruz, A., Gonzalez, T., & Valenzuela, C. (2012). Effects of n-3 polyunsaturated fatty acids on cardiac ion channels. *Frontiers in Physiology*, *3*, 245.

Morimura, K., Yamamura, H., Ohya, S., & Imaizumi, Y. (2006). Voltage-dependent Ca^{2+}-channel block by openers of intermediate and small conductance Ca^{2+}-activated K^+ channels in urinary bladder smooth muscle cells. *Journal of Pharmaceutical Sciences*, *100*(3), 237–241.

Morton, S. K., Chaston, D. J., Howitt, L., Heisler, J., Nicholson, B. J., Fairweather, S., et al. (2015). Loss of functional endothelial connexin40 results in exercise-induced hypertension in mice. *Hypertension*, *65*(3), 662–669.

Nielsen, G., Wandall-Frostholm, C., Sadda, V., Olivan-Viguera, A., Lloyd, E. E., Bryan, R. M., Jr., et al. (2013). Alterations of N-3 polyunsaturated fatty acid-activated K2P channels in hypoxia-induced pulmonary hypertension. *Basic & Clinical Pharmacology & Toxicology*, *113*(4), 250–258.

Olivan-Viguera, A., Valero, M. S., Coleman, N., Brown, B. M., Laria, C., Murillo, M. D., et al. (2015). A novel pan-negative-gating modulator of KCa2/3 channels, the fluoro-di-benzoate, RA-2, inhibits EDH-type relaxation in coronary artery and produces bradycardia in vivo. *Molecular Pharmacology*, *87*, 1–12.

Olivan-Viguera, A., Valero, M. S., Murillo, M. D., Wulff, H., Garcia-Otin, A. L., Arbones-Mainar, J. M., et al. (2013). Novel phenolic inhibitors of small/intermediate-conductance Ca(2)(+)-activated K(+) channels, KCa3.1 and KCa2.3. *PLoS One*, *8*(3), e58614.

Olivan-Viguera, A., Valero, M. S., Pinilla, E., Amor, S., Garcia-Villalon, A. L., Coleman, N., et al. (2016). Vascular reactivity profile of novel KCa3.1-selective positive-gating modulators in the coronary artery bed. *Basic & Clinical Pharmacology & Toxicology*, http://dx.doi.org/10.1111/bcpt.12560 (Epub ahead of print).

Ortega, A., Tarazon, E., Rosello-Lleti, E., Gil-Cayuela, C., Lago, F., Gonzalez-Juanatey, J. R., et al. (2015). Patients with dilated cardiomyopathy and sustained monomorphic ventricular tachycardia show up-regulation of KCNN3 and KCNJ2 genes and CACNG8-linked left ventricular dysfunction. *PLoS One*, *10*(12), e0145518.

Pandit, L. M., Lloyd, E. E., Reynolds, J. O., Lawrence, W. S., Reynolds, C., Wehrens, X. H., et al. (2014). TWIK-2 channel deficiency leads to pulmonary hypertension through a rho-kinase-mediated process. *Hypertension*, *64*(6), 1260–1265.

Panyi, G., Beeton, C., & Felipe, A. (2014). Ion channels and anti-cancer immunity. *Philosophical Transactions of the Royal Society of London. Series B, Biological Sciences*, *369*(1638), 20130106.

Pierce, S. L., & England, S. K. (2010). SK3 channel expression during pregnancy is regulated through estrogen and Sp factor-mediated transcriptional control of the KCNN3 gene. *American Journal of Physiology. Endocrinology and Metabolism*, *299*(4), E640–E646.

Pierce, S. L., Kresowik, J. D., Lamping, K. G., & England, S. K. (2008). Overexpression of SK3 channels dampens uterine contractility to prevent preterm labor in mice. *Biology of Reproduction*, *78*(6), 1058–1063.

Qian, X., Francis, M., Kohler, R., Solodushko, V., Lin, M., & Taylor, M. S. (2014). Positive feedback regulation of agonist-stimulated endothelial Ca^{2+} dynamics by KCa3.1 channels in mouse mesenteric arteries. *Arteriosclerosis, Thrombosis, and Vascular Biology*, *34*(1), 127–135.

Rabjerg, M., Oliván-Viguera, A., Hansen, L. K., Jensen, L., Sevelsted-Møller, L., Walter, S., et al. (2015). High expression of KCa3.1 in patients with clear cell renal carcinoma predicts high metastatic risk and poor survival. *PLoS One*, *10*(4), e0122992.

Radomski, M. W., Palmer, R. M., & Moncada, S. (1987). The anti-aggregating properties of vascular endothelium: Interactions between prostacyclin and nitric oxide. *British Journal of Pharmacology*, *92*(3), 639–646.

Radtke, J., Schmidt, K., Wulff, H., Kohler, R., & de Wit, C. (2013). Activation of KCa3.1 by SKA-31 induces arteriolar dilatation and lowers blood pressure in normo- and hypertensive connexin40-deficient mice. *British Journal of Pharmacology*, *170*(2), 293–303.

Rahbek, M., Nazemi, S., Odum, L., Gupta, S., Poulsen, S. S., Hay-Schmidt, A., et al. (2014). Expression of the small conductance Ca(2)(+)-activated potassium channel subtype 3 (SK3) in rat uterus after stimulation with 17beta-estradiol. *PLoS One*, *9*(2), e87652.

Rapetti-Mauss, R., Lacoste, C., Picard, V., Guitton, C., Lombard, E., Loosveld, M., et al. (2015). A mutation in the Gardos channel is associated with hereditary xerocytosis. *Blood*, *126*(11), 1273–1280.

Roach, K. M., Wulff, H., Feghali-Bostwick, C., Amrani, Y., & Bradding, P. (2014). Increased constitutive αSMA and Smad2/3 expression in idiopathic pulmonary fibrosis myofibroblasts is KCa3.1-dependent. *Respiratory Research*, *15*(1), 155.

Rosa, J. C., Galanakis, D., Ganellin, C. R., Dunn, P. M., & Jenkinson, D. H. (1998). Bis-quinolinium cyclophanes: 6,10-Diaza-3(1,3),8(1,4)-dibenzena-1,5(1,4)-diquinolinacyclodecaphane (UCL 1684), the first nanomolar, non-peptidic blocker of the apamin-sensitive Ca^{2+}-activated K^+ channel. *Journal of Medicinal Chemistry*, *41*(1), 2–5.

Sandow, S. L., Neylon, C. B., Chen, M. X., & Garland, C. J. (2006). Spatial separation of endothelial small- and intermediate-conductance calcium-activated potassium channels (K_{Ca}) and connexins: Possible relationship to vasodilator function? *Journal of Anatomy*, *209*(5), 689–698.

Sankaranarayanan, A., Raman, G., Busch, C., Schultz, T., Zimin, P. I., Hoyer, J., et al. (2009). Naphtho[1,2-d]thiazol-2-ylamine (SKA-31), a new activator of KCa2 and KCa3.1 potassium channels, potentiates the endothelium-derived hyperpolarizing factor response and lowers blood pressure. *Molecular Pharmacology*, *75*(2), 281–295.

Schach, C., Resch, M., Schmid, P. M., Riegger, G. A., & Endemann, D. H. (2014). Type 2 diabetes: Increased expression and contribution of IKCa channels to vasodilation in small mesenteric arteries of ZDF rats. *American Journal of Physiology. Heart and Circulatory Physiology*, *307*(8), H1093–H1102.

Sevelsted-Moeller, L. (2015). *Potential roles of the calcium-activated potassium channel, KCa3.1, in murine and human hepatic fibrogenesis*. PhD thesis, University of Southern Denmark.

Shimokawa, H., & Morikawa, K. (2005). Hydrogen peroxide is an endothelium-derived hyperpolarizing factor in animals and humans. *Journal of Molecular and Cellular Cardiology*, *39*(5), 725–732.

Si, H., Grgic, I., Heyken, W. T., Maier, T., Hoyer, J., Reusch, H. P., et al. (2006). Mitogenic modulation of Ca^{2+}-activated K^+ channels in proliferating A7r5 vascular smooth muscle cells. *British Journal of Pharmacology*, *148*(7), 909–917.

Si, H., Heyken, W. T., Wölfle, S. E., Tysiac, M., Schubert, R., Grgic, I., et al. (2006). Impaired endothelium-derived hyperpolarizing factor-mediated dilations and increased blood pressure in mice deficient of the intermediate-conductance Ca^{2+}-activated K^+ channel. *Circulation Research*, *99*(5), 537–544.

Soder, R. P., Parajuli, S. P., Hristov, K. L., Rovner, E. S., & Petkov, G. V. (2013). SK channel-selective opening by SKA-31 induces hyperpolarization and decreases contractility in human urinary bladder smooth muscle. *American Journal of Physiology. Regulatory, Integrative and Comparative Physiology*, *304*(2), R155–R163.

Srivastava, S., Zhdanova, O., Di, L., Li, Z., Albaqumi, M., Wulff, H., et al. (2008). Protein histidine phosphatase 1 negatively regulates CD4 T cells by inhibiting the K^+ channel KCa3.1. *Proceedings of the National Academy of Sciences of the United States of America*, *105*(38), 14442–14446.

Stott, J. B., Jepps, T. A., & Greenwood, I. A. (2014). K(V)7 potassium channels: A new therapeutic target in smooth muscle disorders. *Drug Discovery Today*, *19*(4), 413–424.

Strobaek, D., Brown, D. T., Jenkins, D. P., Chen, Y. J., Coleman, N., Ando, Y., et al. (2013). NS6180, a new K(Ca) 3.1 channel inhibitor prevents T-cell activation and inflammation in a rat model of inflammatory bowel disease. *British Journal of Pharmacology*, *168*(2), 432–444.

Suematsu, M., Suzuki, H., Delano, F. A., & Schmid-Schonbein, G. W. (2002). The inflammatory aspect of the microcirculation in hypertension: Oxidative stress, leukocytes/endothelial interaction, apoptosis. *Microcirculation*, *9*(4), 259–276.

Takai, J., Santu, A., Zheng, H., Koh, S. D., Ohta, M., Filimban, L. M., et al. (2013). Laminar shear stress upregulates endothelial Ca(2)(+)-activated K(+) channels KCa2.3 and KCa3.1 via a Ca(2)(+)/calmodulin-dependent protein kinase kinase/Akt/p300 cascade. *American Journal of Physiology. Heart and Circulatory Physiology*, *305*(4), H484–H493.

Taylor, M. S., Bonev, A. D., Gross, T. P., Eckman, D. M., Brayden, J. E., Bond, C. T., et al. (2003). Altered expression of small-conductance Ca^{2+}-activated K^{+} (SK3) channels modulates arterial tone and blood pressure. *Circulation Research, 93*(2), 124–131.

Tharp, D. L., Wamhoff, B. R., Wulff, H., Raman, G., Cheong, A., & Bowles, D. K. (2008). Local delivery of the KCa3.1 blocker, TRAM-34, prevents acute angioplasty-induced coronary smooth muscle phenotypic modulation and limits stenosis. *Arteriosclerosis, Thrombosis, and Vascular Biology, 28*(6), 1084–1089.

Toyama, K., Wulff, H., Chandy, K. G., Azam, P., Raman, G., Saito, T., et al. (2008). The intermediate-conductance calcium-activated potassium channel KCa3.1 contributes to atherogenesis in mice and humans. *Journal of Clinical Investigation, 118*(9), 3025–3037.

Tsai, C. T., Hsieh, C. S., Chang, S. N., Chuang, E. Y., Juang, J. M., Lin, L. Y., et al. (2015). Next-generation sequencing of nine atrial fibrillation candidate genes identified novel de novo mutations in patients with extreme trait of atrial fibrillation. *Journal of Medical Genetics, 52*(1), 28–36.

Waeckel, L., Bertin, F., Clavreul, N., Damery, T., Kohler, R., Paysant, J., et al. (2014). Preserved regulation of renal perfusion pressure by small and intermediate conductance K channels in hypertensive mice with or without renal failure. *Pflügers Archiv/European Journal of Physiology, 467*(4), 817–831.

Wandall-Frostholm, C., Dalsgaard, T., Bajoriunas, V., Olivan-Viguera, A., Sadda, V., Beck, L., et al. (2015). Genetic deficit of K 3.1 channels protects against pulmonary circulatory collapse induced by TRPV4 channel activation. *British Journal of Pharmacology*, http://dx.doi.org/10.1111/bph.13234 (Epub ahead of print).

Wandall-Frostholm, C., Skaarup, L. M., Sadda, V., Nielsen, G., Hedegaard, E. R., Mogensen, S., et al. (2014). Pulmonary hypertension in wild type mice and animals with genetic deficit in KCa2.3 and KCa3.1 channels. *PLoS One, 9*(5), e97687.

Wang, R. (2002). Two's company, three's a crowd: Can H_2S be the third endogenous gaseous transmitter? *The FASEB Journal, 16*(13), 1792–1798.

Wang, R. X., Chai, Q., Lu, T., & Lee, H. C. (2011). Activation of vascular BK channels by docosahexaenoic acid is dependent on cytochrome P450 epoxygenase activity. *Cardiovascular Research, 90*(2), 344–352.

Wang, K., Mateos-Aparicio, P., Hönigsperger, C., Raghuram, V., Wu, W. W., Ridder, M. C., et al. (2016). IK1 channels do not contribute to the slow after hyperpolarization in pyramidal neurons. *eLife. 5*, pii: e11206. http://dx.doi.org/10.7554/eLife.11206.

Watanabe, H., Davis, J. B., Smart, D., Jerman, J. C., Smith, G. D., Hayes, P., et al. (2002). Activation of TRPV4 channels (hVRL-2/mTRP12) by phorbol derivatives. *Journal of Biological Chemistry, 277*(16), 13569–13577.

Wei, A. D., Gutman, G. A., Aldrich, R., Chandy, K. G., Grissmer, S., & Wulff, H. (2005). International Union of Pharmacology. LII. Nomenclature and molecular relationships of calcium-activated potassium channels. *Pharmacological Reviews, 57*(4), 463–472.

Weisbrod, D., Khun, S. H., Bueno, H., Peretz, A., & Attali, B. (2016). Mechanisms underlying the cardiac pacemaker: The role of SK4 calcium-activated potassium channels. *Acta Pharmacolologica Sinica, 37*(1), 82–97.

Weston, A. H., Absi, M., Ward, D. T., Ohanian, J., Dodd, R. H., Dauban, P., et al. (2005). Evidence in favor of a calcium-sensing receptor in arterial endothelial cells: Studies with calindol and Calhex 231. *Circulation Research, 97*(4), 391–398.

Wolfle, S. E., Schmidt, V. J., Hoyer, J., Kohler, R., & de Wit, C. (2009). Prominent role of KCa3.1 in endothelium-derived hyperpolarizing factor-type dilations and conducted responses in the microcirculation in vivo. *Cardiovascular Research, 82*(3), 476–483.

Wulff, H., & Castle, N. A. (2010). Therapeutic potential of KCa3.1 blockers: Recent advances and promising trends. *Expert Review of Clinical Pharmacology, 3*(3), 385–396.

Wulff, H., Gutman, G. A., Cahalan, M. D., & Chandy, K. G. (2001). Delineation of the clotrimazole/TRAM-34 binding site on the intermediate conductance calcium-activated potassium channel, IKCa1. *Journal of Biological Chemistry, 276*(34), 32040–32045.

Wulff, H., & Köhler, R. (2013). Endothelial small-conductance and intermediate-conductance KCa channels: An update on their pharmacology and usefulness as cardiovascular targets. *Journal of Cardiovascular Pharmacology, 61*(2), 102–112.

Wulff, H., Miller, M. J., Hansel, W., Grissmer, S., Cahalan, M. D., & Chandy, K. G. (2000). Design of a potent and selective inhibitor of the intermediate-conductance Ca^{2+}-activated K^+ channel, IKCa1: A potential immunosuppressant. *Proceedings of the National Academy of Sciences of the United States of America, 97*(14), 8151–8156.

Yamaguchi, M., Nakayama, T., Fu, Z., Naganuma, T., Sato, N., Soma, M., et al. (2009). Relationship between haplotypes of KCNN4 gene and susceptibility to human vascular diseases in Japanese. *Medical Science Monitor, 15*(8), CR389–CR397.

Yap, F. C., Weber, D. S., Taylor, M. S., Townsley, M. I., Comer, B. S., Maylie, J., et al. (2016). Endothelial SK3 channel-associated Ca^{2+} microdomains modulate blood pressure. *American Journal of Physiology. Heart and Circulatory Physiology*. ajpheart.00787.2015, Epub ahead of print.

Young, E. J., Hill, M. A., Wiehler, W. B., Triggle, C. R., & Reid, J. J. (2008). Reduced EDHF responses and connexin activity in mesenteric arteries from the insulin-resistant obese Zucker rat. *Diabetologia, 51*(5), 872–881.

Zhang, M., Pascal, J. M., Schumann, M., Armen, R. S., & Zhang, J. F. (2012). Identification of the functional binding pocket for compounds targeting small-conductance Ca(2)(+)-activated potassium channels. *Nature Communications, 3*(3), 1021–1040.

Zhao, L. M., Wang, L. P., Wang, H. F., Ma, X. Z., Zhou, D. X., & Deng, X. L. (2015). The role of KCa3.1 channels in cardiac fibrosis induced by pressure overload in rats. *Pflügers Archiv/European Journal of Physiology, 467*(11), 2275–2285.

Zhou, X. B., Feng, Y. X., Sun, Q., Lukowski, R., Qiu, Y., Spiger, K., et al. (2015). Nucleoside diphosphate kinase B-activated intermediate conductance potassium channels are critical for neointima formation in mouse carotid arteries. *Arteriosclerosis, Thrombosis, and Vascular Biology, 35*(8), 1852–1861.

CHAPTER FOUR

Epoxyeicosatrienoic Acids and 20-Hydroxyeicosatetraenoic Acid on Endothelial and Vascular Function

J.D. Imig[1]
Medical College of Wisconsin, Milwaukee, WI, United States
[1]Corresponding author: e-mail address: jdimig@mcw.edu

Contents

Abstract

Endothelial and vascular smooth cells generate cytochrome P450 (CYP) arachidonic acid metabolites that can impact endothelial cell function and vascular homeostasis. The objective of this review is to focus on the physiology and pharmacology of endothelial CYP metabolites. The CYP pathway produces two types of eicosanoid products: epoxyeicosatrienoic acids (EETs), formed by CYP epoxygenases, and hydroxyeicosatetraenoic acids (HETEs), formed by CYP hydroxylases. Advances in CYP enzymes, EETs, and 20-HETE by pharmacological and genetic means have led to a more complete understanding of how these eicosanoids impact on endothelial cell function. Endothelial-derived EETs were initially described as endothelial-derived hyperpolarizing

Advances in Pharmacology, Volume 77
ISSN 1054-3589
http://dx.doi.org/10.1016/bs.apha.2016.04.003

factors. It is now well recognized that EETs importantly contribute to numerous endothelial cell functions. On the other hand, 20-HETE is the predominant CYP hydroxylase synthesized by vascular smooth muscle cells. Like EETs, 20-HETE acts on endothelial cells and impacts importantly on endothelial and vascular function. An important aspect for EETs and 20-HETE endothelial actions is their interactions with hormonal and paracrine factors. These include interactions with the renin–angiotensin system, adrenergic system, puringeric system, and endothelin. Alterations in CYP enzymes, 20-HETE, or EETs contribute to endothelial dysfunction and cardiovascular diseases such as ischemic injury, hypertension, and atherosclerosis. Recent advances have led to the development of potential therapeutics that target CYP enzymes, 20-HETE, or EETs. Thus, future investigation is required to obtain a more complete understanding of how CYP enzymes, 20-HETE, and EETs regulate endothelial cell function.

ABBREVIATIONS

DHETs dihydroxyeicosatrienoic acids
EETs epoxyeicosatrienoic acids
EPC endothelial progenitor cell
HETEs hydroxyeicosatetraenoic acids
sEH soluble epoxide hydrolase

1. INTRODUCTION

Endothelial cells are recognized as important modulators of vascular function and critical for maintaining hemodynamic homeostasis. The endothelium interfaces with plasma and blood cells to respond to physical forces, blood cells, and endocrine and paracrine circulating factors. The endothelial cell can activate cell signaling pathways and release of autocrine and paracrine factors in response. These endothelial factors can regulate vascular inflammation, platelet aggregation, vascular permeability, vascular smooth muscle cell function and blood flow, and angiogenesis.

It is well recognized that endothelial cells release three primary paracrine factors in response to shear stress and hormones to regulate vascular smooth muscle cell function. These endothelial-derived relaxing factors include nitric oxide synthase (NOS) generation of nitric oxide (NO), cyclooxygenase (COX) prostacyclin (PGI_2) generation, and cytochrome P450 (CYP) generation of epoxyeicosatrienoic acids (EETs) (Campbell & Fleming, 2010; Furchgott & Vanhoutte, 1989). Although NO is a major vasodilator factor in large arteries, EETs and NO have similar contributions with a lesser contribution from PGI_2 to endothelial-derived relaxation of smaller resistance arteries and arterioles (Campbell & Fleming, 2010; Imig, 2012). Over the past decades these endothelial-derived factors have been demonstrated to have a number of other activities that maintain vascular homeostasis (Fleming, 2001; Imig, 2012).

The number of endothelial-derived cell signaling pathways, endothelial-derived factors, and endothelial cell physiological roles has been greatly expanding. Endothelial cells are a major focus of investigation and pathological roles in cardiovascular diseases are evaluated for potential therapeutic intervention. A significant role for the endothelium is now recognized for immune diseases, diabetes, Alzheimer's disease, and cancer (Bellien & Joannides, 2013; Tacconelli & Patrignani, 2014). Pharmacological manipulation of endothelial NO and COX metabolites has been extensively evaluated and demonstrated promise in many of these diseases (Bellien & Joannides, 2013; Tacconelli & Patrignani, 2014). An emerging area for pharmacological therapeutics is the endothelial-derived CYP metabolites. This review will focus on the physiology and pharmacology of endothelial CYP metabolites.

2. GENERATION AND PRODUCTION OF EETs AND 20-HETE

Endothelial cells produce a large number of hormonal, paracrine, and autocrine factors to regulate cardiovascular function. Identification of CYP-derived EETs as endothelial-derived hyperpolarizing factors (EDHFs) resulted in concentrated efforts to evaluate their contribution to vascular function (Campbell, Gebremedhin, Pratt, & Harder, 1996; Fisslthaler et al., 1999). Another CYP-derived metabolite, 20-hydroxyeicosatetraenoic acid (20-HETE), was initially thought to be generated by and act on vascular smooth muscle cells and contribute to the myogenic response and blood flow autoregulatory responses (Imig, Zou, Ortiz de Montellano, Sui, & Roman, 1994; Zou et al., 1996; Zou, Imig, Ortiz de Montellano, Sui, & Roman, 1994). More recently, a contribution of vascular smooth muscle cell-derived 20-HETE to endothelial cell function has emerged (Hoopes, Garcia, Edin, Schwartzman, & Zeldin, 2015). Thus, the generation and regulation of EETs and 20-HETE can greatly contribute to endothelial and cardiovascular function.

EETs and 20-HETE are generated from arachidonic acid by distinct enzymatic CYP pathways (Fig. 1). In general, endothelial CYP2C and CYP2J enzymes generate EETs that can be hydrolyzed to dihydroxyeicosatrienoic acids (DHETs) by the enzyme soluble epoxide hydrolase (sEH; Capdevila & Falck, 2001; Imig, 2013). Endothelial CYP2C and CYP2J epoxygenase enzymes vary depending on the species and organ (Capdevila & Falck, 2001). In addition, CYP epoxygenase enzymes generate different portions of the regioisomeric EETs (5,6-EET; 8,9-EET; 11,12-EET; 14,15-EET) (Capdevila & Falck, 2001; Karara et al., 1993). Likewise, the arachidonic acid metabolite, 20-HETE, is generated by different CYP enzymes of the CYP4A and CYP4F enzymatic families (Roman, 2002). CYP4A and CYP4F enzymes also vary depending on sex, species, and organ (Capdevila & Falck, 2001). There are also potential interactions between endothelial cell EET- and 20-HETE-generating enzymes (Imig, 2013). This complex regulation of EETs and 20-HETE requires understanding if their pharmacological manipulation is to be used as a therapeutic.

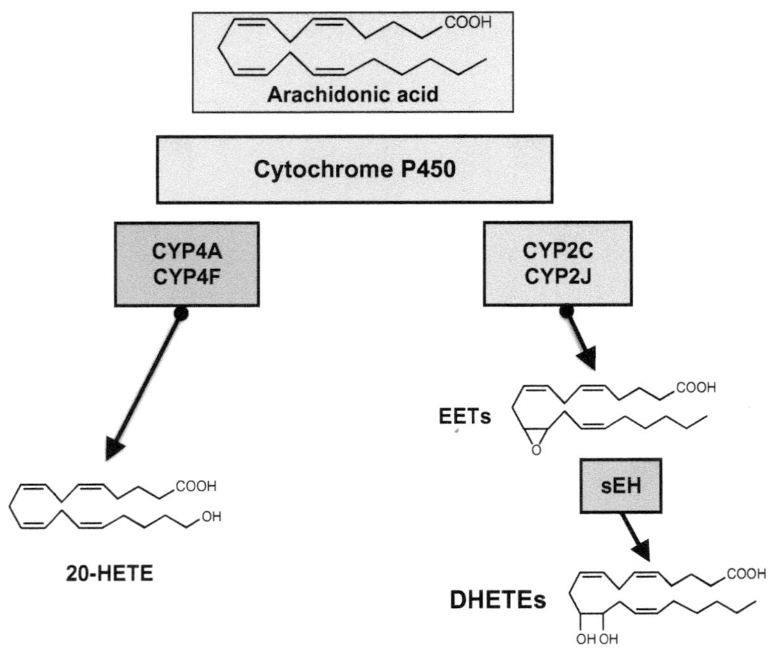

Fig. 1 Diagram depicting pathways of cytochrome P450 (CYP) arachidonic metabolism. CYP 4A and CYP4F enzymes can generate 20-hydroxyeicosatraenoic acid (20-HETE). CYP2C and CYP2C enzymes generate epoxyeicosatrienoic acids (EETs). Soluble epoxide hydrolase (sEH) can hydrate EETs to form dihydroxyeicosatrienoic acids (DHETEs).

The generation of EETs by CYP2C and CYP2J enzymes has been extensively evaluated in a number of species and organ systems (Capdevila & Falck, 2001). These enzymes are localized to the endoplasmic reticulum and convert arachidonic acid that is released from phospholipid membranes to EETs (Spector, Fang, Snyder, & Weintraub, 2004). CYP epoxygenases add an epoxide group across one of four double bones to form 5,6-EET, 8,9-EET, 11,12-EET, and 14,15-EET (Spector et al., 2004). EETs are generated as *cis*-epoxides with a high degree of enantiofacial selectivity of (*R*,*S*) and (*S*,*R*) enantiomers (Capdevila & Falck, 2001; Spector et al., 2004). CYP2C enzymatic proteins are responsible for the majority of endothelial cell EET generation (Fleming, 2007). Species variation exists among the CYP2C isoforms; however, a significant degree of homology and catalytic activity exists between mouse (CYP2c40 and CYP2c44), rat (CYP2C11 and CYP2C23), and human (CYP2C8) epoxygenase enzymes (Capdevila & Falck, 2001). These CYP2C isoforms generate all four regioisomeric EETs with 11,12-EET and 14,15-EET being at least 70–80% of the total EETs formed (Capdevila & Falck, 2001; Zeldin, 2001). The CYP2J family also is capable of converting arachidonic acid to EETs (Zeldin, 2001). Although CYP2J enzymes preferentially generate 14,15-EET, there are species differences between mouse (Cyp2j5), rat (CYP2J3), and human (CYP2J2) enzymes. CYP2C and CYP2J epoxygenase enzymes have been found in the vasculature and endothelial cells of mice, rats, and humans (Zeldin, 2001). Nevertheless, there is still limited information on the contribution of these enzymes to endothelial and vascular smooth muscle cell function.

An important regulator of EET vascular function is the sEH enzyme. The sEH enzyme carries out catabolic conversions of EETs to their corresponding DHETs (Imig & Hammock, 2009; Morisseau & Hammock, 2013). Interestingly, 14,15-EET, the major regioisomeric EET formed by CYP2C and CYP2J isoforms, is the preferred substrate for sEH (Morisseau & Hammock, 2013). Other metabolic pathways become prominent when sEH activity is low or inhibited (Spector et al., 2004). Chain elongation to form 22-carbon products or β-oxidation to generate 16-carbon epoxy fatty acids are the primary alternative 11,12-EET and 14,15-EET metabolic pathways (Spector et al., 2004). EETs can also be regulated by localization to cell membranes and binding to proteins (Spector et al., 2004). Although EETs in phospholipids represent a very small fraction, EETs could impact function through actions at localized domains (Ellinsworth, Earley, Murphy, & Sandow, 2014; Spector et al., 2004).

Binding of EETs to fatty acid-binding proteins could traffic EETs to specific intracellular organelles (Spector et al., 2004). Thus, EET metabolic pathways, EET localization, and EET binding can regulate endothelial cell generated 11,12-EET and 14,15-EET and influence their activity.

Arachidonic acid is acted upon by CYP ω-hydroxylases of the CYP4A and CYP4F enzymatic families to generate 20-HETE (Roman, 2002). CYP ω-hydroxylase enzymes insert a hydroxyl group at the terminal carbon of arachidonic acid. There are several CYP4A and CYP4F isoforms that vary among mice (Cyp4a10 and Cyp4a12), rats (CYP4A1, CYP4A2, CYP4A3, CYP4A8, and CYP4F1), and humans (CYP4A11, CYP4A22, and CYP4F2; Capdevila & Falck, 2001; Roman, 2002). CYP4F2 is the predominant 20-HETE synthase in humans followed by CYP4A11 (Roman, 2002). The CYP4F expression varies between species and has been found in kidney, liver, and vascular smooth muscle cells (Roman, 2002). Likewise, CYP4A enzymes have been localized to the kidney and renal vasculature (Imig et al., 1994; Zou et al., 1994). Endothelial progenitor cells (EPCs) have been demonstrated to produce 20-HETE but vascular endothelial cells are devoid of 20-HETE synthesis capacity (Wu & Schwartzman, 2011). Even though vascular synthesis of 20-HETE occurs primarily in the vascular smooth muscle cell, myeloid cells in the peripheral blood, neutrophils, platelets, and bone marrow cells produce 20-HETE that could impact on endothelial cell function (Hoopes et al., 2015; Wu & Schwartzman, 2011).

CYP4A enzymes are regulated by hormones and can metabolize other substrates, while 20-HETE can be further metabolized. A key feature of CYP4A enzymes is their sex-dependent expression and regulation by androgen (Holla et al., 2001; Wu et al., 2013). Cyp4a12 is expressed in male mice kidneys but is near undetectable levels in the female kidney (Holla et al., 2001). In addition, studies have demonstrated that the murine Cyp4a12 is androgen regulated and this regulation can increase vascular resistance (Holla et al., 2001; Wu et al., 2013). Androgen is a potent inducer of 20-HETE synthesis and has been implicated in increased vascular resistance and development of hypertension (Holla et al., 2001; Wu & Schwartzman, 2011). CYP4A enzymes can metabolize other fatty acids including laurate and ω-hydroxylate EETs (Capdevila & Falck, 2001). CYP4A and CYP4F enzymes also generate other carbons 16–19 on arachidonic acid to produce subterminal HETEs (Capdevila & Falck, 2001). Like EETs, 20-HETE can be metabolized by β-oxidation to generate shorter chain length products (Roman, 2002). In addition, COX enzymes metabolize 20-HETE to produce vasoconstrictor endoperoxides or vasodilator prostanoids (Carroll,

Garcia, Falck, & McGiff, 1992). Thus, these other CYP4A and CYP4F metabolites and 20-HETE metabolites have been demonstrated to have vascular actions.

This review will focus on CYP metabolites of arachidonic acid that are the most abundant and well-characterized metabolites. An emerging area for CYP enzymes is the potential to act on ω-3 polyunsaturated fatty acids, eicosapentaenoic acid and docosahexanoic acid, to generate metabolites that have endothelial cell actions (Ulu et al., 2014; Westphal, Konkel, & Schunck, 2011). Epoxyeicosatrienoylglycerols are a novel class of arachidonic acid metabolites that have been described and demonstrated to cause vasodilation through the activation of cannabinoid CB1 receptors (Chen et al., 2008). Undoubtedly, these and other novel CYP epoxygenase and hydroxylase metabolites will be the focus of future investigations.

3. PHARMACOLOGICAL AND GENETIC TOOLS

A major driving force responsible for our understanding of the contribution of EETs and 20-HETE to endothelial function and vascular homeostasis is the development of pharmacological and genetic tools. CYP enzymatic inhibitors have evolved from nonselective CYP inhibitors to selective epoxygenase and hydroxylase inhibitors (Imig, 2013). Inhibitors of sEH have also been utilized to prevent the conversion of EETs to DHETs (Imig, 2013). 20-HETE and EET structure–activity relationship studies have led to the development of selective agonists and antagonists (Gauthier, Falck, Reddy, & Campbell, 2004). Many of these pharmacological agents can now be used in animals. Genetic manipulation has also been done at the cellular and whole animal levels. CYP2C, CYP2J, CYP4A, CYP4F, and sEH enzymes can be overexpressed or deleted in cellular systems, rats, and mice at the whole animal levels or in an endothelial and vascular smooth muscle cell-specific manner.

CYP enzymatic inhibitors were the first major breakthrough that allowed for evaluation of endothelial and vascular function. Initially, general CYP inhibitors like 17-ODYA were utilized and compared with phospholipase A_2 (PLA_2), COX, and lipoxygenase (LOX) inhibitors (Imig, 2013). Unfortunately, 17-ODYA could not discriminate between epoxygenase and hydroxylase pathways. Antifungal agents such as miconazole were initially used as more selective epoxygenase inhibitors; however, conclusions made with studies employing these agents were not clear-cut. More selective inhibitors were made possible because great strides were made in understanding the biochemistry of epoxygenase and hydroxylase enzymes.

Selective CYP epoxygenase and hydroxylase inhibitors developed included small-molecule mechanism-based inhibitors (Wang et al., 1998). The most widely utilized small molecules act as competitive inhibitors (Wang et al., 1998). An oleinic compound, DDMS, exhibits a high degree of selectivity for inhibiting ω-hydroxylation of arachidonic acid with no selectivity for CYP4A isoforms (Wang et al., 1998). Another widely used and characterized CYP hydroxylase inhibitor is HET0016. HET0016 administered chronically to rats can selectively inhibit 20-HETE formation (Roman, 2002). Likewise, epoxygenase inhibitors PPOH and MS-PPOH selectively inhibit microsomal epoxygenase with little effect on hydroxylase activity at high micromolar concentrations (Wang et al., 1998). Other enzymatic inhibitors were developed to prevent the conversion of EETs to DHETs by sEH (Imig & Hammock, 2009). Potent transition state sEH inhibitors were developed based on the knowledge of the catalytic mechanism of the enzyme (Imig & Hammock, 2009). The most widely used sEH inhibitor is AUDA that can be administered to rodents chronically (Imig & Hammock, 2009). Development of sEH inhibitors for use in humans has advanced and clinical trials in the areas of hypertension, diabetes, and chronic obstructive pulmonary disease conducted (Bellien & Joannides, 2013; Marino, 2009). Investigators have utilized these enzymatic inhibitors to determine the contribution of CYP epoxygenase, CYP hydroxylase, and sEH on endothelial and vascular function.

A key to our understanding of the endothelial actions for EETs was a full understanding to the structure–activity relationship for EETs. EET regioisomer-specific agonists were synthesized to improve solubility and resist autooxidation and metabolism by sEH (Gauthier et al., 2004; Imig, 2012). Initial steps were the replacement of the carboxylic acid with a sulfonamide to resist β-oxidation (Imig, 2012). A structure–activity relationship for EETs emerged as additional modifications were made to the various EET regioisomers. It became evident that the endothelial and biological actions for 11,12-EET and 14,15-EET had three main components. 11,12-EET and 14,15-EET required an acidic carboxyl group, Δ [8] olefin bond, and a *cis*-epoxide (Imig, 2012). EET mimetics/analogs have now been developed to point where they can be used in endothelial cell systems, isolated vascular studies, and administered orally to rodents (Imig, 2012). Additional information from EET structure relationship studies and endothelial cell signaling studies has led to the development of EET antagonists.

Development of EET antagonists has been another essential pharmacological tool to determine EET endothelial actions. EET antagonists are based on the general EET structure with 14,15-EEZE being the most widely utilized EET antagonist (Gauthier et al., 2004). A key component to 14,15-EEZE is that it lacks an 8,9- and 11,12-olefin bond (Gauthier et al., 2004). 14,15-EEZE is a general nonselective EET antagonist that cannot discern the actions of the various regioisomeric EETs (Gauthier et al., 2002). More recently, EET antagonists that selectively inhibit the vascular actions of 11,12-EET or 14,15-EET have been developed (Bukhari et al., 2012). Experimental studies with these selective 11,12-EET and 14,15-EET antagonists have determined unique contributions for 14,15-EET to flow-mediated dilation in mesenteric resistance arteries (Bukhari et al., 2012). Thus, we now have the pharmacological means to determine the endothelial cell functions for specific EET regioisomers.

20-HETE agonists and antagonists have been developed to better understand the contribution of 20-HETE to vascular function (Roman, 2002). Development of 20-HETE analogs has determined that the presence of a hydroxyl group at C20 or C21 is critical for vascular activity (Alonso-Galicia, Falck, Reddy, & Roman, 1999). In addition, the retention of olefin bonds in the 5,6- and 14,15-carbons is necessary for 20-HETE analogs to retain agonist activity (Alonso-Galicia et al., 1999). These structure–activity relationship findings led to the development of 20-5,14-HEDGE which is the most widely utilized 20-HETE analog. 20-5,14-HEDGE has been successfully administered in vivo to rodents; however, β-oxidation and water solubility limit potential for clinical applications (Williams, Murphy, Burke, & Roman, 2010). On the other side of the coin is the development of 20-HETE antagonists (Fernandez, Gonzalez, Williams, Roman, & Nowicki, 2012; Williams et al., 2010). Early studies found that 5-HETE, 15-HETE, and 19-HETE could antagonize the vascular actions of 20-HETE (Alonso-Galicia et al., 1999). This finding led to the development of 20-HEDE that lacks double bonds at the 8,9- and 11,12-positions to resist COX metabolism (Roman, 2002). More recently, a 20-HETE antagonist, SOLA, with greater solubility and oral activity has been used in chronic animal studies (Chen, Ackerman, & Guo, 2012; Gangadhariah et al., 2015). Accordingly, 20-HETE agonists and antagonists have progressed to a state where there cardiovascular therapeutic potential can be evaluated.

Genetic manipulation has provided interesting insight as to the contribution of CYP4A, CYP2C, CYP2J, and sEH enzymes to endothelial function and vascular homeostasis. Approaches have decreased or increased enzymatic protein expression in endothelial cell cultures and endothelial cells at the whole animal level (Capdevila, Falck, & Imig, 2007; Imig, 2012). These genetic manipulations discovered that arachidonic acid metabolism by these enzymes is very complex and identified previously unknown regulation and interactions of these enzymes. It has also greatly enhanced our understanding of EETs and 20-HETE to endothelial cell biology and pathology. More importantly, it has found novel therapeutic targets for endothelial and vascular diseases.

Epoxygenase and sEH enzymes have been genetically manipulated by whole animal gene depletion, and endothelial cell-specific gene deletion or overexpression (Imig, 2012). Endothelial cell culture systems lose expression of epoxygenase enzymes in early passages (Fleming, 2007). This has provided an opportunity to determine the contribution of specific CYP2C and CYP2J enzymes to endothelial cell function. Although epoxygenase enzymes of many species have been evaluated, human CYP2C8 and CYP2J2 expression in endothelial cells has provided excellent insight on their contribution to endothelial cell biology (Imig, 2012). Murine genetic manipulation for Ephx2, CYP2C, and CYP2J enzymes has included global deletion, endothelial cell deletion, and endothelial cell overexpression (Imig, 2012). These genetic-manipulated mice have provided considerable insight on the impact of these enzymes and their metabolites to blood pressure regulation, vascular remodeling, vascular inflammation, and atherosclerosis.

Endothelial cell and vascular 20-HETE biosynthesis can be regulated by genetically manipulating CYP4A and CYP4F expression (Wu & Schwartzman, 2011). CYP4A and CYP4F enzymes have been knocked down or overexpressed in cell culture systems (Roman, 2002; Wu & Schwartzman, 2011). Expression of CYP4A2 in vivo in rat endothelial cells has been accomplished by adenovirus intravenous injection (Wu et al., 2013). Antisense nucleotide CYP4A1/CYP4A2 inhibition in spontaneously hypertensive rats (SHRs) resulted in decreased 20-HETE levels (Sodhi et al., 2010). The CYP4A enzyme has been genetically manipulated in the Dahl salt-sensitive (SS) rat by introgression of CYP4A alleles for the normotensive Brown-Norway rat into the Dahl SS genetic background (Lukaszewicz & Lombard, 2013). This manipulation resulted in a decrease in 20-HETE production in the Dahl SS rat (Lukaszewicz & Lombard, 2013). Cyp4a genes have also been manipulated in mice. Cyp4a14 gene deficiency in mice resulted in increased Cyp4a12(a) expression and increased 20-HETE levels (Holla et al., 2001). On the other hand, Cyp4a10 gene deficiency in mice did not change Cyp4a12(a) expression or 20-HETE levels (Capdevila, Pidkovka, Mei, et al., 2014). Interestingly, Cyp4a10 −/− mice had decreased Cyp2c44 epoxygenase expression (Capdevila et al., 2014). These findings highlight the complex regulation of CYP hydroxylase and epoxygenase enzymes and provide evidence to interactions between these enzymatic pathways. The overall impact of interactions between hydroxylase and epoxygenase pathways on vascular and endothelial study is not completely understood.

4. ENDOTHELIAL AND VASCULAR BLOOD FLOW REGULATION

CYP epoxygenase and hydroxylase have actions on endothelial and vascular smooth muscle cells to influence organ blood flow (Imig, 2012; Roman, 2002; Fig. 2). Initial studies evaluated the effects of EETs, DHETs, and 20-HETE on organ blood flow. Experimental studies in isolated blood vessels were also conducted (Imig, 2012; Roman, 2002). Likewise,

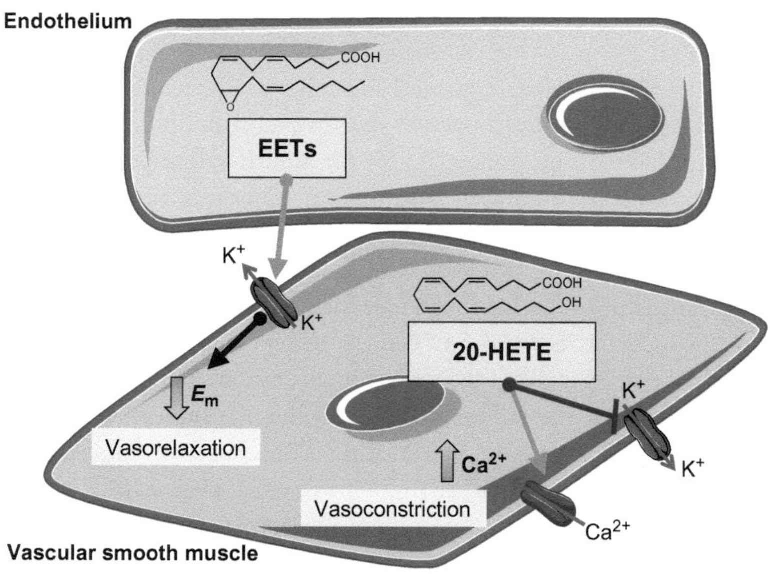

Fig. 2 Diagram depicting 20-hydroxyeicosatraenoic acid (20-HETE) and epoxyeicosatrienoic acids (EETs) cell signaling mechanisms on vascular tone. 20-HETE inhibits large-conductance K^+ channels and activates L-type Ca^{2+} channels leading to an increase in vascular smooth muscle intracellular Ca^{2+} and subsequent vasoconstriction. Endothelial-derived EETs activate vascular smooth muscle cell large-conductance K^+ channels resulting in hyperpolarization and vasorelaxation.

experiments were conducted utilizing pharmacological CYP inhibitors (Imig, 2012; Roman, 2002). Two exciting findings on EETs and 20-HETE and blood flow regulation emerged from these early studies. First, identification of EETs as EDHFs peaked interest in CYP epoxygenase metabolites (Campbell et al., 1996; Fisslthaler et al., 1999). Second, 20-HETE was identified as a critical paracrine factor in blood flow autoregulation (Zou et al., 1994). Numerous investigations have gone on to provide significant detail on EET and 20-HETE endothelial and vascular smooth muscle cell signaling mechanisms. This section will focus on endothelial and vascular regulation by EETs and 20-HETE.

EETs were first described to increase intestinal arteriolar blood flow and subsequent experimental evidence was demonstrated in renal, cerebral, and coronary vasculatures (Imig, 2012; Proctor, Falck, & Capdevila, 1987). EET vasodilation has been demonstrated in rodent and nonrodent species including human blood vessels (Imig, 2012; Larsen et al., 2006). In general, there is agreement that the most abundant EET regioisomers 11,12-EET and 14,15-EET are vasodilators. There is evidence that 5,6-EET and 8,9-EET cause vasoconstriction that is dependent on metabolism by COX enzymes and activation of thromboxane receptors (Carroll, Balazy, Margiotta, Falck, & McGiff, 1993; Imig, Navar, Roman, Reddy, & Falck, 1996; Imig, Zou, et al., 1996). On the other hand, DHETs that are generated by sEH have been demonstrated to be less active than the corresponding EETs (Imig, Navar, et al., 1996; Imig, Zou, et al., 1996). Additional studies demonstrated that nitric oxide-independent dilation in response to bradykinin or acetylcholine was blocked by CYP inhibition (Imig, Falck, Wei, & Capdevila, 2001; Wang, Borrego-Conde, et al., 2003; Wang, Trottier, & Loutzenhiser, 2003). This led to the idea that EETs could be EDHFs responsible for the nitric oxide- and COX-independent dilator response. Ensuing experimental evidence provided definitive proof that EETs were generated in endothelial cells and acted on the adjacent vascular smooth muscle cells to cause hyperpolarization and vasodilation (Campbell & Fleming, 2010; Pfister, Gauthier, & Campbell, 2010). This seminal discovery led to numerous investigations to further delineate EET regioisomeric vascular actions and cell signaling mechanisms.

Endothelial-derived EETs act on vascular smooth muscle cell K^+ channels resulting in vascular smooth muscle cell hyperpolarization and relaxation (Campbell, Holmes, Falck, Capdevila, & Gauthier, 2006; Li, Zhang, Ge, & Campbell, 2002). This vascular smooth muscle cell response appears to be primarily mediated by the large-conductance calcium-activated K^+

(K_{Ca}) channel (Archer et al., 2003; Dimitropoulou et al., 2007; Li et al., 2002). There is also evidence that the ATP-sensitive K^+ (K_{ATP}) participates in the EET-mediated dilation of mesenteric resistance arteries (Chen & Cheung, 1996; Imig, 2012). 11,12-EET and 14,15-EET cell signaling mechanisms have been the most widely studied and EET analogs for these regioisomers have been essential to these experimental studies (Sudhahar, Shaw, & Imig, 2010). These studies have determined that EET-mediated activation of K_{Ca} channels requires the G protein $G_{\alpha s}$ and activation of the cAMP-protein kinase A (PKA) cell signaling pathway (Dimitropoulou et al., 2007; Imig, Inscho, Deichmann, Reddy, & Falck, 1999). Additional evidence points to a contribution of TRP channels and localized increases in calcium resulting in a calcium spark to activate K_{Ca} channels (Fleming et al., 2007; Ma et al., 2015; Vriens et al., 2005; Watanabe et al., 2003). Additional evidence suggests that 11,12-EET can act to hyperpolarize the endothelial cell and this leads to vascular smooth muscle cell hyperpolarization via gap junctions and the ouabain-sensitive Na^+–K^+–ATPase (Ellinsworth et al., 2014; McSherry et al., 2006). EETs also contribute to the conduction of hyperpolarization along an arteriole in response to acetylcholine (McSherry et al., 2006). This finding suggests that EETs can act in a manner other than as an EDHF to cause vasodilation. Overall, there is significant evidence that EETs can act as an EDHF on vascular smooth muscles and in a non-EDHF manner on endothelial cells to cause vasodilation.

Although initial studies were controversial as to the vascular actions of 20-HETE, this CYP hydroxylase metabolite has now clearly been defined as a vasoconstrictor (Alonso-Galicia et al., 1999; Imig, Navar, et al., 1996; Imig, Zou, et al., 1996; Ma et al., 1993). 20-HETE has been demonstrated to be a potent constrictor of cerebral, renal, and mesenteric resistance arteries (Roman, 2002). Vascular smooth muscle cell signaling actions of 20-HETE are in direct opposition to the EET cell signaling actions. 20-HETE inhibits K_{Ca} channels leading to vascular smooth muscle cell depolarization and increases intracellular calcium (Imig, Navar, et al., 1996; Imig, Zou, et al., 1996; Zou et al., 1996). Direct activation of L-type calcium channels and protein kinase C (PKC) also contributes to 20-HETE-mediated vasoconstriction (Ma et al., 1993; Zou et al., 1996). An interesting interaction between 20-HETE and the endothelial-derived relaxing factor, nitric oxide, could also participate in the regulation of vascular tone (Alonso-Galicia, Drummond, Reddy, Falck, & Roman, 1997). Nitric oxide inhibits 20-HETE formation and this decrease in 20-HETE contributes to a cGMP-independent dilator action of nitric oxide (Alonso-Galicia et al., 1997). These 20-HETE vascular smooth muscle cell signaling events are consistent with the concept that 20-HETE is a key contributor to vascular smooth muscle cell autoregulatory responses.

Even though there has been extensive evidence for EETs as an EDHF and 20-HETE as a critical contributor to blood flow autoregulation, the contribution of these CYP metabolites to overall endothelial and vascular function has greatly expanded. EETs and 20-HETE are now recognized to be critically involved in angiogenesis and vascular inflammation.

5. ANGIOGENESIS AND VASCULAR REMODELING

Angiogenesis is one of the few areas where EET and 20-HETE have similar biological actions (Fig. 3). 20-HETE and EETs have proangiogenic activities that are alike in many respects but uniquely regulated depending on the setting (Imig, 2012; Wu & Schwartzman, 2011). These CYP metabolites have overlapping and distinct cell signaling mechanisms that are utilized to promote angiogenesis. Vascular remodeling is also influenced by 20-HETE and EETs and like most biological activities for these CYP metabolites have opposing ying/yang activities (Hoopes et al., 2015; Imig, 2012). Therefore, in physiological and pathological states the contribution of EETs and 20-HETE to angiogenesis and vascular remodeling is complex and depends on the prevailing metabolites.

There is significant evidence that EETs can contribute to angiogenesis and endothelial cell proliferation (Fleming, 2007; Pozzi et al., 2005; Webler et al., 2008; Zhang, Cao, & Rao, 2006). EETs or overexpression of CYP2C9 to cultured endothelial cells resulted in angiogenesis (Fleming, 2007). EETs or overexpression of various CYP epoxygenase enzymes induces angiogenesis in matrigel plugs in rodents and increases the capillary density following in ischemic limb models (Pozzi et al., 2005). EETs appear to contribute to endothelial cell migration and degradation of extracellular matrix (Munzenmaier & Harder, 2000; Pozzi et al., 2005). Vascular endothelial growth factor (VEGF) stimulated endothelial cell proliferation and tube formation involves EET generation (Cheranov et al., 2008; Webler et al., 2008; Yang, Wei, Pozzi, & Capdevila, 2009). Likewise, EETs can increase VEGF expression as a positive feedback mechanism to further increase endothelial cell proliferation

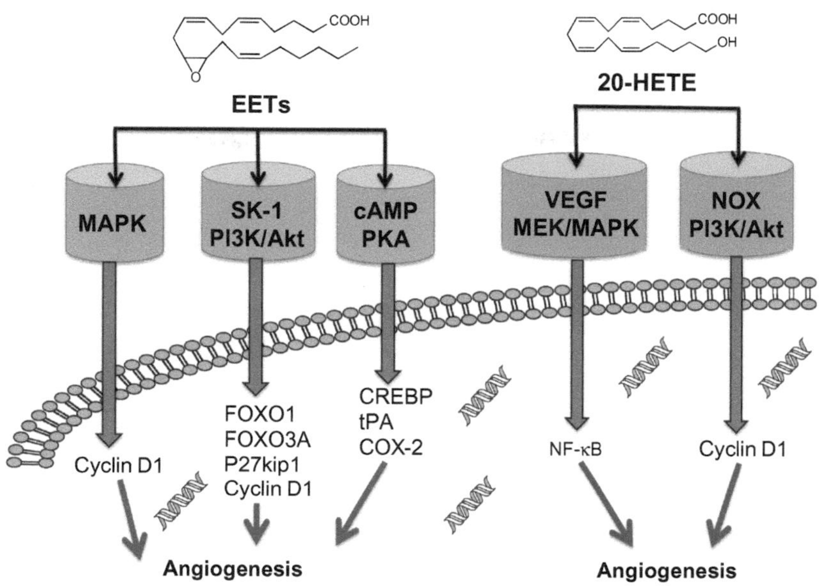

Fig. 3 Diagram depicting 20-hydroxyeicosatraenoic acid (20-HETE) and epoxyeicosatrienoic acids (EETs) on angiogenesis. Angiogenesis is a biological activity where 20-HETE and EETs act in the same manner. EETs promote angiogenesis through mitogen-activated protein kinase (MAPK) activation and nuclear cyclin D1 generation, sphingosine kinase-1 (SK-1), and phosphatidylinositol 3-kinase (PI3K)/Akt pathways to activate transcription factors and the generation of cell cycle modulators, and protein kinase A (PKA) acting via the cAMP/PKA response element-binding protein (CREBP) resulting in COX-2 production that influences angiogenesis. 20-HETE promotes angiogenesis through vascular endothelial growth factor (VEGF) activation of MAPK/extracellular signaling-regulated (ERK) kinase (MEK) to increase the nuclear factor kappa-light-chain enhancer of activated B cells (NF-κB) and through NADPH oxidase (NOX) and PI3K/Akt pathways.

(Cheranov et al., 2008; Webler et al., 2008). These findings demonstrated that EETs contribute importantly to VEGF angiogenic and endothelial cell proliferative actions.

EETs have endothelial cell signaling actions that promote proliferation. Endothelial cell proliferation is increased by CYP2C overexpression or in response to administration of any of the four regioisomeric EETs (Michaelis et al., 2003; Potente, Michaelis, Fisslthaler, Busse, & Fleming, 2002). Two primary endothelial cell signaling pathways have been implicated in EET-induced proliferation. EET activation of MAPK and PI3K/Akt endothelial signaling are major contributors (Potente, Fisslthaler, Busse, & Fleming, 2003; Wang et al., 2005; Webler et al., 2008). Activation of p38MAPK and inactivation of cJun NH2-terminal kinase (JNK) result from CYP2C9 overexpression in human umbilical vein endothelial cells (HUVECs) (Michaelis et al., 2003). 11,12-EET reduces the cyclin D1 inhibitory protein, $P27^{kip1}$, and activates PI3K/Akt to inhibit forkhead transcription factors (FOXO; Potente et al., 2003). EET activation of the EGF receptor and release of heparin-binding EGF-like growth factor (HB-EGF) appear to be upstream of these endothelial cell signaling pathways (Fleming, 2007). Lastly, there are interactions between EETs and COX-2 mediated by the endothelial cell cAMP/PKA pathways. 11,12-EET or CYP2C9 overexpression activates endothelial cell cAMP/PKA pathway to stimulate cAMP response element-binding (CREB) protein to stimulate COX-2 expression (Michaelis et al., 2003). Nevertheless, the coordinated endothelial cell signaling activities by which EETs contribute to endothelial cell proliferation and angiogenesis remain to be explored.

Although many of the biological activities of EETs and 20-HETE are opposing, 20-HETE acts similar to EETs to promote angiogenesis (Chen et al., 2012; Hoopes et al., 2015). 20-HETE can increase endothelial cell VEGF expression via MAPK activation (Hoopes et al., 2015). A major contributor to 20-HETE endothelial cell proliferation appears to be stimulation of reactive oxygen species (ROS) and the production of hypoxia-inducible factor-1α (HIF-1α; Chen et al., 2012). CYP4A overexpression or 20-HETE activates angiogenic factors in endothelial cells (Chen et al., 2012). Intriguingly, endothelial cells from mice with endothelial-specific expression of CYP4F2 exhibited increased growth and tube formation that was VEGF dependent (Cheng et al., 2014). There is also evidence that 20-HETE contributes to angiogenesis associated with tumor growth involves PI3K signaling (Wu & Schwartzman, 2011). Thus, 20-HETE-induced endothelial cell proliferation and angiogenesis have similarities and differences when compared to EETs.

A unique aspect for 20-HETE to regulate angiogenesis is actions on EPC function. EPCs derived from HUVECs express CYP4A11, CYP4A22, and CYP4F2 hydroxylase enzymes and generate 20-HETE (Chen et al., 2014). 20-HETE enhances EPC secretion of key proangiogenic factors such as VEGF, HIF-1α, and stromal-derived factor-1 (SDF-1; Chen et al., 2014). These EPC-secreted factors recruit bone marrow-derived stem cells to promote angiogenesis (Chen et al., 2012). CYP hydroxylase inhibitors or 20-HETE antagonists stop EPC proliferation and migration and EPC-induced endothelial cell differentiation (Chen et al., 2014). Thus, CYP hydroxylases and 20-HETE regulate EPC function and related angiogenic responses.

Vascular remodeling occurs in many cardiovascular disease states and involves structural reorganization of blood vessels in response to stimuli to induce collagen synthesis and reorganization of the vascular wall extracellular matrix. Matrix metalloproteinases (MMPs) and inflammatory signals are key contributors to vascular remodeling that are influenced by 20-HETE and EETs (Hoopes et al., 2015; Imig, 2012). 20-HETE can promote vascular remodeling, whereas endothelial-derived EETs oppose vascular remodeling (Hoopes et al., 2015; Imig, 2012). Renal arteries can be induced to remodeling by 20-HETE (Garcia et al., 2015; Hoopes et al., 2015). In addition, mice with genetic manipulation to induce Cyp4a12 expression resulting in increased 20-HETE levels result in vascular remodeling (Ding et al., 2013). Vascular remodeling in the Cyp4a12 transgenic mice can be blocked by the 20-HETE antagonist, 20-HEDGE (Ding et al., 2013). On the other hand, vascular remodeling is decreased in Ephx2−/− mice or by sEH inhibition (Davis et al., 2002; Simpkins et al., 2010). EETs also have actions on to decrease vascular smooth muscle cell migration and proliferation (Fleming, 2007). Vascular hypertrophic remodeling and collagen deposition in the middle cerebral artery of stroke-prone SHRs are attenuated by sEH inhibition (Simpkins et al., 2009). Likewise, flow-induced remodeling of the carotid artery in rats and mice was reduced by sEH inhibition or Ephx2 gene deficiency (Simpkins et al., 2010). Femoral artery wire-induced endothelial injury and neointimal formation were not decreased by sEH inhibition or Ephx2 gene deficiency (Simpkins et al., 2010). These findings are consistent with the notion that endothelial-derived EETs are required for sEH inhibition to oppose vascular remodeling. Consequently, endothelial-derived EETs will have opposite actions compared to vascular smooth muscle cell-derived 20-HETE on vascular remodeling.

6. ENDOTHELIAL AND VASCULAR INFLAMMATION

Regulation of vascular inflammation by CYP eicosanoids has been an area of extensive investigation. 20-HETE and EETs can influence vascular smooth muscle cell and endothelial cell inflammatory responses (Hoopes et al., 2015; Imig, 2012). In this regard, EETs are antiinflammatory and 20-HETE is inflammatory (Hoopes et al., 2015; Imig, 2012). These opposing vascular inflammatory actions for EETs and 20-HETE on inflammation are mediated via similar cell signaling pathways.

20-HETE promotes vascular inflammation through endothelial cell activation to increase adhesion molecules and proinflammatory cytokines

(Hoopes et al., 2015). In ischemia–reperfusion injury models 20-HETE inhibition decreased vascular inflammation possibly via decreased oxidative stress (Dunn et al., 2008; Hoopes et al., 2015). Decreased 20-HETE synthesis also reduces vascular TNFα, IL-1β, and IL-6 expression (Hoopes et al., 2015). In the other direction, 20-HETE increases vascular ROS and NFκB activation (Ishizuka et al., 2008). Treatment of endothelial cells with 20-HETE leads to activation of NADPH oxidases and NFκB and increased expression of adhesion molecules (Ishizuka et al., 2008). MAPK appears to be involved in the endothelial cell inflammatory response because inhibition of MAPK prevents 20-HETE-induced inflammation (Ishizuka et al., 2008). Taken together, 20-HETE inflammatory actions are dependent on ROS and NFκB activation.

There is significant evidence that EETs and sEH inhibition have vascular antiinflammatory actions (Deng et al., 2011; Imig, 2012; Node et al., 1999). EETs appear to oppose endothelial cell inflammation via inhibition of NFκB activation (Deng et al., 2011). Likewise, EETs and EET analogs inhibit vascular smooth muscle cell TNFα-induced VCAM-1 adhesion molecule expression (Khan et al., 2013; Node et al., 1999). Vascular antiinflammatory actions of EETs or sEH inhibition are mediated by inhibition of phosphor-IKK-derived NFκB activation (Deng et al., 2011; Node et al., 1999). Another mechanism by which EETs are antiinflammatory appears to be through activation of endothelial cell peroxisome proliferator-activated receptor γ (PPARγ) (Liu, Zhang, Schmelzer, et al., 2005). These findings are in agreement with the well-described interactions between PPARs and EETs. Thus, sEH inhibitors and EET analogs are being tested for their therapeutic potential to treat vascular inflammation and atherosclerosis.

7. ENDOCRINE, PARACRINE, AND AUTOCRINE INTERACTIONS

CYP eicosanoids have been demonstrated to contribute to and regulate endothelial and vascular responses evoked by endocrine, paracrine, and autocrine factors (Imig, 2012; Roman, 2002). Interactions between CYP eicosanoids and the renin–angiotensin system, purinergic system, adenosine, nitric oxide, and endothelin have been extensively evaluated. In general, 20-HETE acts as a key mediator for the endothelial and vascular actions of these endocrine, paracrine, and autocrine factors (Roman, 2002). On the other hand, endothelial-derived EETs act to oppose the actions of these factors (Imig, 2012). Thus, it is well recognized that a balance between EETs and 20-HETE and appropriate responses to endocrine, paracrine, and autocrine factors is required for proper endothelial cell function and vascular homeostasis.

20-HETE and EET interactions with the renin–angiotensin system at the level of the endothelial cell and vasculature are required for endothelial cell function (Fig. 4). Endothelial-derived EETs have been demonstrated to oppose angiotensin II vasoconstrictor actions (Imig & Deichmann, 1997; Kohagura et al., 2000). In the other direction, angiotensin II can increase endothelial cell sEH expression to decrease EETs (Ai et al., 2007). Decreased levels of endothelial cell Ephx2 mRNA expression and sEH protein expression in response to angiotensin II are the result of transcriptional factor AP-1 activation (Ai et al., 2007). Angiotensin-converting enzyme (ACE) inhibition can increase EET levels and this is thought to be due to the increase in bradykinin levels (Matsuda et al., 2004). Evidence also suggests that angiotensin type 2 (AT2) receptors increase endothelial EET levels in response to angiotensin II that acts to blunt the angiotensin II vasoconstrictor response (Imig, 2012; Matsuda et al., 2004). On the other hand, 20-HETE appears to

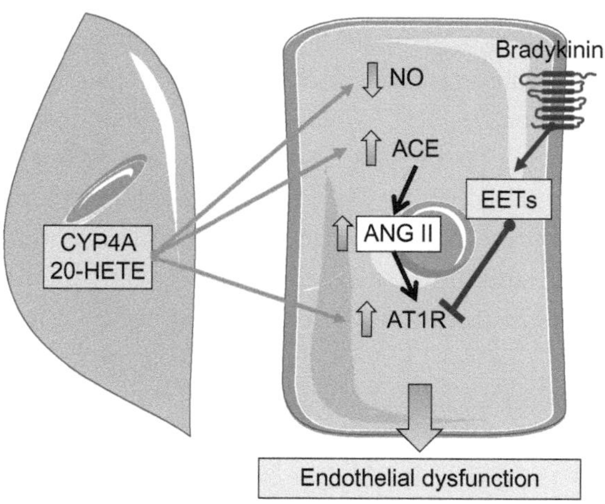

Fig. 4 Diagram depicting 20-hydroxyeicosatraenoic acid (20-HETE) and epoxyeicosatrienoic acids (EETs) interactions on endothelial dysfunction. Vascular smooth muscle cell-derived 20-HETE acts on the endothelial cell to decrease nitric oxide (NO), increase angiotensin type 1 receptor (AT1R), and increase angiotensin-converting enzyme (ACE) to generate angiotensin II (ANG II). Increased ANG II and decreased NO will act to promote endothelial dysfunction. On the other hand, bradykinin stimulates endothelial cell EET generation that acts to interfere with ANG II actions on the AT1R and will oppose decreased NO levels resulting in improved endothelial function.

contribute to angiotensin II vasoconstriction (Alonso-Galicia, Maier, Greene, Cowley, & Roman, 2002). Experimental evidence demonstrates that angiotensin II can increase vascular CYP4A expression and 20-HETE levels (Alonso-Galicia et al., 2002; Croft, McGiff, Sanchez-Mendoza, & Carroll, 2000). 20-HETE also contributes to the ability of angiotensin II to cause artery neointimal thickening (Hoopes et al., 2015). More recently it has been recognized that 20-HETE can increase aspects of the renin–angiotensin system. Cultured endothelial cells incubated with 20-HETE have a big increase in ACE expression (Cheng et al., 2014). Likewise, lentivirus expression of the CYP4A2 cDNA under the control of an endothelial-specific promoter resulted in increased aortic ACE and angiotensin type 1 (AT1) receptor expression (Cheng et al., 2014). These findings strongly implicate 20-HETE as a major contributor to angiotensin II-mediated endothelial and vascular actions. Overall, CYP eicosanoid investigations demonstrate the opposing endothelial interactions between 20-HETE, EETs, and the renin–angiotensin system.

There have also been extensive studies on interactions between CYP eicosanoids and the purinergic system. In this case experimental evidence supports the concept that the purinergic factor adenosine interacts with endothelial cell EETs and that 20-HETE interacts with the purinergic factor ATP. 20-HETE interactions with ATP and the purinergic P2X receptors are important for autoregulation of blood flow (Imig, 2013). 20-HETE synthase inhibition or 20-HETE antagonism significantly attenuates arteriolar constriction to ATP or P2X receptor agonists (Zhao, Falck, Gopal, Inscho, & Imig, 2004; Zhao, Inscho, Bondlela, Falck, & Imig, 2001). This interaction appears to be at the level of the vascular smooth muscle cell because 20-HETE inhibition markedly reduces the increase in vascular smooth muscle cell calcium to P2X receptor but not P2Y receptor agonists (Zhao, Falck, et al., 2004). Thus, 20-HETE appears to be a critical cell signaling factor responsible for ATP P2X receptor-mediated autoregulatory responses. Unlike 20-HETE, EETs do not appear to influence arteriolar or vascular smooth muscle calcium responses to P2X receptor activation (Zhao, Falck, et al., 2004). Interestingly, EETs appear to interact with adenosine receptor-mediated vascular responses. EETs contribute to vasodilation in response to adenosine activation of A2A receptors (Carroll et al., 2006; Cheng et al., 2004). This finding is further supported by the fact that epoxygenase inhibition attenuates vasodilation in response to adenosine or A2A receptor agonists (Cheng et al., 2004). Adenosine through A2A receptor activation and cAMP increases endothelial cell EET generation (Cheng et al., 2004). EETs have also been demonstrated to contribute to adenosine A2B receptor cerebral arteriolar dilation (Alkayed et al., 1997). Taken together, there is clear evidence for 20-HETE interactions with ATP and P2X receptor activation and EET interactions with adenosine and A2A and A2B receptor activation that impact on vasoconstrictor and vasodilator responses.

Major interactions between endothelial factors endothelin-1 and nitric oxide and 20-HETE importantly influence vascular function. Endothelin-1 increases 20-HETE generation by arterioles (Imig, Pham, LeBlanc, Falck, & Inscho, 2000). CYP hydroxylase inhibition or 20-HETE antagonism attenuates endothelin-1 vasoconstriction (Imig et al., 2000). 20-HETE inhibitors also attenuate increases in vascular smooth muscle cell calcium levels in response to endothelin-1 (Imig et al., 2001). Endothelin-1 acts through the endothelin type A (ETA) receptors to increase 20-HETE and vascular smooth muscle cell calcium to cause vasoconstriction (Imig et al., 2001). On the other hand, 20-HETE opposes the vasodilator responses evoked by nitric oxide (Alonso-Galicia et al., 1999). Nitric oxide or superoxide radicals can inhibit vascular 20-HETE generation (Alonso-Galicia et al., 1999). Decreased vascular 20-HETE levels in response to nitric oxide allow for activation of cGMP-independent vasodilation by nitric oxide (Alonso-Galicia et al., 1999). In addition, 20-HETE can act to increase oxidative stress and cause eNOS uncoupling and endothelial dysfunction (Roman, 2002; Williams et al., 2010). Taken as a whole, 20-HETE contributes to the responses to endothelial-derived contracting factor endothelin-1 and opposes dilator responses to endothelial-derived nitric oxide.

As with many biological activities, EETs act as a counterbalance to 20-HETE in regard to vascular responses to the endothelial-derived factors endothelin-1 and nitric oxide. EETs oppose the vasoconstrictor actions of endothelin-1 and have unique interactions with nitric oxide. The epoxygenase inhibitor MS-PPOH enhances endothelin-1 vasoconstriction (Imig et al., 2000). Intriguingly, endothelial expression of human CYP2J2 and CYP2C8 epoxygenases in mice results in attenuated arteriolar constrictor responses to endothelin-1 (Lee et al., 2010). The ability of EETs to oppose endothelin-1 constrictor responses occurs at the endothelial cell because epoxygenase inhibition does not alter endothelin-1-mediated increases in vascular smooth muscle cell calcium (Imig et al., 2000). This finding implies that endothelin-1 increases endothelial generation of EETs to oppose vasoconstriction. In regard to interactions between nitric oxide and EETs, EETs appear to act independent of nitric oxide. On the other hand, endothelial cell EET generation increases in eNOS gene-deficient mice and compensates for the loss of nitric oxide (Huang et al., 2001). This finding demonstrated that EETs minimize the impact of decreased nitric oxide on endothelial cell function and vascular homeostasis. There is also evidence that H_2O_2 can regulate EET bioavailability (Larsen et al., 2008). Evidence has demonstrated that CYP2C9 appears to generate ROS in coronary arteries (Fleming, 2001). CYP2C9 inhibition improves vasodilation and increases nitric oxide levels. Contrary to this finding, EETs can potentiate nitric oxide dilation and EETs increase NOS activity in cultured endothelial cells (Fleming, 2001). Overall, these studies demonstrate endothelial EET generation in response to endothelin-1 opposes vasoconstriction and that interactions between CYP epoxygenases, EETs, nitric oxide, ROS, and eNOS are complex and not fully understood.

These interactions between 20-HETE, EETs, and endocrine, paracrine, and autocrine factors are essential for proper endothelial cell function and vascular homeostasis. This balance is delicate and tightly controlled to maintain overall cardiovascular health. Unfortunately, cardiovascular disease states involve alterations in endocrine, paracrine, and autocrine factors and the ability for CYP eicosanoids to properly maintain endothelial and vascular function. Thus, improper regulation of vascular CYP eicosanoids occurs in cardiovascular disease states and changes in CYP eicosanoids contribute to hypertension, atherosclerosis, stroke, and other cardiovascular diseases.

8. CARDIOVASCULAR DISEASES: HYPERTENSION

Hypertension is a major cardiovascular disease that afflicts one-third of the adult population worldwide. This has led to intensive investigation into the factors contributing to hypertension including CYP eicosanoids. The contribution of CYP eicosanoids to high blood pressure and the associated risk factors has been evaluated in hypertensive animal models as well as in humans (Imig, 2012; Williams et al., 2010). These experimental studies have determined that 20-HETE at the vascular level contributes to hypertension in an androgen-dependent manner (Williams et al., 2010; Wu & Schwartzman, 2011). Decreased endothelial-derived EETs appear to be a contributing factor to increased blood pressure in hypertension (Imig &

Hammock, 2009). In many types of hypertension it is an increase in vascular sEH that is responsible for decreasing EET levels (Imig & Hammock, 2009). Intriguingly, EETs and 20-HETE can also influence renal salt transport to influence blood pressure control in hypertension (Imig, 2012; Roman, 2002). In this regard EETs and 20-HETE increase sodium excretion and are considered to have antihypertensive renal tubular transport actions (Imig, 2012; Roman, 2002). This is in contrast to the EETs and 20-HETE ying/yang actions at the level of the vasculature to influence total peripheral resistance and blood pressure (Imig, 2013). Next, the contribution of CYP eicosanoids to vascular function and blood pressure control in hypertension will be discussed.

20-HETE vascular actions to constrict, increase sensitivity to other vasoconstrictors, enhance angiotensin generation, and interfere with nitric oxide generation are consistent with 20-HETE being prohypertensive (Roman, 2002; Williams et al., 2010). Single nucleotide polymorphisms (SNPs) in CYP4A11 and CYP4F2 enzymes have been linked to increased risk for developing hypertension (Gainer et al., 2005; Williams et al., 2010). There is significant evidence that 20-HETE contributes to angiotensin-dependent hypertension (Wu & Schwartzman, 2011). Human hypertensive patients demonstrate increased 20-HETE levels that correlate with plasma renin activity (Laffer, Laniado-Schwartzman, Wang, Nasjletti, & Elijovich, 2003). 20-HETE inhibition decreases blood pressure when combined with an sEH inhibitor in Ren-2 renin transgenic hypertensive rats (Certikova Chabova et al., 2010). In addition, endothelial expression of CYP4A2 in rats increases ACE activity and AT1 receptor levels to cause angiotensin-dependent hypertension (Cheng et al., 2012). Likewise, vascular dysfunction in the Dahl SS hypertensive rat appears to be due to ROS stimulation of 20-HETE (Lukaszewicz & Lombard, 2013; Williams et al., 2010). Evidence in genetic mice and in rats has demonstrated that increased 20-HETE is a contributor to androgen-dependent hypertension (Holla et al., 2001; Wu et al., 2013). Cyp4a14−/− mice have androgen-mediated hypertension that was a result of increased Cyp4a12 expression and 20-HETE generation (Holla et al., 2001). CYP4A2 transgenic mice demonstrated that human CYP4A2 expression is androgen regulated and that hypertension correlated with 20-HETE levels in these mice (Sodhi et al., 2010; Wu et al., 2013). Even though the contribution of 20-HETE to blood pressure regulation is complex, it is clear that vascular 20-HETE promotes vasoconstriction and vascular dysfunction and contributes to angiotensin II and androgen-dependent hypertension.

Endothelial EET generation has been clearly demonstrated to contribute to many types of hypertension (Imig & Hammock, 2009). Angiotensin-dependent hypertension and SS hypertension have decreased epoxygenase activity or increased sEH activity (Zhao, Pollock, Inscho, Zeldin, & Imig, 2003; Zhao, Yamamoto, et al., 2004). Vascular and kidney CYP2C epoxygenase protein levels are decreased in many types of hypertension (Imig & Hammock, 2009). Likewise, Cyp2c44 genetic-deficient mice develop SS hypertension (Capdevila et al., 2014). Cyp4a10 −/− mice demonstrated decreased Cyp2c44 protein expression, decreased EET generation, and hypertension on a normal salt diet (Nakagawa et al., 2006). There is also evidence for decreased EET levels in human hypertension (Bellien & Joannides, 2013; Tacconelli & Patrignani, 2014). Genetic studies have demonstrated associations between CYP epoxygenase enzyme genetic variants and increased risk for hypertension (Bellien & Joannides, 2013; King et al., 2005). Human studies have also demonstrated decreased EET levels contribute to vascular dysfunction (Lee et al., 2011). Genetic variants in EPHX2 that decreases EET levels result in reduced forearm blood flow responses to bradykinin (Lee et al., 2011). Likewise, decreased renal microvascular CYP2C11 and CYP2C23 expression in obese Zucker rats contributes to increased blood pressure (Zhao et al., 2005). Interestingly, vascular EET levels in obese Zucker rats were further reduced by increased sEH protein expression that contributed endothelial dysfunction (Zhao et al., 2005). Angiotensin-dependent hypertension in rats and mice has also been demonstrated to involve increased vascular sEH protein expression (Imig & Hammock, 2009). Thus, increased sEH and decreased EETs in endothelial cells contribute to the progression of hypertension.

9. CARDIOVASCULAR DISEASES: ATHEROSCLEROSIS AND VASCULAR INFLAMMATION

Atherosclerosis and vascular inflammation are cardiovascular diseases that alterations in CYP eicosanoids contribute importantly to the disease progression. 20-HETE has endothelial cell proinflammatory actions that promote atherosclerosis and vascular remodeling (Hoopes et al., 2015). Proinflammatory endothelial cell changes that occur in response to 20-HETE include increased adhesion molecule expression and cytokine release (Ishizuka et al., 2008). ROS contribute importantly to the vascular inflammatory response to 20-HETE (Ishizuka et al., 2008). 20-HETE increases vascular ROS levels and NF-κB activity to increase endothelial cell adhesion molecules and IL-8 levels (Ishizuka et al., 2008). These findings clearly demonstrate the important contribution of 20-HETE to endothelial and vascular inflammation.

CYP epoxygenases and sEH have been found to influence vascular inflammation and atherosclerosis (Bellien & Joannides, 2013). EETs have been demonstrated to oppose vascular inflammation and decrease endothelial cell adhesion molecule expression (Imig, 2012). Interestingly, EPHX2 polymorphisms have been linked to risk for coronary artery calcification and disease in the Atherosclerosis Risk in Communities and Coronary Artery Risk Development in young adults studies (Bellien & Joannides, 2013). Experimental studies in EPHX2 gene-deficient mice and using sEH inhibitors have demonstrated a contribution for EETs to oppose vascular inflammation, atherosclerosis, and vascular remodeling (Imig, 2012; Imig & Hammock, 2009). Mice with endothelial expression of CYP2C8 or CYP2J2 or EPHX2−/− mice have decreased vascular inflammation and NF-κB activity when exposes to endotoxin (Imig, 2012). EET-positive actions to attenuate atherosclerosis have been associated with decreased adhesion molecules and inflammatory cytokines (Imig, 2012; Node et al., 1999). Thus, EETs and sEH inhibition decrease inflammation and have vascular protective actions that can combat atherosclerosis.

10. CARDIOVASCULAR DISEASES: ISCHEMIC INJURY

The contribution of EETs and 20-HETE to cerebral and cardiac ischemic injury has been extensively investigated. 20-HETE injections into the intracarotid artery can produce cerebral ischemic injury (Roman, 2002; Williams et al., 2010). Likewise, exogenous administration of 20-HETE to coronary arteries increases infarct size in the heart (Roman, 2002; Williams et al., 2010). In agreement with a contribution of 20-HETE to ischemic injury, 20-HETE inhibitors have been demonstrated to reduce cerebral and cardiac ischemia and reperfusion injury (Roman, 2002; Williams et al., 2010). EETs or sEH inhibition protects the brain and heart from damage that occurs following an ischemic event (Alkayed et al., 2002; Imig, 2012; Simpkins et al., 2009). Vasodilator, profibrinolytic, and antiinflammatory actions potentially contribute to the ischemic protective actions for EETs (Imig, 2012). Epidemiological data demonstrating a genetic polymorphism in the EPHX2 gene are associated with increased risk for ischemic stroke (Bellien & Joannides, 2013). EET, EET analogs, and sEH inhibitors have attenuated cerebral and cardiac ischemic injury (Batchu, Lee, Qadhi, et al., 2011; Imig, 2012). This protective action for EETs appears to be multifactorial and includes actions independent of those on the endothelial or vascular smooth muscle cells (Imig, 2012). EETs likely inhibit apoptosis in the cardiac or brain tissue. Brain and cardiac tissue EET cell signaling antiapoptotic mechanisms involve increased Bcl2, ceramide inhibition, and decreased ROS (Alkayed et al., 2002; Batchu et al., 2011; Simpkins et al., 2009). EETs also sequester MAPKs and activate PI3K pathways in brain and cardiac tissue (Alkayed et al., 2002; Batchu et al., 2011; Simpkins et al., 2009). Other studies have demonstrated cardiac tissue protective actions by delaying or inhibiting opening of the mitochondrial transition pore (Katragadda et al., 2009). As a whole, these findings support the idea that decreasing 20-HETE actions or increasing EET actions would be beneficial in cardiac and cerebral ischemia.

Over the past two decades it has become apparent that CYP eicosanoids contribute importantly to endothelial and vascular alterations that occur during cardiovascular disease states. Experimental studies in cardiovascular disease animal models as well as human cardiovascular disease provide convincing evidence that CYP eicosanoids represent a therapeutic target (Imig & Hammock, 2009; Williams et al., 2010). There has been tremendous progress in the development of CYP eicosanoid therapeutics and novel small-molecule eicosanoid-based drugs are currently in clinical trials. Eicosanoid-based therapeutics will be discussed in the next section.

11. EICOSANOID-BASED CARDIOVASCULAR THERAPEUTICS

There has been excellent progress made in the last 15 years to develop CYP eicosanoid-based therapeutics for cardiovascular diseases. Three approaches have been utilized to advance toward human clinical trials: enzymatic inhibitors, agonists, and antagonists. CYP eicosanoid enzymatic inhibitors have progressed the swiftest because detailed information on protein structure for sEH, hydroxylase, and epoxygenase enzymes is available. Development of 20-HETE and EET agonists and antagonists has required structure–activity relationship studies. 20-HETE and EET agonists and antagonists advancement would be greatly accelerated by binding site/receptor protein identification. Another major hurdle was advancing and modifying promising small molecules for chronic administration in animals and humans. There are now novel small molecules that can be administered chronically that target every CYP-eicosanoid metabolite and pathway.

Novel small molecules that target the CYP hydroxylase and 20-HETE that can be chronically administered are emerging (Williams et al., 2010). A number of selective CYP hydroxylase enzymatic inhibitors have been developed (Williams et al., 2010). Experimental studies have extensively evaluated 20-HETE synthase inhibitors such as DDMS, DDBB, and HET0016 in cardiovascular disease models (Williams et al., 2010). 20-HETE agonists and antagonists have been evaluated in ischemia–reperfusion injury models. Agonists of 20-HETE include 5,14,20-HEDE (WIT003) and 5,14,20-HEDGE. Identifying 20-HETE antagonists from agonists in the literature is made difficult because 20-HETE antagonists have very similar names like 6,15,20-HEDE (WIT002) and 6,15,20-HEDGE. A 20-HETE antagonist, SOLA, has been chronically administered in the drinking water and demonstrated antihypertensive actions in diabetic and hypertensive rodents (Cheng et al., 2014; Gangadhariah et al., 2015). More importantly, a CYP hydroxylase inhibitor, TS011, has advanced to human clinical trials for neuroprotection in stroke (Williams et al., 2010). Therefore, novel therapeutics that target CYP hydroxylase and 20-HETE could be used to treat human cardiovascular disease in the near future.

Even though progress of epoxygenase inhibitors and EET antagonists is ongoing, a major focus has been on the developing EET agonists for the treatment of cardiovascular diseases. Selective epoxygenase inhibitors like MS-PPOH are readily available and have been utilized in numerous animal studies (Imig, 2012). Regrettably, these selective epoxygenase inhibitors are not ideal for chronic administration and are currently best delivered by an intraperitoneal pump. A major reason for this slow progress is the absence of a clear clinical indication to use these inhibitors. EET antagonists have been developed based on structure–activity relationship studies (Gauthier, Jagadeesh, Falck, & Campbell, 2003). This has resulted in the development of EET antagonists that are selective for 11,12-EET and 14,15-EET (Bukhari et al., 2012). A nonselective EET antagonist, 14,15-EEZE, has been used in animal studies and could be a precursor to an EET antagonist that could be administered orally and chronically (Gauthier et al., 2003). It is likely that such an EET antagonist could be on the horizon given the advanced development of EET agonists. EET agonists have been rapidly developing over the past 5 years because of their potential to treat cardiovascular diseases (Imig, 2012; Khan et al., 2013). NUDSA was one of the first EET analogs to be administered chronically to rodents and demonstrated antihypertensive actions (Imig et al., 2010; Sodhi et al., 2009). Further development of EET analogs led to two orally active novel small molecules, EET-A and EET-B (Hye Khan et al., 2014; Khan et al., 2013). EET-A and EET-B have been utilized in hypertensive and kidney disease animal models (Hye Khan et al., 2014; Khan et al., 2013). These EET analogs have been demonstrated to decrease blood pressure, reduce inflammation, and decrease end-organ damage in cardiovascular disease animal models (Hye Khan et al., 2014; Khan et al., 2013). The therapeutic potential for EET analogs to treat cardiovascular disease will likely be determined in the next decade.

Development of sEH inhibitors as a means to prevent EET conversion to DHETEs has been tested in human clinical trials for cardiovascular diseases (Imig & Hammock, 2009). Utilization was enhanced by the early development of an orally active sEH inhibitor, AUDA (Imig & Hammock, 2009). AUDA and other sEH inhibitors have been tested in cardiovascular disease animal models such as hypertension, metabolic syndrome, and atherosclerosis (Imig & Hammock, 2009). There was a very rapid pace for developing sEH inhibitors for use in humans because these experimental studies demonstrated great potential to treat cardiovascular diseases. This rapid development resulted in testing sEH inhibitors for hypertension and diabetes in initial human clinical trials (Imig & Hammock, 2009). Currently, Phase I clinical trials evaluating sEH inhibitors for chronic obstructive pulmonary disease have been completed (ClinicalTrials.gov Identifier: NCT01762774, 2012). An aspect of these clinical trials is to evaluate sEH inhibitors on forearm blood flow responses (ClinicalTrials.gov Identifier: NCT01762774, 2012). Thus, the promise for sEH inhibitors to treat cardiovascular disease is progressing and remains an area of active investigation.

12. CONCLUSION

CYP eicosanoid endothelial cell and vascular actions have been extensively studied in physiological and pathological states. These experimental studies have clearly demonstrated opposing endothelial cell actions under the majority of experimental settings for the CYP hydroxylase metabolite 20-HETE with those of CYP epoxygenase EETs. Genetic and pharmacological tools have allowed for extensive evaluation of EET and 20-HETE endothelial cell signaling mechanisms. There have also been numerous studies into CYP eicosanoid interactions with hormones, paracrine factors, and autocrine factors in regard to endothelial cell function and vascular homeostasis. Animal and human experimental studies clearly demonstrate that alteration in CYP eicosanoids at the endothelial and vascular level contributes to cardiovascular diseases. These findings have led to the quest for novel small molecules that act on CYP eicosanoid pathways as therapies for cardiovascular diseases. Accordingly, this is an exciting period for CYP eicosanoids that requires further investigations to reveal new and exciting endothelial cell and vascular actions for the CYP hydroxylase metabolite 20-HETE and CYP epoxygenase EETs.

CONFLICT OF INTEREST STATEMENT

Dr. Imig has patents that cover the composition of matter for EET analogs. There are no other conflicts of interest, financial, or otherwise, declared by the author.

ACKNOWLEDGMENTS

This work was supported by the National Institute of Diabetes and Digestive and Kidney Diseases DK103616 and Dr. Ralph and Marian Falk Medical Research Trust Bank of America, N.A., Trustee. Servier Medical Art was used to generate Figs. 2 and 4 and is licensed by Servier under a Creative Commons Attribution 3.0 Unported License.

REFERENCES

Ai, D., Fu, Y., Guo, D., Tanaka, H., Wang, N., Tang, C., et al. (2007). Angiotensin II up-regulates soluble epoxide hydrolase in vascular endothelium in vitro and in vivo. *Proceedings of the National Academy of Sciences of the United States of America*, *104*, 9018–9023.

Alkayed, N. J., Birks, E. K., Narayanan, J., Petrie, K. A., Kohler-Cabot, A. E., & Harder, D. R. (1997). Role of P-450 arachidonic acid epoxygenase in the response of cerebral blood flow to glutamate in rats. *Stroke*, *28*, 1066–1072.

Alkayed, N. J., Goyagi, T., Joh, H. D., Klaus, J., Harder, D. R., Traystman, R. J., et al. (2002). Neuroprotection and P450 2C11 upregulation after experimental transient ischemic attack. *Stroke*, *33*, 1677–1684.

Alonso-Galicia, M., Drummond, H. A., Reddy, K. K., Falck, J. R., & Roman, R. J. (1997). Inhibition of 20-HETE production contributes to the vascular responses to nitric oxide. *Hypertension*, *29*, 320–325.

Alonso-Galicia, M., Falck, J. R., Reddy, K. M., & Roman, R. J. (1999). 20-HETE agonists and antagonists in the renal circulation. *The American Journal of Physiology*, *277*, F790–F796.

Alonso-Galicia, M., Maier, K. G., Greene, A. S., Cowley, A. W., & Roman, R. J. (2002). Role of 20-hydroxyeicosatetraeinoic acid in the renal and vasoconstrictor actions of angiotensin II. *American Journal of Physiology. Regulatory, Integrative and Comparative Physiology*, *283*, R60–R68.

Archer, S. L., Gragasin, F. S., Wu, X., Wang, S., McMurtry, S., Kim, D. H., et al. (2003). Endothelium-derived hyperpolarizing factor in human internal mammary artery is 11,12-epoxyeicosatrienoic acid and causes relaxation by activating smooth muscle BK(Ca) channels. *Circulation*, *107*, 769–776.

Batchu, S., Lee, S., Qadhi, R., et al. (2011). Cardioprotective effect of a dual acting epoxyeicosatrienoic acid analogue towards ischaemia reperfusion injury. *British Journal of Pharmacology*, *162*(4), 897–907.

Bellien, J., & Joannides, R. (2013). Epoxyeicosatrienoic acid pathway in human health and diseases. *Journal of Cardiovascular Pharmacology, 61*, 188–196.

Bukhari, I. A., Shah, A. J., Gauthier, K. M., Walsh, K. A., Konduru, S. R., Imig, J. D., et al. (2012). 11,12,20-Trihydroxy-eicosa-8(Z)-enoic acid: A selective inhibitor of 11,12-EET induced relaxations of bovine coronary and rat mesenteric arteries. *American Journal of Physiology. Heart and Circulatory Physiology, 302*, H1574–H1583.

Campbell, W. B., & Fleming, I. (2010). Epoxyeicosatrienoic acids and endothelium-dependent responses. *Pflügers Archiv, 459*, 881–895.

Campbell, W. B., Gebremedhin, D., Pratt, P. F., & Harder, D. R. (1996). Identification of epoxyeicosatrienoic acids as endothelium-derived hyperpolarizing factors. *Circulation Research, 78*, 415–423.

Campbell, W. B., Holmes, B. B., Falck, J. R., Capdevila, J. H., & Gauthier, K. M. (2006). Regulation of potassium channels in coronary smooth muscle by adenoviral expression of cytochrome P-450 epoxygenase. *American Journal of Physiology Heart and Circulatory Physiology, 290*, H64–H71.

Capdevila, J. H., & Falck, J. R. (2001). The CYP P450 arachidonic acid monooxygenases: From cell signaling to blood pressure regulation. *Biochemical and Biophysical Research Communications, 285*, 571–576.

Capdevila, J. H., Falck, J. R., & Imig, J. D. (2007). Roles of the cytochrome P450 arachidonic acid monooxygenases in the control of systemic blood pressure and experimental hypertension. *Kidney International, 72*, 683–689.

Capdevila, J. H., Pidkovka, N., Mei, S., et al. (2014). The Cyp2c44 epoxygenase regulates epithelial sodium channel activity and the blood pressure responses to increased dietary salt. *The Journal of Biological Chemistry, 289*(7), 4377–4386.

Carroll, M. A., Balazy, M., Margiotta, P., Falck, J. R., & McGiff, J. C. (1993). Renal vasodilator activity of 5,6-epoxyeicosatrienoic acid depends upon conversion by cyclooxygenase and release of prostaglandins. *The Journal of Biological Chemistry, 268*, 12260–12266.

Carroll, M. A., Doumad, A. B., Li, J., Cheng, M. K., Falck, J. R., & McGiff, J. C. (2006). Adenosine2A receptor vasodilation of rat preglomerular microvessels is mediated by EETs that activate the cAMP/PKA pathway. *American Journal of Physiology. Renal Physiology, 291*, F155–F161.

Carroll, M. A., Garcia, M. P., Falck, J. R., & McGiff, J. C. (1992). Cyclooxygenase dependency of the renovascular actions of cytochrome P450-derived arachidonate metabolites. *The Journal of Pharmacology and Experimental Therapeutics, 260*, 104–109.

Certikova Chabova, V., Walkowska, A., Kompanowska-Jezierska, E., Sadowski, J., Kujal, P., Vernerova, Z., et al. (2010). Combined inhibition of 20-hydroxyeicosatetraenoic acid formation and of epoxyeicosatrienoic acids degradation attenuates hypertension and hypertension-induced end-organ damage in Ren-2 transgenic rats. *Clinical Science (London, England), 118*, 617–632.

Chen, L., Ackerman, R., & Guo, A. M. (2012). 20-HETE in neovascularization. *Prostaglandins & Other Lipid Mediators, 98*(3–4), 63–68.

Chen, L., Ackerman, R., Saleh, M., Gotlinger, K. H., Kessler, M., Mendelowitz, L. G., et al. (2014). 20-HETE regulates the angiogenic functions of human endothelial progenitor cells and contributes to angiogenesis in vivo. *The Journal of Pharmacology and Experimental Therapeutics, 348*(3), 442–451.

Chen, J. K., Chen, J., Imig, J. D., Wei, S., Hachey, D. L., Guthi, J. S., et al. (2008). Identification of novel endogenous cytochrome P450 arachidonate metabolites with high affinity for cannabinoid receptors. *The Journal of Biological Chemistry, 283*, 24514–24524.

Chen, G., & Cheung, D. W. (1996). Modulation of endothelium-dependent hyperpolarization and relaxation to acetylcholine in rat mesenteric artery by cytochrome P450 enzyme activity. *Circulation Research, 79*, 827–833.

Cheng, M. K., Doumad, A. B., Jiang, H., Falck, J. R., McGiff, J. C., & Carroll, M. A. (2004). Epoxyeicosatrienoic acids mediate adenosine-induced vasodilation in rat preglomerular microvessels (PGMV) via A_{2A} receptors. *British Journal of Pharmacology, 141*, 441–448.

Cheng, J., Edin, M. L., Hoopes, S. L., Li, H., Bradbury, J. A., Graves, J. P., et al. (2014). Vascular characterization of mice with endothelial expression of cytochrome P450 4F2. *The FASEB Journal, 28*(7), 2915–2931.

Cheng, J., Garcia, V., Ding, Y., Wu, C. C., Thakar, K., Falck, J. R., et al. (2012). Induction of angiotensin-converting enzyme and activation of the renin-angiotensin system contribute to 20-hydroxyeicosatetraenoic acid-mediated endothelial dysfunction. *Arteriosclerosis, Thrombosis, and Vascular Biology, 32*(8), 1917–1924.

Cheranov, S. Y., Karpurapu, M., Wang, D., Zhang, B., Venema, R. C., & Rao, G. N. (2008). An essential role for SRC-activated STAT-3 in 14,15-EET-induced VEGF expression and angiogenesis. *Blood, 111*, 5581–5591.

ClinicalTrials.gov Identifier: NCT01762774 (2012). *A study to assess the safety, tolerability, pharmacokinetics and pharmacodynamics of single doses of GSK2256294 in healthy volunteers, and single and repeated doses of GSK2256294 in adult male moderately obese smokers.* https://clinicaltrials.gov/ct2/show/NCT01762774?term=soluble+epoxide+hydrolase&rank=2.

Croft, K. D., McGiff, J. C., Sanchez-Mendoza, A., & Carroll, M. A. (2000). Angiotensin II releases 20-HETE from rat renal microvessels. *American Journal of Physiology. Renal Physiology, 279*, F544–F551.

Davis, B. B., Thompson, D. A., Howard, L. L., Morisseau, C., Hammock, B. D., & Weiss, R. H. (2002). Inhibitors of soluble epoxide hydrolase attenuate vascular smooth muscle cell proliferation. *Proceedings of the National Academy of Sciences of the United States of America, 99*, 2222–2227.

Deng, Y., Edin, M. L., Theken, K. N., Schuck, R. N., Flake, G. P., Kannon, M. A., et al. (2011). Endothelial CYP epoxygenase overexpression and soluble epoxide hydrolase disruption attenuate acute vascular inflammatory responses in mice. *The FASEB Journal, 25*, 703–713.

Dimitropoulou, C., West, L., Field, M. B., White, R. E., Reddy, L. M., Falck, J. R., et al. (2007). Protein phosphatase 2A and Ca2+-activated K+ channels contribute to 11,12-epoxyeicosatrienoic acid analog mediated mesenteric arterial relaxation. *Prostaglandins & Other Lipid Mediators, 83*, 50–61.

Ding, Y., Wu, C. C., Garcia, V., Dimitrova, I., Weidenhammer, A., Joseph, G., et al. (2013). 20-HETE induces remodeling of renal resistance arteries independent of blood pressure elevation in hypertension. *American Journal of Physiology. Renal Physiology, 305*, F753–F763.

Dunn, K. M., Renic, M., Flasch, A. K., Harder, D. R., Falck, J., & Roman, R. J. (2008). Elevated production of 20-HETE in the cerebral vasculature contributes to severity of ischemic stroke and oxidative stress in spontaneously hypertensive rats. *American Journal of Physiology. Heart and Circulatory Physiology, 295*(6), H2455–H2465.

Ellinsworth, D. C., Earley, S., Murphy, T. V., & Sandow, S. L. (2014). Endothelial control of vasodilation: Integration of myoendothelial microdomain signalling and modulation by epoxyeicosatrienoic acids. *Pflügers Archiv, 466*(3), 389–405.

Fernandez, M. M., Gonzalez, D., Williams, J. M., Roman, R. J., & Nowicki, S. (2012). Inhibitors of 20-hydroxyeicosatetraenoic acid (20-HETE) formation attenuate the natriuretic effect of dopamine. *European Journal of Pharmacology, 686*, 97–103.

Fisslthaler, B., Popp, R., Kiss, L., Potente, M., Harder, D. R., Fleming, I., et al. (1999). Cytochrome P450 2C is an EDHF synthase in coronary arteries. *Nature, 401*, 493–497.

Fleming, I. (2001). Cytochrome p450 and vascular homeostasis. *Circulation Research, 89*, 753–762.

Fleming, I. (2007). Epoxyeicosatrienoic acids, cell signaling and angiogenesis. *Prostaglandins & Other Lipid Mediators*, *82*, 60–67.

Fleming, I., Rueben, A., Popp, R., Fisslthaler, B., Schrodt, S., Sander, A., et al. (2007). Epoxyeicosatrienoic acids regulate Trp channel dependent Ca2+ signaling and hyperpolarization in endothelial cells. *Arteriosclerosis, Thrombosis, and Vascular Biology*, *27*, 2612–2618.

Furchgott, R. F., & Vanhoutte, P. M. (1989). Endothelium-derived relaxing and contracting factors. *The FASEB Journal*, *3*, 2007–2018.

Gainer, J. V., Bellamine, A., Dawson, E. P., Womble, K. E., Grant, S. W., Wang, Y., et al. (2005). Functional variant of CYP4A11 20-hydroxyeicosatetraenoic acid synthase is associated with essential hypertension. *Circulation*, *111*, 63–69.

Gangadhariah, M. H., Luther, J. M., Garcia, V., Paueksakon, P., Zhang, M. Z., Hayward, S. W., et al. (2015). Hypertension is a major contributor to 20-hydroxyeicosatetraenoic acid-mediated kidney injury in diabetic nephropathy. *Journal of the American Society of Nephrology*, *26*(3), 597–610.

Garcia, V., Joseph, G., Shkolnik, B., Ding, Y., Zhang, F. F., Gotlinger, K., et al. (2015). Angiotensin II receptor blockade or deletion of vascular endothelial ACE does not prevent vascular dysfunction and remodeling in 20-HETE-dependent hypertension. *American Journal of Physiology. Regulatory, Integrative and Comparative Physiology*, *309*(1), R71–R78.

Gauthier, K. M., Deeter, C., Krishna, U. M., Reddy, Y. K., Bondlela, M., Falck, J. R., et al. (2002). 14,15-Epoxyeicosa-5(Z)-enoic acid: A selective epoxyeicosatrienoic acid antagonist that inhibits endothelium-dependent hyperpolarization and relaxation in coronary arteries. *Circulation Research*, *90*(9), 1028–1036. PubMed PMID: 12016270.

Gauthier, K. M., Falck, J. R., Reddy, L. M., & Campbell, W. B. (2004). 14,15-EET analogs: Characterization of structural requirements for agonist and antagonist activity in bovine coronary arteries. *Pharmacological Research*, *49*, 515–524.

Gauthier, K. M., Jagadeesh, S. G., Falck, J. R., & Campbell, W. B. (2003). 14,15-Epoxyeicosa-5(Z)-enoic-mSI: A 14,15- and 5,6-EET antagonist in bovine coronary arteries. *Hypertension*, *42*, 555–561.

Holla, V. R., Adas, F., Imig, J. D., Zhao, X., Price, E., Jr., Olsen, N., et al. (2001). Alterations in the regulation of androgen-sensitive Cyp 4a monooxygenases cause hypertension. *Proceedings of the National Academy of Sciences of the United States of America*, *98*, 5211–5216.

Hoopes, S. L., Garcia, V., Edin, M. L., Schwartzman, M. L., & Zeldin, D. C. (2015). Vascular actions of 20-HETE. *Prostaglandins & Other Lipid Mediators*, *120*, 9–16.

Huang, A., Sun, D., Carroll, M. A., Jiang, H., Smith, C. J., Connetta, J. A., et al. (2001). EDHF mediates flow-induced dilation in skeletal muscle arterioles of female eNOS-KO mice. *American Journal of Physiology. Heart and Circulatory Physiology*, *280*, H2462–H2469.

Hye Khan, M. A., Pavlov, T. S., Christain, S. V., Neckář, J., Staruschenko, A., Gauthier, K. M., et al. (2014). Epoxyeicosatrienoic acid analogue lowers blood pressure through vasodilation and sodium channel inhibition. *Clinical Science (London, England)*, *127*(7), 463–474.

Imig, J. D. (2012). Epoxides and soluble epoxide hydrolase in cardiovascular physiology. *Physiological Reviews*, *92*, 101–130.

Imig, J. D. (2013). Epoxyeicosatrienoic acids, 20-hydroxyeicosatetraenoic acid, and renal microvascular function. *Prostaglandins & Other Lipid Mediators*, *104–105*, 2–7.

Imig, J. D., & Deichmann, P. C. (1997). Afferent arteriolar responses to ANG II involve activation of PLA2 and modulation by lipoxygenase and P-450 pathways. *The American Journal of Physiology*, *273*, F274–F282.

Imig, J. D., Elmarakby, A., Nithipatikom, K., Wei, S., Capdevila, J. H., Tuniki, V. R., et al. (2010). Development of epoxyeicosatrienoic acid analogs with *in vivo* anti-hypertensive actions. *Frontiers in Physiology*, *1*, 157.

Imig, J. D., Falck, J. R., Wei, S., & Capdevila, J. H. (2001). Epoxygenase metabolites contribute to nitric oxide-independent afferent arteriolar vasodilation in response to bradykinin. *Journal of Vascular Research*, *38*, 247–255.

Imig, J. D., & Hammock, B. D. (2009). Soluble epoxide hydrolase as a therapeutic target for cardiovascular diseases. *Nature Reviews. Drug Discovery*, *8*, 794–805.

Imig, J. D., Inscho, E. W., Deichmann, P. C., Reddy, K. M., & Falck, J. R. (1999). Afferent arteriolar vasodilation to the sulfonimide analog of 11,12-epoxyeicosatrienoic acid involves protein kinase A. *Hypertension*, *33*, 408–413.

Imig, J. D., Navar, L. G., Roman, R. J., Reddy, K. K., & Falck, J. R. (1996). Actions of epoxygenase metabolites on the preglomerular vasculature. *Journal of the American Society of Nephrology*, 7, 2364–2370.

Imig, J. D., Pham, B. T., LeBlanc, E. A., Falck, J. R., & Inscho, E. W. (2000). Cytochrome P450 and cyclooxygenase metabolites contribute to the endothelin-1 afferent arteriolar vasoconstrictor and calcium responses. *Hypertension*, *35*, 307–312.

Imig, J. D., Zou, A.-P., Ortiz de Montellano, P. R., Sui, Z., & Roman, R. J. (1994). Cytochrome P450 inhibitors alter the afferent arteriolar response to elevations in perfusion pressure. *American Journal of Physiology. Heart and Circulatory Physiology*, *266*, H1879–H1885.

Imig, J. D., Zou, A. P., Stec, D. E., Harder, D. R., Falck, J. R., & Roman, R. J. (1996). Formation and actions of 20-hydroxyeicosatetraenoic acid in rat renal arterioles. *The American Journal of Physiology*, *270*, R217–R227.

Ishizuka, T., Cheng, J., Singh, H., Vitto, M. D., Manthati, V. L., Falck, J. R., et al. (2008). 20-Hydroxyeicosatetraenoic acid stimulates nuclear factor-kappaB activation and the production of inflammatory cytokines in human endothelial cells. *The Journal of Pharmacology and Experimental Therapeutics*, *324*, 103–110.

Karara, A., Makita, K., Jacobson, H. R., Falck, J. R., Guengerich, F. P., DuBois, R. N., et al. (1993). Molecular cloning, expression, and enzymatic characterization of the rat kidney cytochrome P-450 arachidonic acid epoxygenase. *The Journal of Biological Chemistry*, *268*, 13565–13570.

Katragadda, D., Batchu, S. N., Cho, W. J., Chaudhary, K. R., Falck, J. R., & Seubert, J. M. (2009). Epoxyeicosatrienoic acids limit damage to mitochondrial function following stress in cardiac cells. *Journal of Molecular and Cellular Cardiology*, *46*, 867–875.

Khan, M. A. H., Manthati, V., Errabelli, R., Pavlov, T. S., Staruschenko, A., Falck, J. R., et al. (2013). An orally active epoxyeicosatrienoic acid analog attenuates kidney injury in hypertensive Dahl salt sensitive rat. *Hypertension*, *62*, 905–913.

King, L. M., Gainer, J. V., David, G. L., Dai, D., Goldstein, J. A., Brown, N. J., et al. (2005). Single nucleotide polymorphisms in the CYP2J2 and CYP2C8 genes and the risk of hypertension. *Pharmacogenetics and Genomics*, *15*, 7–13.

Kohagura, K., Endo, Y., Ito, O., Arima, S., Omata, K., & Ito, S. (2000). Endogenous nitric oxide and epoxyeicosatrienoic acids modulate angiotensin II-induced constriction in the rabbit afferent arteriole. *Acta Physiologica Scandinavica*, *168*, 107–112.

Laffer, C. L., Laniado-Schwartzman, M., Wang, M. H., Nasjletti, A., & Elijovich, F. (2003). Differential regulation of natriuresis by 20-hydroxyeicosatetraenoic acid in human salt-sensitive versus salt-resistant hypertension. *Circulation*, *107*, 574–578.

Larsen, B. T., Gutterman, D. D., Sato, A., Toyama, K., Campbell, W. B., Zeldin, D. C., et al. (2008). Hydrogen peroxide inhibits cytochrome p450 epoxygenases: Interaction between two endothelium-derived hyperpolarizing factors. *Circulation Research*, *102*, 59–67.

Larsen, B. T., Miura, H., Hatoum, O. A., Campbell, W. B., Hammock, B. D., Zeldin, D. C., et al. (2006). Epoxyeicosatrienoic and dihydroxyeicosatrienoic acids dilate human coronary arterioles via BK(Ca) channels: Implications for soluble epoxide hydrolase inhibition. *American Journal of Physiology. Heart and Circulatory Physiology*, *290*, H491–H499.

Lee, C. R., Imig, J. D., Edin, M. L., Foley, J., DeGraff, L. M., Bradbury, J. A., et al. (2010). Endothelial expression of human cytochrome P450 epoxygenases lowers blood pressure and attenuates hypertension-induced renal injury in mice. *The FASEB Journal*, *24*, 3770–3781.

Lee, C. R., Pretorius, M., Schuck, R. N., Burch, L. H., Bartlett, J., Williams, S. M., et al. (2011). Genetic variation in soluble epoxide hydrolase (EPHX2) is associated with forearm vasodilator responses in humans. *Hypertension*, *57*, 116–122.

Li, P. L., Zhang, D. X., Ge, Z. D., & Campbell, W. B. (2002). Role of ADP-ribose in 11,12-EET-induced activation of K(Ca) channels in coronary arterial smooth muscle cells. *American Journal of Physiology. Heart and Circulatory Physiology*, *282*, H1229–H1236.

Liu, Y., Zhang, Y., Schmelzer, K., et al. (2005). The antiinflammatory effect of laminar flow: The role of PPARγ, epoxyeicosatrienoic acids, and soluble epoxide hydrolase. *Proceedings of the National Academy of Sciences of the United States of America*, *102*(46), 16747–16752.

Lukaszewicz, K. M., & Lombard, J. H. (2013). Role of the CYP4A/20-HETE pathway in vascular dysfunction of the Dahl salt-sensitive rat. *Clinical Science (London, England)*, *124*(12), 695–700.

Ma, Y. H., Gebremedhin, D., Schwartzman, M. L., Falck, J. R., Clark, J. E., Masters, B. S., et al. (1993). 20-Hydroxyeicosatertraenoic acid is an endogenous vasoconstrictor of canine renal arcuate arteries. *Circulation Research*, *72*, 126–136.

Ma, Y., Zhang, P., Li, J., Lu, J., Ge, J., Zhao, Z., et al. (2015). Epoxyeicosatrienoic acids act through TRPV4-TRPC1-KCa1.1 complex to induce smooth muscle membrane hyperpolarization and relaxation in human internal mammary arteries. *Biochimica et Biophysica Acta*, *1852*, 552–559.

Marino, J. P., Jr. (2009). Soluble epoxide hydrolase, a target with multiple opportunities for cardiovascular drug discovery. *Current Topics in Medicinal Chemistry*, *9*, 452–463.

Matsuda, H., Hayashi, K., Wakino, S., Kubota, E., Honda, M., Tokuyama, H., et al. (2004). Role of endothelium-derived hyperpolarizing factor in ACE inhibitor-induced renal vasodilation in vivo. *Hypertension*, *43*(3), 603–609. Epub 2004 Feb 9. PubMed PMID: 14769805.

McSherry, I. N., Sandow, S. L., Campbell, W. B., Falck, J. R., Hill, M. A., & Dora, K. A. (2006). A role for heterocellular coupling and EETs in dilation of rat cremaster arteries. *Microcirculation*, *13*, 119–130.

Michaelis, U. R., Fisslthaler, B., Medhora, M., Harder, D., Fleming, I., & Busse, R. (2003). Cytochrome P450 2C9-derived epoxyeicosatrienoic acids induce angiogenesis via cross-talk with the epidermal growth factor receptor (EGFR). *The FASEB Journal*, *17*, 770–772.

Morisseau, C., & Hammock, B. D. (2013). Impact of soluble epoxide hydrolase and epoxyeicosanoids on human health. *Annual Review of Pharmacology and Toxicology*, *53*, 37–58.

Munzenmaier, D. H., & Harder, D. R. (2000). Cerebral microvascular endothelial cell tube formation: Role of astrocytic epoxyeicosatrienoic acid release. *American Journal of Physiology. Heart and Circulatory Physiology*, *278*, H1163–H1167.

Nakagawa, K., Holla, V. J., Wei, Y., Wang, W. H., Gatica, A., Wei, S., et al. (2006). Salt-sensitive hypertension is associated with dysfunctional *Cyp4a10* gene and kidney epithelial sodium channel. *The Journal of Clinical Investigation*, *116*, 1696–1702.

Node, K., Huo, Y., Ruan, X., Yang, B., Spiecker, M., Ley, K., et al. (1999). Anti-inflammatory properties of cytochrome P450 epoxygenase-derived eicosanoids. *Science*, *285*, 1276–1279.

Pfister, S. L., Gauthier, K. M., & Campbell, W. B. (2010). Vascular pharmacology of epoxyeicosatrienoic acids. *Advances in Pharmacology*, *60*, 27–59.

Potente, M., Fisslthaler, B., Busse, R., & Fleming, I. (2003). 11,12-Epoxyeicosatrienoic acid-induced inhibition of FOXO factors promotes endothelial proliferation by down-regulating p27Kip1. *The Journal of Biological Chemistry*, *278*, 29619–29625.

Potente, M., Michaelis, U. R., Fisslthaler, B., Busse, R., & Fleming, I. (2002). Cytochrome P450 2C9-induced endothelial cell proliferation involves induction of mitogen-activated protein (MAP) kinase phosphatase-1, inhibition of the c-Jun N-terminal kinase, and up-regulation of cyclin D1. *The Journal of Biological Chemistry*, *277*, 15671–15676.

Pozzi, A., Macias-Perez, I., Abair, T., Wei, S., Su, Y., Zent, R., et al. (2005). Characterization of 5,6- and 8,9-epoxyeicosatrienoic acids (5,6- and 8,9-EET) as potent in vivo angiogenic lipids. *The Journal of Biological Chemistry*, *280*, 27138–27146.

Proctor, K. G., Falck, J. R., & Capdevila, J. (1987). Intestinal vasodilation by epoxyeicosatrienoic acids: Arachidonic acid metabolites produced by a cytochrome P450 monooxygenase. *Circulation Research*, *60*, 50–59.

Roman, R. J. (2002). P-450 metabolites of arachidonic acid in the control of cardiovascular function. *Physiological Reviews*, *82*, 131–185.

Simpkins, A. N., Rudic, R. D., Roy, S., Tsai, H. J., Hammock, B. D., & Imig, J. D. (2010). Soluble epoxide hydrolase inhibition modulates vascular remodeling. *American Journal of Physiology. Heart and Circulatory Physiology*, *298*, H795–H806.

Simpkins, A. N., Rudic, R. D., Schreihofer, D. A., Roy, S., Manhiani, M., Tsai, H. J., et al. (2009). Soluble epoxide inhibition is protective against cerebral ischemia via vascular and neural protection. *The American Journal of Pathology*, *174*, 2086–2095.

Sodhi, K., Inoue, K., Gotlinger, K. H., Canestraro, M., Vanella, L., Kim, D. H., et al. (2009). Epoxyeicosatrienoic acid agonist rescues the metabolic syndrome phenotype of HO-2-null mice. *The Journal of Pharmacology and Experimental Therapeutics*, *331*, 906–916.

Sodhi, K., Wu, C. C., Cheng, J., Gotlinger, K. H., Inoue, K., Goli, M., et al. (2010). CYP4A2-induced hypertension is 20-HETE and angiotensin II-dependent. *Hypertension*, *56*, 871–878.

Spector, A. A., Fang, X., Snyder, G. D., & Weintraub, N. L. (2004). Epoxyeicosatrienoic acids (EETs): Metabolism and biochemical function. *Progress in Lipid Research*, *43*, 55–90.

Sudhahar, V., Shaw, S., & Imig, J. D. (2010). Epoxyeicosatrienoic acid analogs and vascular function. *Current Medicinal Chemistry*, *17*, 1181–1190.

Tacconelli, S., & Patrignani, P. (2014). Inside epoxyeicosatrienoic acids and cardiovascular disease. *Frontiers in Pharmacology*, *5*, 23.

Ulu, A., Stephen Lee, K. S., Miyabe, C., Yang, J., Hammock, B. G., Dong, H., et al. (2014). An omega-3 epoxide of docosahexaenoic acid lowers blood pressure in angiotensin-II-dependent hypertension. *Journal of Cardiovascular Pharmacology*, *64*(1), 87–99.

Vriens, J., Owsianik, G., Fisslthaler, B., Suzuki, M., Janssens, A., Voets, T., et al. (2005). Modulation of the Ca2 permeable cation channel TRPV4 by cytochrome P450 epoxygenases in vascular endothelium. *Circulation Research*, *97*, 908–915.

Wang, D., Borrego-Conde, L. J., Falck, J. R., Sharma, K. K., Wilcox, C. S., & Umans, J. G. (2003). Contributions of nitric oxide, EDHF, and EETs to endothelium-dependent relaxation in renal afferent arterioles. *Kidney International*, *63*, 2187–2193.

Wang, M. H., Brand-Schieber, E., Zand, B. A., Nguyen, X., Falck, J. R., Balu, N., et al. (1998). Cytochrome P450-derived arachidonic acid metabolism in the rat kidney:

Characterization of selective inhibitors. *The Journal of Pharmacology and Experimental Therapeutics*, *284*, 966–973.

Wang, X., Trottier, G., & Loutzenhiser, R. (2003). Determinants of renal afferent arteriolar actions of bradykinin: Evidence that multiple pathways mediate responses attributed to EDHF. *American Journal of Physiology. Renal Physiology*, *285*, F540–F549.

Wang, Y., Wei, X., Xiao, X., Hui, R., Card, J. W., Carey, M. A., et al. (2005). Arachidonic acid epoxygenase metabolites stimulate endothelial cell growth and angiogenesis via mitogen-activated protein kinase and phosphatidylinositol 3-kinase/Akt signaling pathways. *The Journal of Pharmacology and Experimental Therapeutics*, *314*, 522–532.

Watanabe, H., Vriens, J., Prenen, J., Droogmans, G., Voets, T., & Nilius, B. (2003). Anandamide and arachidonic acid use epoxyeicosatrienoic acids to activate TRPV4 channels. *Nature*, *424*, 434–438.

Webler, A. C., Michaelis, U. R., Popp, R., Barbosa-Sicard, E., Murugan, A., Falck, J. R., et al. (2008). Epoxyeicosatrienoic acids are part of the VEGF-activated signaling cascade leading to angiogenesis. *American Journal of Physiology. Cell Physiology*, *295*(5), C1292–C1301. http://dx.doi.org/10.1152/ajpcell.00230.2008. Epub 2008 Sep 11. PubMed PMID: 18787075; PubMed Central PMCID: PMC2584984.

Westphal, C., Konkel, A., & Schunck, W. H. (2011). CYP-eicosanoids—A new link between omega-3 fatty acids and cardiac disease? *Prostaglandins & Other Lipid Mediators*, *96*(1–4), 99–108.

Williams, J. M., Murphy, S., Burke, M., & Roman, R. J. (2010). 20-hydroxyeicosatetraeonic acid: A new target for the treatment of hypertension. *Journal of Cardiovascular Pharmacology*, *56*, 336–344.

Wu, C. C., Mei, S., Cheng, J., Ding, Y., Weidenhammer, A., Garcia, V., et al. (2013). Androgen-sensitive hypertension associates with upregulated vascular CYP4A12-20-HETE synthase. *Journal of the American Society of Nephrology*, *24*, 1288–1296.

Wu, C. C., & Schwartzman, M. L. (2011). The role of 20-HETE in androgen-mediated hypertension. *Prostaglandins & Other Lipid Mediators*, *96*(1–4), 45–53.

Yang, S., Wei, S., Pozzi, A., & Capdevila, J. H. (2009). The arachidonic acid epoxygenase is a component of the signaling mechanisms responsible for VEGF-stimulated angiogenesis. *Archives of Biochemistry and Biophysics*, *489*, 82–91.

Zeldin, D. C. (2001). Epoxygenase pathways of arachidonic acid metabolism. *The Journal of Biological Chemistry*, *276*, 36059–36062.

Zhang, B., Cao, H., & Rao, G. N. (2006). Fibroblast growth factor-2 is a downstream mediator of phosphatidylinositol 3-kinase-Akt signaling in 14,15-epoxyeicosatrienoic acid-induced angiogenesis. *The Journal of Biological Chemistry*, *281*, 905–914.

Zhao, X., Dey, A., Romanko, O. P., Stepp, D. W., Wang, M. H., Zhou, Y., et al. (2005). Decreased epoxygenase and increased epoxide hydrolase expression in the mesenteric artery of obese Zucker rats. *American Journal of Physiology. Regulatory, Integrative and Comparative Physiology*, *288*, R188–R196.

Zhao, X., Falck, J. R., Gopal, V. R., Inscho, E. W., & Imig, J. D. (2004). P2X receptor stimulated calcium responses in preglomerular smooth muscle cells involves 20-hydroxyeicosatetraenoic acid. *The Journal of Pharmacology and Experimental Therapeutics*, *311*, 1211–1217.

Zhao, X., Inscho, E. W., Bondlela, M., Falck, J. R., & Imig, J. D. (2001). The CYP450 hydroxylase pathway contributes to P2X receptor-mediated afferent arteriolar vasoconstriction. *American Journal of Physiology. Heart and Circulatory Physiology*, *281*, H2089–H2096.

Zhao, X., Pollock, D. M., Inscho, E. W., Zeldin, D. C., & Imig, J. D. (2003). Decreased renal cytochrome P450 2C enzymes and impaired vasodilation are associated with salt-sensitive hypertension. *Hypertension*, *41*, 709–714.

Zhao, X., Yamamoto, T., Newman, J. W., Kim, I. H., Watanabe, T., Hammock, B. D., et al. (2004). Soluble epoxide hydrolase inhibition protects the kidney from hypertension-induced damage. *Journal of the American Society of Nephrology*, *15*, 1244–1253.

Zou, A. P., Fleming, J. T., Falck, J. R., Jacobs, E. R., Gebremedhin, D., Harder, D. R., et al. (1996). 20-HETE is an endogenous inhibitor of the large-conductance Ca(2+)-activated K+ channel in renal arterioles. *The American Journal of Physiology*, *270*, R228–R237.

Zou, A.-P., Imig, J. D., Ortiz de Montellano, P. R., Sui, Z., & Roman, R. J. (1994). Inhibition of renal vascular 20-HETE impairs autoregulation of renal blood flow. *American Journal of Physiology. Renal Physiology*, *266*, F275–F282.

CHAPTER FIVE

Endothelin-1: Biosynthesis, Signaling and Vasoreactivity

M. Houde, L. Desbiens, P. D'Orléans-Juste[1]
Université de Sherbrooke, Sherbrooke, QC, Canada
[1]Corresponding author: e-mail address: pedro.dorleans-juste@usherbrooke.ca

Contents

Abstract

Endothelin-1 (ET-1) is an extremely potent vasoconstrictor peptide originally isolated from endothelial cells. Its synthesis, mainly regulated at the gene transcription level, involves processing of a precursor by a furin-type proprotein convertase to an inactive intermediate, big ET-1. The latter peptide can then be cleaved directly by an endothelin-converting enzyme (ECE) into ET-1 or reach the active metabolite through a two-step process involving chymase hydrolyzing big ET-1 to ET-1 (1–31), itself needing conversion

Advances in Pharmacology, Volume 77
ISSN 1054-3589
http://dx.doi.org/10.1016/bs.apha.2016.05.002

to ET-1 by neprilysin (NEP) to exert physiological activity. ET-1 signals through two G protein-coupled receptors, endothelin receptor A (ET_A) and endothelin receptor B (ET_B). Both receptors induce an increase in intracellular Ca^{2+}, mainly from the extracellular space through voltage-independent mechanisms, the receptor-operated channels and store-operated channels. ET-1 also induces signaling through epidermal growth factor receptor transactivation, oxidative stress induction, rho-kinase, and the activation (ET_A) or inhibition (ET_B) of the adenylate cyclase/cyclic adenosine monophosphate pathway. Arterial vasoconstriction is mediated mainly by the ET_A receptor. ET-1, via endothelium-located ET_B, relaxes arteries or constricts vessels following activation of the same receptor type on the smooth muscle, where it can interact with ET_A. In addition, ET_B-dependent vasoconstriction seems more prominent in the venous vasculature. A better understanding of how ET-1 is synthesized and how ET_A and ET_B receptors interact could help design better pharmacological agents in the treatment of cardiovascular diseases where targeting the ET-1 system is indicated.

ABBREVIATIONS

AC adenylyl cyclase
cAMP cyclic adenosine 3′,5′-monophosphate
DAG diacylglycerol
ECE endothelin-converting enzyme
EGFR epidermal growth factor receptor
eNOS endothelial nitric oxide synthase
ET-1 endothelin-1
ET_A/ET_B endothelin receptor A/receptor B
HB-EGF heparin-binding epidermal growth factor
IP_3 inositol 3,4,5-triphosphate
MAPK mitogen-activated protein kinase
mMCP-4 mouse mast cell protease-4
NEP neprilysin
PI-3K phosphatidylinositol-3-kinase
PIP_2 phosphatidylinositol 4,5-bisphosphate
PKA/PKC protein kinase A/protein kinase C
PLC phospholipase C
S6c sarafotoxin 6c
SER sarco/endoplasmic reticulum
TRPC transient receptor protein, canonical subfamily
VOCC/ROCC/SOCC voltage-, receptor-, and store-operated calcium channel
VSMC vascular smooth muscle cell

1. INTRODUCTION

Endothelin-1 (ET-1) was initially reported in 1988 as an extremely potent and long-acting vasoconstrictor (Yanagisawa et al., 1988). It is part of a family of three mammalian isoforms: ET-1, ET-2 (its rodent isoform is alternatively referred to as vasoactive intestinal contract as it differs from human ET-2 by one residue (Saida, Mitsui, & Ishida, 1989)), and ET-3 (Inoue, Yanagisawa, Kimura, et al., 1989). These three 21-amino acid isoforms all possess disulfide bonds between Cys^1–Cys^{15} and Cys^3–Cys^{11} and a tryptophan C-terminal moiety (Inoue, Yanagisawa, Kimura, et al., 1989). The endothelins share their unusual structure with a family of toxins found in the venom of the Middle Eastern mole viper *Atractaspis engaddensis*, termed sarafotoxins (S6a, S6b, S6c, etc.) (Kloog et al., 1988).

Two mammalian ET receptors have been described: endothelin receptor A (ET_A) and endothelin receptor B (ET_B). While pharmacological studies sometimes describe receptor subclasses, only ET_A and ET_B were demonstrated in mammals (Davenport, 2002). ET_A and ET_B are both G protein-coupled receptors (GPCRs) that initiate calcium signaling after their activation via the phospholipase C (PLC) mechanism. ET_B binds with equal affinity to all ET isoforms, while ET-3 possesses very low affinity to ET_A (Rubanyi & Polokoff, 1994). In addition to $G_{q/11}$ signaling, ET receptor activation also influences a variety of second messengers, including the mitogen-activated protein kinase (MAPK) pathway, rho-kinase, and adenylyl cyclase (Ivey, Osman, & Little, 2008). This confers to endothelin peptides potent pharmacological properties beyond blood pressure control, in physiology and pathology. In pathology, ET-1 levels are enhanced in multiple conditions such as heart disease, kidney failure, cancer, and pulmonary hypertension. The latter disease, along with digital ulceration secondary to scleroderma, currently are the two clinical indications for ET receptor antagonists (Davenport et al., 2016).

This review summarizes knowledge to date on ET-1 synthesis and signaling pathways as well as mechanisms by which the potent vasopressor peptide stimulates the responses of endothelial (EC) and vascular smooth muscle cells (VSMCs) as well as in whole blood vessels.

2. ENDOTHELIN BIOSYNTHESIS

2.1 Endothelin Gene

The name of ET peptides derives from their first and best-described source, the endothelium. The vascular wall's inner layer is indeed the most documented source of ET-1. However, other cell types such as VSMCs, renal epithelial cells, and neurons also synthesize ET peptides.

The ET-1 peptide is generated from the End1 gene, 6838 base pairs long, located on chromosome 6. It contains 5 exons, coding for a 2026-nucleotide mRNA (Arinami et al., 1991; Inoue, Yanagisawa, Takuwa, et al., 1989). After proteolytic processing, the N-terminal of active peptide derives from exon 2, while the C-terminal tail of big ET-1 (without the ET-1 sequence) is split between exons 2 and 3. Another endothelin-like sequence, conserving the same N-terminal 4 cysteine residues of ET-1, is located on exon 3, but its role remains unclear (Inoue, Yanagisawa, Takuwa, et al., 1989).

2.2 ET-1 Expression

Numerous factors influence the expression of ET-1. Angiotensin-II, transforming growth factor-β, thrombin, bradykinin, hypoxia, and low-density lipoproteins (either oxidized or acetylated) induce ET-1 expression (Rubanyi & Polokoff, 1994). It can also be regulated by microRNAs (von Brandenstein, Richter, & Fries, 2012). On the other hand, fluid shear stress and nitric oxide (NO) are potent inhibitors of ET-1 gene expression. In fact, nitric oxide synthase (NOS) inhibitor-induced increase in blood pressure can be reversed by ET-1 receptor antagonists, suggesting that NO inhibits both ET-1 gene expression and VSMC contraction (Gratton, Cournoyer, Loffler, Sirois, & D'Orleans-Juste, 1997; Rapoport, 2014).

2.3 Maturation into Big ET-1

As shown in Fig. 1, the End1 gene translated mRNA results in a 212-amino acid protein, preproET-1. A signal peptidase removes the 17 N-terminal residues to produce the 195-residue proET-1. This protein is then cleaved at two dibasic locations by a subtilisin-like proprotein convertase, furin (Denault et al., 1995). Subsequent C-terminal trimming of the dibasic end by carboxypeptidases produces the immediate ET-1 precursor, big ET-1.

2.4 Identification of ECE

The existence of big ET-1 was initially theorized based on analysis of the ET-1 flanking dibasic pairs (Yanagisawa et al., 1988). Big ET-1 is itself inactive and requires the hydrolysis of the Trp^{21}–Val^{22} bond to produce the biologically active metabolite ET-1. Because of the aromatic residue ending the ET-1 sequence, a chymotrypsin-like activity was hypothesized early on. Albeit most probably not involved in vascular ET-1 generation, several types of enzymes were found to produce ET-1 from big ET-1 in vitro, for example, the intestinal chymotrypsin (ET-1 here is actually an intermediate in a three-step degradation process; Takaoka et al.,

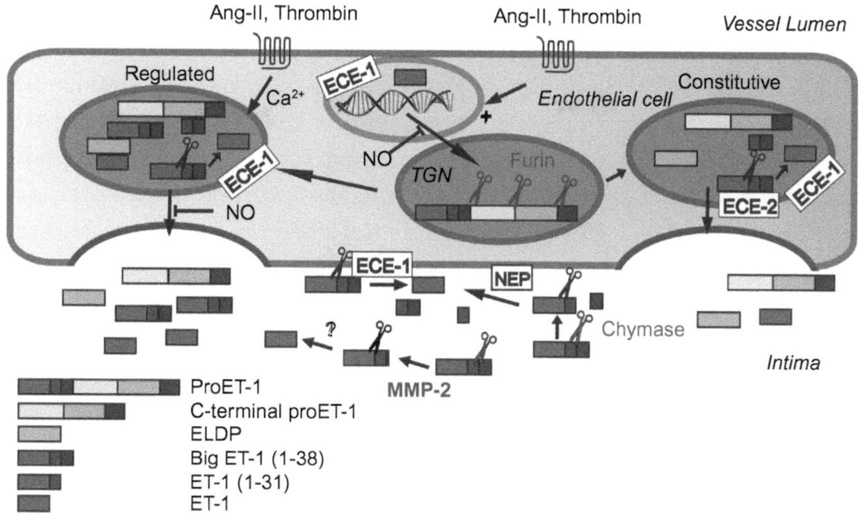

Fig. 1 ET-1 biosynthesis by the endothelial cell. After translation, proET-1 is cleaved in the TGN by furin, which yields big ET-1 and the ELDP or the C-terminal proET-1. Those latter peptides then translocate to secretory vesicles of the constitutive pathway or to storage vesicles (Weibel–Palade bodies in endothelial cells) where the release may be controlled by external factors. Processing of big ET-1 to ET-1 may occur in vesicles, as ECE-1 is present in Weibel–Palade bodies and both ECE-1 and ECE-2 are present in constitutive secretory vesicles. As vesicles fuse with the plasma membrane, on the basolateral side of the endothelial cells, their contents are released in the intimal space. Plasma ET-1 peptides are spilled over from this release. Big ET-1 can be processed in the extracellular space by the ECE-1 on the plasma membrane. In addition, chymase can cleave big ET-1 to produce an intermediate, ET-1 (1–31), which is subsequently hydrolyzed by NEP into its active metabolite, ET-1. Another alternative pathway of unknown physiological significance involves MMP-2-hydrolyzing big ET-1 into ET-1 (1–32). ET-1 (1–32) can elicit ET-1-like actions, but the proteolytic step involved in the mature ET-1 generation remains unclear for this latter pathway.

1990), the ruminant chymosin (Denault et al., 1995), and the bacterial thermolysin (Murphy, Corder, Mallet, & Turner, 1994).

Ohnaka et al. (1990) were the first to suggest that the endothelin-converting enzyme (ECE) was a neutral metalloendopeptidase, later found to be sensitive to phosphoramidon, a neprilysin (NEP) inhibitor (McMahon, Palomo, Moore, McDonald, & Stern, 1991). The same group observed a less potent inhibitory activity in vivo with thiorphan, a more specific NEP inhibitor. NEP itself is unable to directly process big ET-1 to ET-1, and it was originally found to degrade ET-1 instead of synthesizing it (Sokolovsky et al., 1990). The ECE was then isolated from rat lungs and porcine aortic ECs in 1993 (Ohnaka, Takayanagi, Nishikawa,

Haji, & Nawata, 1993; Takahashi, Matsushita, Iijima, & Tanzawa, 1993) and cloned the following year (Schmidt et al., 1994; Xu et al., 1994).

Classified as a heavily glycosylated type-II integral membrane metalloproteinase functioning as a covalently linked dimer, the ECE has a short cytoplasmic tail, a transmembrane region acting as signaling peptide, and a large extracellular/intravesicular domain containing the catalytic site (Turner & Murphy, 1996). Two ECE isoforms are known in humans and rodents, derived from two separate genes, ECE-1 and ECE-2, with each having four splice variants (a–d). ECE-1 is a neural endopeptidase, while ECE-2 prefers an acidic (optimum pH 5.5) environment (Emoto & Yanagisawa, 1995).

2.5 Localization of ECE

The four alternative splicing variants of ECE-1 differ only in their N-terminal cytosolic domain. Their purified forms are enzymatically identical but have diverse subcellular localizations. In Chinese hamster ovary (CHO) cells, ECE-1a and ECE-1c tend to be present on the plasma membrane, while ECE-1b localizes to the interior compartments, colocalizing with a trans-Golgi (TGN) marker (Schweizer et al., 1997). The latter study also found some intracellular ECE-1c staining. In addition, in neuroendocrine cells, ECE-1b and 1d are intracellular, but localized mostly on endosomal vesicles (late endosome for ECE-1b, recycling for ECE-1d), and colocalized weakly with TGN markers. Cells that coexpress multiple ECE-1 isoforms also show heterodimerization, which impact their cell localization (Muller et al., 2003). For example, ECE-1 was also found in the nucleus along with ET-1 (Jafri & Ergul, 2003). As such, ECE-1 can produce ET-1 in the extracellular, cellular granular, and nuclear compartments, contributing to the genesis of different ET-1 pools. The catalytic portion of ECE-1 can also be released in soluble form by ECs and remains active (Kuruppu, Reeve, & Ian Smith, 2007). ECE-1 was isolated from the endothelium, but is also found in a variety of cell types such as VSMCs, macrophages, and neurons (Davenport & Maguire, 2006).

Similar to ECE-1, ECE-2 is expressed as four splice variants (ECE-2a-1, 2a-2, 2b-1, and 2b-2) in multiple cells and tissues including ECs, heart, and lungs. ECE-2a isoforms are more expressed in bovine tissues than the 2b isoforms, which are restricted to cerebral areas and the adrenal medulla (Ikeda, Emoto, Alimsardjono, Yokoyama, & Matsuo, 2002). Using their differences in preferred pH, Russell and Davenport (1999) reported that both ECE-1 and ECE-2 are present on the plasma membrane and in storage granules in human ECs. For both ECE-1 and ECE-2 however, the physiological role for their subcellular localization is not known.

ECs secrete ET-1 and big ET-1 in multiple fashions (Seed et al., 2012). They are constitutively released by ECs at a basal level, but this secretion can be stimulated through multiple pathways. ET-1 and big ET-1 are also stored in ECs' Weibel–Palade bodies, allowing for rapid ET-1 export after stimulation, where big ET-1 can be activated extracellularly. This rapid increase in ET-1 release can be followed by increased gene expression to augment the production of the 21 amino acid peptide (Davenport & Maguire, 2006; Rapoport, 2014). As ET-1 is rapidly removed from the circulation by ET_B receptors and mostly functions in an autocrine/paracrine manner, big ET-1 may act as a circulating pool of ET-1 to act in an endocrine fashion (Johnstrom et al., 2010). Once it is secreted, big ET-1 can then undergo processing to ET-1 by ECE-independent pathways.

2.6 Non-ECE ET-1 Generation

While it does not cleave the Trp^{21}–Val^{22} bond of big ET-1, NEP is involved in ET-1 generation through another pathway. Wypij et al. (1992) demonstrated that a chymostatin-sensitive mast cell serine protease, chymase, was able to generate ET-1 from big ET-1 in vitro. Two groups then independently showed that chymase generated a new peptide, ET-1 (1–31) (the same intermediate was found in the chymotrypsin-dependent processing of big ET-1 to ET-1 (Takaoka et al., 1990)), which expressed ET-1-like activity (Hanson, Andersson, Gyllstedt, Högestätt, & Lindberg, 1997; Nakano et al., 1997). Our group demonstrated that ET-1 (1–31) is itself inactive in vivo, requiring further conversion to ET-1. We showed in three different animal models (guinea pig, rabbit, and mouse) that this action is not mediated by ECE but by NEP (Fecteau et al., 2005; Honore, Plante, Bkaily, Rae, & D'Orleans-Juste, 2002; Simard et al., 2009), demonstrating that the C-terminal tail of big ET-1 has opposite actions on the proteolytic activity of both enzymes. We also demonstrated that the murine chymase isoform responsible for ET-1 (1–31) generation is mouse mast cell protease-4 (Houde et al., 2013). In contrast to chymotrypsin, however, chymase is unable to hydrolyze the Trp^{21}–Val^{22} bond of either big ET-1 or ET-1 (1–31) (Semaan et al., 2015).

Mice knocked-out for ECE-1, or both ECE-1 and ECE-2, die in utero from neural crest defects and cardiac abnormalities. However, they retain over 50% of endogenous ET-1 peptide expression (Yanagisawa et al., 2000, 1998), implying important non-ECE-dependent ET-1 generation, at least in the mouse fetus. Our studies with a chymase inhibitor or mice knocked-out for mouse mast cell protease-4 (mMCP-4) demonstrated reduced pressor response to exogenous big ET-1 administration after chymase activity disruption, similar to the inhibition afforded by the specific ECE inhibitor CGS 35066 (at a non-NEP inhibitory dose) or the specific NEP inhibitor thiorphan (the dual ECE/NEP inhibitor phosphoramidon blocked all big ET-1 pressor activity). Similar results were observed when plasma ET-1 levels were measured after big ET-1 administration (Houde et al., 2013; Simard et al., 2009). The residual response to big ET-1 in mMCP-4 KO mice was sensitive to CGS 35066 but not to thiorphan. Furthermore, mMCP-4 deletion resulted in a 60% decrease of endogenous pulmonary ET-1 levels in mice, indicating that big ET-1 is secreted in the lungs and that mMCP-4 plays a major role in the lung interstitium to convert it to its active metabolite (Houde et al., 2013). Notably, mMCP-4 KO mice are viable, healthy, and do not exhibit ECE-1 KO or End1 KO phenotypes (Tchougounova, Pejler, & Abrink, 2003), indicating that ECE-dependent ET-1 is the main driver of fetal development. Thus, interfering with the chymase-dependent production of ET-1 may be an alternative to the use of ECE inhibitors or ET antagonists which are counterindicated in pregnancy (Taichman et al., 2014).

2.7 Other ET-Like Peptides

The C-terminal tryptophan, shared among all endothelins and the sarafotoxins of the *Atractaspis* snake genus, is crucial in its carboxylic form for endothelin activity. Removal or modification of this C-terminal end results in a 1000-fold decrease in activity (Kimura et al., 1988; Maggi et al., 1989). Any residue remaining after Trp^{21} also prevents any activity, as big ET-1 and ET-1 (1–31) do not bind ET receptors. Furthermore, a new sarafotoxin-m (STX-m) from *Atractaspis microlepidota* contains three residues after Trp^{21}. STX-m is active and lethal to mice, but only weakly binds to both endothelin receptors (Hayashi et al., 2004). A recent study showed that its administration in mice produced a distinct phenotype compared to STX-b (S6b). When mice were administered a truncated version of STX-m, which uncovered Trp^{21}, its distinct activity was lost and it gained STX-b-like activity (Mahjoub et al., 2015).

Yanagisawa et al. (1988) also initially identified an ET-like domain immediately after the big ET-1-Arg-Arg C-terminal. This sequence conserves four cysteines in identical positions to ET-1, but its C-terminal tail differs significantly. This ET-like domain peptide (ELDP) does not exhibit agonistic activity on ET receptors, but recent data suggest that it can act as a potentiator of ET-1 function (Yuzugulen et al., 2012). It has also been suggested, along with the longer C-terminal proET-1, as a plasmatic marker of ET-1 production levels (Dhaun et al., 2015).

Big ET-1 can also be cleaved by gelatinases (matrix metalloproteinase 2 (MMP-2) and MMP-9) to produce the 32-amino acid ET-1 (1–32) (Fernandez-Patron, Radomski, & Davidge, 1999; Fernandez-Patron et al., 2001). This new metabolite is a potent inducer of mesenteric bed vasoconstriction. Remarkably, MMP-2 inhibition only modestly inhibited big ET-1-induced vasoconstriction when the endothelium was present, but almost completely blocked it if the endothelium was removed (Fernandez-Patron et al., 1999). The same group also reported that ET-1 (1–32) induced neutrophil adhesion to ECs, and that this effect was resistant to phosphoramidon but sensitive to ET_A antagonism (Fernandez-Patron et al., 2001). Jeyabalan et al. (2003) reported that the ET_B-mediated relaxin-induced small renal artery vasodilation was unaffected by the dual ECE/NEP inhibitor phosphoramidon, but blocked by MMP-2 inhibition. In contrast, another study found no impact of MMP-2 inhibition on big ET-1-induced contraction of aortic or venous vessels (Watts, Thakali, Smark, Rondelli, & Fink, 2007). Further in vivo evidence is needed to determine whether ET-1 (1–32) needs conversion to ET-1 to exert its ET_A receptor-dependent effects, and whether it is a physiologically relevant big ET-1 metabolite.

2.8 Degradation

Chymase was once thought to be involved in mast cell-dependent extracellular ET-1 degradation (Maurer et al., 2004). However, further studies demonstrated that another protease, carboxypeptidase A3, was responsible for mast cell-dependent ET-1 removal (Schneider, Schlenner, Feyerabend, Wunderlin, & Rodewald, 2007). This protease is coexpressed in the same subset of mast cells as chymase (Pejler, Ronnberg, Waern, & Wernersson, 2010).

Most vascular ET-1 degradation occurs intracellularly. Circulating ET-1 binds rapidly to its ET_B receptor, especially in the lungs, kidneys, and liver (Fukuroda et al., 1994). Apart from the brain, those same organs are the ones with the highest ET_B/ET_A expression ratio (Maguire & Davenport, 2015). The endothelial ET_B receptor is crucial in that regard, as the cell-specific knockout of the latter receptorial moiety increases plasma ET-1 levels similarly to ET_B antagonists (Kelland et al., 2010). The ET-1/ET_B complex then internalizes and is directed to the lysosome for degradation (Bremnes et al., 2000). ET_B is enriched in caveolae, which play a pivotal role in the internalization of those endothelial receptors and thus in the limitation of ET-1-dependent signaling in the endothelium (Oh, Horner, Witkiewicz, & Schnitzer, 2012). Indeed, caveolin-1 is suggested to inhibit ET-1-mediated endothelial nitric oxide synthase (eNOS) activation in ECs (Kwok, Lee, Culberson, Korneszczuk, & Clemens, 2009).

3. ENDOTHELIN-1 RECEPTORS

Endothelins signal via two rhodopsin class G protein-coupled receptors, ET_A and ET_B (Arai, Hori, Aramori, Ohkubo, & Nakanishi, 1990; Sakurai et al., 1990) (Fig. 2). The latter binds all endothelins (and sarafotoxins) with equal affinity. In fact, the linear end of endothelin is all that is required for ET_B binding and activation, as linearized ET-1 analogs $Ala^{1,3,11,15}$-ET-1, ET-1 (16–21), BQ-3020 (Ac-[$Ala^{11,15}$]-ET-1 (6-21)), and IRL-1620 (Suc-[Glu^9, $Ala^{11,15}$]-endothelin-1 (8–21)) can activate it (Ihara et al., 1992; Maggi et al., 1989; Saeki, Ihara, Fukuroda, Yamagiwa, & Yano, 1991; Takai et al., 1992). However, none of these peptides bind ET_A, which is selective for ET-1, ET-2, and some sarafotoxins (Rubanyi & Polokoff, 1994).

In silico analysis shows that the C-terminal tail of endothelins binds deeply in the pocket of the receptor. ET_A activation further requires that the N-terminal of ET-1 and ET-2 binds its extracellular tail, which then opens and directs the C-terminal end of ET-1 in the activation site of the receptor (Lattig, Oksche, Beyermann, Rosenthal, & Krause, 2009). These considerations may explain why no selective agonists of ET_A have been discovered (Watts, 2010).

3.1 Calcium-Dependent Signaling

3.1.1 Canonical Pathway

Both ET_A and ET_B receptors share the $G_{q/11}$ signaling pathway, eliciting Ca^{2+} release into the cytosol. The increase of cytosolic Ca^{2+} is biphasic, with an initial release from intracellular stores followed by an entry from the extracellular fluid. After receptor binding, PLC is activated, processing phosphatidylinositol 4,5-bisphosphate (PIP_2) into diacylglycerol (DAG) and inositol 3,4,5-trisphosphate (IP_3) (Rubanyi & Polokoff, 1994).

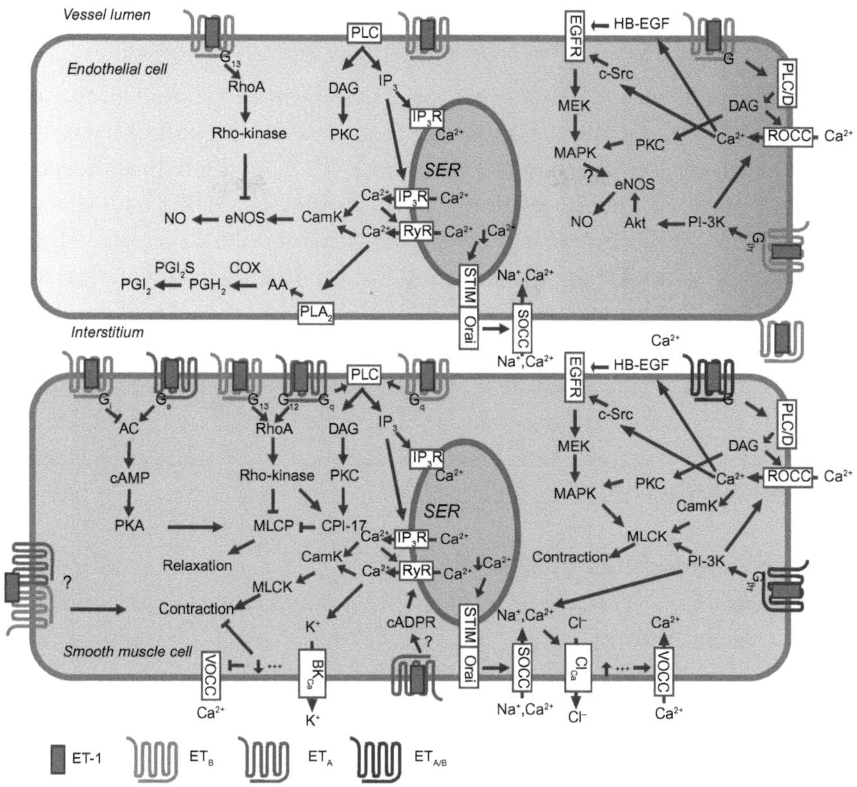

Fig. 2 ET-1 signaling in the vasculature. In the endothelial cell the ET_B receptor is coupled to G_{13}, which activates the Rho-kinase pathway, leading to inhibition of eNOS. In contrast, activation of G_q leads to the activation of PLC, which processes PIP_2 into DAG and IP_3. IP_3 binds its receptor on the sarco/endoplasmic reticulum, which then opens for Ca^{2+} release. This Ca^{2+} signal can be amplified by the ryanodine receptor, which opens upon detection of cytosolic Ca^{2+}. Calcium can also come from the extracellular space, as the emptying of SER stores activates store-operated cation channels, and DAG can activate receptor-operated cation channels, both types of channels permitting Ca^{2+} influx. The cytosolic Ca^{2+} then binds calmodulin, to activate the calmodulin kinase (CamK) to phosphorylate eNOS and initiate NO synthesis. eNOS can also be activated by phosphorylation from Akt (which ET_B activates through a $G_{\beta\gamma}$-mediated activation of PI-3K). ET_B, through EGFR transactivation and DAG-activated PKC, can also activate MAPK. Though the latter can induce eNOS expression, ET_B-mediated eNOS activation through MAPK remains to be confirmed. PI-3K can also mediate ROCC. Intracellular Ca^{2+} also activates PLA_2, leading to arachidonic acid formation, a substrate that COX molds into an endoperoxide (PGH_2), which is then processed into PGI_2 by PGI synthase. Similar signaling pathways are present in the smooth muscle cell, where CamK, MAPK, and PI-3K activate MLCK to induce contraction. Rho-kinase activation (through G_{12} for ET_A) leads to inhibition of MLCP directly and through the activation of CPI-17 (the latter action also mediated by PKC). In a relaxing pathway, ET_A can activate G_s, to initiate cAMP formation and PKA activation, which activate MLCP (through G_i, ET_B inhibits this PKA activation). Activation of ET_A can also activate RyR through cADPR. The influx of Ca^{2+} from the extracellular space is accompanied by Na^+, which, along with Ca^{2+}-activated Cl^- efflux, depolarizes the membrane to open voltage-operated Ca^{2+} channels to help sustain Ca^{2+} influx. In a retrocontrol fashion, Ca^{2+} can also activate K^+ channels to hyperpolarize the membrane and inhibit contraction and VOCC opening. Also illustrated is the yet to be fully validated ET_A/ET_B heterodimerization.

IP_3 then activates its IP_3R receptor channels on the sarco/endoplasmic reticulum (SER) membrane, releasing Ca^{2+} in the cytosol. Meanwhile, DAG is a potent activator of protein kinase C (PKC), which phosphorylates a wide variety of targets including myosin light-chain kinase (MLCK). MLCK is also activated by Ca^{2+}-dependent calmodulin kinase II. MLCK can then phosphorylate the myosin light chain, initiating its interaction with actin and myocyte contraction. In ECs, calmodulin kinase II, activated by the increase in cytosolic Ca^{2+}, and PKC activate the eNOS, phospholipase A_2 (PLA_2), and cyclooxygenase-2 (COX-2), leading to NO and prostacyclin PGI_2 generation. Additionally, the increase in Ca^{2+} may mediate the release of the various endothelium-derived hyperpolarizing factors (Mazzuca & Khalil, 2012).

3.1.2 Calcium-Induced Calcium Release

IP_3-mediated Ca^{2+} from the SER does not account for all Ca^{2+} mobilization from intracellular stores. ET-1 can also activate Ca^{2+}-induced-Ca^{2+}-release from the SER through the ryanodine receptor (RyR), which, after Ca^{2+} binding, opens its pore to release more Ca^{2+} in the cytosolic space. ET-1 can also activate the adenosine diphosphate ribosyl (ADPR) cyclase to form cyclic ADPR, which sensitizes RyR to Ca^{2+} through displacement of FK506 (tacrolimus) binding protein. This contributes to ET-1-mediated reduction of renal blood flow in the rat (Thai & Arendshorst, 2008; Wang et al., 2004).

3.1.3 Calcium Entry

Extracellular Ca^{2+} influx is crucial for ET-1-dependent signaling, especially for its long duration of action and sustained cytosolic Ca^{2+} increasing properties (Rubanyi & Polokoff, 1994). In fact, at low (10^{-10} M) concentration, ET-1 induces arterial constriction and sustained increase of free intracellular Ca^{2+} exclusively from the extracellular medium, without the initial transient peak (Enoki et al., 1995; Zhang et al., 1999).

3.1.3.1 Receptor- and Store-Operated Calcium Channels

A large part of ET-1-mediated Ca^{2+} influx and arterial contraction can be blocked by a broad specter nonvoltage-operated Ca^{2+} channel (non-VOCC) inhibitor, SK&F 96365, which also affects basal vascular tone (Mamo, Angus, Ziogas, Soeding, & Wright, 2014; Xu, Elimban, & Dhalla, 2015). Non-VOCC includes store- and receptor-operated cation channels (SOCC and ROCC). Low-concentration ET-1 seem to elicit

Ca^{2+} influx and contraction through ROCC (Kawanabe, Hashimoto, & Masaki, 2002a; Zhang et al., 1999) in the rat VSMC, while the SOCC component's fraction of Ca^{2+} influx increases with the ET-1 concentration, coherently with the initiation of intracellular Ca^{2+} mobilization (Miwa, Kawanabe, Okamoto, & Masaki, 2005).

Calcium entry in response to ET-1 depends on transient receptor potential channels of the canonical subfamily (TRPCs) (Earley & Brayden, 2015; Horinouchi, Terada, Higashi, & Miwa, 2013). ET-1 can activate TRPCs through multiple pathways, mostly requiring PLC products (IP_3 and/or DAG), although PLC inhibition may not completely inhibit TRPC activation following ET-1 (Peppiatt-Wildman, Albert, Saleh, & Large, 2007), as the latter peptide can also induce DAG synthesis through phospholipase D (Liu, Geisbuhler, & Jones, 1992). DAG is a direct activator of TRPC3, C6, and C7 (Hofmann et al., 1999; Lemonnier, Trebak, & Putney, 2008) and can also induce PKC-dependent inhibition of TRPC3, C4, and C5 (Venkatachalam, Zheng, & Gill, 2003). In rabbit coronary myocytes, the activation of TRPC1 by ET-1 depends on phosphatidylinositol-3-kinase (PI-3K; for ET_A) or PKC (for ET_B) (Saleh, Albert, & Large, 2009; Shi, Ju, Large, & Albert, 2012). In cerebral resistance arterial myocytes, ET-1 can activate TRPC3 in an IP_3-dependent, Ca^{2+}-independent manner. In this model, the IP_3–IP_3R binding causes the receptor to interact directly with TRPC3, causing its activation, Ca^{2+} influx, and cell constriction (Adebiyi et al., 2010). TRPC1 may also be involved in ET-1-mediated contraction in rat caudal and basilar arteries (Bergdahl et al., 2003). Taken together, these results suggest a wide variety of TRPC responses to ET-1, depending on concentration and tissue studied. ECs also express a large number of TRPCs (Earley & Brayden, 2015), but their role in the actions of ET-1 remains unexplored.

3.1.3.2 Voltage-Dependent Calcium Entry

TRPC activation depolarizes the membrane through Ca^{2+} and Na^+ entry. The entry of Ca^{2+} can also activate Cl^- efflux through the Ca^{2+}-activated Cl^- channel to depolarize the membrane further and contribute to ET-1-mediated rabbit basilar artery (Dai & Zhang, 2001) and rat retinal arteriole (McGahon, Needham, Scholfield, McGeown, & Curtis, 2009) constriction. In small arteries, ET-1 can also inhibit the inward rectifier K^+ (Kir2) channels through PKC (Park, Han, & Earm, 2008). In contrast, ET-1-induced Ca^{2+} release can also impair depolarization in rat aortic smooth muscle cells through action of Ca^{2+}-dependent potassium channel

(BK_{Ca})–TRPC1 interaction, stimulating K^+ efflux (Kwan et al., 2009). The membrane voltage resulting from the activity of those channels may then affect ET-1-mediated voltage-operated Ca^{2+} entry.

ET-1-mediated L-type VOCC activation varies widely between models (Rubanyi & Polokoff, 1994). Recent reports also give conflicting data. For instance, nifedipine inhibited ET-1-mediated constriction in murine and rat coronary, but not pulmonary, arteries (Kato et al., 2013; Yang et al., 2016). Pretreatment in rat cerebral arteries did not change the potency of ET-1, but nifedipine terminated the sustained response (Mamo et al., 2014). Coupled with findings that the ET_A receptor can inhibit L-type channels in rabbit cerebral cortex arterioles (Guibert & Beech, 1999), it is difficult to clearly establish a role for L-type channels in the actions of ET-1. R-type and T-type channels on the smooth muscle may also play a role in ET-1 signaling (Benchekroun, Gros-Louis, Bkaily, & D'Orleans-Juste, 1995; Mamo et al., 2014).

3.1.4 Protein Tyrosine Kinase Pathway

ET-1 acts as a mitogen on smooth muscle cells through MAPK and transcription factors c-fos and c-myc activity (Bobik, Grooms, Millar, Mitchell, & Grinpukel, 1990; Fujitani, Ninomiya, Okada, Urade, & Masaki, 1995). MAPK, Src, PI-3K, and Janus tyrosine kinase (Jak) inhibitors can also impair ET-1-mediated, and relax ET-1-induced, constriction in the rabbit basilar artery (Zubkov, Rollins, Parent, Zhang, & Bryan, 2000). ET-1 activates the MAPK pathway through receptor tyrosine kinase mechanisms, such as the activation of the epidermal growth factor receptor (EGFR; Daub, Weiss, Wallasch, & Ullrich, 1996). This transactivation, dependent upon voltage-independent Ca^{2+} channels at lower ET-1 concentrations (1–10 nM) (Kawanabe, Hashimoto, & Masaki, 2002b), involves the metalloproteinase-dependent shedding of heparin-binding EGF (HB-EGF) from the cell membrane, which can then activate EGFR (Prenzel et al., 1999).

In the mouse model, EGFR inhibition, HB-EGF deletion, or PI-3K inhibition blunted ET-1-mediated isolated arterial constriction and in vivo systolic arterial pressure increase (Chansel et al., 2006; Flamant et al., 2003). VSMCs from humans treated with an HB-EGF inhibitor and from HB-EGF KO mice, weak EGFR mice, or WT mice treated with a metalloproteinase inhibitor had reduced BQ-123-sensitive Ca^{2+} currents after ET-1 (100 nM) stimulation (Chansel et al., 2006). The ET_B receptor can also induce EGFR transactivation in A10 rat embryonic VSMCs (Gomez Sandoval, Levesque, Li, & Anand-Srivastava, 2013).

In a mouse brain microvessel EC line, ET-1-induced expression of COX-2 through ET_B-mediated, Ca^{2+} and c-Src dependent, EGFR transactivation, leading to activation of PI-3K and MAPKs (ERK, JUN, p38) (Hsieh, Lin, Chan, Yang, & Yang, 2012). While MAPK can activate eNOS expression (Li, Zhao, Li, Lerea, & Olson, 2007), it does not seem to mediate ET_B-induced NO production, at least in renal medullar collecting duct cells (Hyndman, MacDonell, & Pollock, 2012). The activation of PI-3K, on the other hand, activates Akt, which then activates eNOS to induce NO synthesis (Liu, Premont, Kontos, Huang, & Rockey, 2003). Thus, EGFR transactivation may play a role in ET-1-dependent vasorelaxation.

3.1.5 Oxidative Stress Signaling

ET-1, through both ET_A and ET_B, can also induce oxidative stress signaling, being able to activate the nicotinamide adenine dinucleotide phosphate (NADPH) oxidase through a Ca^{2+}-dependent pathway (Montezano et al., 2010) to produce superoxide $O_2^{\cdot-}$ in ECs (Duerrschmidt, Wippich, Goettsch, Broemme, & Morawietz, 2000), VSMCs (Wedgwood, McMullan, Bekker, Fineman, & Black, 2001), and isolated rat coronary arteries (Li et al., 2003). In mice, ET-1-induced vascular PGI_2 release is impaired in iNOS knockout animals, implicating NO-independent oxidative stress signaling (Carrier, Brochu, de Brum-Fernandes, & D'Orleans-Juste, 2007). In rat arteries, NADPH inhibition impairs ET-1-mediated vasoconstriction (Loomis, Sullivan, Osmond, Pollock, & Pollock, 2005; Meyer, Barton, & Prossnitz, 2014), maybe through impaired Ca^{2+} signaling (Fellner & Arendshorst, 2007). ET-1 can also induce $O_2^{\cdot-}$ production in rat vena cava, but in normal rats, this does not seem to play a role in venoconstriction (Li et al., 2003; Thakali, Demel, Fink, & Watts, 2005). Oxidative stress signaling may also be involved in ET-1-induced EGFR transactivation and VSMC proliferation (Li, Levesque, & Anand-Srivastava, 2010).

3.2 Calcium-Independent Signaling

3.2.1 Rho-Kinase Pathway

ET_A and ET_B receptors can also activate the G_{12}/G_{13} proteins. The G_α subunit can then interact with a Rho guanine nucleotide exchange factor, in this case the leukemia-associated Rho guanine nucleotide exchange factor (LARG), activating in a cascade the small G protein RhoA to induce Rho-kinase (ROCK) phosphorylation of MLCP, inhibiting its action

(Wirth et al., 2008). ROCK can also activate CPI-17, another inhibitor of MLCP activity (Wynne, Chiao, & Webb, 2009). Inhibition of this G_{12}/G_{13} signaling pathway inhibits ET-1-induced smooth muscle cell contraction and vasoconstriction (Gohla, Schultz, & Offermanns, 2000; Wirth et al., 2008). Age-related, ET_A-mediated hypertension in 1-year-old mice involves both the $G_{q/11}$ and G_{12}/G_{13}-LARG pathways, demonstrating that Ca^{2+} sensitization and stress fiber formation are important in ET-1 signaling (Wirth et al., 2016). Rho-kinase inhibition also blocks ET_B-mediated vasoconstriction in the renal microcirculation (Cavarape et al., 2003). In ECs, Rho-kinase activation limits ET_B-mediated NO synthesis, as it inhibits eNOS activation and expression (Rikitake & Liao, 2005). Thus, activation of the rho-kinase pathway by ET-1, in both endothelial and smooth muscle cells, contributes to vasoconstriction.

3.2.2 Adenylyl Cyclase Activity

ET-1 can influence cyclic adenosine monophosphate (cAMP) levels through its receptors, with the VSMC ET_A receptor activating adenylyl cyclase (AC) through G_s and the endothelial ET_B receptor inhibiting AC through G_i (Eguchi, Hirata, Imai, & Marumo, 1993). ET_B can also inhibit AC in ECs through PLC-dependent mechanisms (Ladoux & Frelin, 1991). ET_B activation with S6c induces cell constriction but no change in cAMP levels (El-Mowafy & White, 1998). G_i activation is required before transactivation of EGFR and MAPK sustained signaling by ET_B leading to expression of contractile proteins in VSMCs (Grantcharova et al., 2006). Thus, the AC–cAMP pathway may be a key factor in the differences between ET_A and ET_B signaling.

ET-1 can also activate the downstream target of cAMP, protein kinase A (PKA), through a cAMP-independent manner (Dulin, Niu, Browning, Ye, & Voyno-Yasenetskaya, 2001) or through a complex pathway involving $G_{\beta\gamma\text{-}i}$, MAPK, and PGI_2-dependent G_s activation, which may play a role in VSMC hypertrophic mechanisms (Taurin, Hogarth, Sandbo, Yau, & Dulin, 2007). VSMC activation of AC by ET_A, a relaxing mechanism, possibly indicates a retrocontrol mechanism. The arterial VSMC ET_B receptor may help in limiting the amount of relaxing cAMP production, thus enhancing vascular constriction (Adner, Shankley, & Edvinsson, 2001).

3.3 Endothelin Receptor Dimerization

Similar to other GPCRs, ET-1 receptors can homo- and heterodimerize. In the rat pituitary gland, ET-1 may act as a bivalent ligand, with its

N-terminal loop binding ET_A and its C-terminal end binding and activating ET_B (Harada, Himeno, Shigematsu, Sumikawa, & Niwa, 2002; Himeno, Shigematsu, Taguchi, & Niwa, 1998). In HEK 293 cells, Gregan and colleagues showed that both receptors heterodimerize in a ligand-independent manner but did not observe bivalent binding. Heterodimer dissociation occurred in the endocytic pathway after internalization. The interaction kept the ET_B receptor at the membrane after ET-1 binding longer than if it was expressed alone, and prevented its desensitization (Dai & Galligan, 2006; Gregan et al., 2004). Dimerization affects Ca^{2+} mobilization, as ET_A homodimers elicited a transient Ca^{2+} release from intracellular stores, while ET_B homodimers depended on extracellular Ca^{2+} influx. In cells expressing heterodimers, the increase in cytosolic Ca^{2+} was not transient but sustained and involved intracellular Ca^{2+} stores, and it required dual ET_A/ET_B receptor blockade for inhibition (Evans & Walker, 2008a, 2008b). In a cell-free environment, ET_B homodimerization and ET-1 binding require its first transmembrane segment (Klammt et al., 2007). In CHO cells, a C-terminal peptide from ET_B potentiated ET_A activity upon ET-1 binding, but inhibited ET_B activity. The reverse was observed with an ET_A C-terminal peptide, indicating that homodimerization may inhibit receptor function, but heterodimerization may potentiate it (Yatawara, Wilson, Taylor, Polgar, & Mierke, 2013). Thus, the interaction between ET receptors seems to occur through their C-terminal domains.

Whether this interaction actually occurs in native tissues, with non-overexpressed endogenous receptors, remains to be fully validated. Sauvageau, Thorin, Caron, and Dupuis (2006) observed coimmunoprecipitation of the two receptors in rat small pulmonary arteries, and dual ET receptor antagonism was required to inhibit ET-1-mediated constriction. Using tagged ligands rather than tagged/overexpressed receptors, another report found heterodimerization in rat mesenteric arteries (Kapsokalyvas et al., 2014). To our knowledge, those are the only two reports with endogenously expressed receptors, and more studies are required to determine the importance of dimerization of ET receptors in vivo.

3.4 Nuclear Envelope Receptors

Both the ET_A and ET_B receptors have been located on the nuclear membrane of VSMCs and, notably for the ET_A, on the perinuclear envelope of ECs (Bkaily et al., 2011). They can elicit Ca^{2+} signaling in the nucleus, possibly through cytosolic and nuclear IP_3-dependent mechanisms or R-type VOCC (Bkaily et al., 2015; Merlen et al., 2013). ET-1, either extracellular or injected in the cytosol, can also induce nuclear oxidative stress (Provost et al., 2010). How ET-1 can access its nuclear membrane receptors is unclear. Recent data in cardiomyocytes suggest that nuclear membrane ET_B is not derived from the plasma membrane and is not affected by extracellular ET-1. Rather, it is part of an intracrine signaling system, with ET-1, through ECE-1, coming from intracellular pools (Branco & Allen, 2015; Merlen et al., 2013). The physiopathological role of ET-1 signaling on the nuclear membrane also remains to be determined, particularly in in vivo settings.

4. VASCULAR REACTIVITY TO ET-1

4.1 Arteries

The main explanation for the differences in effects on vascular tone due to activation of both ET receptors resides in their different tissue expressions. Both receptors are present on the arterial vascular smooth muscle, where the ET_A receptor is predominant. This receptor is crucially involved in arterial constriction by ET-1, both in vitro and in vivo (Rubanyi & Polokoff, 1994). What is less clear is the involvement of the ET_B receptor, which is also present on the vascular endothelium, where it is documented to play a clearance role. It may also contribute to arterial relaxation through EDRFs. In vivo, ET_B-selective agonists in rats induce a transient decrease in blood pressure, followed by sustained vasoconstriction (Clozel, Gray, Breu, Loffler, & Osterwalder, 1992). Direct ET_B antagonism or partial repression of that receptor type, probably through impaired ET-1 clearance, increases levels of ET-1, and increases blood pressure which is blocked by an ET_A antagonist (Berthiaume, Yanagisawa, Labonte, & D'Orleans-Juste, 2000; Gratton et al., 1997; Okada & Nishikibe, 2002). Endothelial and VSMC ET_B receptors can compete with one another, leading some investigators to postulate that they are two different receptors, ET_{B1} and ET_{B2} (Warner, Allcock, Corder, & Vane, 1993). However, no evidence have emerged that endothelial and smooth muscle ET_B receptors are distinct except in tissue localization (Davenport, 2002).

In multiple settings, selective blockade of ET_A is sufficient to prevent all ET-1-induced smooth muscle contraction, and the selective ET_B agonist S6c is inactive. There are, on the other hand, a wide variety of other settings in which the ET_B receptor is involved in arterial vasoconstriction (Mazzuca & Khalil, 2012). But the inefficacy of ET_B antagonists to block

ET-1-mediated constriction may mask physiological roles of the ET_B receptor. For example, in rat mesenteric arteries, the constriction-induced ET-1 was resistant to ET_B antagonism (through BQ-788 or ET_B desensitization). However, the ET_B-selective agonist S6c triggers a mild constriction in the same model. Furthermore, low concentrations of ET-1 (up to 0.1 nM) elicited BQ-123-resistant, BQ-788 sensitive constriction (Mickley, Gray, & Webb, 1997). As mentioned earlier, ET_B is linked to G_i. In the same model, pretreatment with a sequence of U46619 (thromboxane A_2 analog) to constrict and then forskolin (AC activator) to relax the artery potentiated S6c response dramatically. These results indicate that basal AC activity may influence the efficacy of the ET_B receptor to induce smooth muscle contraction. The selective ET_B endogenous agonist ET-3 did not show the same profile as S6c and did not require this pretreatment. ET_B desensitization potentiated ET-1, but not S6b, constriction of the endothelium-denuded assay vessels (Adner et al., 2001). Such a phenomenon was observed in rabbit mesenteric arteries, where ET-1 (with BQ-123), ET-3, S6c, and IRL-1620, in a BQ-788 (but not BQ-123)-sensitive manner, relaxed the endothelium-denuded artery in a cyclooxygenase-dependent manner (Iwasaki et al., 1999).

The putative cross talk between ET_A and ET_B in the vascular smooth muscle has been reviewed in detailed recently (Rapoport & Zuccarello, 2012). The discrepancy in ET_B-mediated signaling depends on the animal model, tissue, and concentrations of agonists and antagonists. Basal myogenic tone also plays a role. Using an in situ cranial window to observe a resistance artery (the basilar artery) in vivo in the anesthetized rat, Yoon and colleagues found that S6c, and ET-1 after BQ-123 pretreatment, induced relaxation, consistent with endothelium-dependent vasodilation. ET-1 alone produced sustained vasoconstriction, which ET_A-selective antagonists could relax only partly. An ET_B antagonist could not relax the ET-1-constricted artery, but dual ET_A/ET_B antagonism induced full relaxation to basal levels (Yoon, Zuccarello, & Rapoport, 2012). The same group also recently reported that, in a BQ-788-sensitive manner, S6c could relax (albeit less efficiently) the basilar artery of rescued spotting lethal (*sl*) rats, which do not have a full-length ET_B receptor except under control of the dopamine-β-hydroxylase promoter in the enteric nervous system. The reported cross talk in the vena cava (Thakali, Galligan, Fink, Gariepy, & Watts, 2008) does not occur in these rats. But Yoon, Gariepy, Yanagisawa, Zuccarello, and Rapoport (2016) observed cross talk with the mutant ET_B receptor just as they observed in $ET_B^{+/+}$ animals, although with a lesser potency. Thus, ET-1 binding to ET_B may not be required for ET_A/ET_B cross talk.

4.2 Veins

The endothelin system is also present in the venous vasculature, where its vasoconstrictor activity may be even more potent than in arterial systems (Cocks, Faulkner, Sudhir, & Angus, 1989; D'Orleans-Juste, Finet, de Nucci, & Vane, 1989). ET-1-mediated vasoconstriction in rabbit veins is mediated in large part by the ET_B receptor, with S6c being equipotent to ET-1 to constrict rabbit saphenous vein (Eguchi et al., 1997). Furthermore, ET-1 and S6c constrict the nonchallenged vena cava, but relax it if it was preconstricted with $PGF_{2\alpha}$. Surprisingly, partial relaxation to S6c, but not acetylcholine or ET-1, remained in the vena cava, but not the aorta, after endothelial denudation or NOS inhibition, illustrating that nonendothelial ET_B is involved in venous relaxation, but that the endogenous action from ET-1 is more complex (Tykocki, Gariepy, & Watts, 2009). The same study reported that ET-1 relaxes the $PGF_{2\alpha}$-preconstricted vena cava, but not the aorta (in which relaxation to ET-1 was ET_A antagonist-dependent), up to a certain concentration, after which vasoconstriction was observed.

Similar results were observed in the endothelium-denuded rat mesenteric artery and vein strips. In those vessels, ET-1-mediated constriction is ET_A dependent, and IRL-1620 and S6c are inactive. In contrast, the vein constricts to ET_B-selective agonists (Claing, Shbaklo, Plante, Bkaily, & D'Orleans-Juste, 2002). In addition, both BQ-123 and BQ-788 inhibited the ET-1-induced constriction and intracellular Ca^{2+} increase in veins in overlapping fashion (around 75% inhibition each). A mixed ET_A/ET_B antagonist was unable to reduce the ET-1 response more than the selective agonists. However, a combination of BQ-123 and BQ-788 further inhibited vein constriction (but not free Ca^{2+} increase). Those results also suggest ET_A/ET_B receptor interaction (Claing et al., 2002).

ET_A antagonists are able to block the ET-1-mediated venous constriction only after pretreatment with ET_B receptor ligands (Lodge, Zhang, Halaka, & Moreland, 1995). Desensitization of ET_B receptors was reported in arteries (LaDouceur, Flynn, Keiser, Reynolds, & Haleen, 1993), but S6c-mediated desensitization requires long exposure in veins (Thakali, Fink, & Watts, 2004; Thakali et al., 2008). Intriguingly, the chymase inhibitor chymostatin reduces the constrictor effect of big ET-1 in rat aortic, but not venous, preparations (Watts et al., 2007), reflecting differences in ET-1 processing as well.

4.3 Lymphatic System

In addition to blood vessels, ET-1 contracts lymphatic vessels in vivo in an acetylcholine-, sodium nitroprusside-, and isoproterenol-resistant fashion.

The same study reported those vessels resistant to norepinephrine-induced vasoconstriction, and in which ET-1 was more potent than vasopressin (Fortes, Scivoletto, & Garcia-Leme, 1989). As seen in many blood vessel arteries, lymphatic ECs express ET-1 and the peptide's contractile activity on lymphatic vessels is primarily mediated by its ET_A receptor. Intriguingly, endothelial ET_B activation contributed to vessel constriction, instead of inhibiting it, as BQ-788 provided a small inhibitory effect that was lost with endothelial denudation, whereas NOS inhibition did not affect ET-1-mediated contraction when the endothelium was intact (Reeder & Ferguson, 1996). This is in contrast to a report in vitro in bovine mesenteric lymph vessels, where the endothelial ET_B receptor caused EDRF-dependent vasodilatation and the constriction was only ET_A dependent (Sakai, Ikomi, & Ohhashi, 1999). Those results were also observed in guinea pig mesenteric lymph vessels, where it is also dependent upon Ca^{2+} mobilization (Zhao & van Helden, 2003). Similar to blood vessel ECs, ET-1 can cause lymphatic ECs to invade and migrate, through the ET_B receptor and hypoxia-inducible factor-1 activation, inducing lymphangiogenesis (Spinella et al., 2010, 2009). As such, the lymphatic ET-1–ET_B system might play a role in the development of tumors (Spinella et al., 2014).

5. CONCLUSION

ET-1 gene expression is determinant in its synthesis, but the proteolytic pathways involved in its maturation can offer useful targets to inhibit the endothelin system in disease. As such, the knowledge of the contributions of the different ECE isoforms, chymase, and perhaps MMP-2 needs to be refined, especially since chymase inhibition is seen as a promising strategy in clinical trials for heart failure (such as the CHIARA MIA phase 2 pilot trial, NCT02452515), it is important to consider its effect on the endothelin system in humans.

Knowledge on ET-1 signaling pathways also needs to be refined, by cautious design of precise experimental settings such as the concentration/dose ranges of the ET receptor agonists and antagonists. Multiple Ca^{2+} mobilization mechanisms and other signaling pathways involved in ET-1-induced responses are presented in this review in an attempt to build a comprehensive picture; their roles in physiological settings are however not well documented. There are also pharmacological discrepancies in the literature as ET-1 concentrations used in various biological assays do vary widely (from 0.1 to 200 nM, a 2000-fold range). The overarching contribution of ET_B receptor activation in vascular health or in disease conditions also warrants further investigation. Better understanding of the latter parameters, as well as of the contribution of ET-1 in venous reactivity, arterial pressure, cardiac function and capillary hydrostatic pressures, may contribute in the optimal use of either ET_A-selective or dual ET_A/ET_B antagonists in various clinical settings.

CONFLICT OF INTEREST

The authors have no conflict of interest to declare.

ACKNOWLEDGMENTS

This project was financially supported by the Canadian Institutes for Health Research (MOP-57883) and the Fondation Leducq Grant 13 CVD 03, as well as by the Etienne Lebel Clinical Research Center of the Centre Hospitalier Universitaire de Sherbrooke.

REFERENCES

Adebiyi, A., Zhao, G., Narayanan, D., Thomas-Gatewood, C. M., Bannister, J. P., & Jaggar, J. H. (2010). Isoform-selective physical coupling of TRPC3 channels to IP3 receptors in smooth muscle cells regulates arterial contractility. *Circulation Research*, *106*(10), 1603–1612.

Adner, M., Shankley, N., & Edvinsson, L. (2001). Evidence that ET-1, but not ET-3 and S6b, ET(A)-receptor mediated contractions in isolated rat mesenteric arteries are modulated by co-activation of ET(B) receptors. *British Journal of Pharmacology*, *133*(6), 927–935.

Arai, H., Hori, S., Aramori, I., Ohkubo, H., & Nakanishi, S. (1990). Cloning and expression of a cDNA encoding an endothelin receptor. *Nature*, *348*(6303), 730–732.

Arinami, T., Ishikawa, M., Inoue, A., Yanagisawa, M., Masaki, T., Yoshida, M. C., et al. (1991). Chromosomal assignments of the human endothelin family genes: The endothelin-1 gene (EDN1) to 6p23-p24, the endothelin-2 gene (EDN2) to 1p34, and the endothelin-3 gene (EDN3) to 20q13.2-q13.3. *American Journal of Human Genetics*, *48*(5), 990–996.

Benchekroun, M. T., Gros-Louis, N., Bkaily, G., & D'Orleans-Juste, P. (1995). R-type calcium channel involved in endothelin-1-induced contraction of rabbit aorta. *Journal of Cardiovascular Pharmacology*, *26*(Suppl. 3), S300–S302.

Bergdahl, A., Gomez, M. F., Dreja, K., Xu, S. Z., Adner, M., Beech, D. J., et al. (2003). Cholesterol depletion impairs vascular reactivity to endothelin-1 by reducing store-operated Ca2+ entry dependent on TRPC1. *Circulation Research*, *93*(9), 839–847.

Berthiaume, N., Yanagisawa, M., Labonte, J., & D'Orleans-Juste, P. (2000). Heterozygous knock-out of ETB receptors induces BQ-123-sensitive hypertension in the mouse. *Hypertension*, *36*(6), 1002–1007.

Bkaily, G., Avedanian, L., Al-Khoury, J., Chamoun, M., Semaan, R., Jubinville-Leblanc, C., et al. (2015). Nuclear membrane R-type calcium channels mediate cytosolic ET-1-induced increase of nuclear calcium in human vascular smooth muscle cells. *Canadian Journal of Physiology and Pharmacology*, *93*(4), 291–297.

Bkaily, G., Avedanian, L., Al-Khoury, J., Provost, C., Nader, M., D'Orleans-Juste, P., et al. (2011). Nuclear membrane receptors for ET-1 in cardiovascular function. *American Journal of Physiology Regulatory, Integrative and Comparative Physiology*, *300*(2), R251–R263.

Bobik, A., Grooms, A., Millar, J. A., Mitchell, A., & Grinpukel, S. (1990). Growth factor activity of endothelin on vascular smooth muscle. *American Journal of Physiology, 258*(3 Pt. 1), C408–C415.

Branco, A. F., & Allen, B. G. (2015). G protein-coupled receptor signaling in cardiac nuclear membranes. *Journal of Cardiovascular Pharmacology, 65*(2), 101–109.

Bremnes, T., Paasche, J. D., Mehlum, A., Sandberg, C., Bremnes, B., & Attramadal, H. (2000). Regulation and intracellular trafficking pathways of the endothelin receptors. *Journal of Biological Chemistry, 275*(23), 17596–17604.

Carrier, E., Brochu, I., de Brum-Fernandes, A. J., & D'Orleans-Juste, P. (2007). The inducible nitric-oxide synthase modulates endothelin-1-dependent release of prostacyclin and inhibition of platelet aggregation ex vivo in the mouse. *Journal of Pharmacology and Experimental Therapeutics, 323*(3), 972–978.

Cavarape, A., Endlich, N., Assaloni, R., Bartoli, E., Steinhausen, M., Parekh, N., et al. (2003). Rho-kinase inhibition blunts renal vasoconstriction induced by distinct signaling pathways in vivo. *Journal of the American Society of Nephrology, 14*(1), 37–45.

Chansel, D., Ciroldi, M., Vandermeersch, S., Jackson, L. F., Gomez, A. M., Henrion, D., et al. (2006). Heparin binding EGF is necessary for vasospastic response to endothelin. *The FASEB Journal, 20*(11), 1936–1938.

Claing, A., Shbaklo, H., Plante, M., Bkaily, G., & D'Orleans-Juste, P. (2002). Comparison of the contractile and calcium-increasing properties of platelet-activating factor and endothelin-1 in the rat mesenteric artery and vein. *British Journal of Pharmacology, 135*(2), 433–443.

Clozel, M., Gray, G. A., Breu, V., Loffler, B. M., & Osterwalder, R. (1992). The endothelin ETB receptor mediates both vasodilation and vasoconstriction in vivo. *Biochemical and Biophysical Research Communications, 186*(2), 867–873.

Cocks, T. M., Faulkner, N. L., Sudhir, K., & Angus, J. (1989). Reactivity of endothelin-1 on human and canine large veins compared with large arteries in vitro. *European Journal of Pharmacology, 171*(1), 17–24.

Dai, X., & Galligan, J. J. (2006). Differential trafficking and desensitization of human ET(A) and ET(B) receptors expressed in HEK 293 cells. *Experimental Biology and Medicine (Maywood, N.J.), 231*(6), 746–751.

Dai, Y., & Zhang, J. H. (2001). Role of Cl^- current in endothelin-1-induced contraction in rabbit basilar artery. *American Journal of Physiology. Heart and Circulatory Physiology, 281*(5), H2159–H2167.

Daub, H., Weiss, F. U., Wallasch, C., & Ullrich, A. (1996). Role of transactivation of the EGF receptor in signalling by G-protein-coupled receptors. *Nature, 379*(6565), 557–560.

Davenport, A. P. (2002). International union of pharmacology. XXIX. Update on endothelin receptor nomenclature. *Pharmacological Reviews, 54*(2), 219–226.

Davenport, A. P., Hyndman, K. A., Dhaun, N., Southan, C., Kohan, D. E., Pollock, J. S., et al. (2016). Endothelin. *Pharmacological Reviews, 68*(2), 357–418.

Davenport, A. P., & Maguire, J. J. (2006). Endothelin. *Handbook of Experimental Pharmacology, 176*(Pt. 1), 295–329.

Denault, J. B., Claing, A., D'Orleans-Juste, P., Sawamura, T., Kido, T., Masaki, T., et al. (1995). Processing of proendothelin-1 by human furin convertase. *FEBS Letters, 362*(3), 276–280.

Dhaun, N., Yuzugulen, J., Kimmitt, R. A., Wood, E. G., Chariyavilaskul, P., MacIntyre, I. M., et al. (2015). Plasma pro-endothelin-1 peptide concentrations rise in chronic kidney disease and following selective endothelin A receptor antagonism. *Journal of the American Heart Association, 4*(3), e001624.

D'Orleans-Juste, P., Finet, M., de Nucci, G., & Vane, J. R. (1989). Pharmacology of endothelin-1 in isolated vessels: Effect of nicardipine, methylene blue, hemoglobin, and gossypol. *Journal of Cardiovascular Pharmacology, 13*(Suppl. 5), S19–S22. discussion S45.

Duerrschmidt, N., Wippich, N., Goettsch, W., Broemme, H. J., & Morawietz, H. (2000). Endothelin-1 induces NAD(P)H oxidase in human endothelial cells. *Biochemical and Biophysical Research Communications*, *269*(3), 713–717.

Dulin, N. O., Niu, J., Browning, D. D., Ye, R. D., & Voyno-Yasenetskaya, T. (2001). Cyclic AMP-independent activation of protein kinase A by vasoactive peptides. *Journal of Biological Chemistry*, *276*(24), 20827–20830.

Earley, S., & Brayden, J. E. (2015). Transient receptor potential channels in the vasculature. *Physiological Reviews*, *95*(2), 645–690.

Eguchi, S., Hirata, Y., Imai, T., & Marumo, F. (1993). Endothelin receptor subtypes are coupled to adenylate cyclase via different guanyl nucleotide-binding proteins in vasculature. *Endocrinology*, *132*(2), 524–529.

Eguchi, D., Nishimura, J., Kobayashi, S., Komori, K., Sugimachi, K., & Kanaide, H. (1997). Down-regulation of endothelin B receptors in autogenous saphenous veins grafted into the arterial circulation. *Cardiovascular Research*, *35*(2), 360–367.

El-Mowafy, A. M., & White, R. E. (1998). Evidence for a tyrosine kinase-dependent activation of the adenylyl cyclase/PKA cascade downstream from the G-protein-linked endothelin ETA receptor in vascular smooth muscle. *Biochemical and Biophysical Research Communications*, *251*(2), 494–500.

Emoto, N., & Yanagisawa, M. (1995). Endothelin-converting enzyme-2 is a membrane-bound, phosphoramidon-sensitive metalloprotease with acidic pH optimum. *Journal of Biological Chemistry*, *270*(25), 15262–15268.

Enoki, T., Miwa, S., Sakamoto, A., Minowa, T., Komuro, T., Kobayashi, S., et al. (1995). Long-lasting activation of cation current by low concentration of endothelin-1 in mouse fibroblasts and smooth muscle cells of rabbit aorta. *British Journal of Pharmacology*, *115*(3), 479–485.

Evans, N. J., & Walker, J. W. (2008a). Endothelin receptor dimers evaluated by FRET, ligand binding, and calcium mobilization. *Biophysical Journal*, *95*(1), 483–492.

Evans, N. J., & Walker, J. W. (2008b). Sustained Ca2+ signaling and delayed internalization associated with endothelin receptor heterodimers linked through a PDZ finger. *Canadian Journal of Physiology and Pharmacology*, *86*(8), 526–535.

Fecteau, M.-H., Honoré, J.-C., Plante, M., Labonté, J., Rae, G. A., & D'Orléans-Juste, P. (2005). Endothelin-1 (1-31) is an intermediate in the production of endothelin-1 after big endothelin-1 administration in vivo. *Hypertension*, *46*(1), 87–92.

Fellner, S. K., & Arendshorst, W. (2007). Endothelin-A and -B receptors, superoxide, and Ca2+ signaling in afferent arterioles. *American Journal of Physiology. Renal Physiology*, *292*(1), F175–F184.

Fernandez-Patron, C., Radomski, M. W., & Davidge, S. T. (1999). Vascular matrix metalloproteinase-2 cleaves big endothelin-1 yielding a novel vasoconstrictor. *Circulation Research*, *85*(10), 906–911.

Fernandez-Patron, C., Zouki, C., Whittal, R., Chan, J. S., Davidge, S. T., & Filep, J. G. (2001). Matrix metalloproteinases regulate neutrophil-endothelial cell adhesion through generation of endothelin-1[1-32]. *The FASEB Journal*, *15*(12), 2230–2240.

Flamant, M., Tharaux, P. L., Placier, S., Henrion, D., Coffman, T., Chatziantoniou, C., et al. (2003). Epidermal growth factor receptor trans-activation mediates the tonic and fibrogenic effects of endothelin in the aortic wall of transgenic mice. *The FASEB Journal*, *17*(2), 327–329.

Fortes, Z. B., Scivoletto, R., & Garcia-Leme, J. (1989). Endothelin-1 induces potent constriction of lymphatic vessels in situ. *European Journal of Pharmacology*, *170*(1–2), 69–73.

Fujitani, Y., Ninomiya, H., Okada, T., Urade, Y., & Masaki, T. (1995). Suppression of endothelin-1-induced mitogenic responses of human aortic smooth muscle cells by interleukin-1 beta. *Journal of Clinical Investigation*, *95*(6), 2474–2482.

Fukuroda, T., Fujikawa, T., Ozaki, S., Ishikawa, K., Yano, M., & Nishikibe, M. (1994). Clearance of circulating endothelin-1 by ETB receptors in rats. *Biochemical and Biophysical Research Communications*, *199*(3), 1461–1465.

Gohla, A., Schultz, G., & Offermanns, S. (2000). Role for G(12)/G(13) in agonist-induced vascular smooth muscle cell contraction. *Circulation Research*, *87*(3), 221–227.

Gomez Sandoval, Y. H., Levesque, L. O., Li, Y., & Anand-Srivastava, M. B. (2013). Role of epidermal growth factor receptor transactivation in endothelin-1-induced enhanced expression of Gi protein and proliferation in A10 vascular smooth muscle cells. *Canadian Journal of Physiology and Pharmacology*, *91*(3), 221–227.

Grantcharova, E., Reusch, H. P., Grossmann, S., Eichhorst, J., Krell, H. W., Beyermann, M., et al. (2006). N-terminal proteolysis of the endothelin B receptor abolishes its ability to induce EGF receptor transactivation and contractile protein expression in vascular smooth muscle cells. *Arteriosclerosis, Thrombosis, and Vascular Biology*, *26*(6), 1288–1296.

Gratton, J. P., Cournoyer, G., Loffler, B. M., Sirois, P., & D'Orleans-Juste, P. (1997). ET (B) receptor and nitric oxide synthase blockade induce BQ-123-sensitive pressor effects in the rabbit. *Hypertension*, *30*(5), 1204–1209.

Gregan, B., Jurgensen, J., Papsdorf, G., Furkert, J., Schaefer, M., Beyermann, M., et al. (2004). Ligand-dependent differences in the internalization of endothelin A and endothelin B receptor heterodimers. *Journal of Biological Chemistry*, *279*(26), 27679–27687.

Guibert, C., & Beech, D. J. (1999). Positive and negative coupling of the endothelin ETA receptor to Ca2+-permeable channels in rabbit cerebral cortex arterioles. *Journal of Physiology*, *514*(Pt. 3), 843–856.

Hanson, G. C., Andersson, K. E., Gyllstedt, E., Högestätt, E. D., & Lindberg, B. F. (1997). Hydrolysis of big endothelin-1 by a serine protease in the membrane fraction of human lung. *Regulatory Peptides*, *68*(1), 63–69.

Harada, N., Himeno, A., Shigematsu, K., Sumikawa, K., & Niwa, M. (2002). Endothelin-1 binding to endothelin receptors in the rat anterior pituitary gland: Possible formation of an ETA-ETB receptor heterodimer. *Cellular and Molecular Neurobiology*, *22*(2), 207–226.

Hayashi, M. A., Ligny-Lemaire, C., Wollberg, Z., Wery, M., Galat, A., Ogawa, T., et al. (2004). Long-sarafotoxins: Characterization of a new family of endothelin-like peptides. *Peptides*, *25*(8), 1243–1251.

Himeno, A., Shigematsu, K., Taguchi, T., & Niwa, M. (1998). Endothelin-1 binding to endothelin receptors in the rat anterior pituitary gland: Interaction in the recognition of endothelin-1 between ETA and ETB receptors. *Cellular and Molecular Neurobiology*, *18*(4), 447–452.

Hofmann, T., Obukhov, A. G., Schaefer, M., Harteneck, C., Gudermann, T., & Schultz, G. (1999). Direct activation of human TRPC6 and TRPC3 channels by diacylglycerol. *Nature*, *397*(6716), 259–263.

Honore, J. C., Plante, M., Bkaily, G., Rae, G. A., & D'Orleans-Juste, P. (2002). Pressor and pulmonary responses to ET-1(1-31) in guinea-pigs. *British Journal of Pharmacology*, *136*(6), 819–828.

Horinouchi, T., Terada, K., Higashi, T., & Miwa, S. (2013). Endothelin receptor signaling: New insight into its regulatory mechanisms. *Journal of Pharmacological Sciences*, *123*(2), 85–101.

Houde, M., Jamain, M. D., Labonte, J., Desbiens, L., Pejler, G., Gurish, M., et al. (2013). Pivotal role of mouse mast cell protease 4 in the conversion and pressor properties of Big-endothelin-1. *Journal of Pharmacology and Experimental Therapeutics*, *346*(1), 31–37.

Hsieh, H. L., Lin, C. C., Chan, H. J., Yang, C. M., & Yang, C. M. (2012). c-Src-dependent EGF receptor transactivation contributes to ET-1-induced COX-2 expression in brain microvascular endothelial cells. *Journal of Neuroinflammation*, *9*, 152.

Hyndman, K. A., MacDonell, A. H., & Pollock, J. S. (2012). Extracellular signal-regulated kinases 1/2 signaling pathways are not involved in endothelin regulation of mouse inner medullary collecting duct nitric oxide production. *Life Sciences*, *91*(13–14), 578–582.

Ihara, M., Saeki, T., Fukuroda, T., Kimura, S., Ozaki, S., Patel, A. C., et al. (1992). A novel radioligand [125I]BQ-3020 selective for endothelin (ETB) receptors. *Life Sciences*, *51*(6), L47–L52.

Ikeda, S., Emoto, N., Alimsardjono, H., Yokoyama, M., & Matsuo, M. (2002). Molecular isolation and characterization of novel four subisoforms of ECE-2. *Biochemical and Biophysical Research Communications*, *293*(1), 421–426.

Inoue, A., Yanagisawa, M., Kimura, S., Kasuya, Y., Miyauchi, T., Goto, K., et al. (1989a). The human endothelin family: Three structurally and pharmacologically distinct isopeptides predicted by three separate genes. *Proceedings of the National Academy of Sciences of the United States of America*, *86*(8), 2863–2867.

Inoue, A., Yanagisawa, M., Takuwa, Y., Mitsui, Y., Kobayashi, M., & Masaki, T. (1989b). The human preproendothelin-1 gene. Complete nucleotide sequence and regulation of expression. *Journal of Biological Chemistry*, *264*(25), 14954–14959.

Ivey, M. E., Osman, N., & Little, P. J. (2008). Endothelin-1 signalling in vascular smooth muscle: Pathways controlling cellular functions associated with atherosclerosis. *Atherosclerosis*, *199*(2), 237–247.

Iwasaki, T., Notoya, M., Hayasaki-Kajiwara, Y., Shimamura, T., Naya, N., Ninomiya, M., et al. (1999). Endothelium-independent vascular relaxation mediating ETB receptor in rabbit mesenteric arteries. *American Journal of Physiology*, *276*(2 Pt. 2), H383–H390.

Jafri, F., & Ergul, A. (2003). Nuclear localization of endothelin-converting enzyme-1: Subisoform specificity. *Arteriosclerosis, Thrombosis, and Vascular Biology*, *23*(12), 2192–2196.

Jeyabalan, A., Novak, J., Danielson, L. A., Kerchner, L. J., Opett, S. L., & Conrad, K. P. (2003). Essential role for vascular gelatinase activity in relaxin-induced renal vasodilation, hyperfiltration, and reduced myogenic reactivity of small arteries. *Circulation Research*, *93*(12), 1249–1257.

Johnstrom, P., Fryer, T. D., Richards, H. K., Maguire, J. J., Clark, J. C., Pickard, J. D., et al. (2010). Positron emission tomography of [18F]-big endothelin-1 reveals renal excretion but tissue-specific conversion to [18F]-endothelin-1 in lung and liver. *British Journal of Pharmacology*, *159*(4), 812–819.

Kapsokalyvas, D., Schiffers, P. M., Maij, N., Suylen, D. P., Hackeng, T. M., van Zandvoort, M. A., et al. (2014). Imaging evidence for endothelin ETA/ETB receptor heterodimers in isolated rat mesenteric resistance arteries. *Life Sciences*, *111*(1–2), 36–41.

Kato, K., Okamura, K., Hatta, M., Morita, H., Kajioka, S., Naito, S., et al. (2013). Involvement of IP3-receptor activation in endothelin-1-induced Ca(2+) influx in rat pulmonary small artery. *European Journal of Pharmacology*, *720*(1–3), 255–263.

Kawanabe, Y., Hashimoto, N., & Masaki, T. (2002a). Characterization of Ca2+ channels involved in endothelin-1-induced contraction of rabbit basilar artery. *Journal of Cardiovascular Pharmacology*, *40*(3), 438–447.

Kawanabe, Y., Hashimoto, N., & Masaki, T. (2002b). Characterization of Ca2+ channels involved in ET-1-induced transactivation of EGF receptors. *American Journal of Physiology. Heart and Circulatory Physiology*, *283*(6), H2671–H2675.

Kelland, N. F., Kuc, R. E., McLean, D. L., Azfer, A., Bagnall, A. J., Gray, G. A., et al. (2010). Endothelial cell-specific ETB receptor knockout: Autoradiographic and histological characterisation and crucial role in the clearance of endothelin-1. *Canadian Journal of Physiology and Pharmacology*, *88*(6), 644–651.

Kimura, S., Kasuya, Y., Sawamura, T., Shinmi, O., Sugita, Y., Yanagisawa, M., et al. (1988). Structure-activity relationships of endothelin: Importance of the C-terminal moiety. *Biochemical and Biophysical Research Communications*, *156*(3), 1182–1186.

Klammt, C., Srivastava, A., Eifler, N., Junge, F., Beyermann, M., Schwarz, D., et al. (2007). Functional analysis of cell-free-produced human endothelin B receptor reveals transmembrane segment 1 as an essential area for ET-1 binding and homodimer formation. *FEBS Journal*, *274*(13), 3257–3269.

Kloog, Y., Ambar, I., Sokolovsky, M., Kochva, E., Wollberg, Z., & Bdolah, A. (1988). Sarafotoxin, a novel vasoconstrictor peptide: Phosphoinositide hydrolysis in rat heart and brain. *Science*, *242*(4876), 268–270.

Kuruppu, S., Reeve, S., & Ian Smith, A. (2007). Characterisation of endothelin converting enzyme-1 shedding from endothelial cells. *FEBS Letters*, *581*(23), 4501–4506.

Kwan, H. Y., Shen, B., Ma, X., Kwok, Y. C., Huang, Y., Man, Y. B., et al. (2009). TRPC1 associates with BK(Ca) channel to form a signal complex in vascular smooth muscle cells. *Circulation Research*, *104*(5), 670–678.

Kwok, W., Lee, S. H., Culberson, C., Korneszczuk, K., & Clemens, M. G. (2009). Caveolin-1 mediates endotoxin inhibition of endothelin-1-induced endothelial nitric oxide synthase activity in liver sinusoidal endothelial cells. *American Journal of Physiology. Gastrointestinal and Liver Physiology*, *297*(5), G930–G939.

LaDouceur, D. M., Flynn, M. A., Keiser, J. A., Reynolds, E., & Haleen, S. J. (1993). ETA and ETB receptors coexist on rabbit pulmonary artery vascular smooth muscle mediating contraction. *Biochemical and Biophysical Research Communications*, *196*(1), 209–215.

Ladoux, A., & Frelin, C. (1991). Endothelins inhibit adenylate cyclase in brain capillary endothelial cells. *Biochemical and Biophysical Research Communications*, *180*(1), 169–173.

Lattig, J., Oksche, A., Beyermann, M., Rosenthal, W., & Krause, G. (2009). Structural determinants for selective recognition of peptide ligands for endothelin receptor subtypes ETA and ETB. *Journal of Peptide Science*, *15*(7), 479–491.

Lemonnier, L., Trebak, M., & Putney, J. W., Jr. (2008). Complex regulation of the TRPC3, 6 and 7 channel subfamily by diacylglycerol and phosphatidylinositol-4,5-bisphosphate. *Cell Calcium*, *43*(5), 506–514.

Li, L., Fink, G. D., Watts, S. W., Northcott, C. A., Galligan, J. J., Pagano, P. J., et al. (2003). Endothelin-1 increases vascular superoxide via endothelin(A)-NADPH oxidase pathway in low-renin hypertension. *Circulation*, *107*(7), 1053–1058.

Li, Y., Levesque, L. O., & Anand-Srivastava, M. B. (2010). Epidermal growth factor receptor transactivation by endogenous vasoactive peptides contributes to hyperproliferation of vascular smooth muscle cells of SHR. *American Journal of Physiology. Heart and Circulatory Physiology*, *299*(6), H1959–H1967.

Li, L., Watts, S. W., Banes, A. K., Galligan, J. J., Fink, G. D., & Chen, A. F. (2003). NADPH oxidase-derived superoxide augments endothelin-1-induced venoconstriction in mineralocorticoid hypertension. *Hypertension*, *42*(3), 316–321.

Li, J., Zhao, X., Li, X., Lerea, K. M., & Olson, S. C. (2007). Angiotensin II type 2 receptor-dependent increases in nitric oxide synthase expression in the pulmonary endothelium is mediated via a G alpha i3/Ras/Raf/MAPK pathway. *American Journal of Physiology Cell Physiology*, *292*(6), C2185–C2196.

Liu, Y., Geisbuhler, B., & Jones, A. W. (1992). Activation of multiple mechanisms including phospholipase D by endothelin-1 in rat aorta. *American Journal of Physiology*, *262*(4 Pt. 1), C941–C949.

Liu, S., Premont, R. T., Kontos, C. D., Huang, J., & Rockey, D. C. (2003). Endothelin-1 activates endothelial cell nitric-oxide synthase via heterotrimeric G-protein betagamma subunit signaling to protein kinase B/Akt. *Journal of Biological Chemistry*, *278*(50), 49929–49935.

Lodge, N. J., Zhang, R., Halaka, N. N., & Moreland, S. (1995). Functional role of endothelin ETA and ETB receptors in venous and arterial smooth muscle. *European Journal of Pharmacology*, *287*(3), 279–285.

Loomis, E. D., Sullivan, J. C., Osmond, D. A., Pollock, D. M., & Pollock, J. S. (2005). Endothelin mediates superoxide production and vasoconstriction through activation of NADPH oxidase and uncoupled nitric-oxide synthase in the rat aorta. *Journal of Pharmacology and Experimental Therapeutics*, *315*(3), 1058–1064.

Maggi, C. A., Giuliani, S., Patacchini, R., Rovero, P., Giachetti, A., & Meli, A. (1989). The activity of peptides of the endothelin family in various mammalian smooth muscle preparations. *European Journal of Pharmacology*, *174*(1), 23–31.

Maguire, J. J., & Davenport, A. P. (2015). Endothelin receptors and their antagonists. *Seminars in Nephrology*, *35*(2), 125–136.

Mahjoub, Y., Malaquin, S., Mourier, G., Lorne, E., Abou Arab, O., Massy, Z. A., et al. (2015). Short- versus long-sarafotoxins: Two structurally related snake toxins with very different in vivo haemodynamic effects. *PloS One*, *10*(7), e0132864.

Mamo, Y. A., Angus, J. A., Ziogas, J., Soeding, P. F., & Wright, C. E. (2014). The role of voltage-operated and non-voltage-operated calcium channels in endothelin-induced vasoconstriction of rat cerebral arteries. *European Journal of Pharmacology*, *742*, 65–73.

Maurer, M., Wedemeyer, J., Metz, M., Piliponsky, A. M., Weller, K., Chatterjea, D., et al. (2004). Mast cells promote homeostasis by limiting endothelin-1-induced toxicity. *Nature*, *432*(7016), 512–516.

Mazzuca, M. Q., & Khalil, R. A. (2012). Vascular endothelin receptor type B: Structure, function and dysregulation in vascular disease. *Biochemical Pharmacology*, *84*(2), 147–162.

McGahon, M. K., Needham, M. A., Scholfield, C. N., McGeown, J. G., & Curtis, T. M. (2009). Ca2+-activated Cl- current in retinal arteriolar smooth muscle. *Investigative Ophthalmology and Visual Science*, *50*(1), 364–371.

McMahon, E. G., Palomo, M. A., Moore, W. M., McDonald, J. F., & Stern, M. K. (1991). Phosphoramidon blocks the pressor activity of porcine big endothelin-1-(1-39) in vivo and conversion of big endothelin-1-(1-39) to endothelin-1-(1-21) in vitro. *Proceedings of the National Academy of Sciences of the United States of America*, *88*(3), 703–707.

Merlen, C., Farhat, N., Luo, X., Chatenet, D., Tadevosyan, A., Villeneuve, L. R., et al. (2013). Intracrine endothelin signaling evokes IP3-dependent increases in nucleoplasmic Ca(2)(+) in adult cardiac myocytes. *Journal of Molecular and Cellular Cardiology*, *62*, 189–202.

Meyer, M. R., Barton, M., & Prossnitz, E. R. (2014). Functional heterogeneity of NADPH oxidase-mediated contractions to endothelin with vascular aging. *Life Sciences*, *118*(2), 226–231.

Mickley, E. J., Gray, G. A., & Webb, D. J. (1997). Activation of endothelin ETA receptors masks the constrictor role of endothelin ETB receptors in rat isolated small mesenteric arteries. *British Journal of Pharmacology*, *120*(7), 1376–1382.

Miwa, S., Kawanabe, Y., Okamoto, Y., & Masaki, T. (2005). Ca2+ entry channels involved in endothelin-1-induced contractions of vascular smooth muscle cells. *Journal of Smooth Muscle Research*, *41*(2), 61–75.

Montezano, A. C., Burger, D., Paravicini, T. M., Chignalia, A. Z., Yusuf, H., Almasri, M., et al. (2010). Nicotinamide adenine dinucleotide phosphate reduced oxidase 5 (Nox5) regulation by angiotensin II and endothelin-1 is mediated via calcium/calmodulin-dependent, rac-1-independent pathways in human endothelial cells. *Circulation Research*, *106*(8), 1363–1373.

Muller, L., Barret, A., Etienne, E., Meidan, R., Valdenaire, O., Corvol, P., et al. (2003). Heterodimerization of endothelin-converting enzyme-1 isoforms regulates the subcellular distribution of this metalloprotease. *Journal of Biological Chemistry*, *278*(1), 545–555.

Murphy, L. J., Corder, R., Mallet, A. I., & Turner, A. J. (1994). Generation by the phosphoramidon-sensitive peptidases, endopeptidase-24.11 and thermolysin, of endothelin-1 and c-terminal fragment from big endothelin-1. *British Journal of Pharmacology, 113*(1), 137–142.

Nakano, A., Kishi, F., Minami, K., Wakabayashi, H., Nakaya, Y., & Kido, H. (1997). Selective conversion of big endothelins to tracheal smooth muscle-constricting 31-amino acid-length endothelins by chymase from human mast cells. *Journal of Immunology, 159*(4), 1987–1992.

Oh, P., Horner, T., Witkiewicz, H., & Schnitzer, J. E. (2012). Endothelin induces rapid, dynamin-mediated budding of endothelial caveolae rich in ET-B. *Journal of Biological Chemistry, 287*(21), 17353–17362.

Ohnaka, K., Takayanagi, R., Nishikawa, M., Haji, M., & Nawata, H. (1993). Purification and characterization of a phosphoramidon-sensitive endothelin-converting enzyme in porcine aortic endothelium. OFF. *Journal of Biological Chemistry, 268*(35), 26759–26766.

Ohnaka, K., Takayanagi, R., Yamauchi, T., Okazaki, H., Ohashi, M., Umeda, F., et al. (1990). Identification and characterization of endothelin converting activity in cultured bovine endothelial cells. *Biochemical and Biophysical Research Communications, 168*(3), 1128–1136.

Okada, M., & Nishikibe, M. (2002). BQ-788, a selective endothelin ET(B) receptor antagonist. *Cardiovascular Drug Reviews, 20*(1), 53–66.

Park, W. S., Han, J., & Earm, Y. E. (2008). Physiological role of inward rectifier K(+) channels in vascular smooth muscle cells. *Pflügers Archiv: European Journal of Physiology, 457*(1), 137–147.

Pejler, G., Ronnberg, E., Waern, I., & Wernersson, S. (2010). Mast cell proteases: Multifaceted regulators of inflammatory disease. *Blood, 115*(24), 4981–4990.

Peppiatt-Wildman, C. M., Albert, A. P., Saleh, S. N., & Large, W. A. (2007). Endothelin-1 activates a Ca2+-permeable cation channel with TRPC3 and TRPC7 properties in rabbit coronary artery myocytes. *Journal of Physiology, 580*(Pt. 3), 755–764.

Prenzel, N., Zwick, E., Daub, H., Leserer, M., Abraham, R., Wallasch, C., et al. (1999). EGF receptor transactivation by G-protein-coupled receptors requires metalloproteinase cleavage of proHB-EGF. *Nature, 402*(6764), 884–888.

Provost, C., Choufani, F., Avedanian, L., Bkaily, G., Gobeil, F., & Jacques, D. (2010). Nitric oxide and reactive oxygen species in the nucleus revisited. *Canadian Journal of Physiology and Pharmacology, 88*(3), 296–304.

Rapoport, R. M. (2014). Acute nitric oxide synthase inhibition and endothelin-1-dependent arterial pressure elevation. *Frontiers in Pharmacology, 5*, 57.

Rapoport, R. M., & Zuccarello, M. (2012). Endothelin(A)-endothelin(B) receptor cross talk in endothelin-1-induced contraction of smooth muscle. *Journal of Cardiovascular Pharmacology, 60*(5), 483–494.

Reeder, L. B., & Ferguson, M. K. (1996). Endothelin-1 synthesis and receptor-mediated activity in porcine lymph vessels. *Journal of Surgical Research, 63*(1), 215–219.

Rikitake, Y., & Liao, J. K. (2005). Rho GTPases, statins, and nitric oxide. *Circulation Research, 97*(12), 1232–1235.

Rubanyi, G. M., & Polokoff, M. A. (1994). Endothelins: Molecular biology, biochemistry, pharmacology, physiology, and pathophysiology. *Pharmacological Reviews, 46*(3), 325–415.

Russell, F. D., & Davenport, A. P. (1999). Evidence for intracellular endothelin-converting enzyme-2 expression in cultured human vascular endothelial cells. *Circulation Research, 84*(8), 891–896.

Saeki, T., Ihara, M., Fukuroda, T., Yamagiwa, M., & Yano, M. (1991). [Ala1,3,11,15] endothelin-1 analogs with ETB agonistic activity. *Biochemical and Biophysical Research Communications, 179*(1), 286–292.

Saida, K., Mitsui, Y., & Ishida, N. (1989). A novel peptide, vasoactive intestinal contractor, of a new (endothelin) peptide family. Molecular cloning, expression, and biological activity. *Journal of Biological Chemistry, 264*(25), 14613–14616.

Sakai, H., Ikomi, F., & Ohhashi, T. (1999). Effects of endothelin on spontaneous contractions in lymph vessels. *American Journal of Physiology*, *277*(2 Pt. 2), H459–H466.

Sakurai, T., Yanagisawa, M., Takuwa, Y., Miyazaki, H., Kimura, S., Goto, K., et al. (1990). Cloning of a cDNA encoding a non-isopeptide-selective subtype of the endothelin receptor. *Nature*, *348*(6303), 732–735.

Saleh, S. N., Albert, A. P., & Large, W. A. (2009). Activation of native TRPC1/C5/C6 channels by endothelin-1 is mediated by both PIP3 and PIP2 in rabbit coronary artery myocytes. *Journal of Physiology*, *587*(Pt. 22), 5361–5375.

Sauvageau, S., Thorin, E., Caron, A., & Dupuis, J. (2006). Evaluation of endothelin-1-induced pulmonary vasoconstriction following myocardial infarction. *Experimental Biology and Medicine (Maywood, NJ)*, *231*(6), 840–846.

Schmidt, M., Kroger, B., Jacob, E., Seulberger, H., Subkowski, T., Otter, R., et al. (1994). Molecular characterization of human and bovine endothelin converting enzyme (ECE-1). *FEBS Letters*, *356*(2–3), 238–243.

Schneider, L. A., Schlenner, S. M., Feyerabend, T. B., Wunderlin, M., & Rodewald, H. R. (2007). Molecular mechanism of mast cell mediated innate defense against endothelin and snake venom sarafotoxin. *Journal of Experimental Medicine*, *204*(11), 2629–2639.

Schweizer, A., Valdenaire, O., Nelbock, P., Deuschle, U., Dumas Milne Edwards, J. B., Stumpf, J. G., et al. (1997). Human endothelin-converting enzyme (ECE-1): Three isoforms with distinct subcellular localizations. *Biochemical Journal*, *328*(Pt. 3), 871–877.

Seed, A., Kuc, R. E., Maguire, J. J., Hillier, C., Johnston, F., Essers, H., et al. (2012). The dual endothelin converting enzyme/neutral endopeptidase inhibitor SLV-306 (daglutril), inhibits systemic conversion of big endothelin-1 in humans. *Life Sciences*, *91*(13–14), 743–748.

Semaan, W., Desbiens, L., Houde, M., Labonte, J., Gagnon, H., Yamamoto, D., et al. (2015). Chymase inhibitor-sensitive synthesis of endothelin-1 (1-31) by recombinant mouse mast cell protease 4 and human chymase. *Biochemical Pharmacology*, *94*(2), 91–100.

Shi, J., Ju, M., Large, W. A., & Albert, A. P. (2012). Pharmacological profile of phosphatidylinositol 3-kinases and related phosphatidylinositols mediating endothelin(A) receptor-operated native TRPC channels in rabbit coronary artery myocytes. *British Journal of Pharmacology*, *166*(7), 2161–2175.

Simard, E., Jin, D., Takai, S., Miyazaki, M., Brochu, I., & D'Orleans-Juste, P. (2009). Chymase-dependent conversion of Big endothelin-1 in the mouse in vivo. *Journal of Pharmacology and Experimental Therapeutics*, *328*(2), 540–548.

Sokolovsky, M., Galron, R., Kloog, Y., Bdolah, A., Indig, F. E., Blumberg, S., et al. (1990). Endothelins are more sensitive than sarafotoxins to neutral endopeptidase: Possible physiological significance. *Proceedings of the National Academy of Sciences of the United States of America*, *87*(12), 4702–4706.

Spinella, F., Caprara, V., Cianfrocca, R., Rosano, L., Di Castro, V., Garrafa, E., et al. (2014). The interplay between hypoxia, endothelial and melanoma cells regulates vascularization and cell motility through endothelin-1 and vascular endothelial growth factor. *Carcinogenesis*, *35*(4), 840–848.

Spinella, F., Caprara, V., Garrafa, E., Di Castro, V., Rosano, L., Natali, P. G., et al. (2010). Endothelin axis induces metalloproteinase activation and invasiveness in human lymphatic endothelial cells. *Canadian Journal of Physiology and Pharmacology*, *88*(8), 782–787.

Spinella, F., Garrafa, E., Di Castro, V., Rosano, L., Nicotra, M. R., Caruso, A., et al. (2009). Endothelin-1 stimulates lymphatic endothelial cells and lymphatic vessels to grow and invade. *Cancer Research*, *69*(6), 2669–2676.

Taichman, D. B., Ornelas, J., Chung, L., Klinger, J. R., Lewis, S., Mandel, J., et al. (2014). Pharmacologic therapy for pulmonary arterial hypertension in adults: CHEST guideline and expert panel report. *Chest*, *146*(2), 449–475.

Takahashi, M., Matsushita, Y., Iijima, Y., & Tanzawa, K. (1993). Purification and characterization of endothelin-converting enzyme from rat lung. *Journal of Biological Chemistry, 268*(28), 21394–21398.

Takai, M., Umemura, I., Yamasaki, K., Watakabe, T., Fujitani, Y., Oda, K., et al. (1992). A potent and specific agonist, Suc-[Glu9, Ala11,15]-endothelin-1(8-21), IRL 1620, for the ETB receptor. *Biochemical and Biophysical Research Communications, 184*(2), 953–959.

Takaoka, M., Miyata, Y., Takenobu, Y., Ikegawa, R., Matsumura, Y., & Morimoto, S. (1990). Mode of cleavage of pig big endothelin-1 by chymotrypsin. Production and degradation of mature endothelin-1. *Biochemical Journal, 270*(2), 541–544.

Taurin, S., Hogarth, K., Sandbo, N., Yau, D. M., & Dulin, N. O. (2007). Gbetagamma-mediated prostacyclin production and cAMP-dependent protein kinase activation by endothelin-1 promotes vascular smooth muscle cell hypertrophy through inhibition of glycogen synthase kinase-3. *Journal of Biological Chemistry, 282*(27), 19518–19525.

Tchougounova, E., Pejler, G., & Abrink, M. (2003). The chymase, mouse mast cell protease 4, constitutes the major chymotrypsin-like activity in peritoneum and ear tissue. A role for mouse mast cell protease 4 in thrombin regulation and fibronectin turnover. *Journal of Experimental Medicine, 198*(3), 423–431.

Thai, T. L., & Arendshorst, W. J. (2008). ADP-ribosyl cyclase and ryanodine receptors mediate endothelin ETA and ETB receptor-induced renal vasoconstriction in vivo. *American Journal of Physiology Renal Physiology, 295*(2), F360–F368.

Thakali, K., Demel, S. L., Fink, G. D., & Watts, S. W. (2005). Endothelin-1-induced contraction in veins is independent of hydrogen peroxide. *American Journal of Physiology Heart and Circulatory Physiology, 289*(3), H1115–H1122.

Thakali, K., Fink, G. D., & Watts, S. W. (2004). Arteries and veins desensitize differently to endothelin. *Journal of Cardiovascular Pharmacology, 43*(3), 387–393.

Thakali, K., Galligan, J. J., Fink, G. D., Gariepy, C. E., & Watts, S. W. (2008). Pharmacological endothelin receptor interaction does not occur in veins from ET(B) receptor deficient rats. *Vascular Pharmacology, 49*(1), 6–13.

Turner, A. J., & Murphy, L. J. (1996). Molecular pharmacology of endothelin converting enzymes. *Biochemical Pharmacology, 51*(2), 91–102.

Tykocki, N. R., Gariepy, C. E., & Watts, S. W. (2009). Endothelin ET(B) receptors in arteries and veins: Multiple actions in the vein. *Journal of Pharmacology and Experimental Therapeutics, 329*(3), 875–881.

Venkatachalam, K., Zheng, F., & Gill, D. L. (2003). Regulation of canonical transient receptor potential (TRPC) channel function by diacylglycerol and protein kinase C. *Journal of Biological Chemistry, 278*(31), 29031–29040.

von Brandenstein, M., Richter, C., & Fries, J. W. (2012). MicroRNAs: Small but amazing, and their association with endothelin. *Life Sciences, 91*(13–14), 475–489.

Wang, Y. X., Zheng, Y. M., Mei, Q. B., Wang, Q. S., Collier, M. L., Fleischer, S., et al. (2004). FKBP12.6 and cADPR regulation of Ca2+ release in smooth muscle cells. *American Journal of Physiology. Cell Physiology, 286*(3), C538–C546.

Warner, T. D., Allcock, G. H., Corder, R., & Vane, J. R. (1993). Use of the endothelin antagonists BQ-123 and PD 142893 to reveal three endothelin receptors mediating smooth muscle contraction and the release of EDRF. *British Journal of Pharmacology, 110*(2), 777–782.

Watts, S. W. (2010). Endothelin receptors: What's new and what do we need to know? *American Journal of Physiology Regulatory, Integrative and Comparative Physiology, 298*(2), R254–R260.

Watts, S. W., Thakali, K., Smark, C., Rondelli, C., & Fink, G. D. (2007). Big ET-1 processing into vasoactive peptides in arteries and veins. *Vascular Pharmacology, 47*(5–6), 302–312.

Wedgwood, S., McMullan, D. M., Bekker, J. M., Fineman, J. R., & Black, S. M. (2001). Role for endothelin-1-induced superoxide and peroxynitrite production in rebound pulmonary hypertension associated with inhaled nitric oxide therapy. *Circulation Research*, *89*(4), 357–364.
Wirth, A., Benyo, Z., Lukasova, M., Leutgeb, B., Wettschureck, N., Gorbey, S., et al. (2008). G12-G13-LARG-mediated signaling in vascular smooth muscle is required for salt-induced hypertension. *Nature Medicine*, *14*(1), 64–68.
Wirth, A., Wang, S., Takefuji, M., Tang, C., Althoff, T. F., Schweda, F., et al. (2016). Age-dependent blood pressure elevation is due to increased vascular smooth muscle tone mediated by G-protein signalling. *Cardiovascular Research*, *109*(1), 131–140.
Wynne, B. M., Chiao, C. W., & Webb, R. C. (2009). Vascular smooth muscle cell signaling mechanisms for contraction to angiotensin II and endothelin-1. *Journal of the American Society of Hypertension*, *3*(2), 84–95.
Wypij, D. M., Nichols, J. S., Novak, P. J., Stacy, D. L., Berman, J., & Wiseman, J. S. (1992). Role of mast cell chymase in the extracellular processing of big-endothelin-1 to endothelin-1 in the perfused rat lung. *Biochemical Pharmacology*, *43*(4), 845–853.
Xu, Y. J., Elimban, V., & Dhalla, N. S. (2015). Reduction of blood pressure by store-operated calcium channel blockers. *Journal of Cellular and Molecular Medicine*, *19*(12), 2763–2770.
Xu, D., Emoto, N., Giaid, A., Slaughter, C., Kaw, S., deWit, D., et al. (1994). ECE-1: A membrane-bound metalloprotease that catalyzes the proteolytic activation of big endothelin-1. *Cell*, *78*(3), 473–485.
Yanagisawa, H., Hammer, R. E., Richardson, J. A., Emoto, N., Williams, S. C., Takeda, S., et al. (2000). Disruption of ECE-1 and ECE-2 reveals a role for endothelin-converting enzyme-2 in murine cardiac development. *Journal of Clinical Investigation*, *105*(10), 1373–1382.
Yanagisawa, M., Kurihara, H., Kimura, S., Tomobe, Y., Kobayashi, M., Mitsui, Y., et al. (1988). A novel potent vasoconstrictor peptide produced by vascular endothelial cells. *Nature*, *332*(6163), 411–415.
Yanagisawa, H., Yanagisawa, M., Kapur, R. P., Richardson, J. A., Williams, S. C., Clouthier, D. E., et al. (1998). Dual genetic pathways of endothelin-mediated intercellular signaling revealed by targeted disruption of endothelin converting enzyme-1 gene. *Development*, *125*(5), 825–836.
Yang, H., Kuang, S. J., Rao, F., Xue, Y. M., Liu, X. Y., Shan, Z. X., et al. (2016). Species-specific differences in the role of L-type Ca(2)(+) channels in the regulation of coronary arterial smooth muscle contraction. *Naunyn-Schmiedebergs Archives of Pharmacology*, *389*(2), 151–157.
Yatawara, A., Wilson, J. L., Taylor, L., Polgar, P., & Mierke, D. F. (2013). C-terminus of ETA/ETB receptors regulate endothelin-1 signal transmission. *Journal of Peptide Science*, *19*(4), 257–262.
Yoon, S., Gariepy, C. E., Yanagisawa, M., Zuccarello, M., & Rapoport, R. M. (2016). Functional ETA-ETB receptor cross-talk in basilar artery in situ from ETB receptor deficient rats. *Journal of Cardiovascular Pharmacology*, *67*(3), 212–217.
Yoon, S., Zuccarello, M., & Rapoport, R. M. (2012). EndothelinA-endothelinB receptor cross-talk in rat basilar artery in situ. *Naunyn-Schmiedebergs Archives of Pharmacology*, *385*(4), 437–441.
Yuzugulen, J., Wood, E. G., Douthwaite, J. A., Villar, I. C., Patel, N. S., Jegard, J., et al. (2012). *Abstract 18898: Characterization of proendothelin-1 derived peptides identifies a Co-secreted modulator of ET-1 vasoconstriction, and provides insights for biomarker measurements: Paper presented at the American Heart Association Scientific Sessions, Los Angeles, CA, USA.*
Zhang, X. F., Iwamuro, Y., Enoki, T., Okazawa, M., Lee, K., Komuro, T., et al. (1999). Pharmacological characterization of Ca2+ entry channels in endothelin-1-induced

contraction of rat aorta using LOE 908 and SK&F 96365. *British Journal of Pharmacology*, *127*(6), 1388–1398.

Zhao, J., & van Helden, D. F. (2003). ET-1-associated vasomotion and vasospasm in lymphatic vessels of the guinea-pig mesentery. *British Journal of Pharmacology*, *140*(8), 1399–1413.

Zubkov, A. Y., Rollins, K. S., Parent, A. D., Zhang, J., & Bryan, R. M., Jr. (2000). Mechanism of endothelin-1-induced contraction in rabbit basilar artery. *Stroke*, *31*(2), 526–533.

CHAPTER SIX

Endothelium-Dependent Contractions: Prostacyclin and Endothelin-1, Partners in Crime?

O. Baretella, P.M. Vanhoutte[1]
State Key Laboratory of Pharmaceutical Biotechnology, The University of Hong Kong, Hong Kong S.A.R., China
[1]Corresponding author: e-mail address: vanhoutt@hku.hk

Contents

Abstract

Both the lipid prostacyclin and the peptide endothelin-1 are endothelium-derived substances. Endothelin-1 is one of the most powerful endogenous vasoconstrictors, while prostacyclin is a potent antiaggregatory and vasodilator mediator upon activation of prostaglandin I_2 (IP) receptors. During endothelium-dependent, prostanoid-mediated contractions/constrictions, however, prostacyclin appears to be a major endothelium-derived contracting factor (EDCF). Such cyclooxygenase-dependent responses, whether measured ex vivo or in vivo, are exacerbated by aging, obesity, diabetes, or hypertension. On the background of such cardiovascular risk factors, endothelin-1 may potentiate these contractions by promoting prostacyclin production. The latter is reduced by endothelin-A (ET_A) receptor antagonists. This receptor subtype is recognized for mediating contractions of smooth muscle cells to endothelin-1. However, it is present also on endothelial cells, where its activation increases intracellular calcium concentration with subsequent initiation of phospholipase A_2 that provides arachidonic acid for

Advances in Pharmacology, Volume 77
ISSN 1054-3589
http://dx.doi.org/10.1016/bs.apha.2016.04.006

metabolism by cyclooxygenases. Thus, endothelin-1 favors cyclooxygenase-dependent vasoconstrictor prostanoid formation, including prostacyclin. Activation of endothelial endothelin-B (ET_B) receptors promotes the release of nitric oxide, which opposes both EDCF and endothelin-1. This is less pronounced in disease promoting ET_A- and smooth muscle ET_B receptor-dependent as well as prostanoid-mediated contractions. In addition, the thromboxane prostanoid (TP) receptors on vascular smooth muscle cells become hyperresponsive to EDCF under pathophysiological conditions, while IP receptor responsiveness diminishes. A better understanding of the interaction between prostacyclin and endothelin-1 and the determination of the roles of the TP and IP receptors involved in prostanoid-mediated contractions in health and during disease will help to define advanced pharmacological strategies for the therapy of cardiovascular disorders.

ABBREVIATIONS

5-HT 5-hydroxytryptamine = serotonin
AP-1 activator protein 1
cAMP cyclic adenosine monophosphate
cGMP cyclic guanosine monophosphate
COX cyclooxygenase(s)
ECE endothelin-converting enzyme(s)
EDCF endothelium-derived contracting factor(s)
EDH endothelium-dependent hyperpolarization
eNOS endothelial nitric oxide synthase
ET-1 endothelin-1
ET_A endothelin-A receptor
ET_B endothelin-B receptor
IP prostaglandin I_2 = prostacyclin receptor
M_3 muscarinic receptor 3
NFκB nuclear factor kappa B
NO nitric oxide
PGI_2 prostaglandin I_2
PLA_2 phospholipase A_2
PPAR peroxisome proliferator-activated receptor
TP thromboxane prostanoid receptor

1. INTRODUCTION: THE DISCOVERY OF ENDOTHELIAL FUNCTION

The existence of endothelial cells as the innermost layer of blood vessels has been known since the mid-19th century when the Swiss anatomist Wilhelm His coined the term *endothelium* for this continuous lining (Aird, 2007; His, 1865). More than a century later, endothelial cells were demonstrated to be essential for relaxation of isolated arteries induced by acetylcholine (Furchgott, 1993; Furchgott & Zawadzki, 1980). This originally serendipitous finding has become one of the foundations of modern vascular biology leading to the introduction of nitric oxide as the major endothelium-derived relaxing factor, activating as downstream target soluble guanylyl cyclase with increased production of cyclic guanosine monophosphate (cGMP) (Furchgott, 1988; Furchgott & Vanhoutte, 1989; Ignarro, Buga, Wood, Byrns, & Chaudhuri, 1987; Ignarro, Byrns, & Wood, 1988; Rapoport & Murad, 1983; Vanhoutte, 2009). The concept of endothelium-dependent control of vascular tone has not remained limited to relaxations/dilatations, as initially veins and later certain isolated arteries have been found to contract to certain stimuli in an endothelium-dependent manner (De Mey & Vanhoutte, 1982; Katušić, Shepherd, & Vanhoutte, 1988; Lüscher & Vanhoutte, 1986; Miller & Vanhoutte, 1985a). Such endothelium-dependent contractions require the metabolism of arachidonic acid by cyclooxygenases, since such responses are eliminated by inhibitors of these enzymes (Katušić et al., 1988; Lüscher & Vanhoutte, 1986; Miller & Vanhoutte, 1985a).

Since the very beginning, acetylcholine (activating endothelial muscarinic M_3 receptors; Boulanger, Morrison, & Vanhoutte, 1994) has been the gold standard to induce endothelium-dependent responses (Furchgott & Zawadzki, 1980; Lüscher & Vanhoutte, 1986; Vanhoutte, Shimokawa, Tang, & Félétou, 2009). Thus, for the study of such responses, the muscarinic agonist is most widely used pharmacologically, although endothelium-dependent relaxations also are induced physiologically by increases in flow (Rubanyi, Romero, & Vanhoutte, 1986), while endothelium-dependent contractions are elicited by augmenting physical stretch (Katušić, Shepherd, & Vanhoutte, 1987) or inducing hypoxia (De Mey & Vanhoutte, 1983; Furchgott & Vanhoutte, 1989; Rubanyi & Vanhoutte, 1985; Vanhoutte et al., 2009). The presence of enzymes instrumental for the genesis of such endothelium-dependent relaxations or -contractions [endothelial nitric oxide synthase (eNOS) and cyclooxygenases (COX), respectively] has been demonstrated in the endothelium (DeWitt, Day, Sonnenburg, & Smith, 1983; Félétou, Huang, & Vanhoutte, 2010; Palmer & Moncada, 1989; Vanhoutte et al., 2009; Yang, Félétou, Levens, Zhang, & Vanhoutte, 2003).

2. ENDOTHELIUM-DEPENDENT RELAXATIONS

In addition to acetylcholine (Furchgott & Zawadzki, 1980), endothelium-dependent relaxations also are evoked, among others

(Vanhoutte, Shimokawa, Félétou, & Tang, 2015; Vanhoutte et al., 2009), by catecholamines or serotonin subject to the vascular preparation studied (Cocks & Angus, 1983; Cohen, Shepherd, & Vanhoutte, 1983; Miller & Vanhoutte, 1985b). These eNOS-dependent, nitric oxide-mediated relaxations are caused by activation of endothelial α_2-adrenoceptors (Miller & Vanhoutte, 1985b; Vanhoutte & Miller, 1989) or 5-HT_{1D} receptors (Schoeffter & Hoyer, 1990), respectively. Mostly, nitric oxide-dependent responses to these agonists are studied in larger arteries such as the aorta (Furchgott & Zawadzki, 1980) or coronary arteries (Cocks & Angus, 1983; Cohen et al., 1983; Schoeffter & Hoyer, 1990), but they also occur in preparations like femoral arteries and veins (Miller & Vanhoutte, 1985b; Vanhoutte & Miller, 1989; Vanhoutte et al., 2009).

In smaller preparations and even in large coronary arteries, acetylcholine (and other stimuli) not only induces the release of nitric oxide- but also evokes endothelium-dependent hyperpolarization (EDH) and subsequent relaxations of the vascular smooth muscle cells (Chen, Suzuki, & Weston, 1988; Cohen & Vanhoutte, 1995; Félétou & Vanhoutte, 1988, 2013; Nagao, Illiano, & Vanhoutte, 1992; Nagao & Vanhoutte, 1991). This pathway is attenuated chronically by nitric oxide and thus is regarded as a backup mechanism when the bioavailability of the latter is reduced (Bauersachs et al., 1996; Chadha et al., 2010; Félétou, Tang & Vanhoutte, 2008; Mokhtar et al., 2016; Najibi, Cowan, Palacino, & Cohen, 1994; Olmos, Mombouli, Illiano, & Vanhoutte, 1995).

Both endothelium-dependent NO- and EDH-mediated relaxations are physiologically relevant for humans in vivo as demonstrated by experiments in the forearm (Halcox, Narayanan, Cramer-Joyce, Mincemoyer, & Quyyumi, 2001; Linder, Kiowski, Bühler, & Lüscher, 1990; Panza, Quyyumi, Brush, & Epstein, 1990; Taddei, Ghiadoni, Virdis, Buralli, & Salvetti, 1999).

3. ENDOTHELIUM-DEPENDENT CONTRACTIONS

Contractions in isolated preparations with endothelium to thrombin and arachidonic acid were observed originally in veins but not in arteries (De Mey & Vanhoutte, 1982) and found to be COX-dependent (Miller & Vanhoutte, 1985a). Subsequent experiments with acetylcholine have revealed endothelium-dependent, prostanoid-mediated contractions in rodent- (Lüscher & Vanhoutte, 1986) and canine arteries (Katušić et al., 1988). The endothelium-derived COX products (endothelium-derived contracting factors, EDCFs) activate thromboxane prostanoid (TP) receptors on the vascular smooth muscle cells (Auch-Schwelk, Katušić, & Vanhoutte, 1990; Tesfamariam, Jakubowski, & Cohen, 1989), as the final step in the endothelium-dependent contractions in most arteries (Vanhoutte et al., 2009). However, endothelium-dependent contractions not due to endothelium-derived prostanoids (although they may involve activation of cyclooxygenases in vascular smooth muscle cells; Auch-Schwelk, Katušić & Vanhoutte, 1989) is triggered, for example, by anoxia (hypoxia) or oxidative stress (Chan, Mak, Gao, Man, & Vanhoutte, 2011; Chen et al., 2014; Gao & Vanhoutte, 2014; Gräser & Vanhoutte, 1991; Katušić, Schugel, Cosentino, & Vanhoutte, 1993; Katušić & Vanhoutte, 1989; Pearson, Lin, Schaff, & Vanhoutte, 1996; Rubanyi & Vanhoutte, 1985; Tesfamariam, 1994; Yang et al., 2002).

As EDH-mediated relaxations (Bauersachs et al., 1996; Olmos et al., 1995), the COX-dependent EDCF pathway is inhibited, both acutely and long-term, by nitric oxide (Auch-Schwelk, Katušić, & Vanhoutte, 1992; Félétou et al., 2008; Tang, Félétou, Huang, Man, & Vanhoutte, 2005). Of the COX products, prostacyclin appears to be the major EDCF released from isolated arteries (at least in rodents) stimulated with acetylcholine (Auch-Schwelk et al., 1990; Gluais, Lonchampt, Morrow, Vanhoutte, & Félétou, 2005; Rapoport & Williams, 1996; Tesfamariam et al., 1989). Cyclooxygenase 1 (COX-1) is the main prostanoid source in these rodent preparations (Ge et al., 1995; Gluais et al., 2006; Tang, Ku, et al., 2005; Traupe, Lang, et al., 2002). In certain species (including humans) cyclooxygenase 2 (COX-2) may contribute to altered acetylcholine-induced responses (Virdis et al., 2013; Wong et al., 2009). The identity of the COX isoform responsible for prostanoid production, as well as its location (being restricted to the endothelium or extended to the vascular smooth muscle layer) is of additional interest for the understanding of pathophysiological situations in patients (Taddei, Virdis, Mattei, & Salvetti, 1993; Vanhoutte, 2013).

3.1 Role in Cardiovascular Disease

COX- and endothelium-dependent contractions are prominent in aortic rings of spontaneously hypertensive rats (Lüscher & Vanhoutte, 1986), the prototypical animal model for essential hypertension in humans. Furthermore, such contractions are augmented in arteries of aged- (Koga et al., 1989), diabetic- (Tesfamariam et al., 1989), or obese animals

(Traupe, Lang, et al., 2002). By contrast, endothelium-dependent, nitric oxide-mediated relaxations in preparations from these animals are preserved despite the presence of cardiovascular risk factors (Koga et al., 1989; Lüscher & Vanhoutte, 1986; Tesfamariam et al., 1989; Traupe, Lang, et al., 2002). Besides acetylcholine, several other stimulators of the endothelium can promote COX-dependent contractions under cardiovascular risk conditions (Vanhoutte et al., 2009). The known substances include platelet products [adenosine di- or triphosphate activating purinoceptors (Mombouli & Vanhoutte, 1993) and serotonin (Shimokawa & Vanhoutte, 1989) acting through 5-HT_2 receptors (Park, Lee, Kirchengast, Boulanger, & Vanhoutte, 1995)], ergonovine (also activating 5-HT_2 receptors; Shimokawa, Flavahan, Shepherd, & Vanhoutte, 1989), as well as endothelin-1 (ET-1) activating endothelin-A (ET_A) receptors (Park, Lee, & Vanhoutte, 1999; Taddei & Vanhoutte, 1993; Fig. 1). Particularly the contractions to aggregating platelets are apparent in the context of a regenerated endothelium and/or atherosclerosis (Vanhoutte, 2016). Thus, endothelium-dependent, prostanoid-mediated contractions appear to be of importance for endothelial- and vascular dysfunction under pathophysiological conditions.

Mechanistically, the cause for the prominent EDCF-mediated responses in disease can be found in the augmented production of prostanoids and/or higher sensitivity of the smooth muscle cells to these agents (Auch-Schwelk et al., 1990; Ge et al., 1995; Gluais et al., 2005; Rapoport & Williams, 1996). The former has been demonstrated in the aorta of spontaneously hypertensive rats (Auch-Schwelk et al., 1990; Ge et al., 1995; Gluais et al., 2005; Rapoport & Williams, 1996), but corresponding preparations of diabetic- and control animals released similar prostanoid amounts upon stimulation with acetylcholine (Tesfamariam et al., 1989). Despite an increased generation of certain prostanoids in arteries incubated in very high levels of glucose (Tesfamariam, Brown, Deykin, & Cohen, 1990), an augmented responsiveness of their smooth muscle to these vasoconstrictors appears to be the more relevant reason for potentiated EDCF-mediated contractions in arteries from animals with cardiovascular risk factors other than hypertension. Indeed, an increased TP receptor responsiveness is present particularly in the renal vasculature of both type I- (Michel, Simonet, et al., 2008) and type II diabetic rodents (Matsumoto, Watanabe, Kawamura, Taguchi, & Kobayashi, 2014) as well as obese mice (Baretella, Chung, Barton, Xu, & Vanhoutte, 2014). By contrast, in renal arterial smooth muscle full TP receptor activation is comparable in preparations of spontaneously hypertensive- and normotensive Wistar-Kyoto rats (Michel, Man, Man, & Vanhoutte, 2008). In corresponding

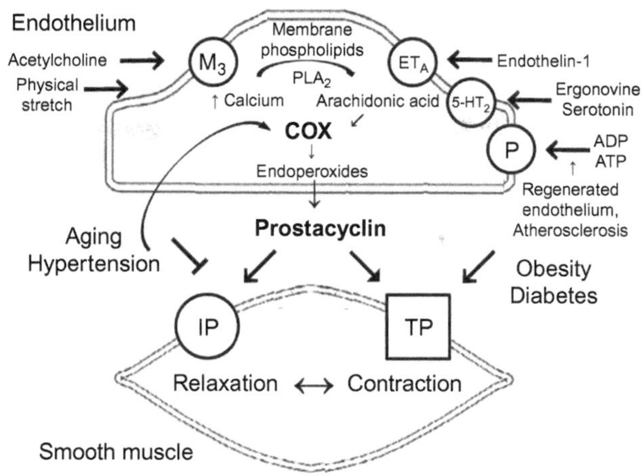

Fig. 1 In endothelial cells, prostacyclin (prostaglandin I_2/PGI_2) is generated by cyclooxygenases (COX) via endoperoxides from phospholipase A_2 (PLA_2)-liberated membrane lipid-derived arachidonic acid upon exposure to stimulators such as acetylcholine activating muscarinic (M_3) receptors, physical stretch, endothelin-1 activating endothelin-A (ET_A) receptors, ergonovine, and the platelet products serotonin acting through 5-HT_2 receptors as well as adenosine diphosphate (ADP) or –triphosphate (ATP) activating purinoceptors (P). Prostacyclin induces relaxation of vascular smooth muscle cells by activating prostaglandin I_2 (IP) receptors, or at high concentration causes contraction through activation of thromboxane prostanoid (TP) receptors. The responsiveness of the latter is augmented by obesity or diabetes, while aging and hypertension reduce the relaxing capacity of prostacyclin via IP receptors. In addition, hypertension promotes cyclooxygenase-dependent production of prostacyclin. Thus, aging and cardiovascular risk factors induce an imbalance by reducing relaxing IP- but augmenting procontractile TP receptor function resulting in increased endothelium-dependent vasoconstrictor tone particularly in the face of greater prostacyclin production.

preparations of these animals, COX-dependent and TP receptor-mediated contractions to acetylcholine are evoked also in the absence of hypertension although the latter aggravates them (Michel, Man, et al., 2008). In the forearm of essential hypertensive patients, vasodilatations to acetylcholine are impaired in a COX-dependent manner compared to normotensive control subjects indicating a pathophysiological relevance of EDCF for humans in vivo (Taddei et al., 1993).

4. PROSTACYCLIN

Prostaglandin I_2 (PGI_2), or prostacyclin has been discovered as an unstable substance inhibiting platelet aggregation and originally recognized for its propensity to relax isolated blood vessels (Moncada, Gryglewski, Bunting, & Vane, 1976), which both are mediated via activation of prostaglandin I_2 (IP) receptors (Armstrong, Lawrence, Jones, Wilson, & Collier, 1989; Boie et al., 1994; Nakagawa et al., 1994; Fig. 1). By contrast to the initial findings in rabbit arteries (Bunting, Gryglewski, Moncada, & Vane, 1976; Moncada et al., 1976) and bovine coronaries (Dusting, Moncada, & Vane, 1977a), vasoconstrictor responses have been demonstrated subsequently for prostacyclin in the porcine coronary circulation (Dusting, Moncada, & Vane, 1977b) and in venous and aortic preparations of the rat (Levy, 1978, 1980). These contractions are due to the activation of TP receptors of the vascular smooth muscle (Williams, Dorn, & Rapoport, 1994; Fig. 1). The stable prostacyclin metabolite 6-keto prostaglandin $F_{1\alpha}$ (Johnson et al., 1976) is released from endothelial cells during acetylcholine-induced, endothelium-dependent contractions in aortic rings of experimental animals (Auch-Schwelk et al., 1990; Tesfamariam et al., 1989), and prostacyclin eventually has been proposed as an endothelium-derived contracting factor (Rapoport & Williams, 1996). When added exogenously, however, this prostanoid induces relatively weak and slow in onset increases in tension. This is in contrast to the larger amplitude of contractions in response to endogenously released prostacyclin, which underlies the endothelium- and COX-dependent contractions evoked by acetylcholine (Baretella, Xu, & Vanhoutte, 2014; Gluais et al., 2005; Tesfamariam et al., 1989). The explanation for this discrepancy remains speculative, but is likely to reflect different potencies between substances added exogenously or produced endogenously and released by the endothelium in preferential sites, as also seen for NO whether released by acetylcholine or administered in the form of an NO donor (Furchgott, 1988). Indeed, prostacyclin is unstable and degraded rapidly in aqueous solutions (Cho & Allen, 1978; Johnson et al., 1976; Moncada et al., 1976); this rapid decay may be of less significance if the mediator is produced locally by the endothelium and diffuses directly to the underlying smooth muscle cells (Furchgott, 1988). The comparable transient nature of contractions induced by acetylcholine and exogenous prostacyclin suggests the release of the latter during EDCF-mediated responses upon stimulation with the former (Gluais et al., 2005). The determination of the production of its stable breakdown product (6-keto prostaglandin $F_{1\alpha}$), has confirmed that prostacyclin is an EDCF released from murine arteries upon stimulation with acetylcholine (Baretella, Xu, et al., 2014; Liu et al., 2012a). In both porcine- (Liu et al., 2012b) and murine renal arteries (Liu et al., 2013), COX-1-derived prostacyclin increases endothelium-dependent vasomotor tone. These findings seem to be at variance with the original notion that a lack of this prostaglandin could be responsible for elevations in arterial blood pressure (Moncada et al., 1976). However, most of the studies in isolated preparations of experimental animals have mainly focused on the TP receptor dependency (Williams et al., 1994) of these contractions (Auch-Schwelk et al., 1990; Gluais et al., 2005; Rapoport & Williams, 1996; Tang, Ku, et al., 2005; Traupe, Lang, et al., 2002). The possible functional importance of IP receptors for these responses has been largely neglected so far.

4.1 Role in Cardiovascular Disease

Contractions to prostacyclin in aortic preparations of spontaneously hypertensive- and normotensive Wistar-Kyoto rats are similar (Levy, 1980) or only slightly facilitated by hypertension (Gluais et al., 2005). With aging and particularly with sustained increases in systolic arterial blood pressure, relaxations of spontaneously hypertensive rat aortae to prostacyclin diminish (Gluais et al., 2005; Gomez et al., 2008; Fig. 1). By contrast, prostacyclin synthase expression increases in this context particularly in endothelial cells (Tang & Vanhoutte, 2008), while IP receptors are less expressed already in prehypertensive spontaneously hypertensive rat aortae (Numaguchi et al., 1999).

While the absence of TP receptors or deletion of COX-1 attenuate the development of high arterial blood pressure (Francois, Athirakul, Mao, Rockman, & Coffman, 2004; Sparks et al., 2013), the role of IP receptors in hypertension is inconclusive to judge from studies with knockout mice (Francois, et al., 2005; Fujino et al., 2004). The presence of prostacyclin-induced renin release from the kidney (Jensen, Schmid, & Kurtz, 1996; Whorton et al., 1977) explains why lack of IP receptors attenuates the development of renovascular hypertension (Fujino et al., 2004). By contrast, IP receptor knockout aggravates high salt-induced elevations in arterial blood pressure, emphasizing the protective effects of prostacyclin (Francois et al., 2005). This is further illustrated by the accelerated cardiac hypertrophy following pressure overload in IP receptor knockout mice (Hara et al., 2005). IP receptor deletion also aggravates cardiac ischemia injury (Xiao et al., 2001) and atherosclerosis development (Kobayashi et al., 2004). Despite differences in the signaling pathways between mice and men (Miggin & Kinsella, 2002), dysfunction of IP receptors may contribute to cardiovascular disease also in humans (Arehart et al., 2008; Patrignani et al., 2008; Stitham et al., 2011). Furthermore, the concentration of prostacyclin determines whether it induces relaxation via IP- or contraction via TP receptor activation in the rat aorta (Williams et al., 1994) as well as in human kidney arteries (Eskildsen et al., 2014). Several molecular mechanisms and cellular events cause impairments of IP receptor signaling (Majed & Khalil, 2012). The interaction of prostacyclin with IP receptors is defined by transmembrane domains determining ligand binding or receptor activity (Narumiya, Sugimoto, & Ushikubi, 1999; Stitham et al., 2007). Vasodilator effects of prostacyclin depend on increases in adenylyl cyclase activity that is initiated by coupling of the ligand to binding proteins (G_s) on the G protein-coupled IP receptors (Majed & Khalil, 2012; Wright et al., 2001). By contrast, other IP receptor binding protein subtypes (G_i and G_q) inhibit adenylyl cyclase and favor contraction (Majed & Khalil, 2012; Narumiya et al., 1999). Moreover, posttranslational modification of the IP receptor via glycosylation (a hallmark of hyperglycaemia in diabetes) decreases prostacyclin-induced adenylyl cyclase activation, and thus cyclic adenosine monophosphate (cAMP) production (Zhang, Austin, & Smyth, 2001). IP Receptor desensitization is caused by epigenetic receptor modulation such as phosphorylation, and internalization may occur separately (Hata & Breyer, 2004; Majed & Khalil, 2012). Particularly in diabetes, oxidative stress inactivates prostacyclin synthase via nitration (Giacco & Brownlee, 2010; Majed & Khalil, 2012), in addition to a reduced expression of IP receptors on platelets favoring their aggregation in diabetic patients (Knebel, Sprague, & Stephenson, 2015).

Taken in conjunction, the available data suggest that prostacyclin is a major contributor to vascular dysfunction in disease states when it may no longer be able of mediating its protective effects and much higher levels are produced/released compared to other prostanoids such as prostaglandin $F_{2\alpha}$, prostaglandin E_2, and thromboxane A_2 that have known vasoconstrictor effects (Gluais et al., 2005).

5. ENDOTHELIN-1

Following several studies investigating endothelium-dependent vasoconstrictors (De Mey & Vanhoutte, 1982; Gillespie, Owasoyo, McMurtry, & O'Brien, 1986; Hickey, Rubanyi, Paul, & Highsmith, 1985; O'Brien, Robbins, & McMurtry, 1987), a 21 amino acid peptide has been identified in cultured porcine aortic endothelial cells and named endothelin-1 (ET-1) (Barton & Yanagisawa, 2008; De Mey &

Vanhoutte, 2014; Yanagisawa et al., 1988). Two subtypes of receptors for the ET-1 peptide have been discovered subsequently (Arai, Hori, Aramori, Ohkubo, & Nakanishi, 1990; Sakurai et al., 1990), and termed endothelin-A (ET_A) and endothelin-B (ET_B) receptors (Masaki, Vane, & Vanhoutte, 1994). Both receptor subtypes initiate contraction of vascular smooth muscle cells (Sumner, Cannon, Mundin, White, & Watts, 1992; Fig. 2). Hence, ET-1 has been proposed as a potential EDCF released during endothelium-dependent contractions (Vanhoutte & Katušić, 1988). Indeed, stimulated endothelial cells produce increased amounts of the ET-1 peptide contributing to vasoconstrictor tone (Goel et al., 2010) via activation of ET_A receptors on vascular smooth muscle cells (Wirth et al., 2016). However, activation of the ET_B subtype located on endothelial cells causes the release of nitric oxide (Hirata et al., 1993), which inhibits both the production and action of ET-1 (Boulanger & Lüscher, 1990; De Mey & Vanhoutte, 2014; de Nucci et al., 1988; Miller, Komori, Burnett, & Vanhoutte, 1989; Rapoport, 2014; Thorin & Clozel, 2010; Vanhoutte, 2000; Fig. 2).

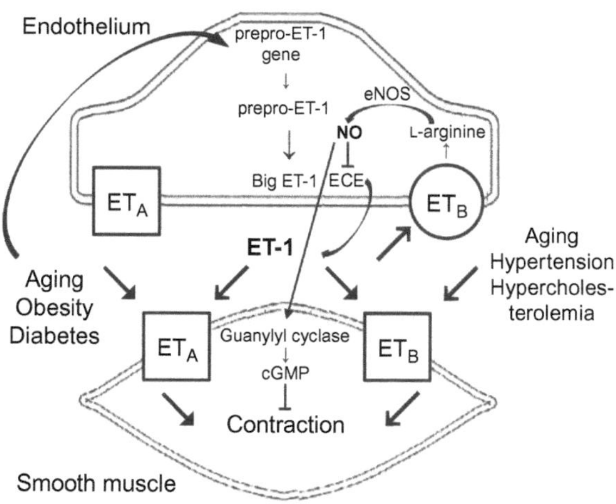

Fig. 2 Endothelin-1 (ET-1) originates from its precursor *prepro ET-1* via big ET-1 processed by endothelin-converting enzymes (ECE) into the active peptide. Vascular smooth muscle contains both endothelin-A (ET_A) and endothelin-B (ET_B) receptors which initiate contraction, while ET_B receptors on endothelial cells promote the generation of nitric oxide (NO) from L-arginine by endothelial NO synthase (eNOS). NO inhibits the production of ET-1 in endothelial cells and in vascular smooth muscle prevents the action of the peptide by activating soluble guanylyl cyclase producing cyclic guanosine monophosphate (cGMP). During aging, endothelial ET-1 levels increase and both smooth muscle ET_A- as well as ET_B receptor-mediated vasoconstrictions are augmented. The latter augmentation is promoted further by hypercholesterolemia.

Moreover, ET-1 clearance in the lungs is an ET_B receptor-mediated process (Fukuroda et al., 1994). ET_B receptors on smooth muscle cells contribute to vasoconstriction in canine coronary arteries (Teerlink, Breu, Sprecher, Clozel, & Clozel, 1994) and human arteries and veins (Seo, Oemar, Siebenmann, von Segesser, & Lüscher, 1994). Besides initiating contraction of vascular smooth muscle (Masaki et al., 1994; Sumner et al., 1992; Taddei & Vanhoutte, 1993), ET_A receptors also are present on endothelial cells (Amano, Hori, Ozaki, Matsuda, & Karaki, 1994; Avedanian et al., 2010; Nishimura et al., 1995; Vigne & Frelin, 1994).

Plasma ET-1 levels in humans depend on ethnic and cardiovascular risk factors such as obesity, diabetes, and hypertension (Ergul, Parish, Puett, & Ergul, 1996; Ferri et al., 1995; Saito, Nakao, Mukoyama, & Imura, 1990; Schneider et al., 2002; Takahashi, Ghatei, Lam, O'Halloran, & Bloom, 1990). However, tissue levels of the vasoconstrictor peptide appear more relevant as secretion occurs mainly towards the smooth muscle layer located at the abluminal side of the endothelial cells (Wagner et al., 1992).

5.1 Role in Cardiovascular Disease

Both *prepro ET-1* gene expression (Larivière, Day, & Schiffrin, 1993) and presence of the peptide are increased in arteries of animals with salt-sensitive hypertension (Larivière, Thibault, & Schiffrin, 1993) and humans at cardiovascular risk (Bacon, Cary, & Davenport, 1996; Rossi et al., 1999). In order to better understand the role of ET-1 in physiology and disease, transgenic animals have been engineered. Mice with heterozygous deletion of the peptide exhibit a paradoxical elevation in arterial blood pressure and if the peptide is completely lacking, do not survive due to respiratory failure, indicating an essential role of ET-1 in embryonal development (Kurihara et al., 1994; Vanhoutte, 1994). Overexpressing the peptide's gene globally leads to glomerulosclerosis, kidney- and lung fibrosis without a necessarily maintained increase in ET-1 plasma- and/or tissue levels (Hocher et al., 2000, 1997; Shindo et al., 2002). A rise in arterial blood pressure in these animals is observed only under high salt diet (Shindo et al., 2002) and during inhibition- (Hocher et al., 2004) or absence of endothelial nitric oxide synthase (Quaschning et al., 2007). Heterozygous endothelium-restricted ET-1 overexpression does not cause hypertension under basal conditions (Amiri, Paradis, Reudelhuber, & Schiffrin, 2008; Amiri et al., 2004; Javeshghani, Barhoumi, Idris-Khodja, Paradis, & Schiffrin, 2013), but results in a salt-dependent elevation in systolic arterial blood pressure sensitive to

ET_A receptor blockade (Amiri, Ko, Javeshghani, Reudelhuber, & Schiffrin, 2010). Homozygous overexpression of the murine *prepro ET-1* gene in endothelial cells elevates systolic arterial blood pressure in an ET_A receptor-dependent manner without pulmonary- or renal fibrosis (Leung, Ho, Lo, Chung, & Chung, 2004; Leung et al., 2011). This ET_A receptor dependency of an increase in arterial blood pressure is further observed in mice with inducible heterozygous overexpression of human ET-1 (Rautureau et al., 2015), while induction of ET_A receptor deficiency in vascular smooth muscle cells prevents age-dependent blood pressure increases (Wirth et al., 2016). Similarly in humans, ET-1 has ET_A receptor-mediated vasoconstrictor activity in vivo particularly under pathophysiological conditions such as hypertension (Cardillo, Kilcoyne, Waclawiw, Cannon, & Panza, 1999), obesity, and diabetes (Cardillo, Campia, Bryant, & Panza, 2002; Cardillo, Campia, Iantorno, & Panza, 2004; Mather, Mirzamohammadi, Lteif, Steinberg, & Baron, 2002; Weil et al., 2011). A higher cardiovascular risk as in obesity or aging is further associated with upregulation of the *prepro ET-1* gene expression (Goel et al., 2010; Traupe, Lang, et al., 2002; Wirth et al., 2016) and ET_A receptor presence (Mundy et al., 2007; Fig. 2). ET_A receptor blockade attenuates endothelial dysfunction and reduces atherosclerotic lesion formation in experimental hyperlipidemia (Barton et al., 1998). ET_B receptors on vascular smooth muscle cells contribute to vasoconstriction in renal arteries of spontaneously hypertensive rats particularly with aging (Seo & Lüscher, 1995), as well as mesenteric arteries of hypercholesterolemic rats (Xu et al., 2014), and porcine coronaries (Hasdai et al., 1997; Fig. 2). As regards the human, ET_B receptor-mediated vasomotor tone plays a role in vivo in essential hypertensive- (Cardillo et al., 1999), but not in hypercholesterolemic patients (Cardillo, Kilcoyne, Cannon, & Panza, 2000). Additional ET_B receptor blockade does not have an effect compared to the ET_A receptor antagonist alone in attenuating the increased ET-1-mediated vascular tone in diabetic humans (Cardillo et al., 2002). Similarly, ET_A receptor blockade not only improves vascular function but also prevents remodeling throughout vascular beds relevant in diabetes-related complications (Ergul, 2011).

Contractions to endogenously processed ET-1 following administration of its precursor big ET-1 are augmented in atherosclerotic human coronary arteries indicating that the endothelin-converting enzyme (ECE) activity is increased in disease (Maguire & Davenport, 1998). This ECE activity is higher in patients with atherosclerosis also in vivo (Böhm, Johansson, Hedin, Alving, & Pernow, 2002). At the molecular basis, the inflammatory potency of ET-1 is illustrated by an increased activation of the transcription factors activator protein 1 (AP-1) and nuclear factor kappa B (NFκB) in arteries of ET-1 overexpressing mice (Amiri et al., 2008). These transcription factors are further important for ET-1 expression itself (Stow, Jacobs, Wingo, & Cain, 2011). Inhibition of AP-1 by activation of peroxisome proliferator-activated receptor (PPAR) types α and γ decreases stimulated ET-1 production in endothelial cells (Delerive et al., 1999). PPAR activity is of particular importance in disease states such as obesity, diabetes, and atherosclerosis (Kersten, Desvergne, & Wahli, 2000; Semple, Chatterjee, & O'Rahilly, 2006). Activation of PPARγ has beneficial cardiovascular effects by attenuating ET-1-driven inflammation (Montezano, Amiri, Tostes, Touyz, & Schiffrin, 2007), while PPARα activation reduces end organ damage possibly via suppression of AP-1-mediated augmentation of ET-1 expression (Ogata et al., 2002). Particularly in diabetes, PPARγ agonists represent a valuable therapeutic approach to address among others the augmented ET-1 activity (Matsumoto, Kobayashi, & Kamata, 2008).

Deletion of the *prepro ET-1* gene specifically in endothelial cells confirms that the endothelium is the main source of the peptide and results in a decrease in systolic arterial blood pressure (Kisanuki et al., 2010), further supporting a role for ET-1 in the maintenance of basal vasomotor tone (Haynes & Webb, 1994). Under chronic pressure overload, endothelium-derived ET-1 is required to preserve cardiac function (Heiden et al., 2014).

6. INTERACTION BETWEEN PROSTACYCLIN AND ET-1

Besides releasing nitric oxide opposing its action (Boulanger & Lüscher, 1990; De Mey & Vanhoutte, 2014; Rapoport, 2014; Thorin & Clozel, 2010; Vanhoutte, 2000), ET-1 promotes the COX-dependent generation of prostacyclin and thromboxane A_2 (de Nucci et al., 1988). Moreover, intrarenal- or systemic administration of ET-1 augments plasma levels of prostacyclin (Carrier, Brochu, de Brum-Fernandes, & D'Orléans-Juste, 2007; Chou, Dahhan, & Porush, 1990). Thromboxane A_2 has been identified as an EDCF augmenting contractions in ET-1 stimulated aortae with endothelium of spontaneously hypertensive rats (Auch-Schwelk & Vanhoutte, 1992; Taddei & Vanhoutte, 1993). Prostacyclin, however, attenuates the vasoconstrictor responses to ET-1 in human veins in vivo (Haynes & Webb, 1993) and inhibits its production and secretion from endothelial cells (Prins et al., 1994). By contrast, contractions to ET-1 are partially mediated by prostanoids distinct from thromboxane A_2 in isolated human placental veins with endothelium (Le, Wasserstrum, Mombouli & Vanhoutte, 1993). Indeed, endoperoxides (the precursors of prostanoids, including prostacyclin) contribute to the vasoconstriction induced by ET-1 in porcine coronary arteries with regenerated endothelium (Park et al., 1999), while prostacyclin attenuates ET-1-induced vascular tone in this preparation from healthy young animals (Suzuki, Kajikuri, Suzuki, & Itoh, 1991). Blockade of endothelin receptors decreases the formation of prostanoids, including prostacyclin, in isolated kidney- and lung preparations (Télémaque, Gratton, Claing, & D'Orléans-Juste, 1993; Uhlig, von Bethmann, Featherstone, & Wendel, 1995; Warner, Battistini, Allcock, & Vane, 1993). Indeed, activation of endothelial ET_A receptors by ET-1 causes an increase in cytosolic calcium (Amano et al., 1994; Avedanian et al., 2010; Nishimura et al., 1995) activating phospholipase A_2 which then liberates membrane lipids to form arachidonic acid (N'Diaye et al., 1997; Vigne & Frelin, 1994), the essential substrate for COX to synthesize vasoconstrictor prostanoids in the endothelium (Vanhoutte et al., 2009; Fig. 3).

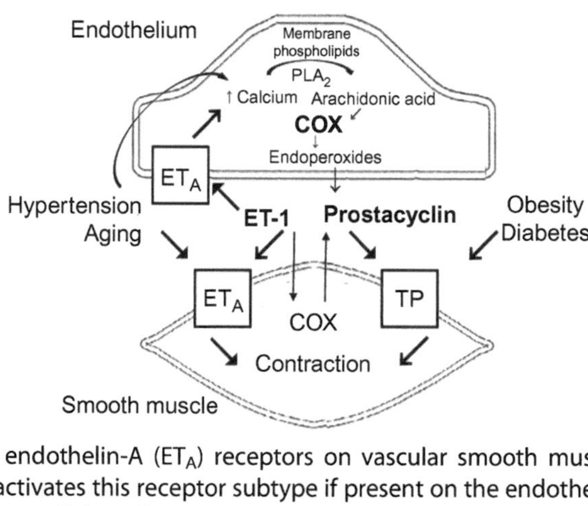

Fig. 3 Besides endothelin-A (ET_A) receptors on vascular smooth muscle cells, endothelin-1 (ET-1) activates this receptor subtype if present on the endothelium. This leads to increase in intracellular calcium activating phospholipase A_2 (PLA_2) to liberate membrane phospholipids form arachidonic acid, which acts as substrate for cyclooxygenases (COX) generating endoperoxides that are transformed further into prostacyclin (PGI_2). Moreover, ET-1 augments COX activity in vascular smooth muscle cells, which become another source of vasoconstrictor prostaglandins activating TP receptors. With aging and additional cardiovascular risk factors (hypertension, obesity, diabetes) the facilitated endothelium-dependent, prostanoid-mediated contractions are potentiated further by the peptide.

Acute- (Moreau, Takase, & Lüscher, 1996) as well as chronic (Traupe, D'Uscio, et al., 2002) blockade of ET_A receptors curtail endothelium-dependent, prostanoid-mediated contractions, suggesting a causal link between ET-1's action and these COX-dependent responses. In addition to activating phospholipase A_2 (Resink, Scott-Burden, & Bühler, 1989), ET-1 augments cyclooxygenase (COX-2) gene expression and protein presence in vascular smooth muscle (Chen, Balyakina, Lawrence, Christman, & Meyrick, 2003; Fig. 3). In the absence of nitric oxide, such increased COX-2 expression contributes to the facilitated ET_A receptor-dependent contractions to ET-1 (Zhou et al., 2006). At the molecular level, both AP-1 and NFκB are implicated in ET-1-driven COX-2 expression in brain microvascular endothelial cells (Hsieh, Lin, Chan, Yang, & Yang, 2012; Lin et al., 2013). By contrast, ET-1-induced hypertrophy of aortic vascular smooth muscle is COX-1-dependent (Taurin, Hogarth, Sandbo, Yau, & Dulin, 2007), while in pulmonary vascular smooth muscle cells the peptide activates both COX-1 and COX-2 (Deacon & Knox, 2010). The peptide stimulates ET_A receptors coupled to calcium-dependent PLA_2 activation providing the arachidonic acid needed for prostacyclin synthesis, and to a lesser extent that of prostaglandin E_2 (Deacon & Knox, 2010). Taken in conjunction, these findings prompt the conclusion that prostacyclin and ET-1 may well be partners in crime in exacerbating endothelial dysfunction due to the prominence of endothelium-dependent contractions.

7. THERAPEUTIC APPROACHES

Despite encouraging results from experimental studies (Barton et al., 2000, 1998; Cardillo et al., 2004, 1999; Mather et al., 2002; Moreau et al., 1996; Weil et al., 2011), endothelin receptor antagonists have so far not been established as a widely applied therapy for cardiovascular- and related renal diseases in humans. This is possibly due to the nonhuman characteristics of the animal models used as well as to issues concerning the timing and dosing of the treatment (Barton, 2011; Kohan, Cleland, Rubin, Theodorescu, & Barton, 2012). Further disappointing have been attempts to inhibit cyclooxygenases (which are important not only for inflammation but also essential for prostanoid-mediated contractions) since the use of compounds inhibiting these enzymes is associated with renal damage (Perneger, Whelton, & Klag, 1994; Spühler & Zollinger, 1953) and more frequent cardiovascular events (FitzGerald, 2004; Jüni et al., 2004). TP receptor antagonists may have therapeutic value especially against end organ damage including atherosclerosis (Félétou, Cohen, Vanhoutte, & Verbeuren, 2010; Francois et al., 2005), while orally active IP receptor agonists show promising effects in pulmonary arterial hypertension (Simonneau et al., 2012; Sitbon et al., 2015) and may open the field for the treatment of systemically increased systolic arterial blood pressure. Such IP receptor agonists appear to be more efficacious in inducing relaxations in pulmonary arteries than conventional prostacyclin analogs (Morrison et al., 2012). The latter have additional benefits for pulmonary arterial hypertension patients under treatment with an endothelin receptor antagonist in terms of a decreased pulmonary arterial pressure and vascular resistance (McLaughlin et al., 2006). This is in line with animal studies showing not only that endothelin receptor blockade and prostacyclin analogs are equivalent in treating pulmonary hypertension (Ueno, Miyauchi, Sakai, Goto, & Yamaguchi, 2000), but also that the combination of these two classes of drugs is better than the administration of each of them individually (Ueno et al., 2002). Since prostacyclin analogs decrease ET-1 levels during lung injury (Kawashima et al., 2003) and reduce the cardiomyocyte hypertrophy caused by the peptide (Ritchie et al., 2004), restoration of IP receptor-mediated decreases in pressure may also be useful in the treatment of systemic arterial hypertension as prostacyclin counteracts renal constrictor effects of ET-1 (Chou et al., 1990). Therefore, potent and effective IP receptor activation may be a valid pharmacological target for the treatment of high arterial blood pressure (Simonneau et al., 2012; Sitbon et al., 2015).

8. CONCLUSION

Prostacyclin has been established as the major EDCF involved in prostanoid-mediated contractions leading to activation of TP receptors of vascular smooth muscle (Auch-Schwelk et al., 1990; Gluais et al., 2005; Rapoport & Williams, 1996; Tesfamariam et al., 1989). ET-1 modulates such responses by further promoting the generation of prostacyclin (de Nucci et al., 1988) potentially leading to aggravated prostanoid-mediated contractions under pathophysiological conditions (Park et al., 1999; Taddei & Vanhoutte, 1993). The vasoconstrictor activity of high levels of prostacyclin in the face of reduced IP receptor responsiveness, in the renal vasculature (Eskildsen et al., 2014; Liu et al., 2012b, 2013) can become instrumental in the development and maintenance of hypertension, for which a decreased blood supply to the kidney can be held responsible (Garovic & Textor, 2005; Goldblatt, Lynch, Hanzal, & Summerville,

1934; Guyton et al., 1972; Oparil & Haber, 1974; Ponnuchamy & Khalil, 2009). In order to preserve the beneficial effects of moderate levels of prostacyclin on the vasculature (Moncada et al., 1976), activation of its IP receptors rather than synthesis inhibition of prostacyclin could be of therapeutic value. Indeed, a new class of orally available, selective agonists have promising effects in pulmonary arterial hypertension by restoring IP receptor function (Simonneau et al., 2012; Sitbon et al., 2015) and its usefulness in the treatment of systemic increases in arterial blood pressure warrants further investigation.

CONFLICT OF INTEREST

The authors have no conflicts of interest to declare.

ACKNOWLEDGMENTS

O.B. was supported by the Swiss National Science Foundation (Grants 138 754 and 143 672) and by a HKU Postgraduate Fellowship in Integrative Medicine.

REFERENCES

Aird, W. C. (2007). *Endothelial biomedicine*. Cambridge, United Kingdom: Cambridge University Press.

Amano, K., Hori, M., Ozaki, H., Matsuda, Y., & Karaki, H. (1994). Ca^{++} mobilization mediated by endothelin ET_A receptor in endothelium of rabbit aortic valve. *Journal of Pharmacology and Experimental Therapeutics*, *271*(3), 1359–1364.

Amiri, F., Ko, E. A., Javeshghani, D., Reudelhuber, T. L., & Schiffrin, E. L. (2010). Deleterious combined effects of salt-loading and endothelial cell restricted endothelin-1 overexpression on blood pressure and vascular function in mice. *Journal of Hypertension*, *28*(6), 1243–1251.

Amiri, F., Paradis, P., Reudelhuber, T. L., & Schiffrin, E. L. (2008). Vascular inflammation in absence of blood pressure elevation in transgenic murine model overexpressing endothelin-1 in endothelial cells. *Journal of Hypertension*, *26*(6), 1102–1109.

Amiri, F., Virdis, A., Neves, M. F., Iglarz, M., Seidah, N. G., Touyz, R. M., et al. (2004). Endothelium-restricted overexpression of human endothelin-1 causes vascular remodeling and endothelial dysfunction. *Circulation*, *110*(15), 2233–2240.

Arai, H., Hori, S., Aramori, I., Ohkubo, H., & Nakanishi, S. (1990). Cloning and expression of a cDNA encoding an endothelin receptor. *Nature*, *348*(6303), 730–732.

Arehart, E., Stitham, J., Asselbergs, F. W., Douville, K., MacKenzie, T., Fetalvero, K. M., et al. (2008). Acceleration of cardiovascular disease by a dysfunctional prostacyclin receptor mutation: Potential implications for cyclooxygenase-2 inhibition. *Circulation Research*, *102*(8), 986–993.

Armstrong, R. A., Lawrence, R. A., Jones, R. L., Wilson, N. H., & Collier, A. (1989). Functional and ligand binding studies suggest heterogeneity of platelet prostacyclin receptors. *British Journal of Pharmacology*, *97*(3), 657–668.

Auch-Schwelk, W., Katušić, Z. S., & Vanhoutte, P. M. (1989). Contractions to oxygen-derived free radicals are augmented in aorta of the spontaneously hypertensive rat. *Hypertension*, *13*(6 Pt. 2), 859–864.

Auch-Schwelk, W., Katušić, Z. S., & Vanhoutte, P. M. (1990). Thromboxane A_2 receptor antagonists inhibit endothelium-dependent contractions. *Hypertension, 15*(6 Pt. 2), 699–703.

Auch-Schwelk, W., Katušić, Z. S., & Vanhoutte, P. M. (1992). Nitric oxide inactivates endothelium-derived contracting factor in the rat aorta. *Hypertension, 19*(5), 442–445.

Auch-Schwelk, W., & Vanhoutte, P. M. (1992). Contractions to endothelin in normotensive and spontaneously hypertensive rats: Role of endothelium and prostaglandins. *Blood Pressure, 1*(1), 45–49.

Avedanian, L., Riopel, J., Bkaily, G., Nader, M., D'Orléans-Juste, P., & Jacques, D. (2010). ET_A receptors are present in human aortic vascular endothelial cells and modulate intracellular calcium. *Canadian Journal of Physiology and Pharmacology, 88*(8), 817–829.

Bacon, C. R., Cary, N. R., & Davenport, A. P. (1996). Endothelin peptide and receptors in human atherosclerotic coronary artery and aorta. *Circulation Research, 79*(4), 794–801.

Baretella, O., Chung, S. K., Barton, M., Xu, A., & Vanhoutte, P. M. (2014). Obesity and heterozygous endothelial overexpression of prepro-endothelin-1 modulate responsiveness of mouse main and segmental renal arteries to vasoconstrictor agents. *Life Sciences, 118*(2), 206–212.

Baretella, O., Xu, A., & Vanhoutte, P. M. (2014). Acidosis prevents and alkalosis augments endothelium-dependent contractions in mouse arteries. *Pflugers Archiv: European Journal of Physiology, 466*(2), 295–305.

Barton, M. (2011). Endothelin antagonism and reversal of proteinuric renal disease in humans. *Contributions to Nephrology, 172*, 210–222.

Barton, M., Carmona, R., Morawietz, H., d'Uscio, L. V., Goettsch, W., Hillen, H., et al. (2000). Obesity is associated with tissue-specific activation of renal angiotensin-converting enzyme *in vivo*: Evidence for a regulatory role of endothelin. *Hypertension, 35*(1 Pt. 2), 329–336.

Barton, M., Haudenschild, C. C., d'Uscio, L. V., Shaw, S., Münter, K., & Lüscher, T. F. (1998). Endothelin ET_A receptor blockade restores NO-mediated endothelial function and inhibits atherosclerosis in apolipoprotein E-deficient mice. *Proceedings of the National Academy of Sciences of the United States of America, 95*(24), 14367–14372.

Barton, M., & Yanagisawa, M. (2008). Endothelin: 20 years from discovery to therapy. *Canadian Journal of Physiology and Pharmacology, 86*(8), 485–498.

Bauersachs, J., Popp, R., Hecker, M., Sauer, E., Fleming, I., & Busse, R. (1996). Nitric oxide attenuates the release of endothelium-derived hyperpolarizing factor. *Circulation, 94*(12), 3341–3347.

Böhm, F., Johansson, B. L., Hedin, U., Alving, K., & Pernow, J. (2002). Enhanced vasoconstrictor effect of big endothelin-1 in patients with atherosclerosis: Relation to conversion to endothelin-1. *Atherosclerosis, 160*(1), 215–222.

Boie, Y., Rushmore, T. H., Darmon-Goodwin, A., Grygorczyk, R., Slipetz, D. M., Metters, K. M., et al. (1994). Cloning and expression of a cDNA for the human prostanoid IP receptor. *Journal of Biological Chemistry, 269*(16), 12173–12178.

Boulanger, C., & Lüscher, T. F. (1990). Release of endothelin from the porcine aorta. Inhibition by endothelium-derived nitric oxide. *Journal of Clinical Investigation, 85*(2), 587–590.

Boulanger, C. M., Morrison, K. J., & Vanhoutte, P. M. (1994). Mediation by M_3-muscarinic receptors of both endothelium-dependent contraction and relaxation to acetylcholine in the aorta of the spontaneously hypertensive rat. *British Journal of Pharmacology, 112*(2), 519–524.

Bunting, S., Gryglewski, R., Moncada, S., & Vane, J. R. (1976). Arterial walls generate from prostaglandin endoperoxides a substance (prostaglandin X) which relaxes strips of mesenteric and coeliac ateries and inhibits platelet aggregation. *Prostaglandins, 12*(6), 897–913.

Cardillo, C., Campia, U., Bryant, M. B., & Panza, J. A. (2002). Increased activity of endogenous endothelin in patients with type II diabetes mellitus. *Circulation, 106*(14), 1783–1787.

Cardillo, C., Campia, U., Iantorno, M., & Panza, J. A. (2004). Enhanced vascular activity of endogenous endothelin-1 in obese hypertensive patients. *Hypertension, 43*(1), 36–40.

Cardillo, C., Kilcoyne, C. M., Cannon, R. O., 3rd, & Panza, J. A. (2000). Increased activity of endogenous endothelin in patients with hypercholesterolemia. *Journal of the American College of Cardiology, 36*(5), 1483–1488.

Cardillo, C., Kilcoyne, C. M., Waclawiw, M., Cannon, R. O., 3rd, & Panza, J. A. (1999). Role of endothelin in the increased vascular tone of patients with essential hypertension. *Hypertension, 33*(2), 753–758.

Carrier, E., Brochu, I., de Brum-Fernandes, A. J., & D'Orléans-Juste, P. (2007). The inducible nitric-oxide synthase modulates endothelin-1-dependent release of prostacyclin and inhibition of platelet aggregation *ex vivo* in the mouse. *Journal of Pharmacology and Experimental Therapeutics, 323*(3), 972–978.

Chadha, P. S., Haddock, R. E., Howitt, L., Morris, M. J., Murphy, T. V., Grayson, T. H., et al. (2010). Obesity up-regulates intermediate conductance calcium-activated potassium channels and myoendothelial gap junctions to maintain endothelial vasodilator function. *Journal of Pharmacology and Experimental Therapeutics, 335*(2), 284–293.

Chan, C. K., Mak, J., Gao, Y., Man, R. Y., & Vanhoutte, P. M. (2011). Endothelium-derived NO, but not cyclic GMP, is required for hypoxic augmentation in isolated porcine coronary arteries. *American Journal of Physiology. Heart and Circulatory Physiology, 301*(6), H2313–H2321.

Chen, D., Balyakina, E. V., Lawrence, M., Christman, B. W., & Meyrick, B. (2003). Cyclooxygenase is regulated by ET-1 and MAPKs in peripheral lung microvascular smooth muscle cells. *American Journal of Physiology. Lung Cellular and Molecular Physiology, 284*(4), L614–L621.

Chen, G., Suzuki, H., & Weston, A. H. (1988). Acetylcholine releases endothelium-derived hyperpolarizing factor and EDRF from rat blood vessels. *British Journal of Pharmacology, 95*(4), 1165–1174.

Chen, Z., Zhang, X., Ying, L., Dou, D., Li, Y., Bai, Y., et al. (2014). cIMP synthesized by sGC as a mediator of hypoxic contraction of coronary arteries. *American Journal of Physiology. Heart and Circulatory Physiology, 307*(3), H328–H336.

Cho, M. J., & Allen, M. A. (1978). Chemical stability of prostacyclin (PGI_2) in aqueous solutions. *Prostaglandins, 15*(6), 943–954.

Chou, S. Y., Dahhan, A., & Porush, J. G. (1990). Renal actions of endothelin: Interaction with prostacyclin. *American Journal of Physiology, 259*(4 Pt. 2), F645–F652.

Cocks, T. M., & Angus, J. A. (1983). Endothelium-dependent relaxation of coronary arteries by noradrenaline and serotonin. *Nature, 305*(5935), 627–630.

Cohen, R. A., Shepherd, J. T., & Vanhoutte, P. M. (1983). 5-Hydroxytryptamine can mediate endothelium-dependent relaxation of coronary arteries. *American Journal of Physiology, 245*(6), H1077–H1080.

Cohen, R. A., & Vanhoutte, P. M. (1995). Endothelium-dependent hyperpolarization. Beyond nitric oxide and cyclic GMP. *Circulation, 92*(11), 3337–3349.

Deacon, K., & Knox, A. J. (2010). Endothelin-1 (ET-1) increases the expression of remodeling genes in vascular smooth muscle through linked calcium and cAMP pathways: Role of a phospholipase A_2 ($cPLA_2$)/cyclooxygenase-2 (COX-2)/prostacyclin receptor-dependent autocrine loop. *Journal of Biological Chemistry, 285*(34), 25913–25927.

Delerive, P., Martin-Nizard, F., Chinetti, G., Trottein, F., Fruchart, J. C., Najib, J., et al. (1999). Peroxisome proliferator-activated receptor activators inhibit thrombin-induced endothelin-1 production in human vascular endothelial cells by inhibiting the activator protein-1 signaling pathway. *Circulation Research, 85*(5), 394–402.

De Mey, J. G., & Vanhoutte, P. M. (1982). Heterogeneous behavior of the canine arterial and venous wall. Importance of the endothelium. *Circulation Research*, *51*(4), 439–447.
De Mey, J. G., & Vanhoutte, P. M. (1983). Anoxia and endothelium-dependent reactivity of the canine femoral artery. *The Journal of Physiology*, *335*, 65–74.
De Mey, J. G., & Vanhoutte, P. M. (2014). End o' the line revisited: Moving on from nitric oxide to CGRP. *Life Sciences*, *118*(2), 120–128.
de Nucci, G., Thomas, R., D'Orléans-Juste, P., Antunes, E., Walder, C., Warner, T. D., et al. (1988). Pressor effects of circulating endothelin are limited by its removal in the pulmonary circulation and by the release of prostacyclin and endothelium-derived relaxing factor. *Proceedings of the National Academy of Sciences of the United States of America*, *85*(24), 9797–9800.
DeWitt, D. L., Day, J. S., Sonnenburg, W. K., & Smith, W. L. (1983). Concentrations of prostaglandin endoperoxide synthase and prostaglandin I_2 synthase in the endothelium and smooth muscle of bovine aorta. *Journal of Clinical Investigation*, *72*(6), 1882–1888.
Dusting, G. J., Moncada, S., & Vane, J. R. (1977a). Prostacyclin (PGX) is the endogenous metabolite responsible for relaxation of coronary arteries induced by arachindonic acid. *Prostaglandins*, *13*(1), 3–15.
Dusting, G. J., Moncada, S., & Vane, J. R. (1977b). Prostacyclin (PGI_2) is a weak contractor of coronary arteries of the pig. *European Journal of Pharmacology*, *45*(3), 301–304.
Ergul, A. (2011). Endothelin-1 and diabetic complications: Focus on the vasculature. *Pharmacological Research*, *63*(6), 477–482.
Ergul, S., Parish, D. C., Puett, D., & Ergul, A. (1996). Racial differences in plasma endothelin-1 concentrations in individuals with essential hypertension. *Hypertension*, *28*(4), 652–655.
Eskildsen, M. P., Hansen, P. B., Stubbe, J., Toft, A., Walter, S., Marcussen, N., et al. (2014). Prostaglandin I_2 and prostaglandin E_2 modulate human intrarenal artery contractility through prostaglandin E_2-EP_4, prostacyclin-IP, and thromboxane A_2-TP receptors. *Hypertension*, *64*(3), 551–556.
Félétou, M., Cohen, R. A., Vanhoutte, P. M., & Verbeuren, T. J. (2010). TP receptors and oxidative stress hand in hand from endothelial dysfunction to atherosclerosis. *Advances in Pharmacology*, *60*, 85–106.
Félétou, M., Huang, Y., & Vanhoutte, P. M. (2010). Vasoconstrictor prostanoids. *Pflugers Archiv: European Journal of Physiology*, *459*(6), 941–950.
Félétou, M., Tang, E. H., & Vanhoutte, P. M. (2008). Nitric oxide the gatekeeper of endothelial vasomotor control. *Frontiers in Bioscience*, *13*, 4198–4217.
Félétou, M., & Vanhoutte, P. M. (1988). Endothelium-dependent hyperpolarization of canine coronary smooth muscle. *British Journal of Pharmacology*, *93*(3), 515–524.
Félétou, M., & Vanhoutte, P. M. (2013). Endothelium-dependent hyperpolarization: No longer an F-word! *Journal of Cardiovascular Pharmacology*, *61*(2), 91–92.
Ferri, C., Bellini, C., Desideri, G., Di Francesco, L., Baldoncini, R., Santucci, A., et al. (1995). Plasma endothelin-1 levels in obese hypertensive and normotensive men. *Diabetes*, *44*(4), 431–436.
FitzGerald, G. A. (2004). Coxibs and cardiovascular disease. *New England Journal of Medicine*, *351*(17), 1709–1711.
Francois, H., Athirakul, K., Howell, D., Dash, R., Mao, L., Kim, H. S., et al. (2005). Prostacyclin protects against elevated blood pressure and cardiac fibrosis. *Cell Metabolism*, *2*(3), 201–207.
Francois, H., Athirakul, K., Mao, L., Rockman, H., & Coffman, T. M. (2004). Role for thromboxane receptors in angiotensin-II-induced hypertension. *Hypertension*, *43*(2), 364–369.
Fujino, T., Nakagawa, N., Yuhki, K., Hara, A., Yamada, T., Takayama, K., et al. (2004). Decreased susceptibility to renovascular hypertension in mice lacking the prostaglandin I_2 receptor IP. *Journal of Clinical Investigation*, *114*(6), 805–812.

Fukuroda, T., Fujikawa, T., Ozaki, S., Ishikawa, K., Yano, M., & Nishikibe, M. (1994). Clearance of circulating endothelin-1 by ET_B receptors in rats. *Biochemical and Biophysical Research Communications*, *199*(3), 1461–1465.

Furchgott, R. F. (1988). Studies on relaxation of rabbit aorta by sodium nitrite: The basis for the proposal that the acid-activatable inhibitory factor from bovine retractor penis is inorganic nitrite and the endothelium-derived relaxing factor is nitric oxide. In P. M. Vanhoutte (Ed.), *Vasodilatation: Vascular smooth muscle, peptides, autonomic nerves and endothelium: Vol. 4.* (pp. 401–414). New York: Raven Press.

Furchgott, R. F. (1993). The discovery of endothelium-dependent relaxation. *Monograph-American Heart Association*, *87*(5), V3–V8.

Furchgott, R. F., & Vanhoutte, P. M. (1989). Endothelium-derived relaxing and contracting factors. *FASEB Journal*, *3*(9), 2007–2018.

Furchgott, R. F., & Zawadzki, J. V. (1980). The obligatory role of endothelial cells in the relaxation of arterial smooth muscle by acetylcholine. *Nature*, *288*(5789), 373–376.

Gao, Y., & Vanhoutte, P. M. (2014). Tissues cIMPly do not lie. *Naunyn-Schmiedebergs Archives of Pharmacology*, *387*(9), 901–903.

Garovic, V. D., & Textor, S. C. (2005). Renovascular hypertension and ischemic nephropathy. *Circulation*, *112*(9), 1362–1374.

Ge, T., Hughes, H., Junquero, D. C., Wu, K. K., Vanhoutte, P. M., & Boulanger, C. M. (1995). Endothelium-dependent contractions are associated with both augmented expression of prostaglandin H synthase-1 and hypersensitivity to prostaglandin H_2 in the SHR aorta. *Circulation Research*, *76*(6), 1003–1010.

Giacco, F., & Brownlee, M. (2010). Oxidative stress and diabetic complications. *Circulation Research*, *107*(9), 1058–1070.

Gillespie, M. N., Owasoyo, J. O., McMurtry, I. F., & O'Brien, R. F. (1986). Sustained coronary vasoconstriction provoked by a peptidergic substance released from endothelial cells in culture. *Journal of Pharmacology and Experimental Therapeutics*, *236*(2), 339–343.

Gluais, P., Lonchampt, M., Morrow, J. D., Vanhoutte, P. M., & Félétou, M. (2005). Acetylcholine-induced endothelium-dependent contractions in the SHR aorta: The Janus face of prostacyclin. *British Journal of Pharmacology*, *146*(6), 834–845.

Gluais, P., Paysant, J., Badier-Commander, C., Verbeuren, T., Vanhoutte, P. M., & Félétou, M. (2006). In SHR aorta, calcium ionophore A-23187 releases prostacyclin and thromboxane A_2 as endothelium-derived contracting factors. *American Journal of Physiology. Heart and Circulatory Physiology*, *291*(5), H2255–H2264.

Goel, A., Su, B., Flavahan, S., Lowenstein, C. J., Berkowitz, D. E., & Flavahan, N. A. (2010). Increased endothelial exocytosis and generation of endothelin-1 contributes to constriction of aged arteries. *Circulation Research*, *107*(2), 242–251.

Goldblatt, H., Lynch, J., Hanzal, R. F., & Summerville, W. W. (1934). Studies on experimental hypertension: I. The production of persistent elevation of systolic blood pressure by means of renal ischemia. *Journal of Experimental Medicine*, *59*(3), 347–379.

Gomez, E., Schwendemann, C., Roger, S., Simonet, S., Paysant, J., Courchay, C., et al. (2008). Aging and prostacyclin responses in aorta and platelets from WKY and SHR rats. *American Journal of Physiology. Heart and Circulatory Physiology*, *295*(5), H2198–H2211.

Gräser, T., & Vanhoutte, P. M. (1991). Hypoxic contraction of canine coronary arteries: Role of endothelium and cGMP. *American Journal of Physiology*, *261*(6 Pt. 2), H1769–H1777.

Guyton, A. C., Coleman, T. G., Cowley, A. V., Jr., Scheel, K. W., Manning, R. D., Jr., & Norman, R. A., Jr. (1972). Arterial pressure regulation. Overriding dominance of the kidneys in long-term regulation and in hypertension. *American Journal of Medicine*, *52*(5), 584–594.

Halcox, J. P., Narayanan, S., Cramer-Joyce, L., Mincemoyer, R., & Quyyumi, A. A. (2001). Characterization of endothelium-derived hyperpolarizing factor in the human forearm microcirculation. *American Journal of Physiology. Heart and Circulatory Physiology, 280*(6), H2470–H2477.
Hara, A., Yuhki, K., Fujino, T., Yamada, T., Takayama, K., Kuriyama, S., et al. (2005). Augmented cardiac hypertrophy in response to pressure overload in mice lacking the prostaglandin I_2 receptor. *Circulation, 112*(1), 84–92.
Hasdai, D., Mathew, V., Schwartz, R. S., Smith, L. A., Holmes, D. R., Jr., Katušić, Z. S., et al. (1997). Enhanced endothelin-B-receptor-mediated vasoconstriction of small porcine coronary arteries in diet-induced hypercholesterolemia. *Arteriosclerosis, Thrombosis, and Vascular Biology, 17*(11), 2737–2743.
Hata, A. N., & Breyer, R. M. (2004). Pharmacology and signaling of prostaglandin receptors: Multiple roles in inflammation and immune modulation. *Pharmacology and Therapeutics, 103*(2), 147–166.
Haynes, W. G., & Webb, D. J. (1993). Endothelium-dependent modulation of responses to endothelin-I in human veins. *Clinical Science (London, England), 84*(4), 427–433.
Haynes, W. G., & Webb, D. J. (1994). Contribution of endogenous generation of endothelin-1 to basal vascular tone. *Lancet, 344*(8926), 852–854.
Heiden, S., Vignon-Zellweger, N., Masuda, S., Yagi, K., Nakayama, K., Yanagisawa, M., et al. (2014). Vascular endothelium derived endothelin-1 is required for normal heart function after chronic pressure overload in mice. *PLoS One, 9*(2), e88730.
Hickey, K. A., Rubanyi, G., Paul, R. J., & Highsmith, R. F. (1985). Characterization of a coronary vasoconstrictor produced by cultured endothelial cells. *American Journal of Physiology, 248*(5 Pt. 1), C550–C556.
Hirata, Y., Emori, T., Eguchi, S., Kanno, K., Imai, T., Ohta, K., et al. (1993). Endothelin receptor subtype B mediates synthesis of nitric oxide by cultured bovine endothelial cells. *Journal of Clinical Investigation, 91*(4), 1367–1373.
His, W. (1865). *Academisches Programm: Die Häute und Höhlen des Körpers [academic program: the membranes and cavities of the body].* Basel, Switzerland: Schweighauser.
Hocher, B., Schwarz, A., Fagan, K. A., Thöne-Reineke, C., El-Hag, K., Kusserow, H., et al. (2000). Pulmonary fibrosis and chronic lung inflammation in ET-1 transgenic mice. *American Journal of Respiratory Cell and Molecular Biology, 23*(1), 19–26.
Hocher, B., Schwarz, A., Slowinski, T., Bachmann, S., Pfeilschifter, J., Neumayer, H. H., et al. (2004). *In-vivo* interaction of nitric oxide and endothelin. *Journal of Hypertension, 22*(1), 111–119.
Hocher, B., Thöne-Reineke, C., Rohmeiss, P., Schmager, F., Slowinski, T., Burst, V., et al. (1997). Endothelin-1 transgenic mice develop glomerulosclerosis, interstitial fibrosis, and renal cysts but not hypertension. *Journal of Clinical Investigation, 99*(6), 1380–1389.
Hsieh, H. L., Lin, C. C., Chan, H. J., Yang, C. M., & Yang, C. M. (2012). c-Src-dependent EGF receptor transactivation contributes to ET-1-induced COX-2 expression in brain microvascular endothelial cells. *Journal of Neuroinflammation, 9*, 152.
Ignarro, L. J., Buga, G. M., Wood, K. S., Byrns, R. E., & Chaudhuri, G. (1987). Endothelium-derived relaxing factor produced and released from artery and vein is nitric oxide. *Proceedings of the National Academy of Sciences of the United States of America, 84*(24), 9265–9269.
Ignarro, L., Byrns, R., & Wood, K. (1988). Biochemical and pharmacological properties of endothelium-derived relaxing factor and its similarity to nitric oxide radical. In P. M. Vanhoutte (Ed.), *Vasodilatation: Vascular smooth muscle, peptides, autonomic nerves and endothelium* (pp. 427–435). New York: Raven Press.
Javeshghani, D., Barhoumi, T., Idris-Khodja, N., Paradis, P., & Schiffrin, E. L. (2013). Reduced macrophage-dependent inflammation improves endothelin-1-induced vascular injury. *Hypertension, 62*(1), 112–117.

Jensen, B. L., Schmid, C., & Kurtz, A. (1996). Prostaglandins stimulate renin secretion and renin mRNA in mouse renal juxtaglomerular cells. *American Journal of Physiology, 271*(3 Pt. 2), F659–F669.

Johnson, R. A., Morton, D. R., Kinner, J. H., Gorman, R. R., McGuire, J. C., Sun, F. F., et al. (1976). The chemical structure of prostaglandin X (prostacyclin). *Prostaglandins, 12*(6), 915–928.

Jüni, P., Nartey, L., Reichenbach, S., Sterchi, R., Dieppe, P. A., & Egger, M. (2004). Risk of cardiovascular events and rofecoxib: Cumulative meta-analysis. *Lancet, 364*(9450), 2021–2029.

Katušić, Z. S., Schugel, J., Cosentino, F., & Vanhoutte, P. M. (1993). Endothelium-dependent contractions to oxygen-derived free radicals in the canine basilar artery. *American Journal of Physiology, 264*(3 Pt. 2), H859–H864.

Katušić, Z. S., Shepherd, J. T., & Vanhoutte, P. M. (1987). Endothelium-dependent contraction to stretch in canine basilar arteries. *American Journal of Physiology, 252*(3 Pt. 2), H671–H673.

Katušić, Z. S., Shepherd, J. T., & Vanhoutte, P. M. (1988). Endothelium-dependent contractions to calcium ionophore A23187, arachidonic acid, and acetylcholine in canine basilar arteries. *Stroke, 19*(4), 476–479.

Katušić, Z. S., & Vanhoutte, P. M. (1989). Superoxide anion is an endothelium-derived contracting factor. *American Journal of Physiology, 257*(1 Pt. 2), H33–H37.

Kawashima, M., Nakamura, T., Schneider, S., Vollmar, B., Lausberg, H. F., Bauer, M., et al. (2003). Iloprost ameliorates post-ischemic lung reperfusion injury and maintains an appropriate pulmonary ET-1 balance. *Journal of Heart and Lung Transplantation, 22*(7), 794–801.

Kersten, S., Desvergne, B., & Wahli, W. (2000). Roles of PPARs in health and disease. *Nature, 405*(6785), 421–424.

Kisanuki, Y. Y., Emoto, N., Ohuchi, T., Widyantoro, B., Yagi, K., Nakayama, K., et al. (2010). Low blood pressure in endothelial cell-specific endothelin 1 knockout mice. *Hypertension, 56*(1), 121–128.

Knebel, S. M., Sprague, R. S., & Stephenson, A. H. (2015). Prostacyclin receptor expression on platelets of humans with type 2 diabetes is inversely correlated with hemoglobin A1c levels. *Prostaglandins and Other Lipid Mediators, 116–117*, 131–135.

Kobayashi, T., Tahara, Y., Matsumoto, M., Iguchi, M., Sano, H., Murayama, T., et al. (2004). Roles of thromboxane A_2 and prostacyclin in the development of atherosclerosis in apoE-deficient mice. *Journal of Clinical Investigation, 114*(6), 784–794.

Koga, T., Takata, Y., Kobayashi, K., Takishita, S., Yamashita, Y., & Fujishima, M. (1989). Age and hypertension promote endothelium-dependent contractions to acetylcholine in the aorta of the rat. *Hypertension, 14*(5), 542–548.

Kohan, D. E., Cleland, J. G., Rubin, L. J., Theodorescu, D., & Barton, M. (2012). Clinical trials with endothelin receptor antagonists: What went wrong and where can we improve? *Life Sciences, 91*(13–14), 528–539.

Kurihara, Y., Kurihara, H., Suzuki, H., Kodama, T., Maemura, K., Nagai, R., et al. (1994). Elevated blood pressure and craniofacial abnormalities in mice deficient in endothelin-1. *Nature, 368*(6473), 703–710.

Larivière, R., Day, R., & Schiffrin, E. L. (1993). Increased expression of endothelin-1 gene in blood vessels of deoxycorticosterone acetate-salt hypertensive rats. *Hypertension, 21*(6 Pt. 2), 916–920.

Larivière, R., Thibault, G., & Schiffrin, E. L. (1993). Increased endothelin-1 content in blood vessels of deoxycorticosterone acetate-salt hypertensive but not in spontaneously hypertensive rats. *Hypertension, 21*(3), 294–300.

Le, S. Q., Wasserstrum, N., Mombouli, J. V., & Vanhoutte, P. M. (1993). Contractile effect of endothelin in human placental veins: Role of endothelium prostaglandins and thromboxane. *American Journal of Obstetrics and Gynecology, 169*(4), 919–924.

Leung, J. W., Ho, M. C., Lo, A. C., Chung, S. S., & Chung, S. K. (2004). Endothelial cell-specific over-expression of endothelin-1 leads to more severe cerebral damage following transient middle cerebral artery occlusion. *Journal of Cardiovascular Pharmacology, 44*(Suppl. 1), S293–S300.

Leung, J. W., Wong, W. T., Koon, H. W., Mo, F. M., Tam, S., Huang, Y., et al. (2011). Transgenic mice over-expressing ET-1 in the endothelial cells develop systemic hypertension with altered vascular reactivity. *PLoS One, 6*(11), e26994.

Levy, J. V. (1978). Contractile responses to prostacyclin (PGI_2) of isolated human saphenous and rat venous tissue. *Prostaglandins, 16*(1), 93–97.

Levy, J. V. (1980). Prostacyclin-induced contraction of isolated aortic strips from normal and spontaneously hypertensive rats (SHR). *Prostaglandins, 19*(4), 517–525.

Lin, C. C., Hsieh, H. L., Shih, R. H., Chi, P. L., Cheng, S. E., & Yang, C. M. (2013). Up-regulation of COX-2/PGE_2 by endothelin-1 via MAPK-dependent NF-kB pathway in mouse brain microvascular endothelial cells. *Cell Communication and Signaling: CCS, 11*(1), 8.

Linder, L., Kiowski, W., Bühler, F. R., & Lüscher, T. F. (1990). Indirect evidence for release of endothelium-derived relaxing factor in human forearm circulation *in vivo*. Blunted response in essential hypertension. *Circulation, 81*(6), 1762–1767.

Liu, B., Luo, W., Zhang, Y., Li, H., Zhu, N., Huang, D., et al. (2012a). Involvement of cyclo-oxygenase-1-mediated prostacyclin synthesis in the vasoconstrictor activity evoked by ACh in mouse arteries. *Experimental Physiology, 97*(2), 277–289.

Liu, B., Luo, W., Zhang, Y., Li, H., Zhu, N., Huang, D., et al. (2012b). Role of cyclooxygenase-1-mediated prostacyclin synthesis in endothelium-dependent vasoconstrictor activity of porcine interlobular renal arteries. *American Journal of Physiology. Renal Physiology, 302*(9), F1133–F1140.

Liu, B., Zhang, Y., Zhu, N., Li, H., Luo, W., & Zhou, Y. (2013). A vasoconstrictor role for cyclooxygenase-1-mediated prostacyclin synthesis in mouse renal arteries. *American Journal of Physiology. Renal Physiology, 305*(9), F1315–F1322.

Lüscher, T. F., & Vanhoutte, P. M. (1986). Endothelium-dependent contractions to acetylcholine in the aorta of the spontaneously hypertensive rat. *Hypertension, 8*(4), 344–348.

Maguire, J. J., & Davenport, A. P. (1998). Increased response to big endothelin-1 in atherosclerotic human coronary artery: Functional evidence for up-regulation of endothelin-converting enzyme activity in disease. *British Journal of Pharmacology, 125*(2), 238–240.

Majed, B. H., & Khalil, R. A. (2012). Molecular mechanisms regulating the vascular prostacyclin pathways and their adaptation during pregnancy and in the newborn. *Pharmacological Reviews, 64*(3), 540–582.

Masaki, T., Vane, J. R., & Vanhoutte, P. M. (1994). International Union of Pharmacology nomenclature of endothelin receptors. *Pharmacological Reviews, 46*(2), 137–142.

Mather, K. J., Mirzamohammadi, B., Lteif, A., Steinberg, H. O., & Baron, A. D. (2002). Endothelin contributes to basal vascular tone and endothelial dysfunction in human obesity and type 2 diabetes. *Diabetes, 51*(12), 3517–3523.

Matsumoto, T., Kobayashi, T., & Kamata, K. (2008). Relationships among ET-1, PPARgamma, oxidative stress and endothelial dysfunction in diabetic animals. *Journal of Smooth Muscle Research, 44*(2), 41–55.

Matsumoto, T., Watanabe, S., Kawamura, R., Taguchi, K., & Kobayashi, T. (2014). Enhanced uridine adenosine tetraphosphate-induced contraction in renal artery from type 2 diabetic Goto-Kakizaki rats due to activated cyclooxygenase/thromboxane receptor axis. *Pflugers Archiv: European Journal of Physiology, 466*(2), 331–342.

McLaughlin, V. V., Oudiz, R. J., Frost, A., Tapson, V. F., Murali, S., Channick, R. N., et al. (2006). Randomized study of adding inhaled iloprost to existing bosentan in pulmonary

arterial hypertension. *American Journal of Respiratory and Critical Care Medicine, 174*(11), 1257–1263.
Michel, F. S., Man, G. S., Man, R. Y., & Vanhoutte, P. M. (2008). Hypertension and the absence of EDHF-mediated responses favour endothelium-dependent contractions in renal arteries of the rat. *British Journal of Pharmacology, 155*(2), 217–226.
Michel, F., Simonet, S., Vayssettes-Courchay, C., Bertin, F., Sansilvestri-Morel, P., Bernhardt, F., et al. (2008). Altered TP receptor function in isolated, perfused kidneys of nondiabetic and diabetic ApoE-deficient mice. *American Journal of Physiology. Renal Physiology, 294*(1), F120–F129.
Miggin, S. M., & Kinsella, B. T. (2002). Investigation of the mechanisms of G protein: Effector coupling by the human and mouse prostacyclin receptors. Identification of critical species-dependent differences. *Journal of Biological Chemistry, 277*(30), 27053–27064.
Miller, V. M., Komori, K., Burnett, J. C., Jr., & Vanhoutte, P. M. (1989). Differential sensitivity to endothelin in canine arteries and veins. *American Journal of Physiology, 257*(4 Pt. 2), H1127–H1131.
Miller, V. M., & Vanhoutte, P. M. (1985a). Endothelium-dependent contractions to arachidonic acid are mediated by products of cyclooxygenase. *American Journal of Physiology, 248*(4 Pt. 2), H432–H437.
Miller, V. M., & Vanhoutte, P. M. (1985b). Endothelial α_2-adrenoceptors in canine pulmonary and systemic blood vessels. *European Journal of Pharmacology, 118*(1–2), 123–129.
Mokhtar, S. S., Vanhoutte, P. M., Leung, S. W. S., Yusof, M. I., Wan Sulaiman, W. A., Mat Saad, A. Z., et al. (2016). Endothelium dependent hyperpolarization-type relaxation compensates for attenuated nitric oxide-mediated responses in subcutaneous arteries of diabetic patients. *Nitric Oxide, 53*, 35–44.
Mombouli, J. V., & Vanhoutte, P. M. (1993). Purinergic endothelium-dependent and -independent contractions in rat aorta. *Hypertension, 22*(4), 577–583.
Moncada, S., Gryglewski, R., Bunting, S., & Vane, J. R. (1976). An enzyme isolated from arteries transforms prostaglandin endoperoxides to an unstable substance that inhibits platelet aggregation. *Nature, 263*(5579), 663–665.
Montezano, A. C., Amiri, F., Tostes, R. C., Touyz, R. M., & Schiffrin, E. L. (2007). Inhibitory effects of PPAR-gamma on endothelin-1-induced inflammatory pathways in vascular smooth muscle cells from normotensive and hypertensive rats. *Journal of the American Society of Hypertension, 1*(2), 150–160.
Moreau, P., Takase, H., & Lüscher, T. F. (1996). Effect of endothelin antagonists on the responses to prostanoid endothelium-derived contracting factor. *British Journal of Pharmacology, 118*(6), 1429–1432.
Morrison, K., Studer, R., Ernst, R., Haag, F., Kauser, K., & Clozel, M. (2012). Differential effects of Selexipag and prostacyclin analogs in rat pulmonary artery. *Journal of Pharmacology and Experimental Therapeutics, 343*(3), 547–555.
Mundy, A. L., Haas, E., Bhattacharya, I., Widmer, C. C., Kretz, M., Hofmann-Lehmann, R., et al. (2007). Fat intake modifies vascular responsiveness and receptor expression of vasoconstrictors: Implications for diet-induced obesity. *Cardiovascular Research, 73*(2), 368–375.
Nagao, T., Illiano, S., & Vanhoutte, P. M. (1992). Heterogeneous distribution of endothelium-dependent relaxations resistant to N^G-nitro-L-arginine in rats. *American Journal of Physiology, 263*(4 Pt. 2), H1090–H1094.
Nagao, T., & Vanhoutte, P. M. (1991). Hyperpolarization contributes to endothelium-dependent relaxations to acetylcholine in femoral veins of rats. *American Journal of Physiology, 261*(4 Pt. 2), H1034–H1037.
Najibi, S., Cowan, C. L., Palacino, J. J., & Cohen, R. A. (1994). Enhanced role of potassium channels in relaxations to acetylcholine in hypercholesterolemic rabbit carotid artery. *American Journal of Physiology, 266*(5 Pt. 2), H2061–H2067.

Nakagawa, O., Tanaka, I., Usui, T., Harada, M., Sasaki, Y., Itoh, H., et al. (1994). Molecular cloning of human prostacyclin receptor cDNA and its gene expression in the cardiovascular system. *Circulation, 90*(4), 1643–1647.
Narumiya, S., Sugimoto, Y., & Ushikubi, F. (1999). Prostanoid receptors: Structures, properties, and functions. *Physiological Reviews, 79*(4), 1193–1226.
N'Diaye, N., Pueyo, M. E., Battle, T., Ossart, C., Guédin, D., & Michel, J. B. (1997). Conversion of big-endothelin-1 elicits an endothelin ET_A receptor-mediated response in endothelial cells. *European Journal of Pharmacology, 321*(3), 387–396.
Nishimura, J., Aoki, H., Chen, X., Shikasho, T., Kobayashi, S., & Kanaide, H. (1995). Evidence for the presence of endothelin ET_A receptors in endothelial cells *in situ* on the aortic side of porcine aortic valve. *British Journal of Pharmacology, 115*(8), 1369–1376.
Numaguchi, Y., Harada, M., Osanai, H., Hayashi, K., Toki, Y., Okumura, K., et al. (1999). Altered gene expression of prostacyclin synthase and prostacyclin receptor in the thoracic aorta of spontaneously hypertensive rats. *Cardiovascular Research, 41*(3), 682–688.
O'Brien, R. F., Robbins, R. J., & McMurtry, I. F. (1987). Endothelial cells in culture produce a vasoconstrictor substance. *Journal of Cellular Physiology, 132*(2), 263–270.
Ogata, T., Miyauchi, T., Sakai, S., Irukayama-Tomobe, Y., Goto, K., & Yamaguchi, I. (2002). Stimulation of peroxisome-proliferator-activated receptor alpha (PPAR alpha) attenuates cardiac fibrosis and endothelin-1 production in pressure-overloaded rat hearts. *Clinical Science (London, England), 103*(Suppl. 48), 284S–288S.
Olmos, L., Mombouli, J. V., Illiano, S., & Vanhoutte, P. M. (1995). cGMP mediates the desensitization to bradykinin in isolated canine coronary arteries. *American Journal of Physiology, 268*(2 Pt. 2), H865–H870.
Oparil, S., & Haber, E. (1974). The renin-angiotensin system (first of two parts). *New England Journal of Medicine, 291*(8), 389–401.
Palmer, R. M., & Moncada, S. (1989). A novel citrulline-forming enzyme implicated in the formation of nitric oxide by vascular endothelial cells. *Biochemical and Biophysical Research Communications, 158*(1), 348–352.
Panza, J. A., Quyyumi, A. A., Brush, J. E., Jr., & Epstein, S. E. (1990). Abnormal endothelium-dependent vascular relaxation in patients with essential hypertension. *New England Journal of Medicine, 323*(1), 22–27.
Park, S. J., Lee, J. J., Kirchengast, M., Boulanger, C. M., & Vanhoutte, P. M. (1995). The combined 5-HT_2 receptor and calcium channel inhibitor LU49938 restores endothelium dependent responses in the regenerated endothelium of the porcine coronary artery. *Cardiovascular Research, 29*(1), 95–101.
Park, S. J., Lee, J. J., & Vanhoutte, P. M. (1999). Endothelin-1 releases endothelium-derived endoperoxides and thromboxane A_2 in porcine coronary arteries with regenerated endothelium. *Acta Pharmacologica Sinica, 20*(10), 872–878.
Patrignani, P., Di Febbo, C., Tacconelli, S., Douville, K., Guglielmi, M. D., Horvath, R. J., et al. (2008). Differential association between human prostacyclin receptor polymorphisms and the development of venous thrombosis and intimal hyperplasia: A clinical biomarker study. *Pharmacogenetics and Genomics, 18*(7), 611–620.
Pearson, P. J., Lin, P. J., Schaff, H. V., & Vanhoutte, P. M. (1996). Augmented endothelium-dependent constriction to hypoxia early and late following reperfusion of the canine coronary artery. *Clinical and Experimental Pharmacology and Physiology, 23*(8), 634–641.
Perneger, T. V., Whelton, P. K., & Klag, M. J. (1994). Risk of kidney failure associated with the use of acetaminophen, aspirin, and nonsteroidal antiinflammatory drugs. *New England Journal of Medicine, 331*(25), 1675–1679.
Ponnuchamy, B., & Khalil, R. A. (2009). Cellular mediators of renal vascular dysfunction in hypertension. *American Journal of Physiology. Regulatory, Integrative and Comparative Physiology, 296*(4), R1001–R1018.

Prins, B. A., Hu, R. M., Nazario, B., Pedram, A., Frank, H. J., Weber, M. A., et al. (1994). Prostaglandin E_2 and prostacyclin inhibit the production and secretion of endothelin from cultured endothelial cells. *Journal of Biological Chemistry*, *269*(16), 11938–11944.

Quaschning, T., Voss, F., Relle, K., Kalk, P., Vignon-Zellweger, N., Pfab, T., et al. (2007). Lack of endothelial nitric oxide synthase promotes endothelin-induced hypertension: Lessons from endothelin-1 transgenic/endothelial nitric oxide synthase knockout mice. *Journal of the American Society of Nephrology*, *18*(3), 730–740.

Rapoport, R. M. (2014). Nitric oxide inhibition of endothelin-1 release in the vasculature: *In vivo* relevance of *in vitro* findings. *Hypertension*, *64*(5), 908–914.

Rapoport, R. M., & Murad, F. (1983). Agonist-induced endothelium-dependent relaxation in rat thoracic aorta may be mediated through cGMP. *Circulation Research*, *52*(3), 352–357.

Rapoport, R. M., & Williams, S. P. (1996). Role of prostaglandins in acetylcholine-induced contraction of aorta from spontaneously hypertensive and Wistar-Kyoto rats. *Hypertension*, *28*(1), 64–75.

Rautureau, Y., Coelho, S. C., Fraulob-Aquino, J. C., Huo, K. G., Rehman, A., Offermanns, S., et al. (2015). Inducible human endothelin-1 overexpression in endothelium raises blood pressure via endothelin type A receptors. *Hypertension*, *66*(2), 347–355.

Resink, T. J., Scott-Burden, T., & Bühler, F. R. (1989). Activation of phospholipase A_2 by endothelin in cultured vascular smooth muscle cells. *Biochemical and Biophysical Research Communications*, *158*(1), 279–286.

Ritchie, R. H., Rosenkranz, A. C., Huynh, L. P., Stephenson, T., Kaye, D. M., & Dusting, G. J. (2004). Activation of IP prostanoid receptors prevents cardiomyocyte hypertrophy via cAMP-dependent signaling. *American Journal of Physiology. Heart and Circulatory Physiology*, *287*(3), H1179–H1185.

Rossi, G. P., Colonna, S., Pavan, E., Albertin, G., Della Rocca, F., Gerosa, G., et al. (1999). Endothelin-1 and its mRNA in the wall layers of human arteries *ex vivo*. *Circulation*, *99*(9), 1147–1155.

Rubanyi, G. M., Romero, J. C., & Vanhoutte, P. M. (1986). Flow-induced release of endothelium-derived relaxing factor. *American Journal of Physiology*, *250*(6 Pt. 2), H1145–H1149.

Rubanyi, G. M., & Vanhoutte, P. M. (1985). Hypoxia releases a vasoconstrictor substance from the canine vascular endothelium. *The Journal of Physiology*, *364*, 45–56.

Saito, Y., Nakao, K., Mukoyama, M., & Imura, H. (1990). Increased plasma endothelin level in patients with essential hypertension. *New England Journal of Medicine*, *322*(3), 205.

Sakurai, T., Yanagisawa, M., Takuwa, Y., Miyazaki, H., Kimura, S., Goto, K., et al. (1990). Cloning of a cDNA encoding a non-isopeptide-selective subtype of the endothelin receptor. *Nature*, *348*(6303), 732–735.

Schneider, J. G., Tilly, N., Hierl, T., Sommer, U., Hamann, A., Dugi, K., et al. (2002). Elevated plasma endothelin-1 levels in diabetes mellitus. *American Journal of Hypertension*, *15*(11), 967–972.

Schoeffter, P., & Hoyer, D. (1990). 5-Hydroxytryptamine (5-HT)-induced endothelium-dependent relaxation of pig coronary arteries is mediated by 5-HT receptors similar to the 5-HT_{1D} receptor subtype. *Journal of Pharmacology and Experimental Therapeutics*, *252*(1), 387–395.

Semple, R. K., Chatterjee, V. K., & O'Rahilly, S. (2006). PPAR gamma and human metabolic disease. *Journal of Clinical Investigation*, *116*(3), 581–589.

Seo, B., & Lüscher, T. F. (1995). ET_A and ET_B receptors mediate contraction to endothelin-1 in renal artery of aging SHR. Effects of FR139317 and bosentan. *Hypertension*, *25*(4 Pt. 1), 501–506.

Seo, B., Oemar, B. S., Siebenmann, R., von Segesser, L., & Lüscher, T. F. (1994). Both ET_A and ET_B receptors mediate contraction to endothelin-1 in human blood vessels. *Circulation*, *89*(3), 1203–1208.

Shimokawa, H., Flavahan, N. A., Shepherd, J. T., & Vanhoutte, P. M. (1989). Endothelium-dependent inhibition of ergonovine-induced contraction is impaired in porcine coronary arteries with regenerated endothelium. *Circulation, 80*(3), 643–650.

Shimokawa, H., & Vanhoutte, P. M. (1989). Impaired endothelium-dependent relaxation to aggregating platelets and related vasoactive substances in porcine coronary arteries in hypercholesterolemia and atherosclerosis. *Circulation Research, 64*(5), 900–914.

Shindo, T., Kurihara, H., Maemura, K., Kurihara, Y., Ueda, O., Suzuki, H., et al. (2002). Renal damage and salt-dependent hypertension in aged transgenic mice overexpressing endothelin-1. *Journal of Molecular Medicine (Berlin, Germany), 80*(2), 105–116.

Simonneau, G., Torbicki, A., Hoeper, M. M., Delcroix, M., Karlócai, K., Galiè, N., et al. (2012). Selexipag: An oral, selective prostacyclin receptor agonist for the treatment of pulmonary arterial hypertension. *European Respiratory Journal, 40*(4), 874–880.

Sitbon, O., Channick, R., Chin, K. M., Frey, A., Gaine, S., Galiè, N., et al. (2015). Selexipag for the treatment of pulmonary arterial hypertension. *New England Journal of Medicine, 373*(26), 2522–2533.

Sparks, M. A., Makhanova, N. A., Griffiths, R. C., Snouwaert, J. N., Koller, B. H., & Coffman, T. M. (2013). Thromboxane receptors in smooth muscle promote hypertension, vascular remodeling, and sudden death. *Hypertension, 61*(1), 166–173.

Spühler, O., & Zollinger, H. (1953). Die chronisch-interstitielle Nephritis [chronic interstitial nephritis]. *Zeitschrift für klinische Medizin, 151*(1), 1–50.

Stitham, J., Arehart, E., Elderon, L., Gleim, S. R., Douville, K., Kasza, Z., et al. (2011). Comprehensive biochemical analysis of rare prostacyclin receptor variants: Study of association of signaling with coronary artery obstruction. *Journal of Biological Chemistry, 286*(9), 7060–7069.

Stitham, J., Arehart, E., Gleim, S. R., Li, N., Douville, K., & Hwa, J. (2007). New insights into human prostacyclin receptor structure and function through natural and synthetic mutations of transmembrane charged residues. *British Journal of Pharmacology, 152*(4), 513–522.

Stow, L. R., Jacobs, M. E., Wingo, C. S., & Cain, B. D. (2011). Endothelin-1 gene regulation. *FASEB Journal, 25*(1), 16–28.

Sumner, M. J., Cannon, T. R., Mundin, J. W., White, D. G., & Watts, I. S. (1992). Endothelin ET_A and ET_B receptors mediate vascular smooth muscle contraction. *British Journal of Pharmacology, 107*(3), 858–860.

Suzuki, S., Kajikuri, J., Suzuki, A., & Itoh, T. (1991). Effects of endothelin-1 on endothelial cells in the porcine coronary artery. *Circulation Research, 69*(5), 1361–1368.

Taddei, S., Ghiadoni, L., Virdis, A., Buralli, S., & Salvetti, A. (1999). Vasodilation to bradykinin is mediated by an ouabain-sensitive pathway as a compensatory mechanism for impaired nitric oxide availability in essential hypertensive patients. *Circulation, 100*(13), 1400–1405.

Taddei, S., & Vanhoutte, P. M. (1993). Role of endothelium in endothelin-evoked contractions in the rat aorta. *Hypertension, 21*(1), 9–15.

Taddei, S., Virdis, A., Mattei, P., & Salvetti, A. (1993). Vasodilation to acetylcholine in primary and secondary forms of human hypertension. *Hypertension, 21*(6 Pt. 2), 929–933.

Takahashi, K., Ghatei, M. A., Lam, H. C., O'Halloran, D. J., & Bloom, S. R. (1990). Elevated plasma endothelin in patients with diabetes mellitus. *Diabetologia, 33*(5), 306–310.

Tang, E. H., Félétou, M., Huang, Y., Man, R. Y., & Vanhoutte, P. M. (2005). Acetylcholine and sodium nitroprusside cause long-term inhibition of EDCF-mediated contractions. *American Journal of Physiology. Heart and Circulatory Physiology, 289*(6), H2434–H2440.

Tang, E. H., Ku, D. D., Tipoe, G. L., Félétou, M., Man, R. Y., & Vanhoutte, P. M. (2005). Endothelium-dependent contractions occur in the aorta of wild-type and COX2-/- knockout but not COX1-/- knockout mice. *Journal of Cardiovascular Pharmacology, 46*(6), 761–765.

Tang, E. H., & Vanhoutte, P. M. (2008). Gene expression changes of prostanoid synthases in endothelial cells and prostanoid receptors in vascular smooth muscle cells caused by aging and hypertension. *Physiological Genomics*, *32*(3), 409–418.

Taurin, S., Hogarth, K., Sandbo, N., Yau, D. M., & Dulin, N. O. (2007). $G\beta\gamma$-mediated prostacyclin production and cAMP-dependent protein kinase activation by endothelin-1 promotes vascular smooth muscle cell hypertrophy through inhibition of glycogen synthase kinase-3. *Journal of Biological Chemistry*, *282*(27), 19518–19525.

Teerlink, J. R., Breu, V., Sprecher, U., Clozel, M., & Clozel, J. P. (1994). Potent vasoconstriction mediated by endothelin ET_B receptors in canine coronary arteries. *Circulation Research*, *74*(1), 105–114.

Télémaque, S., Gratton, J. P., Claing, A., & D'Orléans-Juste, P. (1993). Endothelin-1 induces vasoconstriction and prostacyclin release via the activation of endothelin ET_A receptors in the perfused rabbit kidney. *European Journal of Pharmacology*, *237*(2–3), 275–281.

Tesfamariam, B. (1994). Free radicals in diabetic endothelial cell dysfunction. *Free Radical Biology and Medicine*, *16*(3), 383–391.

Tesfamariam, B., Brown, M. L., Deykin, D., & Cohen, R. A. (1990). Elevated glucose promotes generation of endothelium-derived vasoconstrictor prostanoids in rabbit aorta. *Journal of Clinical Investigation*, *85*(3), 929–932.

Tesfamariam, B., Jakubowski, J. A., & Cohen, R. A. (1989). Contraction of diabetic rabbit aorta caused by endothelium-derived PGH_2-TxA_2. *American Journal of Physiology*, *257*(5 Pt. 2), H1327–H1333.

Thorin, E., & Clozel, M. (2010). The cardiovascular physiology and pharmacology of endothelin-1. *Advances in Pharmacology*, *60*, 1–26.

Traupe, T., D'Uscio, L. V., Münter, K., Morawietz, H., Vetter, W., & Barton, M. (2002). Effects of obesity on endothelium-dependent reactivity during acute nitric oxide synthase inhibition: Modulatory role of endothelin. *Clinical Science (London, England)*, *103*(Suppl. 48), 13S–15S.

Traupe, T., Lang, M., Goettsch, W., Münter, K., Morawietz, H., Vetter, W., et al. (2002). Obesity increases prostanoid-mediated vasoconstriction and vascular thromboxane receptor gene expression. *Journal of Hypertension*, *20*(11), 2239–2245.

Ueno, M., Miyauchi, T., Sakai, S., Goto, K., & Yamaguchi, I. (2000). Endothelin-A-receptor antagonist and oral prostacyclin analog are comparably effective in ameliorating pulmonary hypertension and right ventricular hypertrophy in rats. *Journal of Cardiovascular Pharmacology*, *36*(5 Suppl. 1), S305–S310.

Ueno, M., Miyauchi, T., Sakai, S., Yamauchi-Kohno, R., Goto, K., & Yamaguchi, I. (2002). A combination of oral endothelin-A receptor antagonist and oral prostacyclin analogue is superior to each drug alone in ameliorating pulmonary hypertension in rats. *Journal of the American College of Cardiology*, *40*(1), 175–181.

Uhlig, S., von Bethmann, A. N., Featherstone, R. L., & Wendel, A. (1995). Pharmacologic characterization of endothelin receptor responses in the isolated perfused rat lung. *American Journal of Respiratory and Critical Care Medicine*, *152*(5 Pt. 1), 1449–1460.

Vanhoutte, P. M. (1994). Endothelin-1. A matter of life and breath. *Nature*, *368*(6473), 693–694.

Vanhoutte, P. M. (2000). Say NO to ET. *Journal of the Autonomic Nervous System*, *81*(1–3), 271–277.

Vanhoutte, P. M. (2009). How we learned to say NO. *Arteriosclerosis, Thrombosis, and Vascular Biology*, *29*(8), 1156–1160.

Vanhoutte, P. M. (2013). One or two, does it matter as long as the arterial wall is coxygenated? *Hypertension*, *62*(2), 244–246.

Vanhoutte, P. M. (2016). Regenerated endothelium and its senescent response to aggregating platelets. *Circulation Journal*, *80*(4), 783–790.

Vanhoutte, P. M., & Katušić, Z. S. (1988). Endothelium-derived contracting factor: Endothelin and/or superoxide anion? *Trends in Pharmacological Sciences*, *9*(7), 229–230.

Vanhoutte, P. M., & Miller, V. M. (1989). Alpha 2-adrenoceptors and endothelium-derived relaxing factor. *The American Journal of Medicine*, *87*(3C), 1S–5S.

Vanhoutte, P. M., Shimokawa, H., Félétou, M., & Tang, E. H. (2015). Endothelial dysfunction and vascular disease—A thirtieth anniversary update. *Acta physiologica*. http://dx.doi.org/10.1111/apha.12646. [Epub ahead of print].

Vanhoutte, P. M., Shimokawa, H., Tang, E. H., & Félétou, M. (2009). Endothelial dysfunction and vascular disease. *Acta Physiologica*, *196*(2), 193–222.

Vigne, P., & Frelin, C. (1994). Endothelins activate phospholipase A_2 in brain capillary endothelial cells. *Brain Research*, *651*(1–2), 342–344.

Virdis, A., Bacca, A., Colucci, R., Duranti, E., Fornai, M., Materazzi, G., et al. (2013). Endothelial dysfunction in small arteries of essential hypertensive patients: Role of cyclooxygenase-2 in oxidative stress generation. *Hypertension*, *62*(2), 337–344.

Wagner, O. F., Christ, G., Wojta, J., Vierhapper, H., Parzer, S., Nowotny, P. J., et al. (1992). Polar secretion of endothelin-1 by cultured endothelial cells. *Journal of Biological Chemistry*, *267*(23), 16066–16068.

Warner, T. D., Battistini, B., Allcock, G. H., & Vane, J. R. (1993). Endothelin ET_A and ET_B receptors mediate vasoconstriction and prostanoid release in the isolated kidney of the rat. *European Journal of Pharmacology*, *250*(3), 447–453.

Weil, B. R., Westby, C. M., Van Guilder, G. P., Greiner, J. J., Stauffer, B. L., & DeSouza, C. A. (2011). Enhanced endothelin-1 system activity with overweight and obesity. *American Journal of Physiology. Heart and Circulatory Physiology*, *301*(3), H689–H695.

Whorton, A. R., Misono, K., Hollifield, J., Frolich, J. C., Inagami, T., & Oates, J. A. (1977). Prostaglandins and renin release: I. Stimulation of renin release from rabbit renal cortical slices by PGI_2. *Prostaglandins*, *14*(6), 1095–1104.

Williams, S. P., Dorn, G. W., 2nd, & Rapoport, R. M. (1994). Prostaglandin I_2 mediates contraction and relaxation of vascular smooth muscle. *American Journal of Physiology*, *267*(2 Pt. 2), H796–H803.

Wirth, A., Wang, S., Takefuji, M., Tang, C., Althoff, T. F., Schweda, F., et al. (2016). Age-dependent blood pressure elevation is due to increased vascular smooth muscle tone mediated by G-protein signalling. *Cardiovascular Research*, *109*(1), 131–140.

Wong, S. L., Leung, F. P., Lau, C. W., Au, C. L., Yung, L. M., Yao, X., et al. (2009). Cyclooxygenase-2-derived prostaglandin $F_{2\alpha}$ mediates endothelium-dependent contractions in the aortae of hamsters with increased impact during aging. *Circulation Research*, *104*(2), 228–235.

Wright, D. H., Abran, D., Bhattacharya, M., Hou, X., Bernier, S. G., Bouayad, A., et al. (2001). Prostanoid receptors: Ontogeny and implications in vascular physiology. *American Journal of Physiology. Regulatory, Integrative and Comparative Physiology*, *281*(5), R1343–R1360.

Xiao, C. Y., Hara, A., Yuhki, K., Fujino, T., Ma, H., Okada, Y., et al. (2001). Roles of prostaglandin I_2 and thromboxane A_2 in cardiac ischemia-reperfusion injury: A study using mice lacking their respective receptors. *Circulation*, *104*(18), 2210–2215.

Xu, C. B., Zheng, J. P., Zhang, W., Liu, E., Edvinsson, L., & Zhang, Y. (2014). Low density lipoprotein induces upregulation of vasoconstrictive endothelin type B receptor expression. *Vascular Pharmacology*, *60*(1), 42–48.

Yanagisawa, M., Kurihara, H., Kimura, S., Tomobe, Y., Kobayashi, M., Mitsui, Y., et al. (1988). A novel potent vasoconstrictor peptide produced by vascular endothelial cells. *Nature*, *332*(6163), 411–415.

Yang, D., Félétou, M., Boulanger, C. M., Wu, H. F., Levens, N., Zhang, J. N., et al. (2002). Oxygen-derived free radicals mediate endothelium-dependent contractions to acetylcholine in aortas from spontaneously hypertensive rats. *British Journal of Pharmacology, 136*(1), 104–110.

Yang, D., Félétou, M., Levens, N., Zhang, J. N., & Vanhoutte, P. M. (2003). A diffusible substance(s) mediates endothelium-dependent contractions in the aorta of SHR. *Hypertension, 41*(1), 143–148.

Zhang, Z., Austin, S. C., & Smyth, E. M. (2001). Glycosylation of the human prostacyclin receptor: Role in ligand binding and signal transduction. *Molecular Pharmacology, 60*(3), 480–487.

Zhou, Y., Mitra, S., Varadharaj, S., Parinandi, N., Zweier, J. L., & Flavahan, N. A. (2006). Increased expression of cyclooxygenase-2 mediates enhanced contraction to endothelin ET_A receptor stimulation in endothelial nitric oxide synthase knockout mice. *Circulation Research, 98*(11), 1439–1445.

CHAPTER SEVEN

Communication Through Gap Junctions in the Endothelium

K. Schmidt*,†, R. Windler*,†, C. de Wit*,†,1
*Institut für Physiologie, Universität zu Lübeck, Lübeck, Germany
†Deutsches Zentrum für Herz-Kreislauf-Forschung (DZHK) e.V. (German Center for Cardiovascular Research), Partner Site Hamburg/Kiel/Lübeck, Lübeck, Germany
[1]Corresponding author: e-mail address: dewit@uni-luebeck.de

Contents

Abstract

A swarm of fish displays a collective behavior (swarm behavior) and moves "en masse" despite the huge number of individual animals. In analogy, organ function is supported by a huge number of cells that act in an orchestrated fashion and this applies also to vascular cells along the vessel length. It is obvious that communication is required to achieve this vital goal. Gap junctions with their modular bricks, connexins (Cxs), provide channels that interlink the cytosol of adjacent cells by a pore sealed against the extracellular space. This allows the transfer of ions and charge and thereby the travel of membrane potential changes along the vascular wall. The endothelium provides a low-resistance pathway that depends crucially on connexin40 which is required for long-distance conduction of dilator signals in the microcirculation. The experimental evidence for membrane potential changes synchronizing vascular behavior is manifold but the functional verification of a physiologic role is still open. Other molecules may also be exchanged that possibly contribute to the synchronization (eg, Ca^{2+}). Recent data suggest that vascular Cxs have more functions than just facilitating

Advances in Pharmacology, Volume 77
ISSN 1054-3589
http://dx.doi.org/10.1016/bs.apha.2016.04.004

communication. As pharmacological tools to modulate gap junctions are lacking, Cx-deficient mice provide currently the standard to unravel their vascular functions. These include arteriolar dilation during functional hyperemia, hypoxic pulmonary vasoconstriction, vascular collateralization after ischemia, and feedback inhibition on renin secretion in the kidney.

ABBREVIATIONS

ACh acetylcholine
$[Ca^{2+}]_i$ intracellular Ca^{2+}
Cx connexin
CD73 cluster of differentiation 73 (or ecto-5′-nucleotidase)
EDH endothelium-derived hyperpolarization
eNOS endothelial NO synthase
HPV hypoxic pulmonary vasoconstriction
K_{Ca} Ca^{2+}-dependent K^+-channels
K_{ATP} ATP-dependent K^+-channel
K_{IR} inwardly rectifying K^+-channel
NO nitric oxide
IP_3 inositol trisphosphate
TRPV vanilloid-receptor transient receptor potential channel

1. INTRODUCTION

Communication between cells is an important task to ensure organ function since their function requires individual cells to act in a concerted synchronized fashion. The most well-known example is the heart in which individual cardiomyocytes synchronize their initiation of contraction through intercellular communication and only this achieves the expulsion of blood from the cardiac cavities. The cells communicate through specialized cell–cell junctions (gap junctions) that interlink the cytosol of adjacent cells by a pore which permits the transfer of ions along differences in membrane potential and thereby electrotonic spread of membrane potential changes. In the same manner, communication is put into effect in other organs in which the activity of a substantial number of cells is required to achieve physiologic effects, eg, in secretory glands (Bosco, Haefliger, & Meda, 2011). A concerted behavior is likewise required in blood vessels: Resistance along a vessel can only be substantially altered if the vessel constricts or dilates along its entire length (de Wit, 2004; Murrant & Sarelius, 2000; Sarelius & Pohl, 2010). Consequently, longitudinal signaling is present in the vascular wall and due to their tight coupling the endothelial cell layer provides a preferential communication pathway. A syncytium of interconnected cells is created that allows concerted activity. Smooth muscle cells also form gap junctions with their neighbors, however, their coupling is less efficient. Vascular cells are not only coupled homocellularly (between similar cell types) but also heterocellularly, ie, gap junctions connect endothelial with smooth muscle cells. This heterocellular coupling was suggested to underly the endothelium-dependent dilation that is associated with smooth muscle hyperpolarization. This so-called EDH-type dilation is independent of the endothelial mediators nitric oxide (NO) and prostaglandins and may be due to simple electrotonic spread of an endothelial hyperpolarization into the smooth muscle cell layer through gap junctional communication (de Wit & Griffith, 2010; Feletou & Vanhoutte, 2009; Figueroa & Duling, 2009; Garland, Hiley, & Dora, 2011; Grgic, Kaistha, Hoyer, & Kohler, 2009), although some evidence indicates otherwise (Boettcher & de Wit, 2011; de Wit, Boettcher, & Schmidt, 2008; Dora, Gallagher, McNeish, & Garland, 2008; Just et al., 2009; Radtke, Schmidt, Wulff, Kohler, & de Wit, 2013).

The longitudinal communication pathway is experimentally verified by the demonstration of so-called conducted or propagated vasomotor responses, ie, the spread of dilator or constrictor responses in arterioles up- and downstream from an initiation site which is stimulated in a spatially confined manner. Remote vascular responses were initially identified by August Krogh who investigated mechanisms controlling capillary blood flow. He demonstrated the recruitment of nonperfused capillaries during exercise (awarded with the Nobel Prize in Physiology or Medicine in 1920) as well as a vasomotor influence from the capillary level into arterioles (Krogh, 1920). Such conducted responses were rediscovered by Brian Duling in 1970 (Duling & Berne, 1970), are independent of flow (Hoepfl, Rodenwaldt, Pohl, & de Wit, 2002; Segal & Duling, 1986; Segal, Welsh, & Kurjiaka, 1999), are related to changes of membrane potential changes (Emerson & Segal, 2000a, 2000b; Siegl, Koeppen, Wolfle, Pohl, & de Wit, 2005), and depend on gap junction proteins (Cohen, Berg, & Sarelius, 2000; de Wit et al., 2000; Wolfle et al., 2007). Here we discuss longitudinal signaling along the endothelial cell layer, the underlying cellular signals and mechanisms, and the functional role of this pathway in vascular tone regulation. In addition, we elucidate the molecular bricks of gap junctional coupling in endothelial cells and potential other functions of these proteins.

2. GAP JUNCTION CHANNELS ARE FORMED BY CONNEXINS

Gap junctions are specialized areas at cell borders where the plasma membranes of adjacent cells closely approach each other. A remaining gap is bridged by intercellular channels. Each channel is composed of two separate hemichannels that dock face-to-face such that they interconnect their central pore without obstruction (Johnstone, Isakson, & Locke, 2009; Oshima, 2014). Thus, an intercellular channel is formed which creates a communication pathway between these neighboring cells (Fig. 1). Since in these areas hundreds of channels are clustered, the gap junction forms a low-resistance connection between adjacent cells (Koval, 2006; Saez, Berthoud, Branes, Martinez, & Beyer, 2003). The channels are built by connexins (Cxs), a protein family consisting of 20 members in mice and 21 members in humans. Their genes have been classified according to sequence homology into five groups (α, β, γ, δ, ε) and named accordingly (*GJA*, *GJB*, *GJC*, *GJD*, *GJE*). Within these groups, the genes are numbered in the order of their discovery. The proteins are often termed according to the predicted molecular weight and both terminologies are used, eg, Cx43 has a molecular weight of 43 kDa and is encoded by the gene *GJA1* since it is the first member of the *GJA* group (Sohl & Willecke, 2004).

Cxs are synthetized in the endoplasmic reticulum in which they already assemble as hexameric ring structure named connexon consisting of six individual proteins. This process is finished in the Golgi apparatus from where they are transferred along microtubules to the plasma membrane (Segretain & Falk, 2004; Thevenin et al., 2013). The individual Cx exhibits four transmembrane spanning domains (M1–M4) and the central pore is suggested in model systems to be lined by the M3 and M1 transmembrane domains (Unger, Kumar, Gilula, & Yeager, 1999). The N- and C-terminal regions are located in the cytoplasm and in most cases length differences of the C-terminal cytoplasmic domain are specifically responsible for the distinct predicted molecular masses and thus for the diversity of Cx proteins (Yeager & Harris, 2007). The connections of the transmembrane domains give rise to one cytoplasmic (that adds also to the diversity of different Cxs) and two extracellular loops (E1 and E2). Three cysteine residues are contained within each extracellular loop, which are separated by two single amino acids (Fig. 1). This structure is highly conserved and the cysteine residues may contribute to and stabilize the intercellular connection by forming

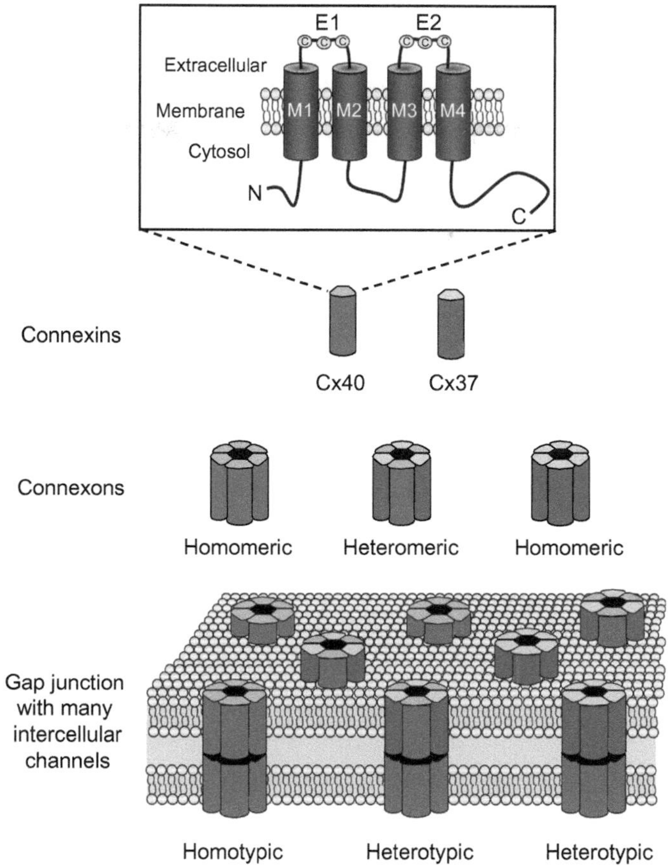

Fig. 1 Molecular structure of gap junctions. Connexin proteins are the molecular bricks of gap junctions and each connexin exhibits four transmembrane spanning domains (inset on *top*: M1–M4). The N- and C-terminal regions are located in the cytoplasm and their length varies between different connexins which determine differences in molecular masses, eg, between Cx40 (*blue* (*dark gray* in the print version)) and Cx37 (*red* (*gray* in the print version)) which are abundantly expressed in endothelial cells. Each of the two extracellular loops (E1, E2) contains three cysteine residues (inset on *top*) which may contribute to and stabilize the intercellular connection by forming disulfide bonds. Six connexins form a hemichannel (*middle*: connexon) with a central pore, which may consist of similar (homomeric) or different connexins (heteromeric). The intercellular channel is composed of two connexons that dock face-to-face such that they interconnect their central pore without obstruction. Hundreds of such channels cluster at specific areas in the cell membrane forming a gap junction (*bottom*), a low-resistance connection between adjacent cells.

disulfide bonds (Martin & Evans, 2004). Cxs from adjacent cells dock together face-to-face forming a dodecameric channel (12 subunits). The central pore has to be shielded against the extracellular environment which is achieved by the extracellular loops E1 and E2 and by the staggering (rotation) of the two hemichannels against each other and their respective interdigitation (Kovacs, Baker, Altenberg, Abagyan, & Yeager, 2007; Yeager, 2009). The tight interaction seals the central pore against the extracellular space and prevents a leakage of molecules to the extracellular fluid as well as an influx of substances into the cytoplasmic bridge. The dimension of the pore allows transfer of ions as well as water and other polar molecules up to a size of $\sim$1000 Da (Harris & Locke, 2009). Therefore, gap junctions create not only an electrical continuity that exhibits a certain conductance but also enable the diffusion of signaling molecules such as calcium ions (Ca^{2+}), inositol trisphosphate (IP_3), cyclic adenosine monophosphate, or cyclic guanosine monophosphate (Bedner et al., 2006; Harris, 2007; Rackauskas, Verselis, & Bukauskas, 2007). Additionally, other metabolites as well as small linear peptides or microRNAs are able to pass through such channels (Lemcke, Steinhoff, & David, 2015; Zong, Zhu, Liang, & Zhao, 2016). Together, gap junction channels provide a pathway for the exchange of a multitude of molecules and thus allow versatile communication. Recently, it was shown that Cxs may also function as plasma membrane channels since a hemichannel traverses the membrane and establishes a pore between the intracellular and extracellular space (Goodenough & Paul, 2003; Zhou & Jiang, 2014). It has been suggested that signaling molecules (such as ATP) are released through this pathway (Baroja-Mazo, Barbera-Cremades, & Pelegrin, 2013; Lohman & Isakson, 2014). This function is also supported by pannexins, a closely related protein family, which are reviewed elsewhere (Bond & Naus, 2014; Gaete, Lillo, & Figueroa, 2014; Isakson & Thompson, 2014; Penuela, Gehi, & Laird, 2013).

3. CXS IN THE VESSEL WALL

Gap junctional coupling is crucial in many different tissues and multicellular organs not only for homeostasis and function but also for embryonic development and growth (Levin, 2002). The diversity in the Cx protein family can partially be related to different cell types in which specific Cx members serve coupling according to specific tissue needs. In line with this, each Cx displays a tissue- or cell-type specific expression, but, on the other hand, most cells express more than one member of the Cx family.

Certain Cxs (Cx43 [*GJA1*] and Cx32 [*GJB1*]) are expressed in many different cell types as well as tissues and enable possibly a more general type of coupling whereas other Cxs are expressed only in a limited number of cell types suggesting a specific function in these cells.

In the cardiovascular system four members of the 21 human Cxs are primarily expressed, namely Cx43 (*GJA1*), Cx40 (*GJA5*), Cx37 (*GJA4*), and Cx45 (*GJC1*). Orthologs for these human Cxs exist in other species that are likewise the predominant Cxs in cells of the cardiovascular system. Across this system, the expression of the four dominant Cxs is heterogeneous (ie, in heart, arteries, veins, and lymphatic vessels). In vessels, as a general rule, Cx40 and Cx37 are found in endothelial cells, whereas smooth muscle cells express Cx43 and, to a lower extent, Cx45 (Brisset, Isakson, & Kwak, 2009; de Wit, Wolfle, & Hopfl, 2006; Hakim, Jackson, & Segal, 2008; Johnstone et al., 2009; Meens, Pfenniger, Kwak, & Delmar, 2013; Schmidt et al., 2012). The presence of two distinct Cxs in the different vascular cells opens the possibility of a variety of different channels. A connexon (hemichannel) may be composed of just a single Cx (consequently the connexon is named homomeric) but may also be constituted of different Cxs (named heteromeric). Two connexons that dock together and form a functional channel may be composed of two similar or distinct connexons (homotypic or heterotypic). Consequently, a full intercellular channel is homomeric–homotypic, homomeric–heterotypic, heteromeric–homotypic, or heteromeric–heterotypic. The multifaceted composition of gap junction channels may relate to different function because conductances, permeabilities, and gating (open vs closed) characteristics of the intercellular channels vary with the Cxs forming them. Importantly, varying channels may be subject to variable regulation because the C-terminal region that is very heterogeneous for distinct Cx proteins is targeted by regulatory phosphorylation processes (Haefliger, Nicod, & Meda, 2004; Johnstone et al., 2009).

4. LONGITUDINAL COMMUNICATION ALONG THE VESSEL WALL

As outlined in Section 1, the formation of a syncytium by gap junctional coupling ensures that vascular cells act in a concerted fashion. Experimentally, confined stimulation of arterioles in the microcirculation using micropipettes exerts a dilation at the site of application (Fig. 2). However, this dilation is not restricted to the stimulation site but can also be observed at remote, distant up- and downstream sites of the arteriole (Cohen et al., 2000;

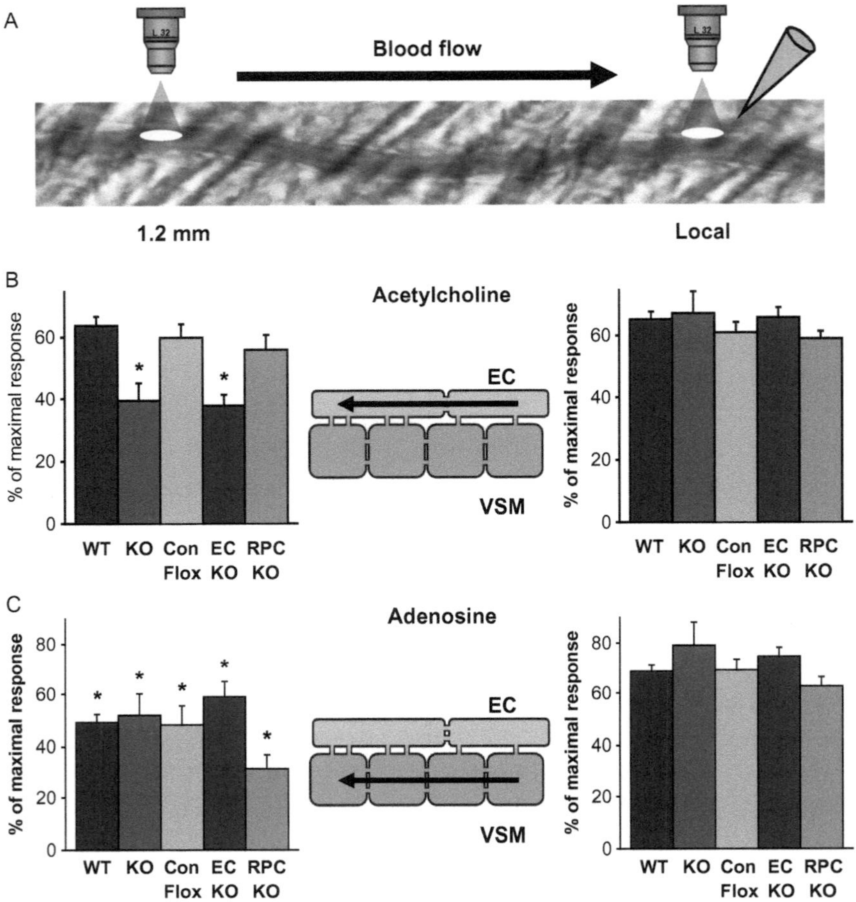

Fig. 2 Conducted dilations in the microcirculation. Arterioles in the microcirculation in vivo are stimulated by application of acetylcholine or adenosine through micropipettes and diameter changes assessed at the site of stimulation (local) and at an upstream site at a distance of 1.2 mm (scheme in A). Responses to acetylcholine are depicted in (B): local dilations (amplitude in % of maximal response, *right columns*) are comparable in different Cx40-deficient mice. At the remote upstream site (*left columns*), the amplitudes are not different from the local site in wild-type mice (WT), in control mice carrying homozygously a floxed Cx40 gene (Con Flox), or in mice deficient for Cx40 in renin-producing cells (RPC KO). In contrast, the amplitude was significantly reduced in global Cx40-deficient mice (KO) as well as in endothelial-specific Cx40-deficient mice (EC KO). This suggests that the conducting signal is transmitted along endothelial cells in a Cx40-dependent manner (B, inset in the *middle*). Adenosine induced a local dilation (C, *right columns*) and also initiated a conducted response (C, *left columns*). However, the amplitude was reduced in all genotypes compared to the local dilation in a similar fashion. This suggests that the conduction process is independent of Cx40 and the smooth muscle layer is the likely conduction pathway (C, inset in the *middle*). For more details, see text. *Data reproduced from de Wit, C. (2010). Different pathways with distinct properties conduct dilations in the microcirculation in vivo.* Cardiovascular Research, 85, *604–613 and Jobs, A., Schmidt, K., Schmidt, V. J., Lubkemeier, I., van Veen, T. A. B., Kurtz, A., et al. (2012). Defective Cx40 maintains Cx37 expression but intact Cx40 is crucial for conducted dilations irrespective of hypertension.* Hypertension, 60, *1422–1429.*

Emerson & Segal, 2000a; Hoepfl et al., 2002; Segal & Duling, 1986; Segal et al., 1999; Wolfle & de Wit, 2005). In fact, the dilation spreads very quickly to such remote sites, giving the impression that the arteriole dilates simultaneously along its entire length in a concerted manner. This so-called conducted response (or propagated response) can also be exerted by stimuli that elicit a constriction and was initially identified by Krogh (1920) and rediscovered by Duling and Berne (1970). The original idea (postulated nearly a century ago) that a signal generated in the periphery of the vascular system (eg, in capillaries) spreads in an ascending fashion against the streaming blood into feeding arterioles and arteries (so-called ascending dilation) (Fleisch, 1935; Hilton, 1959; Krogh, 1920; Schretzenmayr, 1933) has therefore been revived (Segal, 2005; Segal & Duling, 1986). The potential physiologic impact of signaling along the vessel has spurred considerable interest in this field of vascular research. Such a mechanism may help to match blood flow and oxygen delivery with oxygen demand in, for example, working skeletal muscle. In addition, other mechanisms such as flow-induced dilation that relies on local increases in shear stress may support this matching (Pohl, Holtz, Busse, & Bassenge, 1986). Both mechanisms, although mechanistically very distinct (signal being conducted in the vascular wall vs a change in flow that elicits a stimulus at all sites along the vessel), may help that vascular cells act in synchrony (de Wit, 2004; Pohl & de Wit, 1999).

A conducted response is elicited only by some vasodilators, whereas others induce solely a local dilation that does not conduct to remote sites (Segal et al., 1999). This suggests that dilation per se is not sufficient and argues against a flow-mediated upstream response elicited by distal dilation (and flow increase). In fact, shear rate did not increase during conducted dilations (Hoepfl et al., 2002). Acetylcholine (ACh) elicits a conducted response, whereas exogenous application of NO does not. In line with this, the blockade of NO synthase does not change the capability of ACh to do so (Hoepfl et al., 2002; Welsh & Segal, 2000; Wolfle & de Wit, 2005). This suggests that endothelial NO is not responsible for initiating a conducted response. Other agents (bradykinin, adenosine) that are able to initiate the response have with ACh in common that they elicit a dilation by the activation of K^+-channels (in case of ACh and bradykinin Ca^{2+}-dependent K^+-channels, KCa) and subsequent hyperpolarization (Domeier & Segal, 2007; Garland, Yarova, Jimenez-Altayo, & Dora, 2011; Hoepfl et al., 2002). This suggests that membrane potential changes are a crucial requirement for the response and, in fact, membrane potential changes have been measured at local and remote sites using sharp electrodes impaled into endothelial or smooth muscle cells of the vascular wall (Emerson, Neild, & Segal, 2002; Emerson & Segal, 2000a, 2001; Siegl et al., 2005; Takano, Dora, Spitaler, & Garland, 2004; Welsh & Segal, 2000). Conversely, a local depolarization and constriction initiated by, for example, high local concentrations of potassium elicit similarly a conducted response (in this case a constriction) further verifying this concept (de Wit et al., 2000). Such an electronic conduction of membrane potential changes through low-resistance gap junction channels as the underlying mechanism is also supported by the speed of the conduction of the dilation (in the range of mm/s).

Endothelial cells express abundant amounts of Cxs (Looft-Wilson, Payne, & Segal, 2004; Sandow et al., 2003), forming robust gap junctions, and exhibit a morphology with their long axes in the direction of the length of the vessel (Haas & Duling, 1997). Both structural features favor longitudinal electrotonic conduction. In contrast, smooth muscle cells are wrapped with their long axis circularly around the vessel wall and longitudinal conduction has to cross a considerably higher number of cell membranes (Fig. 3). These exert even in the presence of gap junctions that also link smooth muscle cells a substantial resistance. Experimental evidence suggests that both cell layers act as a signaling pathway that are possibly independent of each other. A conducted response was always observed up to but never beyond the second site of damage in skeletal muscle arterioles, if the endothelial and the smooth muscle cell layer are destroyed in succession along the vessel in the microcirculation in vivo. This was independent of the fact whether the endothelial or the smooth muscle cells were impaired at the first site and the respective other layer at the second site. The authors concluded from these elegant experiments that a dilator signal is initiated at the site of stimulation upon ACh in both layers and conducts in each layer up to the site of damage (Budel, Bartlett, & Segal, 2003). This also implies that the signal does not cross between layers while traveling. However, the data did not allow to compare the efficacy of conduction in these supposedly separate and independent layers. Interestingly, some endothelium-dependent dilators (bradykinin) seem to use only the endothelial layer as a transduction pathway (Budel et al., 2003). In other vessels, the sole damage of endothelial cells along the conduction pathway prevented the passage of the dilation (Emerson & Segal, 2000a, 2001; Segal & Jacobs, 2001; Takano et al., 2004). These small arteries that were located outside of the tissue were different from arterioles within the tissue since they lacked a smooth muscle pathway. This discrepancy was suggested to be related to their rich

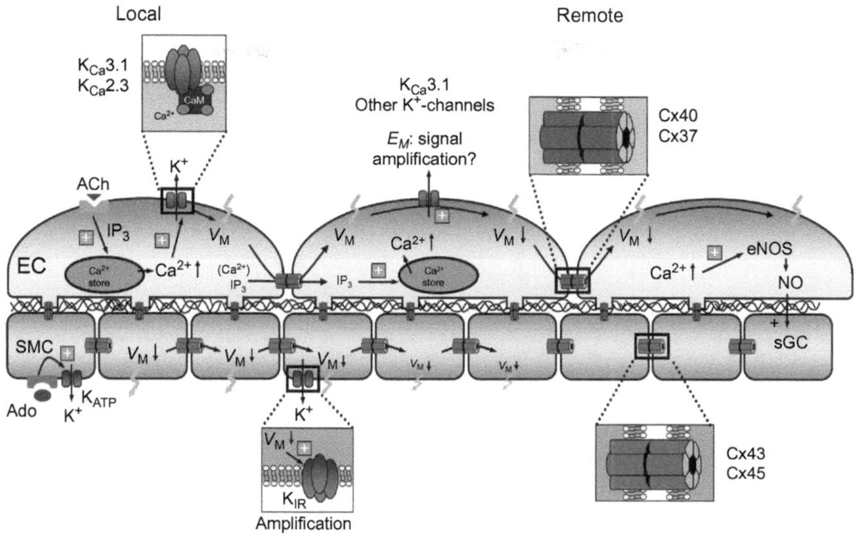

Fig. 3 Signals travel along the vessel wall. Local stimulation of the arteriole using acetylcholine (ACh) activates endothelial KCa channels and hyperpolarizes endothelial cells (EC). The hyperpolarization is transmitted through gap junctions connecting endothelial cells to each other that are composed of Cx40 and Cx37. The hyperpolarization is boosted while traveling through activation of K^+-channels including KCa3.1 but it remains to be clarified which channels are how activated. Additionally, a wave of calcium is initiated that travels at a slower speed along the endothelial cell layer supposedly by gap junctional transfer of IP_3. This calcium increase may further dilate arterioles at remote sites through the release of endothelial autacoids (eg, NO). Other dilators (eg, adenosine) elicit a hyperpolarization in smooth muscle cells, which are likewise connected through gap junctions, mainly composed of Cx43 and Cx45. The smooth muscle cell layer possesses a higher resistance and the dilation decays more quickly with distance. For further details, see text.

sympathetic innervation (Haug, Welsh, & Segal, 2003). These findings indicate that different vessels (depending on their location in the network and the supplied tissue) exert somewhat different signaling mechanisms but the endothelial cell layer provides the most important pathway along which membrane potential changes conduct to initiate a concerted behavior of vascular cells.

5. DISTANCE AND MODULATION OF THE CONDUCTION OF ELECTRICAL SIGNALS ALONG THE ENDOTHELIUM

An important aspect is the distance that responses are able to conduct. Theoretically, the amplitude of an electronically conducting signal should exponentially decrease and be related to the longitudinal resistance along the pathway. Accordingly, endothelial cells provide a more suitable pathway as outlined earlier. Depolarization-induced conducted constrictions decay indeed quickly with distance and the decline of the amplitude can be fitted to an exponential function (Rodenwaldt, Pohl, & de Wit, 2007). This decline is modified by the presence of NO (Rodenwaldt et al., 2007) suggesting that NO modulates resistance most likely due to an alteration of gap junction channel conductivity. However, such responses travel along the smooth muscle layer as suggested by experiments in Cx-deficient mice (see later). In contrast, the amplitude of conducted dilations decays considerably less, which is even more pronounced if endothelium-dependent agonists are used to initiate the response (Emerson et al., 2002). In fact, a decay of dilatory amplitudes with distance (for those distances studied, up to 1.5 mm) is almost not observed if ACh or bradykinin is employed to elicit conducted responses. Simulation of vascular conduction underline the important role of the endothelium (Diep, Vigmond, Segal, & Welsh, 2005) but also suggest that the coupling efficiency of endothelial cells to smooth muscle modulates the distance that the signal is traveling (Hald, Jacobsen, Sandow, Holstein-Rathlou, & Welsh, 2014; Tran, Vigmond, Plane, & Welsh, 2009; Tran & Welsh, 2010).

Alternatively, these observations suggest that the conducting signal is amplified by active mechanisms such as ion channel opening while traveling (Crane, Neild, & Segal, 2004). A suitable candidate which is activated by hyperpolarization is the inwardly rectifying K^+-channel (K_{IR}) that may augment the hyperpolarization during its travel, a process which justifies a term like "propagated response" (Smith et al., 2008). Such amplifying processes may vary between smooth muscle which expresses K_{IR} channels and endothelial cells which express additionally KCa (KCa3.1 and KCa2.3). In fact, stimulation of the KCa3.1 channel by pharmacological activation elicits a conducted response and KCa3.1 is also involved in the initiation of the response upon ACh stimulation. However, conducted responses can still be elicited by ACh in mice deficient for KCa3.1. This remaining response is abrogated after pharmacologic blockade of KCa2.3 indicating that both channels are able to initiate an endothelial hyperpolarization and a resulting conducted response (Wolfle, Schmidt, Hoyer, Kohler, & de Wit, 2009). It seems that KCa3.1 has additionally a role in the augmentation of the signal during traveling in the sense that it boosts the hyperpolarization during conduction since the dilatory amplitude was decreased in KCa3.1-deficient animals but not in wild-type controls in a distance of 1.1 mm. On the contrary, pharmacologic blockade of KCa2.3 alone (in wild-type mice) did not alter dilatory amplitudes at remote sites (Wolfle et al., 2009). These experiments suggest that both, KCa3.1 and KCa2.3, contribute to initiate the response whereas only KCa3.1 exerts an augmentation of the propagating signal (Fig. 3). Whether this is related to concomitant changes of intracellular Ca^{2+} ($[Ca^{2+}]_i$) (see also later) remains to be clarified. Interestingly, ACh exerts a specific effect and is able to engage powerful amplification mechanisms that are independent of KCa3.1 activation: Conducted responses upon ACh did not decay up to a distance of 1.1 mm, whereas the dilatory amplitude of conducted dilations upon KCa3.1 stimulation was decreased at this distance (Wolfle et al., 2009). It may seem favorable to amplify conducted responses as the dilations cover larger distances but such augmentation processes exhibit also a downside: the signal is perturbed in its amplitude and a graded and possibly additive behavior (eg, at a parent vessel receiving signals from many daughter arterioles) is difficult to envision under such circumstances. In fact, regenerative processes have also been questioned and the great distances that are covered by conducted dilations in experimental settings have been explained by supramaximal hyperpolarization at the stimulation site. While the dilation did not decay along the vessel, the electrical signal (ie, the hyperpolarization) indeed decayed but due to its large amplitude at the local site this decay did not translate into an attenuated dilation. The authors postulate that there is a limit for the change in membrane potential to translate into a mechanical response and beyond this threshold (or border) a further dilation was not observed (Wolfle et al., 2011).

Charge may also dissipate through the cell membrane and not be available for driving a membrane potential change at remote sites. This idea relies on cable theory, which predicts that the length constant (lambda) determines the distance that a voltage change spreads. The length constant, in turn, is proportional to the square root of the membrane resistance (divided by axial resistance which is mainly dependent on gap junctional conductance). Membrane resistance itself depends on the number of active ion channels and their respective conductance. It seems therefore obvious that ion channel activation modulates the change of membrane potential at remote sites. Indeed, the conduction efficiency of electrical signals upon current injection was modulated by activation or inhibition of endothelial KCa2.3 and KCa3.1 channels in isolated endothelial cell tubes. Activation of KCa channels decreased efficiency, ie, the changes of the membrane potential at distant sites upon local current injection were attenuated, and conversely, blockade of these channels enhanced conduction efficiency (Behringer & Segal, 2012). This demonstrates that endothelial cell tubes behave indeed as predicted by cable theory which is often used in theoretical simulations of electrical behavior of coupled cells in vessels (Diep et al., 2005). However, it cannot be deduced from these experiments that KCa channels do not contribute as amplifiers during signal propagation. Channel activation during traveling of the signal would, in fact, increase the available voltage for further signal propagation as long as a driving force across the membrane is present. This is the case if the prevailing membrane potential is different from the equilibrium potential of the channel since the difference between these two parameters drives current through an open channel. However, if the cell layer is already hyperpolarized before a conducting signal is initiated or if the potassium equilibrium potential is artificially altered (by raising the extracellular potassium concentration) the driving force for potassium ions across the membrane and, consequently, the current through KCa channels is minimized. At such circumstances, KCa channel opening does not drive a further membrane potential change but rather "clamps" the membrane potential at the equilibrium potential of the channel by generating short-circuit currents. These currents dissipate the change of membrane potential induced by current injection as done in those experiments (Behringer & Segal, 2012) and distant potential changes are reduced. Therefore, the conduction of hyperpolarizations (initiated by current injection) is indeed reduced under conditions of augmented KCa channel activity as suggested to occur with aging (Behringer, Shaw, Westcott, Socha, & Segal, 2013), but it has to be considered that such vessels are also more dilated just due to this fact. The conclusion that KCa channel activation contributes to flow limitations with advanced age (Behringer et al., 2013) needs to be vigorously verified. Membrane potential changes beyond the potassium equilibrium are hard to envision just as a dilation beyond the vessel's maximal diameter. If the arteriole is already close to its maximal diameter the small additional (remaining) change is "as good as it gets".

6. CALCIUM AS A SECONDARY CONDUCTED SIGNAL: CALCIUM WAVE ALONG THE ENDOTHELIUM

Recently, another signaling molecule, namely $[Ca^{2+}]_i$ was implicated in conducted responses. Stimulation of endothelial cells with agonists used to elicit conducted responses such as ACh that are coupled to G-protein receptors are well known to increase $[Ca^{2+}]_i$. This involves activation of phospholipase C which hydrolyzes phosphatidyl-4,5-bisphosphate to diacylglycerol and inositol-1,4,5-trisphosphate (IP_3). The latter releases Ca^{2+} from the endoplasmic reticulum through activation of IP_3 receptors. Endothelial increase of $[Ca^{2+}]_i$ is an important step to activate endothelial NO synthase (eNOS) and release NO to induce dilation. Focal application of ACh onto isolated small arteries of hamsters using micropipettes indeed initiated a local $[Ca^{2+}]_i$ increase and an ensuing Ca^{2+} wave that propagated along endothelial cells in both directions (Takano et al., 2004; Uhrenholt, Domeier, & Segal, 2007). The propagation of the Ca^{2+} wave was considerably slower (0.1 mm/s) than the conduction of the membrane potential change and of the dilation (in the range of cm/s or mm/s). However, blockade of KCa channels that abrogated the initiation of the fast spreading hyperpolarization and dilation uncovered a rather slowly propagating dilation (0.02 mm/s) which was preceded by the Ca^{2+} wave (Fig. 3). This slow conducting dilation was abrogated upon blockade of NO synthase and cyclooxygenase indicating that it is mediated through NO (and prostaglandin) release (Uhrenholt et al., 2007). A similar "two-step" conducted dilation was identified in the microcirculation of mice in vivo: A rapid conducted dilation without any changes of Ca^{2+} and a slower endothelial Ca^{2+}-wave that enhanced the dilation slightly. Furthermore, this secondary slower component covered only smaller distances (Tallini et al., 2007). This Ca^{2+} wave is certainly not just Ca^{2+} diffusion from the initialized local elevation of $[Ca^{2+}]_i$. More likely, it involves a propagating Ca^{2+} increase that is amplified during traveling. An unresolved question is which molecule diffuses actually from one cell to its neighbor. Candidates are Ca^{2+} itself which may elicit Ca^{2+}-induced Ca^{2+} release or, more likely, the second messenger IP_3 that invokes an increase in $[Ca^{2+}]_i$ in the adjacent cell through $[Ca^{2+}]_i$ release by IP_3-receptor stimulation (Kameritsch, Pogoda, Ritter, Munzing, & Pohl, 2012). The interaction of the hyperpolarization and $[Ca^{2+}]_i$ in endothelial cells is complex and, in fact, a matter of debate (Dora & Garland, 2013). Endothelial cells lack classical voltage-dependent Ca^{2+}-channels and thus a depolarization induced increase of $[Ca^{2+}]_i$ is not likely. Twenty-five years ago, the idea was put forward that a hyperpolarization would enhance the driving force for Ca^{2+}-influx as deduced from experiments on isolated cultured endothelial cells (Luckhoff & Busse, 1990). More recent experiments, however, did not verify these findings in endothelial cells of arterioles (Marrelli, Eckmann, & Hunte, 2003; Rahman, Hughes, Matchkov, Nilsson, & Aalkjaer, 2007; Takano et al., 2004). Conversely, increases in $[Ca^{2+}]_i$ may promote opening of KCa channels and hyperpolarization involving a transient receptor potential channel (TRPV4) (Bagher et al., 2012; Ma et al., 2013; Sonkusare et al., 2012). However, the difference in conduction velocity argues against the idea that the conducting Ca^{2+} wave drives and augments the much faster conduction of the change in membrane potential leaving it unresolved on how the signal is able to travel such great distances.

7. ENDOTHELIAL CONNEXIN40 EXERTS A DECISIVE FUNCTION IN LONGITUDINAL COUPLING

Gap junctions in the vessel wall are formed by Cx37, Cx40, Cx43, and Cx45 with a cell-specific preference, namely Cx37 and Cx40 in the endothelium and Cx43 and Cx45 in smooth muscle cells as outlined earlier (Fig. 3). Specific inhibition of gap junctions is difficult, most substances (eg, 18alpha- or 18beta-glycyrrhetinic acid, carbenoxolone) that are employed as gap junction inhibitors are notorious for affecting other targets such as ion channels (Behringer, Socha, Polo-Parada, & Segal, 2012; Guan, Si, & Jiang, 2007; Matchkov, Rahman, Peng, Nilsson, & Aalkjaer, 2004; Tare, Coleman, & Parkington, 2002). A specific blockade of Cx subtypes is even more difficult, in fact, pharmacologic substances that are useful for this purpose are not available. However, small peptides that act by modulating the gating of gap junction channels were shown to be effective blockers of gap junctions. The modification of their amino acid sequence enables to target the extracellular loops of specific Cxs (so-called gap peptides or Cx-mimetic peptides; Evans & Leybaert, 2007) and such peptides were successfully used in cells or isolated vessels to interfere (supposedly) with specific Cxs (Chaytor, Bakker, Edwards, & Griffith, 2005; Chaytor, Martin, Edwards, & Griffith, 2001; de Wit & Griffith, 2010). However, this strategy is not very useful to be employed in vivo or in the microcirculation since the delivery to the target side is difficult to obtain in the required concentrations. In some cases, these peptides have been reported to exert specific effects in vivo (De Vriese, Van de Voorde, & Lameire, 2002). We have also used this approach in the cremaster microcirculation but did not achieve inhibition of effects known to be gap junction dependent (unpublished observations).

A more promising approach to analyze the contribution of specific Cxs in longitudinal communication is studying Cx-deficient animals. Mice which lack either Cx37 or Cx40 globally are viable and fertile (except for female Cx37-deficient mice) (Kirchhoff et al., 1998; Simon,

Goodenough, Li, & Paul, 1997; Simon, Goodenough, & Paul, 1998) whereas combined deficiency of both endothelial Cxs is embryonically lethal due to vascular malformation (Simon & McWhorter, 2002) pointing to the crucial role of these Cxs in the development of the vasculature. Longitudinal communication has been studied in the microcirculation of these Cx-deficient mice by stimulation of an arteriole in a confined manner and the assessment of responses at the local and remote upstream sites as outlined earlier. Global deletion of Cx37 did not alter the spread of conducted dilations initiated by ACh compared to wild-type controls (Figueroa & Duling, 2008). In contrast, Cx40-deficient mice exhibited an attenuation of dilations at remote sites despite a similar amplitude of the local dilation if the response was initiated by endothelium-dependent dilators (eg, ACh or bradykinin) (Fig. 2) (de Wit et al., 2000; Figueroa & Duling, 2008). Likewise, the amplitude of a conducting dilation that was elicited by electrical stimulation was reduced at distant, upstream sites in Cx40-deficient mice (Figueroa et al., 2003). In contrast, conducted responses triggered by other dilators, which induce dilation through a direct effect on the vascular smooth muscle (such as adenosine or the ATP-dependent K^+-channel (K_{ATP})-channel opener pinacidil) remained unaffected in Cx40-deficient mice (de Wit, 2010; Figueroa & Duling, 2008). These substances do, in fact, elicit a hyperpolarization and thus are able to produce a communicating dilator signal, but the hyperpolarization is initiated in this case most likely in the vascular smooth muscle and not in endothelial cells. At least, adenosine does not elicit an endothelial hyperpolarization in these vessels in mice and the dilation is mediated through activation of K_{ATP}-channels (de Wit, 2010) which are expressed in most vessels only in smooth muscle cells. Reports for expression of these channels in endothelial cells are scarce (Aziz et al., 2014; Casalini et al., 2014; Jackson, 2005; Nichols, Singh, & Grange, 2013). A further distinction between conducted dilations triggered by ACh (or bradykinin) and those upon adenosine is the change of the amplitude during propagation. Whereas the amplitude does not substantially diminish within a distance of 1–1.5 mm upon ACh, it decreases significantly with distance after adenosine. Interestingly, the behavior of conducted dilations upon ACh and adenosine (or a pharmacologic K_{ATP}-channel opener) resembled each other in Cx40-deficient mice with respect to amplitude (de Wit, 2010; Figueroa & Duling, 2008) suggesting that the regenerative, amplifying mechanisms initiated by ACh requires Cx40 (and the endothelium). In the absence of Cx40 remains only the pathway that enables a mere passive nonregenerating spread of the dilations. This latter pathway is probably the smooth muscle cell layer, which is coupled less efficiently and a higher number of cell membranes have to be traversed, highlighting the efficiency of the endothelial longitudinal signaling pathway that relies on Cx40. These experimental data are supported by mathematical modeling that suggests the endothelial cell layer as the main conducting pathway and the smooth muscle cell layer as a less efficient backup system (Diep et al., 2005).

It has to be considered that the conduction deficit might be related to the considerable arterial hypertension that is found in Cx40-deficient mice and the hypertension itself may impact on the endothelial cell pathway through modulation of Cx expression and conduction (Dal-Ros et al., 2009; Goto, Rummery, Grayson, & Hill, 2004; Kurjiaka, Bender, Nye, Wiehler, & Welsh, 2005; Rummery & Hill, 2004; Rummery, McKenzie, Whitworth, & Hill, 2002). This possibility was tested after generating mice carrying cell-specific deletions of Cx40 which is expressed, in addition to endothelial cells, in cardiomyocytes, and in renin-producing cells. Surprisingly, only mice that lack Cx40 in renin-producing cells were hypertensive, whereas endothelial deficiency of Cx40 was not associated with exaggerated arterial pressure (Wagner et al., 2010). This finding is in line with the lack of feedback inhibition of pressure on renin secretion as observed in globally Cx40-deficient mice previously (Krattinger et al., 2007; Wagner et al., 2007) and implicates Cx40 as a crucial protein in the renal baroreceptor pathway that leads to the inverse relationship of renin secretion and pressure (Castrop et al., 2010; Wagner & Kurtz, 2013). Further evidence for this hypothesis was achieved by reexpression of Cx40 selectively in renin-producing cells, which reduces hypertension (Le Gal et al., 2014). Conducted dilations triggered by endothelium-dependent dilators remained unaltered in hypertensive renin-producing cell-specific Cx40-deficient mice (Fig. 2) (Jobs et al., 2012). Likewise, conduction was not affected in eNOS-deficient mice (de Wit et al., 2000) that exhibit a moderate hypertension suggesting that hypertension per se does not alter longitudinal signalling along the endothelial cell layer. Conversely, endothelial-specific Cx40 deficient arterioles exhibited a conduction deficit comparable to global Cx40-deficient mice although endothelial-specific Cx40 deficient mice are normotensive (Fig. 2) (Jobs et al., 2012). This verifies that the lack of Cx40 in endothelial cells indeed alters longitudinal signaling and this defect is not causally related to hypertension. Further confirmation was obtained in mice that express Cx45 in place of Cx40. These animals exhibit only a moderate hypertension and hyperreninemia is prevented (Schweda et al., 2009) but the conduction deficit persisted

(Wolfle et al., 2007). These data indicate that Cx45 which exhibits a considerably lower conductance than Cx40 (35 vs 200 pS) (van Veen, van Rijen, & Jongsma, 2000) cannot support the decisive function of Cx40 despite its expression in endothelial cells in these genetically modified animals.

In all these Cx40-deficient murine models a conducted response persisted albeit the amplitude at distant sites was diminished. The remaining underlying signal may travel along the smooth muscle cell layer or may be supported by the second endothelial Cx (Cx37) which is also prominently expressed. Interestingly, the expression of the two endothelial Cxs (Cx37 and Cx40) is interlinked, the deletion of one reduced the expression of the nontargeted Cx. Deletion of Cx40 reduced the expression of Cx37 dramatically in aortic endothelial cells and, conversely, deletion of Cx37 reduced Cx40 protein expression, although this latter effect was only moderate (Simon & McWhorter, 2003). In arterioles of Cx40-deficient mice, Cx37 could not be detected by immunostaining verifying the crucial importance of Cx40 for proper plasma membrane location of Cx37 (de Wit, 2010). Again, hypertension was not an influential factor: Mice carrying a deletion of Cx40 specifically in renin-producing cells expressed Cx37 in the endothelium of arterioles at the level of controls despite their hypertension, whereas mice with endothelial cell-specific Cx40 deletion that are normotensive did not express Cx37 in endothelial cells (as detected by immunostaining in arterioles). In contrast to protein levels, transcriptional Cx37 activity was similar in all genotypes, suggesting a specific role for Cx40 in the translocation of Cx37 toward the membrane, its insertion, or, alternatively, preventing its removal from the membrane and degradation (Jobs et al., 2012). Interestingly, Cx45 expressed instead of Cx40 in the endothelium was likewise incapable of normalizing Cx37 expression in arterioles (Wolfle et al., 2007). The mechanisms for this strong dependency are still obscure. One may speculate that heteromeric hemichannels are formed in endothelial cells and Cx40 is required to do so. However, it has to be kept in mind that Cx37 formed functional homotypic channels in cell expression systems (Banach, Ramanan, & Brink, 2000). Seeing that the presence of Cx37 is strongly reduced in endothelial cell membranes in Cx40-ablated endothelium it is tempting to speculate that the smooth muscle cell layer indeed provides the passage for longitudinal conduction in these conditions and serves as a backup system in case the endothelial pathway is compromised. Final support for this hypothesis comes from experiments in mice that express a human nonconducting mutated Cx40 that was identified in a patient group suffering from atrial fibrillation (Gollob et al., 2006). The conduction of dilations initiated by endothelium-dependent stimuli was impaired in mice that carried the mutated Cx with an amino acid exchange at position 96. The conducting dilatory amplitude in these mice was attenuated to a degree that was comparable to animals which lacked endothelial Cx40. This occurred despite the proper localization of Cx37 in the membrane suggesting that Cx37 expression does not maintain longitudinal signaling along the endothelial cell layer. However, the presence of mutated Cx40 proteins may have blocked conduction by integration of such nonconducting proteins into hemichannels. If this is indeed the case a dominant-negative effect of mutated Cx40 should be expected. Although we did not study conduction in heterozygous mice we did obtain arterial pressure measurements and pressure was only elevated in mice expressing the mutated Cx40 in a homozygous fashion (Jobs et al., 2012). Again the pressure increase was most likely related to defects in renin secretion (Lubkemeier et al., 2011) and not to dysfunctional endothelial communication. Taken together, the endothelial cell layer provides an important signalling pathway along the length of the vessel and Cx40 exerts a crucial function in this pathway. Cx40 forms the structural basis in the endothelium that enables the cells of the vascular wall to act jointly in a synchronized fashion.

8. EVIDENCE FROM KNOCKOUT MICE: PHYSIOLOGIC FUNCTION OF VASCULAR CXS

The idea that Cx-dependent cellular signaling is an important requirement for vascular function has been proposed repeatedly. It is supported by experiments studying coupling in cell cultures, isolated vessels, or even in the microcirculation in vivo using intravital microscopy as outlined in detail earlier. While these experiments highlight that Cx-mediated cellular coupling is existent and effects relying on such mechanisms can be demonstrated, more direct evidence for a role of longitudinal signaling in vessels in the intact organism is sparse. Functional hyperemia in response to active skeletal muscle contraction is a classical example of arteriolar regulation that requires a coordination of cellular behavior along the vascular tree (Bagher & Segal, 2011; de Wit, 2004; Murrant & Sarelius, 2000; Pohl & de Wit, 1999; Segal, 2005). Skeletal muscle contraction by stimulation of the motor nerve is a convenient approach to trigger functional hyperemia and can be applied in the microcirculation while measuring arteriolar diameters

(Murrant et al., 2014; VanTeeffelen & Segal, 2006). The amplitude of the arteriolar dilation is dependent on stimulus frequency (muscle twitching vs tetanic contraction) and duration of the contraction (Armstrong, Dua, & Murrant, 2007; Dua, Dua, & Murrant, 2009). A tetanic stimulation (100 Hz) for 60 s induces a near maximal arteriolar dilation in the cremaster muscle of mice (~80% of maximal diameter). This dilation was impaired in global Cx40-deficient mice (to 60%) but also dilations upon less vigorous muscle stimulation were reduced in these mice indicating that indeed propagation of signals along the vessel is contributing to functional hyperemic responses (Milkau, Kohler, & de Wit, 2010).

Recently, signaling along endothelial cells in a Cx40-dependent fashion was also demonstrated in the pulmonary vascular bed (Wang et al., 2012). Vasoconstriction in pulmonary arteries in response to low oxygen pressure in lung alveoli is an important physiologic regulation to redirect blood flow to lung areas with high oxygen content to ensure full oxygenation of blood during its passage through the pulmonary circulation. This so-called hypoxic pulmonary vasoconstriction (HPV) prevents perfusion of insufficiently ventilated areas and avoids functional right-left shunts. It requires signalling from down- to upstream vessel sites much like the responses in functional hyperemia outlined earlier. Interestingly, global deficiency of Cx40 abrogated HPV in isolated lungs. Since Cx40 was expressed abundantly in endothelial cells in pulmonary vessels but not in smooth muscle cells the endothelium seems to be the crucial signalling pathway. Indeed, hypoxia depolarized endothelial cells originating in alveolar capillaries. The depolarization propagated upstream into arterioles in wild-type but not in Cx40-deficient mice as measured using fluorescence imaging. The transformation of endothelial depolarization into a functional vasoconstriction involved endothelial voltage-dependent T-type Ca^{2+}-channels ($Ca_V3.1$), phospholipase A2, and epoxyeicosatrienoic acids (Wang et al., 2012). Functionally, a partial blockade or obstruction of lung ventilation, thereby challenging HPV, induced severe hypoxemia in Cx40-deficient mice but not in wild-type mice indicating the requirement of Cx40 for this important response in vivo. Interestingly, the vascular maladaptation to chronic exposure to hypoxia which consists of muscularization of pulmonary arterioles and right ventricular hypertrophy was also absent in Cx40-deficient mice (Wang et al., 2012).

Signals from the periphery may also have a functional role in ischemia-induced arteriogenesis. In fact, mice lacking Cx40 exhibited reduced arteriogenesis and experienced severe reduction in limb perfusion compared to wild-type mice after induction of limb ischemia by vessel ligation

(Buschmann et al., 2010; Fang, Angelov, Simon, & Burt, 2012). This compromised tissue perfusion upon ischemic damage was unrelated to the angiotensin II excess observed in Cx40-deficient mice (provoked by the lack of pressure-mediated inhibition of renin secretion as mentioned earlier) because blockade of the angiotensin receptor 1 did not improve perfusion after ligation (Fang, Angelov, Simon, & Burt, 2013). Surprisingly, deletion of Cx37 improved recovery and enhanced limb flow in this setting. The causal mechanisms for the deleterious effect of the presence of Cx37 in this pathologic model are not clear. The authors speculated that Cx37 may exert an antiproliferative effect (Fang, Angelov, Simon, & Burt, 2011) or interact with endothelial NO synthase (Pfenniger et al., 2010). Interestingly, the presence of Cx37 reduced platelet reactivity and thereby limits thrombus propensity (Angelillo-Scherrer et al., 2011). However, these data have recently been questioned by others who demonstrated exactly the opposite, namely reduced platelet function at multiple levels of platelet activation, as well as attenuated clot retraction in Cx37-deficient mice (Vaiyapuri et al., 2012). Similarly, deletion of Cx40 reduced platelet aggregation, an effect that was independent of Cx37 (Vaiyapuri et al., 2013). Such effects in platelets (or in other circulating blood cells) may also contribute during ischemic challenges. For example, Cx40 may exert its beneficial effects in ischemia also independently of gap junctional communication: In cardiac tissue endothelial Cx40 limited myocardial damage upon ischemia/reperfusion, ie, lack of Cx40 resulted in exaggerated myocardial infarction. This protective effect was linked to activation of the cluster of differentiation 73 (or ecto-5′-nucleotidase) (CD73) and required the presence of leukocytes (Morel et al., 2014). This signaling pathway (Cx40, CD73) that prevented inflammation and leukocyte infiltration was also implicated in the protective effect of endothelial Cx40 in atherosclerosis (Chadjichristos et al., 2010) indicating that Cx40 is much more than just a molecule promoting longitudinal signaling in the endothelium. Finally, it has been clearly demonstrated that Cx expression is modulated in atherosclerosis and they exert distinct promoting and inhibiting effects in this disease (Brisset et al., 2009; Morel, 2014; Pfenniger, Chanson, & Kwak, 2013) further highlighting this fact.

9. CONCLUSION

Experiments in Cx-deficient mice have uncovered a multiplicity of defects in these mice, some being unexpected and surprising. Thus, these mice provide valuable models to obtain mechanistic insight into Cx function in the cardiovascular system. Unfortunately, the physiological function of Cxs in the vascular wall, aside from pathophysiologic challenges, is far from being clear. Specifically, the hypothesis that longitudinal signaling in the vessel wall is a crucial factor for functional hyperemia requires further investigation and experimental verification although there is no doubt that an endothelial communication pathway exists that is, in principle, able to concert vascular cell behavior. Currently, conducted dilations have not been examined in humans, even not, to the best of our knowledge, in isolated human vessels. Thus, it is currently unknown whether this important physiologic response is altered in human diseases, such as atherosclerosis. Since atherosclerosis affects primarily larger arteries whereas conduction of vasomotor signals have been demonstrated to be functionally important in small arteries and in the microcirculation an alteration in atherosclerosis seems to be unlikely. This view is supported by unhindered conducted dilations in hypercholesterolemic mice (Beleznai et al., 2011; Wolfle & de Wit, 2005). Other human diseases that also affect small arteries (eg, diabetic angiopathy) have not been examined with regard to this aspect in experimental animal studies. On the other hand, mutations in Cxs have been demonstrated to be causal for a multitude of human diseases affecting the skin, myelinated neurons, hearing, the lens, or even leading to complex syndromes (Abrams & Scherer, 2012; Cotrina & Nedergaard, 2012; Laird, 2014; Pfenniger, Wohlwend, & Kwak, 2011; Xu & Nicholson, 2013). In the cardiovascular system, Cx mutations have been implicated in cardiac arrhythmias (Delmar & Makita, 2012) including atrial fibrillation (Molica, Meens, Morel, & Kwak, 2014). Conduction deficits in the cardiac atria may also be acquired (Chaldoupi, Loh, Hauer, de Bakker, & van Rijen, 2009) or related to cardiac diseases (eg, myocardial infarction) in the ventricles and modulation of Cx function is an emerging target for drug development (Dhein et al., 2010). Considering the importance of Cxs in atherosclerosis (Pfenniger et al., 2013), they may also be a future target in this disease. The interesting role of Cx40 in blood pressure control led to the speculation that mutations in *GJA5* may also underlie human hypertension which was, however, not verified (Firouzi et al., 2006; Schmidt, Kaiser, Erdmann, & de Wit, 2015).

CONFLICT OF INTEREST

The authors have no conflicts of interest to declare.

REFERENCES

Abrams, C. K., & Scherer, S. S. (2012). Gap junctions in inherited human disorders of the central nervous system. *Biochimica et Biophysica Acta*, *1818*, 2030–2047.

Angelillo-Scherrer, A., Fontana, P., Burnier, L., Roth, I., Sugamele, R., Brisset, A., et al. (2011). Connexin 37 limits thrombus propensity by downregulating platelet reactivity. *Circulation*, *124*, 930–939.

Armstrong, M. L., Dua, A. K., & Murrant, C. L. (2007). Time course of vasodilation at the onset of repetitive skeletal muscle contractions. *American Journal of Physiology. Regulatory, Integrative and Comparative Physiology*, *292*, R505–R515.

Aziz, Q., Thomas, A. M., Gomes, J., Ang, R., Sones, W. R., Li, Y., et al. (2014). The ATP-sensitive potassium channel subunit, Kir6.1, in vascular smooth muscle plays a major role in blood pressure control. *Hypertension*, *64*, 523–529.

Bagher, P., Beleznai, T., Kansui, Y., Mitchell, R., Garland, C. J., & Dora, K. A. (2012). Low intravascular pressure activates endothelial cell TRPV4 channels, local Ca^{2+} events, and IK_{Ca} channels, reducing arteriolar tone. *Proceedings of the National Academy of Sciences of the United States of America*, *109*, 18174–18179.

Bagher, P., & Segal, S. S. (2011). Regulation of blood flow in the microcirculation: Role of conducted vasodilation. *Acta Physiologica (Oxford, England)*, *202*, 271–284.

Banach, K., Ramanan, S. V., & Brink, P. R. (2000). The influence of surface charges on the conductance of the human connexin37 gap junction channel. *Biophysical Journal*, *78*, 752–760.

Baroja-Mazo, A., Barbera-Cremades, M., & Pelegrin, P. (2013). The participation of plasma membrane hemichannels to purinergic signaling. *Biochimica et Biophysica Acta*, *1828*, 79–93.

Bedner, P., Niessen, H., Odermatt, B., Kretz, M., Willecke, K., & Harz, H. (2006). Selective permeability of different connexin channels to the second messenger cyclic amp. *Journal of Biological Chemistry*, *281*, 6673–6681.

Behringer, E. J., & Segal, S. S. (2012). Tuning electrical conduction along endothelial tubes of resistance arteries through Ca(2+)-activated K(+) channels. *Circulation Research*, *110*, 1311–1321.

Behringer, E. J., Shaw, R. L., Westcott, E. B., Socha, M. J., & Segal, S. S. (2013). Aging impairs electrical conduction along endothelium of resistance arteries through enhanced Ca^{2+}-activated K^{+} channel activation. *Arteriosclerosis, Thrombosis, and Vascular Biology*, *33*, 1892–1901.

Behringer, E. J., Socha, M. J., Polo-Parada, L., & Segal, S. S. (2012). Electrical conduction along endothelial cell tubes from mouse feed arteries: Confounding actions of glycyrrhetinic acid derivatives. *British Journal of Pharmacology*, *166*, 774–787.

Beleznai, T., Takano, H., Hamill, C., Yarova, P., Douglas, G., Channon, K., et al. (2011). Enhanced K(+)-channel-mediated endothelium-dependent local and conducted dilation of small mesenteric arteries from ApoE(−/−) mice. *Cardiovascular Research*, *92*, 199–208.

Boettcher, M., & de Wit, C. (2011). Distinct endothelium-derived hyperpolarizing factors emerge in vitro and in vivo and are mediated in part via connexin 40-dependent myoendothelial coupling. *Hypertension*, *57*, 802–808.

Bond, S. R., & Naus, C. C. (2014). The pannexins: Past and present. *Frontiers in Physiology*, *5*, 58.

Bosco, D., Haefliger, J. A., & Meda, P. (2011). Connexins: Key mediators of endocrine function. *Physiological Reviews*, *91*, 1393–1445.

Brisset, A. C., Isakson, B. E., & Kwak, B. R. (2009). Connexins in vascular physiology and pathology. *Antioxidants and Redox Signaling*, *11*, 267–282.

Budel, S., Bartlett, I. S., & Segal, S. S. (2003). Homocellular conduction along endothelium and smooth muscle of arterioles in hamster cheek pouch: Unmasking an no wave. *Circulation Research*, *93*, 61–68.

Buschmann, I., Pries, A., Styp-Rekowska, B., Hillmeister, P., Loufrani, L., Henrion, D., et al. (2010). Pulsatile shear and GJA5 modulate arterial identity and remodeling events during flow-driven arteriogenesis. *Development*, *137*, 2187–2196.
Casalini, E. D., Goodwill, A. G., Owen, M. K., Moberly, S. P., Berwick, Z. C., & Tune, J. D. (2014). Contribution of hydrogen sulfide to the control of coronary blood flow. *Microcirculation*, *21*, 104–111.
Castrop, H., Hocherl, K., Kurtz, A., Schweda, F., Todorov, V., & Wagner, C. (2010). Physiology of kidney renin. *Physiological Reviews*, *90*, 607–673.
Chadjichristos, C. E., Scheckenbach, K. E., van Veen, T. A., Richani Sarieddine, M. Z., de Wit, C., Yang, Z., et al. (2010). Endothelial-specific deletion of connexin40 promotes atherosclerosis by increasing CD73-dependent leukocyte adhesion. *Circulation*, *121*, 123–131.
Chaldoupi, S. M., Loh, P., Hauer, R. N. W., de Bakker, J. M. T., & van Rijen, H. V. M. (2009). The role of connexin40 in atrial fibrillation. *Cardiovascular Research*, *84*, 15–23.
Chaytor, A. T., Bakker, L. M., Edwards, D. H., & Griffith, T. M. (2005). Connexin-mimetic peptides dissociate electrotonic EDHF-type signalling via myoendothelial and smooth muscle gap junctions in the rabbit iliac artery. *British Journal of Pharmacology*, *144*, 108–114.
Chaytor, A. T., Martin, P. E., Edwards, D. H., & Griffith, T. M. (2001). Gap junctional communication underpins EDHF-type relaxations evoked by ACh in the rat hepatic artery. *American Journal of Physiology. Heart and Circulatory Physiology*, *280*, H2441–H2450.
Cohen, K. D., Berg, B. R., & Sarelius, I. H. (2000). Remote arteriolar dilations in response to muscle contraction under capillaries. *American Journal of Physiology. Heart and Circulatory Physiology*, *278*, H1916–H1923.
Cotrina, M. L., & Nedergaard, M. (2012). Brain connexins in demyelinating diseases: Therapeutic potential of glial targets. *Brain Research*, *1487*, 61–68.
Crane, G. J., Neild, T. O., & Segal, S. S. (2004). Contribution of active membrane processes to conducted hyperpolarization in arterioles of hamster cheek pouch. *Microcirculation*, *11*, 425–433.
Dal-Ros, S., Bronner, C., Schott, C., Kane, M. O., Chataigneau, M., Schini-Kerth, V. B., et al. (2009). Angiotensin II-induced hypertension is associated with a selective inhibition of endothelium-derived hyperpolarizing factor-mediated responses in the rat mesenteric artery. *Journal of Pharmacology and Experimental Therapeutics*, *328*, 478–486.
De Vriese, A. S., Van de Voorde, J., & Lameire, N. H. (2002). Effects of connexin-mimetic peptides on nitric oxide synthase- and cyclooxygenase-independent renal vasodilation. *Kidney International*, *61*, 177–185.
de Wit, C. (2004). Connexins pave the way for vascular communication. *News in Physiological Sciences*, *19*, 148–153.
de Wit, C. (2010). Different pathways with distinct properties conduct dilations in the microcirculation in vivo. *Cardiovascular Research*, *85*, 604–613.
de Wit, C., Boettcher, M., & Schmidt, V. J. (2008). Signaling across myoendothelial gap junctions—Fact or fiction? *Cell Communication & Adhesion*, *15*, 231–245.
de Wit, C., & Griffith, T. M. (2010). Connexins and gap junctions in the EDHF phenomenon and conducted vasomotor responses. *Pflügers Archiv/European Journal of Physiology*, *459*, 897–914.
de Wit, C., Roos, F., Bolz, S. S., Kirchhoff, S., Kruger, O., Willecke, K., et al. (2000). Impaired conduction of vasodilation along arterioles in connexin40 deficient mice. *Circulation Research*, *86*, 649–655.
de Wit, C., Wolfle, S. E., & Hopfl, B. (2006). Connexin-dependent communication within the vascular wall: Contribution to the control of arteriolar diameter. *Advances in Cardiology*, *42*, 268–283.
Delmar, M., & Makita, N. (2012). Cardiac connexins, mutations and arrhythmias. *Current Opinion in Cardiology*, *27*, 236–241.

Dhein, S., Hagen, A., Jozwiak, J., Dietze, A., Garbade, J., Barten, M., et al. (2010). Improving cardiac gap junction communication as a new antiarrhythmic mechanism: The action of antiarrhythmic peptides. *Naunyn-Schmiedeberg's Archives of Pharmacology, 381*, 221–234.
Diep, H. K., Vigmond, E. J., Segal, S. S., & Welsh, D. G. (2005). Defining electrical communication in skeletal muscle resistance arteries: A computational approach. *Journal of Physiology, 568*, 267–281.
Domeier, T. L., & Segal, S. S. (2007). Electromechanical and pharmacomechanical signalling pathways for conducted vasodilatation along endothelium of hamster feed arteries. *Journal of Physiology, 579*, 175–186.
Dora, K. A., Gallagher, N. T., McNeish, A., & Garland, C. J. (2008). Modulation of endothelial cell KCa3.1 channels during endothelium-derived hyperpolarizing factor signaling in mesenteric resistance arteries. *Circulation Research, 102*, 1247–1255.
Dora, K. A., & Garland, C. J. (2013). Linking hyperpolarization to endothelial cell calcium events in arterioles. *Microcirculation, 20*, 248–256.
Dua, A. K., Dua, N., & Murrant, C. L. (2009). Skeletal muscle contraction-induced vasodilator complement production is dependent on stimulus and contraction frequency. *American Journal of Physiology. Heart and Circulatory Physiology, 297*, H433–H442.
Duling, B. R., & Berne, R. M. (1970). Propagated vasodilation in the microcirculation of the hamster cheek pouch. *Circulation Research, 26*, 163–170.
Emerson, G. G., Neild, T. O., & Segal, S. S. (2002). Conduction of hyperpolarization along hamster feed arteries: Augmentation by acetylcholine. *American Journal of Physiology. Heart and Circulatory Physiology, 283*, H102–H109.
Emerson, G. G., & Segal, S. S. (2000a). Endothelial cell pathway for conduction of hyperpolarization and vasodilation along hamster feed artery. *Circulation Research, 86*, 94–100.
Emerson, G. G., & Segal, S. S. (2000b). Electrical coupling between endothelial cells and smooth muscle cells in hamster feed arteries—Role in vasomotor control. *Circulation Research, 87*, 474–479.
Emerson, G. G., & Segal, S. S. (2001). Electrical activation of endothelium evokes vasodilation and hyperpolarization along hamster feed arteries. *American Journal of Physiology. Heart and Circulatory Physiology, 280*, H160–H167.
Evans, W. H., & Leybaert, L. (2007). Mimetic peptides as blockers of connexin channel-facilitated intercellular communication. *Cell Communication & Adhesion, 14*, 265–273.
Fang, J. S., Angelov, S. N., Simon, A. M., & Burt, J. M. (2011). Cx37 deletion enhances vascular growth and facilitates ischemic limb recovery. *American Journal of Physiology. Heart and Circulatory Physiology, 301*, H1872–H1881.
Fang, J. S., Angelov, S. N., Simon, A. M., & Burt, J. M. (2012). Cx40 is required for, and Cx37 limits, postischemic hindlimb perfusion, survival and recovery. *Journal of Vascular Research, 49*, 2–12.
Fang, J. S., Angelov, S. N., Simon, A. M., & Burt, J. M. (2013). Compromised regulation of tissue perfusion and arteriogenesis limit, in an at1r-independent fashion, recovery of ischemic tissue in Cx40(−/−) mice. *American Journal of Physiology. Heart and Circulatory Physiology, 304*, H816–H827.
Feletou, M., & Vanhoutte, P. M. (2009). EDHF: An update. *Clinical Science (London), 117*, 139–155.
Figueroa, X. F., & Duling, B. R. (2008). Dissection of two Cx37-independent conducted vasodilator mechanisms by deletion of Cx40: Electrotonic versus regenerative conduction. *American Journal of Physiology. Heart and Circulatory Physiology, 295*, H2001–H2007.
Figueroa, X. F., & Duling, B. R. (2009). Gap junctions in the control of vascular function. *Antioxidants and Redox Signaling, 11*, 251–266.
Figueroa, X. F., Paul, D. L., Simon, A. M., Goodenough, D. A., Day, K. H., Damon, D. N., et al. (2003). Central role of connexin40 in the propagation of electrically activated vasodilation in mouse cremasteric arterioles in vivo. *Circulation Research, 92*, 793–800.

Firouzi, M., Kok, B., Spiering, W., Busjahn, A., Bezzina, C. R., Ruijter, J. M., et al. (2006). Polymorphisms in human connexin40 gene promoter are associated with increased risk of hypertension in men. *Journal of Hypertension, 24*, 325–330.

Fleisch, A. (1935). Les reflexes nutritifs ascendants producteurs de dilatation arterielle. *Archives Internationales de Physiologie, 41*, 141–167.

Gaete, P. S., Lillo, M. A., & Figueroa, X. F. (2014). Functional role of connexins and pannexins in the interaction between vascular and nervous system. *Journal of Cellular Physiology, 229*, 1336–1345.

Garland, C. J., Hiley, C. R., & Dora, K. A. (2011). EDHF: Spreading the influence of the endothelium. *British Journal of Pharmacology, 164*, 839–852.

Garland, C. J., Yarova, P. L., Jimenez-Altayo, F., & Dora, K. A. (2011). Vascular hyperpolarization to beta-adrenoceptor agonists evokes spreading dilatation in rat isolated mesenteric arteries. *British Journal of Pharmacology, 164*, 913–921.

Gollob, M. H., Jones, D. L., Krahn, A. D., Danis, L., Gong, X. Q., Shao, Q., et al. (2006). Somatic mutations in the connexin 40 gene (GJA5) in atrial fibrillation. *New England Journal of Medicine, 354*, 2677–2688.

Goodenough, D. A., & Paul, D. L. (2003). Beyond the gap: Functions of unpaired connexon channels. *Nature Reviews. Molecular Cell Biology, 4*, 285–294.

Goto, K., Rummery, N. M., Grayson, T. H., & Hill, C. E. (2004). Attenuation of conducted vasodilatation in rat mesenteric arteries during hypertension: Role of inwardly rectifying potassium channels. *Journal of Physiology, 561*, 215–231.

Grgic, I., Kaistha, B. P., Hoyer, J., & Kohler, R. (2009). Endothelial Ca^{2+}-activated K+ channels in normal and impaired EDHF-dilator responses—Relevance to cardiovascular pathologies and drug discovery. *British Journal of Pharmacology, 157*, 509–526.

Guan, B. C., Si, J. Q., & Jiang, Z. G. (2007). Blockade of gap junction coupling by glycyrrhetinic acids in guinea pig cochlear artery: A whole-cell voltage- and current-clamp study. *British Journal of Pharmacology, 151*, 1049–1060.

Haas, T. L., & Duling, B. R. (1997). Morphology favors an endothelial cell pathway for longitudinal conduction within arterioles. *Microvascular Research, 53*, 113–120.

Haefliger, J. A., Nicod, P., & Meda, P. (2004). Contribution of connexins to the function of the vascular wall. *Cardiovascular Research, 62*, 345–356.

Hakim, C. H., Jackson, W. F., & Segal, S. S. (2008). Connexin isoform expression in smooth muscle cells and endothelial cells of hamster cheek pouch arterioles and retractor feed arteries. *Microcirculation, 15*, 503–514.

Hald, B. O., Jacobsen, J. C. B., Sandow, S. L., Holstein-Rathlou, N. H., & Welsh, D. G. (2014). Less is more: Minimal expression of myoendothelial gap junctions optimizes cell-cell communication in virtual arterioles. *Journal of Physiology, 592*, 3243–3255.

Harris, A. L. (2007). Connexin channel permeability to cytoplasmic molecules. *Progress in Biophysics and Molecular Biology, 94*, 120–143.

Harris, A. L., & Locke, D. (2009). Permeability of connexin channels. In A. L. Harris & D. Locke (Eds.), *Connexins: A guide* (pp. 165–206). New York City, NY: Humana Press.

Haug, S. J., Welsh, D. G., & Segal, S. S. (2003). Sympathetic nerves inhibit conducted vasodilatation along feed arteries during passive stretch of hamster skeletal muscle. *Journal of Physiology, 552*, 273–282.

Hilton, S. M. (1959). A peripheral arterial conducting mechanism underlying dilatation of the femoral artery concerned in functional vasodilatation in skeletal muscle. *Journal of Physiology, 149*, 93–111.

Hoepfl, B., Rodenwaldt, B., Pohl, U., & de Wit, C. (2002). EDHF, but not NO or prostaglandins, is critical to evoke a conducted dilation upon ach in hamster arterioles. *American Journal of Physiology. Heart and Circulatory Physiology, 283*, H996–H1004.

Isakson, B. E., & Thompson, R. J. (2014). Pannexin-1 as a potentiator of ligand-gated receptor signaling. *Channels (Austin, Texas), 8*, 118–123.

Jackson, W. F. (2005). Potassium channels in the peripheral microcirculation. *Microcirculation, 12,* 113–127.

Jobs, A., Schmidt, K., Schmidt, V. J., Lubkemeier, I., van Veen, T. A. B., Kurtz, A., et al. (2012). Defective Cx40 maintains Cx37 expression but intact Cx40 is crucial for conducted dilations irrespective of hypertension. *Hypertension, 60,* 1422–1429.

Johnstone, S., Isakson, B., & Locke, D. (2009). Biological and biophysical properties of vascular connexin channels. *International Review of Cell and Molecular Biology, 278,* 69–118.

Just, A., Kurtz, L., de Wit, C., Wagner, C., Kurtz, A., & Arendshorst, W. J. (2009). Connexin 40 mediates the tubuloglomerular feedback contribution to renal blood flow autoregulation. *Journal of the American Society of Nephrology, 20,* 1577–1585.

Kameritsch, P., Pogoda, K., Ritter, A., Munzing, S., & Pohl, U. (2012). Gap junctional communication controls the overall endothelial calcium response to vasoactive agonists. *Cardiovascular Research, 93,* 508–515.

Kirchhoff, S., Nelles, E., Hagendorff, A., Kruger, O., Traub, O., & Willecke, K. (1998). Reduced cardiac conduction velocity and predisposition to arrhythmias in connexin40-deficient mice. *Current Biology, 8,* 299–302.

Kovacs, J. A., Baker, K. A., Altenberg, G. A., Abagyan, R., & Yeager, M. (2007). Molecular modeling and mutagenesis of gap junction channels. *Progress in Biophysics and Molecular Biology, 94,* 15–28.

Koval, M. (2006). Pathways and control of connexin oligomerization. *Trends in Cell Biology, 16,* 159–166.

Krattinger, N., Capponi, A., Mazzolai, L., Aubert, J. F., Caille, D., Nicod, P., et al. (2007). Connexin40 regulates renin production and blood pressure. *Kidney International, 72,* 814–822.

Krogh, A. (1920). Studies on the capillaromotor mechanism. I. the reaction to stimuli and the innervation of the blood vessels in the tongue of the frog. *Journal of Physiology, 53,* 399–419.

Kurjiaka, D. T., Bender, S. B., Nye, D. D., Wiehler, W. B., & Welsh, D. G. (2005). Hypertension attenuates cell-to-cell communication in hamster retractor muscle feed arteries. *American Journal of Physiology. Heart and Circulatory Physiology, 288,* H861–H870.

Laird, D. W. (2014). Syndromic and non-syndromic disease-linked cx43 mutations. *FEBS Letters, 588,* 1339–1348.

Le Gal, L., Alonso, F., Wagner, C., Germain, S., Nardelli Haefliger, D., Meda, P., et al. (2014). Restoration of connexin 40 (Cx40) in renin-producing cells reduces the hypertension of cx40 null mice. *Hypertension, 63,* 1198–1204.

Lemcke, H., Steinhoff, G., & David, R. (2015). Gap junctional shuttling of miRNA—A novel pathway of intercellular gene regulation and its prospects in clinical application. *Cellular Signalling, 27,* 2506–2514.

Levin, M. (2002). Isolation and community: A review of the role of gap-junctional communication in embryonic patterning. *Journal of Membrane Biology, 185,* 177–192.

Lohman, A. W., & Isakson, B. E. (2014). Differentiating connexin hemichannels and pannexin channels in cellular atp release. *FEBS Letters, 588,* 1379–1388.

Looft-Wilson, R. C., Payne, G. W., & Segal, S. S. (2004). Connexin expression and conducted vasodilation along arteriolar endothelium in mouse skeletal muscle. *Journal of Applied Physiology, 97,* 1152–1158.

Lubkemeier, I., Machura, K., Kurtz, L., Neubauer, B., Dobrowolski, R., Schweda, F., et al. (2011). The connexin40 A96S mutation causes renin-dependent hypertension. *Journal of the American Society of Nephrology, 22,* 1031–1040.

Luckhoff, A., & Busse, R. (1990). Activators of potassium channels enhance calcium influx into endothelial cells as a consequence of potassium currents. *Naunyn-Schmiedeberg's Archives of Pharmacology, 342,* 94–99.

Ma, X., Du, J., Zhang, P., Deng, J., Liu, J., Lam, F. F. Y., et al. (2013). Functional role of TRPV4-KCa2.3 signaling in vascular endothelial cells in normal and streptozotocin-induced diabetic rats. *Hypertension*, *62*, 134–139.

Marrelli, S. P., Eckmann, M. S., & Hunte, M. S. (2003). Role of endothelial intermediate conductance KCa channels in cerebral EDHF-mediated dilations. *American Journal of Physiology. Heart and Circulatory Physiology*, *285*, H1590–H1599.

Martin, P. E. M., & Evans, W. H. (2004). Incorporation of connexins into plasma membranes and gap junctions. *Cardiovascular Research*, *62*, 378–387.

Matchkov, V. V., Rahman, A., Peng, H., Nilsson, H., & Aalkjaer, C. (2004). Junctional and nonjunctional effects of heptanol and glycyrrhetinic acid derivates in rat mesenteric small arteries. *British Journal of Pharmacology*, *142*, 961–972.

Meens, M. J., Pfenniger, A., Kwak, B. R., & Delmar, M. (2013). Regulation of cardiovascular connexins by mechanical forces and junctions. *Cardiovascular Research*, *99*, 304–314.

Milkau, M., Kohler, R., & de Wit, C. (2010). Crucial importance of the endothelial K^+ channel SK3 and connexin40 in arteriolar dilations during skeletal muscle contraction. *FASEB Journal*, *24*, 3572–3579.

Molica, F., Meens, M. J. P., Morel, S., & Kwak, B. R. (2014). Mutations in cardiovascular connexin genes. *Biology of the Cell*, *106*, 269–293.

Morel, S. (2014). Multiple roles of connexins in atherosclerosis- and restenosis-induced vascular remodelling. *Journal of Vascular Research*, *51*, 149–161.

Morel, S., Braunersreuther, V., Chanson, M., Bouis, D., Rochemont, V., Foglia, B., et al. (2014). Endothelial Cx40 limits myocardial ischaemia/reperfusion injury in mice. *Cardiovascular Research*, *102*, 329–337.

Murrant, C. L., Dodd, J. D., Foster, A. J., Inch, K. A., Muckle, F. R., Ruiz, D. A., et al. (2014). Prostaglandins induce vasodilatation of the microvasculature during muscle contraction and induce vasodilatation independent of adenosine. *Journal of Physiology*, *592*, 1267–1281.

Murrant, C. L., & Sarelius, I. H. (2000). Coupling of muscle metabolism and muscle blood flow in capillary units during contraction. *Acta Physiologica Scandinavica*, *168*, 531–541.

Nichols, C. G., Singh, G. K., & Grange, D. K. (2013). K_{ATP} channels and cardiovascular disease: Suddenly a syndrome. *Circulation Research*, *112*, 1059–1072.

Oshima, A. (2014). Structure and closure of connexin gap junction channels. *FEBS Letters*, *588*, 1230–1237.

Penuela, S., Gehi, R., & Laird, D. W. (2013). The biochemistry and function of pannexin channels. *Biochimica et Biophysica Acta*, *1828*, 15–22.

Pfenniger, A., Chanson, M., & Kwak, B. R. (2013). Connexins in atherosclerosis. *Biochimica et Biophysica Acta*, *1828*, 157–166.

Pfenniger, A., Derouette, J., Verma, V., Lin, X., Foglia, B., Coombs, W., et al. (2010). Gap junction protein Cx37 interacts with endothelial nitric oxide synthase in endothelial cells. *Arteriosclerosis, Thrombosis, and Vascular Biology*, *30*, 827–834.

Pfenniger, A., Wohlwend, A., & Kwak, B. R. (2011). Mutations in connexin genes and disease. *European Journal of Clinical Investigation*, *41*, 103–116.

Pohl, U., & de Wit, C. (1999). A unique role of no in the control of blood flow. *News in Physiological Sciences*, *14*, 74–80.

Pohl, U., Holtz, J., Busse, R., & Bassenge, E. (1986). Crucial role of endothelium in the vasodilator response to increased flow in vivo. *Hypertension*, *8*, 37–44.

Rackauskas, M., Verselis, V. K., & Bukauskas, F. F. (2007). Permeability of homotypic and heterotypic gap junction channels formed of cardiac connexins mCx30.2, Cx40, Cx43, and Cx45. *American Journal of Physiology. Heart and Circulatory Physiology*, *293*, H1729–H1736.

Radtke, J., Schmidt, K., Wulff, H., Kohler, R., & de Wit, C. (2013). Activation of KCa3.1 by SKA-31 induces arteriolar dilation and lowers blood pressure in normo- and hypertensive connexin40-deficient mice. *British Journal of Pharmacology*, *170*, 293–303.

Rahman, A., Hughes, A., Matchkov, V., Nilsson, H., & Aalkjaer, C. (2007). Antiphase oscillations of endothelium and smooth muscle [Ca(2+)](i) in vasomotion of rat mesenteric small arteries. *Cell Calcium*, *42*, 536–547.

Rodenwaldt, B., Pohl, U., & de Wit, C. (2007). Endogenous and exogenous no attenuates conduction of vasoconstrictions along arterioles in the microcirculation. *American Journal of Physiology. Heart and Circulatory Physiology*, *292*, H2341–H2348.

Rummery, N. M., & Hill, C. E. (2004). Vascular gap junctions and implications for hypertension. *Clinical and Experimental Pharmacology and Physiology*, *31*, 659–667.

Rummery, N. M., McKenzie, K. U. S., Whitworth, J. A., & Hill, C. E. (2002). Decreased endothelial size and connexin expression in rat caudal arteries during hypertension. *Journal of Hypertension*, *20*, 247–253.

Saez, J. C., Berthoud, V. M., Branes, M. C., Martinez, A. D., & Beyer, E. C. (2003). Plasma membrane channels formed by connexins: Their regulation and functions. *Physiological Reviews*, *83*, 1359–1400.

Sandow, S. L., Looft-Wilson, R., Doran, B., Grayson, T. H., Segal, S. S., & Hill, C. E. (2003). Expression of homocellular and heterocellular gap junctions in hamster arterioles and feed arteries. *Cardiovascular Research*, *60*, 643–653.

Sarelius, I., & Pohl, U. (2010). Control of muscle blood flow during exercise: Local factors and integrative mechanisms. *Acta Physiologica*, *199*, 349–365.

Schmidt, V. J., Jobs, A., von Maltzahn, J., Worsdorfer, P., Willecke, K., & de Wit, C. (2012). Connexin45 is expressed in vascular smooth muscle but its function remains elusive. *PloS One*, *7*, e42287.

Schmidt, K., Kaiser, F. J., Erdmann, J., & de Wit, C. (2015). Two polymorphisms in the cx40 promoter are associated with hypertension and left ventricular hypertrophy preferentially in men. *Clinical and Experimental Hypertension*, *37*, 580–586.

Schretzenmayr, A. (1933). Ueber kreislaufregulatorische Vorgaenge an den grossen Arterien bei Muskelarbeit. *Pflügers Archiv/European Journal of Physiology*, *232*, 743–748.

Schweda, F., Kurtz, L., de Wit, C., Janssen-Bienhold, U., Kurtz, A., & Wagner, C. (2009). Substitution of connexin40 with connexin45 prevents hyperreninemia and attenuates hypertension. *Kidney International*, *75*, 482–489.

Segal, S. S. (2005). Regulation of blood flow in the microcirculation. *Microcirculation*, *12*, 33–45.

Segal, S. S., & Duling, B. R. (1986). Flow control among microvessels coordinated by intercellular conduction. *Science*, *234*, 868–870.

Segal, S. S., & Jacobs, T. L. (2001). Role for endothelial cell conduction in ascending vasodilatation and exercise hyperaemia in hamster skeletal muscle. *Journal of Physiology*, *536*, 937–946.

Segal, S. S., Welsh, D. G., & Kurjiaka, D. T. (1999). Spread of vasodilatation and vasoconstriction along feed arteries and arterioles of hamster skeletal muscle. *Journal of Physiology*, *516*, 283–291.

Segretain, D., & Falk, M. M. (2004). Regulation of connexin biosynthesis, assembly, gap junction formation, and removal. *Biochimica et Biophysica Acta*, *1662*, 3–21.

Siegl, D., Koeppen, M., Wolfle, S. E., Pohl, U., & de Wit, C. (2005). Myoendothelial coupling is not prominent in arterioles within the mouse cremaster microcirculation in vivo. *Circulation Research*, *97*, 781–788.

Simon, A. M., Goodenough, D. A., Li, E., & Paul, D. L. (1997). Female infertility in mice lacking connexin 37. *Nature*, *385*, 525–529.

Simon, A. M., Goodenough, D. A., & Paul, D. L. (1998). Mice lacking connexin40 have cardiac conduction abnormalities characteristic of atrioventricular block and bundle branch block. *Current Biology*, *8*, 295–298.

Simon, A. M., & McWhorter, A. R. (2002). Vascular abnormalities in mice lacking the endothelial gap junction proteins connexin37 and connexin40. *Developmental Biology*, *251*, 206–220.

Simon, A. M., & McWhorter, A. R. (2003). Decreased intercellular dye-transfer and down-regulation of non-ablated connexins in aortic endothelium deficient in connexin37 or connexin40. *Journal of Cell Science, 116*, 2223–2236.

Smith, P. D., Brett, S. E., Luykenaar, K. D., Sandow, S. L., Marrelli, S. P., Vigmond, E. J., et al. (2008). Kir channels function as electrical amplifiers in rat vascular smooth muscle. *Journal of Physiology, 586*, 1147–1160.

Sohl, G., & Willecke, K. (2004). Gap junctions and the connexin protein family. *Cardiovascular Research, 62*, 228–232.

Sonkusare, S. K., Bonev, A. D., Ledoux, J., Liedtke, W., Kotlikoff, M. I., Heppner, T. J., et al. (2012). Elementary Ca^{2+} signals through endothelial TRPV4 channels regulate vascular function. *Science, 336*, 597–601.

Takano, H., Dora, K. A., Spitaler, M. M., & Garland, C. J. (2004). Spreading dilatation in rat mesenteric arteries associated with calcium-independent endothelial cell hyperpolarization. *Journal of Physiology, 556*, 887–903.

Tallini, Y. N., Brekke, J. F., Shui, B., Doran, R., Hwang, S. M., Nakai, J., et al. (2007). Propagated endothelial Ca^{2+} waves and arteriolar dilation in vivo: Measurements in Cx40BAC GCaMP2 transgenic mice. *Circulation Research, 101*, 1300–1309.

Tare, M., Coleman, H. A., & Parkington, H. C. (2002). Glycyrrhetinic derivatives inhibit hyperpolarization in endothelial cells of guinea pig and rat arteries. *American Journal of Physiology. Heart and Circulatory Physiology, 282*, H335–H341.

Thevenin, A. F., Kowal, T. J., Fong, J. T., Kells, R. M., Fisher, C. G., & Falk, M. M. (2013). Proteins and mechanisms regulating gap-junction assembly, internalization, and degradation. *Physiology (Bethesda), 28*, 93–116.

Tran, C. H. T., Vigmond, E. J., Plane, F., & Welsh, D. G. (2009). Mechanistic basis of differential conduction in skeletal muscle arteries. *Journal of Physiology, 587*, 1301–1318.

Tran, C. H. T., & Welsh, D. G. (2010). The differential hypothesis: A provocative rationalization of the conducted vasomotor response. *Microcirculation, 17*, 226–236.

Uhrenholt, T. R., Domeier, T. L., & Segal, S. S. (2007). Propagation of calcium waves along endothelium of hamster feed arteries. *American Journal of Physiology. Heart and Circulatory Physiology, 292*, H1634–H1640.

Unger, V. M., Kumar, N. M., Gilula, N. B., & Yeager, M. (1999). Three-dimensional structure of a recombinant gap junction membrane channel. *Science, 283*, 1176–1180.

Vaiyapuri, S., Jones, C. I., Sasikumar, P., Moraes, L. A., Munger, S. J., Wright, J. R., et al. (2012). Gap junctions and connexin hemichannels underpin hemostasis and thrombosis. *Circulation, 125*, 2479–2491.

Vaiyapuri, S., Moraes, L. A., Sage, T., Ali, M. S., Lewis, K. R., Mahaut-Smith, M. P., et al. (2013). Connexin40 regulates platelet function. *Nature Communications, 4*, 2564.

van Veen, T. A., van Rijen, H. V., & Jongsma, H. J. (2000). Electrical conductance of mouse connexin45 gap junction channels is modulated by phosphorylation. *Cardiovascular Research, 46*, 496–510.

VanTeeffelen, J. W. G. E., & Segal, S. S. (2006). Rapid dilation of arterioles with single contraction of hamster skeletal muscle. *American Journal of Physiology. Heart and Circulatory Physiology, 290*, H119–H127.

Wagner, C., de Wit, C., Kurtz, L., Grunberger, C., Kurtz, A., & Schweda, F. (2007). Connexin40 is essential for the pressure control of renin synthesis and secretion. *Circulation Research, 100*, 556–563.

Wagner, C., Jobs, A., Schweda, F., Kurtz, L., Kurt, B., Lopez, M. L. S., et al. (2010). Selective deletion of connexin 40 in renin-producing cells impairs renal baroreceptor function and is associated with arterial hypertension. *Kidney International, 78*, 762–768.

Wagner, C., & Kurtz, A. (2013). Distribution and functional relevance of connexins in renin-producing cells. *Pflügers Archiv/European Journal of Physiology, 465*, 71–77.

Wang, L., Yin, J., Nickles, H. T., Ranke, H., Tabuchi, A., Hoffmann, J., et al. (2012). Hypoxic pulmonary vasoconstriction requires connexin 40-mediated endothelial signal conduction. *Journal of Clinical Investigation*, *122*, 4218–4230.

Welsh, D. G., & Segal, S. S. (2000). Role of EDHF in conduction of vasodilation along hamster cheek pouch arterioles in vivo. *American Journal of Physiology. Heart and Circulatory Physiology*, *278*, H1832–H1839.

Wolfle, S. E., Chaston, D. J., Goto, K., Sandow, S. L., Edwards, F. R., & Hill, C. E. (2011). Non-linear relationship between hyperpolarisation and relaxation enables long distance propagation of vasodilatation. *Journal of Physiology*, *589*, 2607–2623.

Wolfle, S. E., & de Wit, C. (2005). Intact endothelium-dependent dilation and conducted responses in resistance vessels of hypercholesterolemic mice in vivo. *Journal of Vascular Research*, *42*, 475–482.

Wolfle, S. E., Schmidt, V. J., Hoepfl, B., Gebert, A., Alcolea, S., Gros, D., et al. (2007). Connexin45 cannot replace the function of connexin40 in conducting endothelium-dependent dilations along arterioles. *Circulation Research*, *101*, 1292–1299.

Wolfle, S. E., Schmidt, V. J., Hoyer, J., Kohler, R., & de Wit, C. (2009). Prominent role of KCa3.1 in endothelium-derived hyperpolarizing factor-type dilations and conducted responses in the microcirculation in vivo. *Cardiovascular Research*, *82*, 476–483.

Xu, J., & Nicholson, B. J. (2013). The role of connexins in ear and skin physiology—Functional insights from disease-associated mutations. *Biochimica et Biophysica Acta*, *1828*, 167–178.

Yeager, M. (2009). Gap junction channel structure. In A. L. Harris & D. Locke (Eds.), *Connexins: A guide* (pp. 27–75). New York, NY: Springer.

Yeager, M., & Harris, A. L. (2007). Gap junction channel structure in the early 21st century: Facts and fantasies. *Current Opinion in Cell Biology*, *19*, 521–528.

Zhou, J. Z., & Jiang, J. X. (2014). Gap junction and hemichannel-independent actions of connexins on cell and tissue functions—An update. *FEBS Letters*, *588*, 1186–1192.

Zong, L., Zhu, Y., Liang, R., & Zhao, H. B. (2016). Gap junction mediated miRNa intercellular transfer and gene regulation: A novel mechanism for intercellular genetic communication. *Scientific Reports*, *6*, 19884.

CHAPTER EIGHT

Molecular Signaling Pathways Controlling Vascular Tube Morphogenesis and Pericyte-Induced Tube Maturation in 3D Extracellular Matrices

S.L.K. Bowers, P.R. Norden, G.E. Davis[1]
Dalton Cardiovascular Research Center, University of Missouri School of Medicine, Columbia, MO, United States
[1]Corresponding author: e-mail address: davisgeo@health.missouri.edu

Contents

Advances in Pharmacology, Volume 77
ISSN 1054-3589
http://dx.doi.org/10.1016/bs.apha.2016.04.005

Abstract

During capillary network formation, ECs establish interconnecting tubes with defined lumens that reside within vascular guidance tunnels (physical spaces generated during EC tubulogenesis). Pericytes are recruited to EC tubes within these tunnels and capillary basement membrane deposition occurs to facilitate tube maturation. Here, we discuss molecular mechanisms controlling EC tubulogenesis demonstrating the involvement of integrins, MT1-MMP, extracellular matrix, Cdc42, Rac1, Rac2, k-Ras, Rap1b, and key downstream effectors including Pak2, Pak4, IQGAP1, MRCKβ, and Rasip1. These molecules activate kinase cascades controlling EC tube formation, in conjunction with growth factor receptor signaling, which involve PKCε, Src family, Raf, Mek, and Erk kinases. These molecules and signaling cascades stimulate EC lumen and tube formation by: regulating MT-MMP-dependent lumen expansion and vascular guidance tunnel formation; generation of intracellular vacuoles/vesicles to create EC apical membranes; and establishing cytoskeletal polarity with acetylated tubulin distributed subapically (and F-actin basally) to facilitate vacuole trafficking/fusion in a polarized, perinuclear region. Using defined serum-free models, we have demonstrated that human EC tubulogenesis and EC–pericyte tube coassembly requires five exogenously applied growth factors which are SCF, IL-3, SDF-1α, FGF-2, and insulin (Factors). Also, we have demonstrated that EC-derived PDGF-BB and HB-EGF are necessary for pericytes to proliferate, recruit to tubes, and induce basement membrane assembly. Finally, we have shown that VEGF fails to directly stimulate EC tubulogenesis. In contrast, it acts as an upstream EC primer of downstream "Factor"-induced tubulogenic and EC–pericyte tube coassembly by upregulating c-Kit, IL-3Rα, and CXCR4 as well as PDGF-BB and HB-EGF expression.

ABBREVIATIONS

Ang-1 angiopoietin-1
BMP bone morphogenic protein
CA constitutively active
CCM cerebral cavernous malformation
DN dominant negative
EC endothelial cell
ECM extracellular matrix
EGF epidermal growth factor
ERK extracellular signal-regulated kinase
FGF/FGFR fibroblast growth factor/receptor
GAP GTPase-activating protein
GEF guanine exchange factor
GFP green fluorescent protein
GIT1 G protein-coupled receptor kinase-interacting ArfGAP1
HB-EGF heparin-binding EGF-like growth factor
HDAC histone deacetylase
HEG heart of glass protein
IL interleukin
iPS induced pluripotent stem

Jak Janus-activated kinase
MMP matrix metalloproteinase
MT1-MMP membrane type 1 matrix metalloproteinase
PDGFR platelet-derived growth factor receptor
PKCε protein kinase C epsilon
RSII reduced serum supplement
SCF stem cell factor
SDF stromal-derived factor
siRNA small interfering RNA
Stat signal transducer and activator of transcription

1. INTRODUCTION

Considerable advances have been made concerning fundamental questions underlying how blood vessels form, how different vascular beds within the blood vs lymphatic vasculatures develop, and how blood vessels are altered and play a pathologic role in major human diseases (eg, cancer, diabetes, and hypertension) (Adams & Alitalo, 2007; Augustin, Koh, Thurston, & Alitalo, 2009; Carmeliet & Jain, 2011; Davis, 2012; Davis, Norden, & Bowers, 2015; Davis & Senger, 2005; Davis, Stratman, Sacharidou, & Koh, 2011; Dejana, Tournier-Lasserve, & Weinstein, 2009; Eelen, de Zeeuw, Simons, & Carmeliet, 2015; Eichmann & Simons, 2012; Eichmann & Thomas, 2013; Fish & Wythe, 2015; Giannotta, Trani, & Dejana, 2013; Hynes, 2007; Jeltsch, Leppanen, Saharinen, & Alitalo, 2013; Marcelo, Goldie, & Hirschi, 2013; Ramasamy, Kusumbe, & Adams, 2015; Rocha & Adams, 2009; Sacharidou, Stratman, & Davis, 2012; Senger & Davis, 2011; Simons & Eichmann, 2015; Stratman & Davis, 2012; Swift & Weinstein, 2009; Tirziu & Simons, 2009). Elucidation of these mechanisms has been performed using a combination of in vivo vs in vitro strategies. Major regulators of vascular morphogenesis are growth factors (including cytokines and chemokines), small molecules such as bioactive peptides and lipids, mural cells such as pericytes and vascular smooth muscle cells, inflammatory cells including macrophages and mast cells, and tissue parenchymal cells, which influence endothelial cells (ECs) to develop unique characteristics in different tissue beds (Adams & Alitalo, 2007; Arroyo & Iruela-Arispe, 2010; Davis, 2012; Davis et al., 2011; Helker et al., 2015; Jeltsch et al., 2013; Klein & Caron, 2015; Tirziu & Simons, 2009; van Hinsbergh, 2012; Zhang et al., 2012). It is becoming increasingly clear that these processes are highly complex, but the underlying cell and molecular biology of vascular cells, morphogenic events, and key states such as vessel maturation and stabilization can be elucidated with systematic investigations in vivo and in vitro using state-of-the-art approaches (Davis, Norden, & Bowers, 2015; Davis et al., 2011; Simons et al., 2015; Stratman & Davis, 2012).

Our laboratory has been focused on determining how ECs form tube structures with a particular emphasis on elucidating how capillary networks assemble since they represent the most abundant blood vessel type in all vascularized tissues (Davis et al., 2011; Sacharidou et al., 2012). Thus, we have also been investigating how ECs interact with pericytes, the other major cell type present in capillary structures (Armulik, Genove, & Betsholtz, 2011; Davis, Norden, & Bowers, 2015; Stratman & Davis, 2012). One of the fascinating questions that underlie our interests is why ECs are able to form tubes, while cells such as pericytes do not. We have made great progress in this pursuit over the years using defined 3D models with various human ECs and have contributed to the general field of lumen formation as well (Davis et al., 2011; Iruela-Arispe & Davis, 2009; Sacharidou et al., 2012) (eg, discovery that Cdc42 is a major regulator of EC lumen formation) (Bayless & Davis, 2002). Later studies revealed Cdc42's role in epithelial cell lumen formation in vitro (Bray, Brakebusch, & Vargo-Gogola, 2011; Bryant et al., 2010; Martin-Belmonte et al., 2007) as well as EC and epithelial lumen formation in vivo (Barry et al., 2015; Hoang, Nagy, & Senger, 2011; Kesavan et al., 2009; Wan et al., 2013). We have focused on a number of aspects of this process including the role of Rho GTPases and their effectors, the role of integrins and matrix metalloproteinases, the role of kinase signaling cascades, and a key role for both the actin and microtubule cytoskeletons (Davis, Bayless, & Mavila, 2002; Davis & Camarillo, 1996; Davis, Koh, & Stratman, 2007; Davis et al., 2011; Sacharidou et al., 2012). We have investigated the mechanisms underlying how ECs become polarized during tubulogenesis including membrane trafficking events and the role of intracellular vacuoles in creation of the apical membrane (Bayless & Davis, 2002; Davis & Camarillo, 1996; Davis, Speichinger, Norden, Kim, & Bowers, 2015; Kamei et al., 2006; Norden, Kim, Barry, Cleaver, & Davis, 2016), and following pericyte recruitment, how the EC abluminal surface is selectively decorated with pericytes and the vascular basement membrane matrix (Smith, Bowers, Stratman, & Davis, 2013; Stratman & Davis, 2012; Stratman, Malotte, Mahan, Davis, & Davis, 2009; Stratman, Schwindt, Malotte, & Davis, 2010). Finally, we have established models of vascular tube morphogenesis and maturation under serum-free defined conditions

(Koh, Stratman, Sacharidou, & Davis, 2008), which allows us to independently determine the growth factor requirements for EC tube assembly and EC–pericyte tube coassembly (mimicking the development of human capillary networks) leading to vascular basement membrane matrix deposition (Bowers, Meng, Davis, & Davis, 2015; Davis et al., 2013; Smith et al., 2013; Stratman, Davis, & Davis, 2011) (Figs. 1 and 2). These studies led to the surprising finding that VEGF or the combination of VEGF and fibroblast growth factor (FGF)-2, two major growth factors that were thought to be the primary factors controlling the development of the vasculature, failed to allow for human EC tube formation or EC–pericyte tube coassembly (Smith et al., 2013; Stratman et al., 2011; Stratman, Malotte, et al., 2009). In contrast, we discovered that five growth factors when added in combination allow for both processes to dramatically occur, and they are stem cell factor (SCF), interleukin-3 (IL-3), stromal-derived factor (SDF)-1α, FGF-2, and insulin (Factors) (Smith et al., 2013; Stratman et al., 2011) (Fig. 1). Of great interest is that we determined that VEGF is capable of acting as an upstream

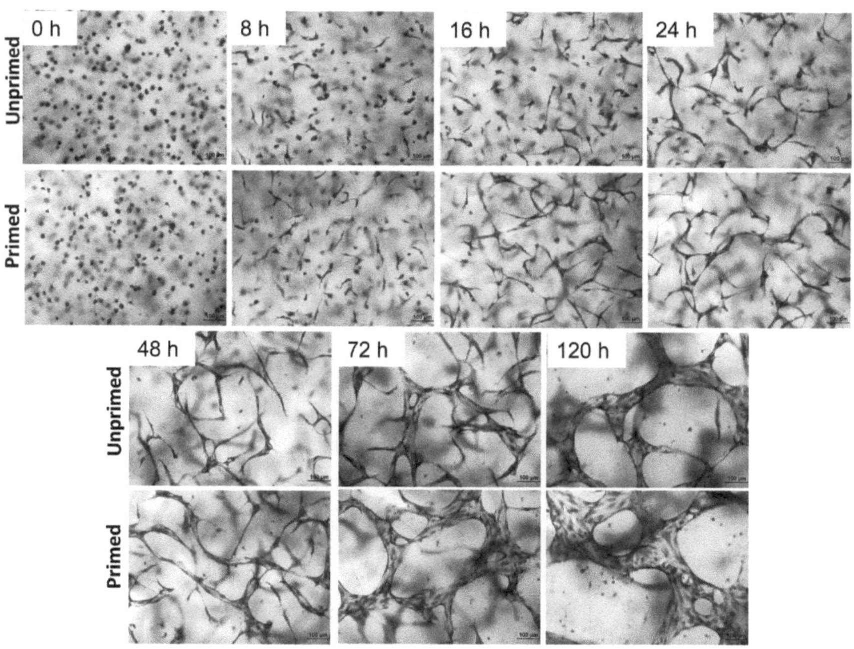

Fig. 1 VEGF priming stimulates EC tubulogenesis in response to the promorphogenic growth factors, SCF, IL-3, SDF-1α, FGF-2, and insulin. ECs were primed or not and then were placed in 3D collagen matrices. EC tube formation was allowed to proceed for indicated time points where cultures were fixed, stained with toluidine blue, and photographed. Bar equals 100 μm.

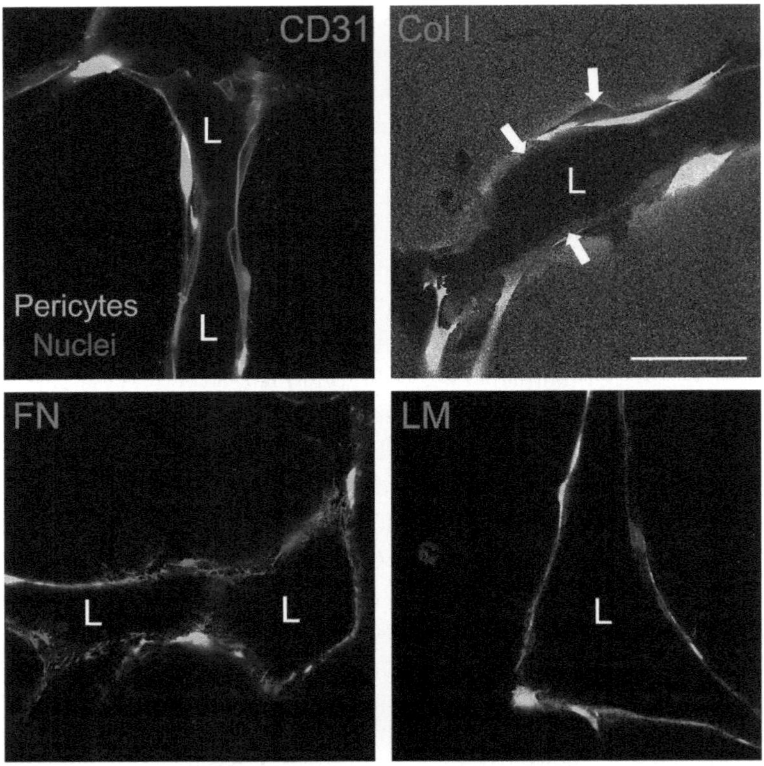

Fig. 2 Polarized EC–pericyte tube coassembly occurs with vascular guidance tunnels and results in vascular basement membrane matrix deposition. ECs were seeded with GFP-pericytes in 3D collagen matrices, and after 5 days were fixed, immunostained with the indicated antigens, and imaged using confocal microscopy. Pericytes are observed to recruit selectively to the EC abluminal tube surface (ie, basal side) as indicated with CD31 staining. Pericytes recruit to the tube surface within vascular guidance tunnels as revealed using anticollagen type I antibodies (Col I) (*arrows* indicate the tunnel borders). Vascular basement membrane matrix deposition is observed between the ECs and pericytes as revealed using antibodies directed to laminin (LM) or fibronectin (FN). The deposited FN matrix appears to be fibrillar in the image shown. Bar equals 25 μm.

priming agent to stimulate tube morphogenic and sprouting responses to these "Factors," but only if added prior to them (Fig. 1). Thus, we believe that VEGF's major effect is to prime ECs and thus, to prepare them for subsequent steps of vascular morphogenesis that require the "Factor" combination that we describe (Stratman et al., 2011). During vascular development, it appears that VEGF may also be essential for specifying the EC lineage as well (Fish & Wythe, 2015; Marcelo et al., 2013; Swift & Weinstein, 2009). Overall, the focus of our work on fundamental questions concerning how ECs perform critical functions such as assemble tube networks and coassemble with pericytes under defined conditions has led to significant advances in our understanding of these processes.

2. MOLECULAR DETERMINANTS CONTROLLING THE ABILITY OF ECs TO FORM TUBE NETWORKS IN 3D TISSUE AND EXTRACELLULAR MATRIX ENVIRONMENTS

A key function of ECs is to form tubes, a property of these cells that is frequently overlooked and is certainly understudied (Figs. 1 and 2). ECs of all types and in each vascularized tissue, including blood-derived and lymphatic ECs, form tube structures. Most EC-lined tubes have mural cells (either pericytes or vascular smooth muscle cells) associated with them, with the exception of lymphatic capillaries. The most abundant blood vessels by far are capillaries, and these critical vessel structures are lined by ECs and pericytes which cover a variable percentage (depending on the tissue) of the EC abluminal surface (Armulik et al., 2011). Our laboratory has focused considerable attention on an investigation of human capillary tube network assembly in three-dimensions with and without pericytes using highly defined in vitro systems (Davis et al., 2013; Smith et al., 2013; Stratman et al., 2011; Stratman, Malotte, et al., 2009). The process of EC tubulogenesis has been investigated in detail over the past 15–20 years and considerable progress has been made, with respect to identifying the critical roles of individual molecules and signaling cascades (Fig. 3). General questions such as the relative importance of extracellular matrix (ECM), tissue cell surfaces, and growth factors/chemokines/cytokines in these processes have also been addressed (Davis & Senger, 2005; Hynes, 2009; Jeltsch et al., 2013; Senger & Davis, 2011). It is also well known that both the ECM and cell surfaces present growth factors to cells so there is an important overlap in how cues that stimulate EC tubulogenesis are presented. In addition, receptors for ECM proteins and growth factor receptors can be coactivated together and can associate in unique ways to control the signaling necessary to control different stages of EC tubulogenesis, maturation and stabilization (Somanath, Ciocea, & Byzova, 2009; Stratman & Davis, 2012; Stratman, Malotte, et al., 2009). Since vascular cells (including both ECs and mural cells) produce growth factors and remodel the ECM during vascular morphogenesis, combined growth factor/ECM-mediated signaling will vary depending on the stage of vessel

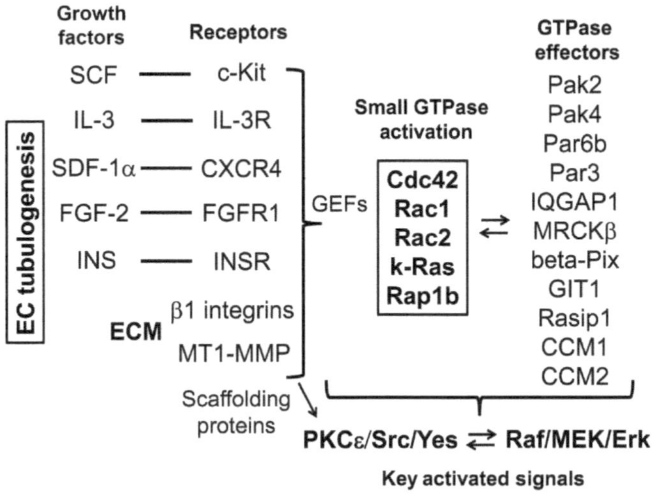

Fig. 3 The defined growth factors, SCF, IL-3, SDF-1α, FGF-2, and insulin, in conjunction with 3D extracellular matrices, stimulate EC tubulogenesis through the indicated molecular regulators. We hypothesize that this combination of growth factors signal in concert with integrins and MT1-MMP to activate kinase cascades including PKCε, Src, Yes, Pak2, Pak4, B-Raf/c-Raf, Mek, and Erk, all of which are necessary for ECs to form lumen and tube networks. We have identified five major small GTPases that control this process including Cdc42, Rac1, Rac2, k-Ras, and Rap1b. Key effectors that have been identified downstream of these GTPases which control EC tubulogenesis, include Pak2, Pak4, Par6b, Par3, IQGAP1, MRCKβ, beta-Pix, GIT1, Rasip1, CCM1, and CCM2. Future work will identify scaffolding proteins that connect the pathways downstream of receptor-mediated signaling, additional kinases that affect these events and GEFs that activate the key GTPases controlling this tubulogenic process.

formation and maturation. For example, it is known that specific growth factors can associate with particular ECM proteins, such as BMP-4 with collagen type IV (Wang, Harris, Bayston, & Ashe, 2008), while VEGF can associate with fibronectin (Wijelath et al., 2006), and keratinocyte growth factor (KGF or FGF-7) can bind to interstitial collagens such as type I and type III (Ruehl et al., 2002). Interestingly, integrin α5β1, a fibronectin-binding integrin, has been found to associate with Tie2, an angiopoietin-1 (Ang-1) receptor (Cascone, Napione, Maniero, Serini, & Bussolino, 2005). These types of specific associations allow for the possibility of unique signaling events for ECs, for example, where integrin α5β1 signals with Tie2, when Ang-1 is presented with fibronectin during vessel maturation events.

Two major events termed vasculogenesis (de novo assembly of individual and small groups of ECs into tube networks during development) and angiogenesis (sprouting and tube formation from pre-existing vessels) are where the EC tubulogenic process is observed (Adams & Alitalo, 2007; Davis et al., 2007, 2011; Xu & Cleaver, 2011). We have demonstrated that EC tubulogenesis in a 3D matrix environment is accompanied by clearing of the ECM through MT1-MMP-mediated proteolysis creating vascular guidance tunnels in which EC tubes reside (Stratman, Saunders, et al., 2009) (Fig. 2). Blockade of MT1-MMP interferes with EC tubulogenesis and the creation of vascular guidance tunnels simultaneously. In fact, when EC tubulogenesis is blocked by inhibitors of Src, PKCε, or anti-α2β1 integrin antibodies, vascular guidance tunnels fail to form as well (Stratman, Saunders, et al., 2009). These vascular guidance tunnels are matrix conduits which allow for EC tube formation, EC tube remodeling and are a physical space in which mural cells such as pericytes are recruited to migrate along the EC tube abluminal surface (Smith et al., 2013; Stratman, Malotte, et al., 2009). Although MT1-MMP activity is required to create vascular guidance tunnels, once they are formed and both ECs and pericytes are present within the tunnels, each cell type can migrate freely in an MMP-independent manner (mimicking the motility of cells on 2D substrates which do not require MMPs for their migratory ability) (Stratman, Malotte, et al., 2009; Stratman, Saunders, et al., 2009). Thus, generation of vascular guidance tunnels is required for EC tubulogenesis to occur and is an essential step in the process. These tunnel spaces are also necessary for EC tube remodeling and polarized EC–pericyte interactions leading to capillary basement membrane formation (Stratman, Malotte, et al., 2009) (Fig. 2).

The presence of pericytes with ECs markedly alters the pattern and remodeling of developing EC tube networks (Senger & Davis, 2011; Stratman & Davis, 2012; Stratman, Malotte, et al., 2009; Stratman et al., 2010). ECs alone are fully capable of making lumen and tube networks; but over time, they become wider and wider without pericyte coverage. In contrast, pericyte addition and recruitment to these tubes restricts EC tube diameter, which in part results from vascular basement membrane matrix assembly, a process that is markedly enhanced when both cell types are present together (Stratman & Davis, 2012; Stratman, Malotte, et al., 2009) (Fig. 2). Disruption of pericyte recruitment or function leads to increased vessel diameters which have been observed both in vitro and in vivo (Hellstrom et al., 2001; Hellstrom, Kalen, Lindahl, Abramsson, & Betsholtz, 1999; Stratman, Malotte, et al., 2009; Stratman et al., 2010). Thus, ECs can be defined by their ability to develop tube networks and specialize into different vascular beds depending on cues from their surrounding environment, while pericytes can be functionally defined by their ability to recruit to EC-lined tubes, affect EC tube remodeling events and coregulate vascular basement membrane matrix assembly (Bowers et al., 2015). More frequently, ECs and pericytes are defined by their expression of various immunomarkers and genes, and although this is very helpful, we propose that a functional definition may be more relevant and useful in our interpretation of studies and the potential use of ECs and pericytes in cellular therapeutic strategies such as tissue engineering applications.

3. MOLECULAR CONTROL OF EC–PERICYTE TUBE COASSEMBLY AND VASCULAR BASEMENT MEMBRANE MATRIX DEPOSITION

We have shown that pericyte recruitment to the abluminal EC tube surface occurs within vascular guidance tunnels, and that ECs and pericytes both migrate in a polarized manner with these matrix spaces to deposit the vascular basement membrane specifically along the EC abluminal surface (Stratman, Malotte, et al., 2009) (Fig. 2). ECs fail to deposit the vascular basement membrane on their own despite their ability to synthesize many of the key basement membrane components (Stratman, Malotte, et al., 2009). Major basement membrane components in the vascular basement membrane are laminins 411/421, laminins 511/521, collagen type IV, fibronectin, nidogens 1 and 2, perlecan, and collagen type XVIII (Davis & Senger, 2005; Senger & Davis, 2011). ECs and pericytes cocontribute these components and then deposit them between the two cell types. When siRNA suppression of collagen type IV was performed in ECs vs pericytes, we observed that disruption of collagen type IV in either cell type had a blocking effect, although the effect was greater in ECs (likely due to the fact that there are more ECs involved) (Stratman, Malotte, et al., 2009). Furthermore, we demonstrated that selective components of the basement membrane matrix including fibronectin, nidogen 1, laminin isoforms, and perlecan were induced when ECs and pericytes were cocultured in 3D collagen matrices compared to either cell type alone. Interestingly, we observed increases in pericyte basement membrane mRNAs for the laminin subunits $\alpha4$, $\alpha5$, $\beta1$, $\beta2$, and $\gamma1$, collagen type IV $\alpha1$ and $\alpha2$ chains, and perlecan in EC–pericyte cocultures compared to pericytes alone (Stratman, Malotte, et al., 2009). Increases in EC basement membrane mRNAs for the laminin chains $\alpha5$ and $\beta1$, nidogen 2, collagen type IV $\alpha1$ and $\alpha2$, perlecan, and fibronectin were observed in EC–pericyte cocultures compared to ECs alone (Stratman, Malotte, et al., 2009). Thus, it is clear that ECs and pericytes synergize in their ability to synthesize basement membrane components when they coassemble and furthermore, they work together to deposit this specialized matrix to create the unique capillary basement membrane.

A final interesting point is that ECs and pericytes migrate along each other within vascular guidance tunnels in a polarized fashion during basement membrane assembly, and we hypothesized that mechanical forces exerted on each other might be a critical signal that is necessary for basement membrane deposition. For example, it is well appreciated that fibronectin is a mechanosensitive protein and that is possesses matricryptic domains that are exposed following force application which allow it to self-assemble during ECM deposition (Davis, Bayless, Davis, & Meininger, 2000; Zhong et al., 1998). Interestingly, blockade of actomyosin force generation in our coculture system with the myosin II inhibitor, blebbistatin, interferes with fibronectin, and collagen type IV deposition, while EC–pericyte coassembly does occur as shown by CD31 staining to label the EC tubes, while pericytes contain green fluorescent protein (GFP) (Fig. 4). Previously, we reported that a 70 kDa N-terminal fibronectin fragment interfered with fibronectin deposition during EC–pericyte coassembly, but also collagen type IV deposition, suggesting that fibronectin regulates collagen type IV assembly during capillary basement membrane formation (Stratman, Malotte, et al., 2009). These data indicate that force generation between ECs and pericytes plays a fundamental regulatory role during capillary basement membrane matrix assembly.

4. INTEGRIN SIGNALING REQUIREMENTS CHANGE DURING EC TUBULOGENESIS VS EC–PERICYTE TUBE COASSEMBLY AND VASCULAR BASEMENT MEMBRANE MATRIX DEPOSITION

A variety of studies indicate that interstitial matrices such as collagen matrices strongly stimulate vascular tube morphogenesis (Davis & Senger, 2005; Senger & Davis, 2011). In addition, provisional matrices such as fibrin matrices, which contain fibronectin and vitronectin incorporated into them in fibrin clots, serve as matrices that stimulate EC tubulogenesis as well. Interestingly, tubulogenesis in collagen matrices appears to primarily depend on the collagen receptor $\alpha2\beta1$ (Davis & Camarillo, 1996), while

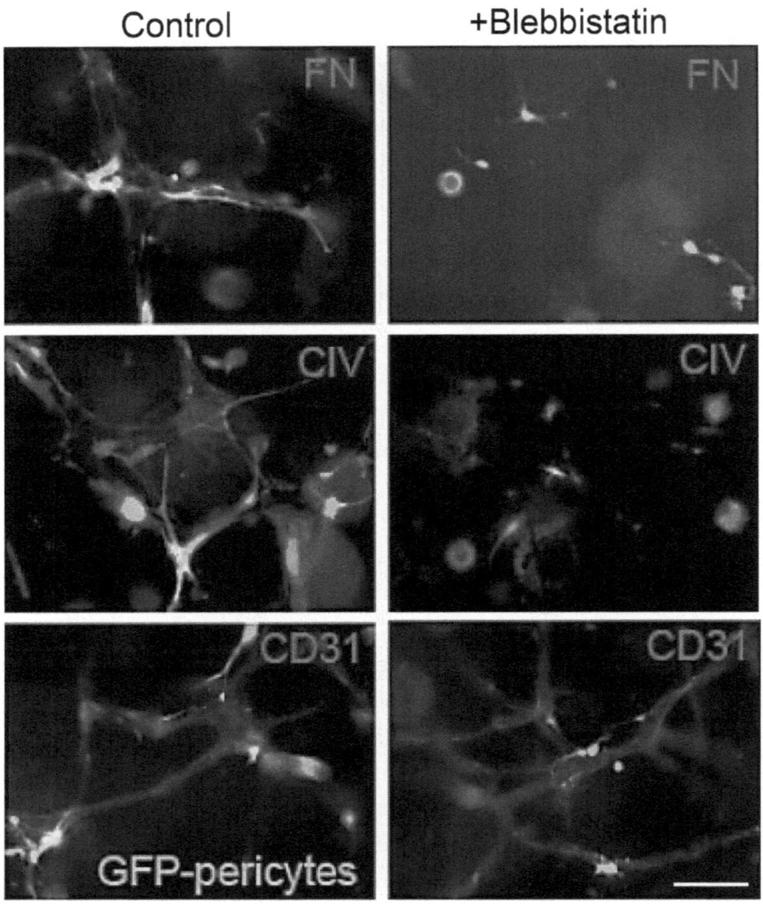

Fig. 4 Blockade of myosin II-dependent actomyosin contractility disrupts vascular basement membrane matrix assembly in 3D matrices. EC–pericyte cocultures were established in the absence or presence of 20 μM blebbistatin which blocks myosin II-dependent cellular contractility and cultures were fixed at 120 h and immunostained with fibronectin (FN), collagen type IV (CIV), or CD31 (to label ECs) antibodies. Pericytes were labeled with GFP. Bar equals 50 μm.

the integrins, α5β1 and αvβ3 control tube formation in fibrin matrices due to their affinity for fibronectin and vitronectin (both present in plasma and/or serum), respectively, which coassemble with fibrin (Bayless, Salazar, & Davis, 2000). Interestingly, we recently reported using a serum-free fibrin model with supplemented fibronectin incorporated into the fibrin matrix that this was exclusively dependent on the α5β1 integrin (Smith et al., 2013). Another study using fibrin matrices, which is supplemented with fibroblasts to support EC tube formation, demonstrated a role for EC α2β1 in this matrix due to the ability of fibroblasts to produce and deposit interstitial collagens, which can then be used by the ECs to form tubes in this remodeled fibrin matrix (Newman, Nakatsu, Chou, Gershon, & Hughes, 2011).

Of great interest is that integrin requirements change in collagen matrices, when cultures of ECs alone are compared to EC–pericyte cocultures (where basement membrane matrix assembly selectively occurs). In this case, EC only cultures depend on α2β1 and its effect can be observed using blocking antibodies, siRNAs, or specific chemical inhibitors (Davis & Camarillo, 1996). α2β1 controls both the formation and maintenance of EC-lined tubes in collagen matrices, while other integrin blocking reagents that interfere with integrins with known affinity for basement membrane components (ie, α1β1, α3β1, α5β1, α6β1) have no effects (Davis & Camarillo, 1996). The influence of α2β1 blocking antibodies has much less effect over time in EC–pericyte cocultures due to the fact that ECs begin to encounter basement membrane matrix proteins as they become deposited. Of great interest is that blocking antibodies directed to α1β1 (affinity for collagen type IV), α3β1 (affinity for laminin isoforms and nidogens), α5β1 (fibronectin), and α6β1 (laminin isoforms) begin to have strong disrupting effects which are selective to EC–pericyte cocultures when these are established in 3D collagen matrices (Stratman, Malotte, et al., 2009). This result strongly suggests that basement membrane matrix deposition results in a marked functional switch between integrin signaling downstream of α2β1 vs integrin signaling mediated by these basement membrane-binding integrins (ie, α1β1, α3β1, α5β1, α6β1) (Stratman, Malotte, et al., 2009). Interestingly, a series of studies indicate that basement membrane components interfere with EC morphogenesis suggests that this switch may prevent further morphogenesis and thus, promote, tube maturation and stabilization. Another key point is that pericytes make the MT1-MMP inhibitor, TIMP-3, which is known to bind ECM, such as perlecan, through heparan sulfate side chains, in the basement membrane and which could inhibit further tube morphogenesis in this manner (Saunders et al., 2006).

A final point of great interest is whether EC- and pericyte-derived growth factors specifically adsorb particular components of the vascular basement membrane which facilitate tube maturation and stabilization by affecting both cell types. Tissue parenchymal cells as well as other stromal cells, which are adjacent to these mature capillaries, will also be exposed to these EC–pericyte factors and such signaling may be fundamentally critical to tissue formation, differentiation, maintenance, and health of the key parenchymal cell types. We recently hypothesized that "capillaries are disease suppressors," in that factors derived from healthy capillaries will inhibit key disease mechanisms such as thrombosis, inflammation, ischemia, fibrosis, hemorrhage, and carcinogenesis (Davis, Norden, & Bowers, 2015). In diseases such as diabetes, capillary dysfunction is markedly observed, which may increase the incidence of such basic disease mechanisms leading to the common clinical complications of diabetes which manifests in many tissues including the retina and kidney.

5. CRITICAL GROWTH FACTOR REQUIREMENTS FOR EC–PERICYTE TUBE COASSEMBLY LEADING TO CAPILLARY TUBE FORMATION AND MATURATION

Considerable information suggests an important role for PDGF isoforms in cardiovascular development. Genetic knockout of PDGF-BB or platelet-derived growth factor receptor β (PDGFRβ) in mice leads to embryonic lethality and the vessel abnormalities that are observed resemble those seen in diabetes (Armulik et al., 2011; Hellstrom et al., 2001, 1999). One major observed phenotype are abnormalities in the extent of mural cell coverage, seen particularly in capillaries, suggesting a defect regarding EC–pericyte interactions. Suggested possible reasons for this defect included abnormalities in pericyte proliferation, differentiation, or recruitment. To address such questions in more detail, we set out to assess the role of PDGF-BB and other possible growth factors during EC–pericyte tube coassembly in our defined serum-free models of the process in 3D collagen or fibrin matrices (Smith et al., 2013; Stratman et al., 2010). We have shown that EC-derived PDGF-BB is important toward the ability of pericytes to migrate, proliferate, and recruit to developing EC-lined tubes in 3D matrices. When pericytes are cultured alone in 3D collagen matrices under serum-free conditions, they fail to migrate or proliferate (Stratman et al., 2010). However, if ECs and pericytes are cocultured together, pericytes proliferate, markedly migrate, and recruit to EC tubes (Stratman et al., 2010). This depends on EC PDGF-BB since blocking antibodies, receptor traps, and chemical inhibition of the PDGFRβ receptor leads to reduced pericyte coverage, widening of EC tubes, and reduced basement membrane formation (Stratman et al., 2010). In addition, we demonstrated that heparin-binding EGF-like growth factor (HB-EGF) is also critical to this process and combined blockade of both PDGF-BB and HB-EGF or their receptors leads to more marked inhibitory effects on pericyte proliferation, pericyte recruitment, and basement membrane matrix deposition. We were also able to demonstrate these phenomena in vitro and in vivo in developing quail embryos (Stratman et al., 2010). Thus, our work and that of others suggests a fundamental role for both PDGF-BB and HB-EGF in capillary tube assembly. Future work will need to identify other critical growth factors that direct this process. One of the fascinating questions is how many growth factors are actually necessary to control these events.

6. RHO GTPase-DEPENDENT CONTROL OF EC TUBULOGENESIS AND KEY FUNCTIONAL ROLES FOR GTPase EFFECTORS AND GTPase-ACTIVATING PROTEINS

A major direction in our laboratory has been to investigate the role of master regulators of the cytoskeleton, the Rho GTPases during EC tube morphogenesis (Fig. 4). We first demonstrated that Cdc42 and Rac1 were required for EC lumen formation by expressing dominant negative (DN) mutants in ECs and performing tube formation assays (Bayless & Davis, 2002). We observed that DN Cdc42 and Rac1, blocked lumen formation, while DN RhoA did not. In support of this conclusion, the addition of C3 exoenzyme, which selectively inhibits Rho, had no ability to inhibit tube formation (Bayless & Davis, 2002). Interestingly, constitutively active (CA) Rac1-stimulated lumen and tube formation, while CA RhoA markedly inhibited the process (Bayless & Davis, 2002). We also demonstrated that pinocytic intracellular vacuoles were critical to the EC lumen formation process (Bayless & Davis, 2002; Davis et al., 2002; Davis & Camarillo, 1996) and that they did not form well if Cdc42 or Rac1 were inhibited, while RhoA inhibition had no influence (Bayless & Davis, 2002). Importantly, we showed that GFP fusion proteins of Rac1 and Cdc42 were able to highlight the vacuoles and developing lumen surface during this process (Bayless & Davis, 2002). Here, we illustrate this vacuole formation and fusion process over time using ECs labeled with GFP-RalA (and mCherry-Cdc42) (Fig. 5) and GFP-Rac1 (Fig. 6). The vacuole trafficking and fusion events occur in a polarized, perinuclear position that defines the apical membrane domain. Also, in a collaborative study using zebrafish embryos expressing GFP-Cdc42 in ECs, it was shown that intracellular vacuoles played a key role during EC lumen formation in vivo (Kamei et al., 2006). We have observed abundant EC intracellular vacuoles

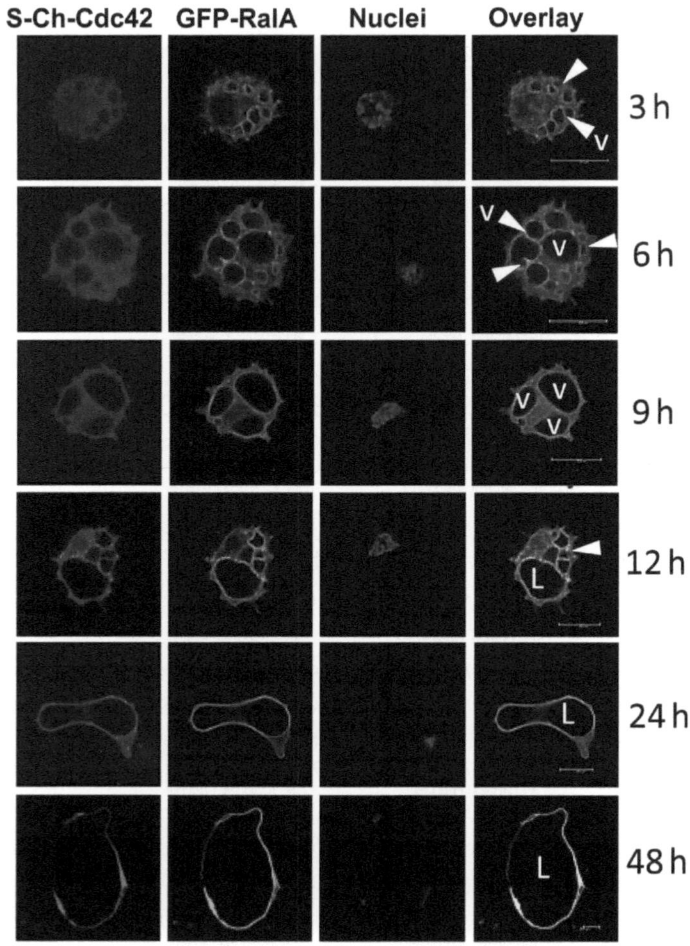

Fig. 5 Intracellular vacuole formation and fusion controls EC lumen formation in 3D collagen matrices. ECs were labeled with S tag-Cherry-Cdc42 (S-Ch-Cdc42) and GFP-RalA fusion proteins and were seeded in 3D collagen matrices. At the indicated time points, cultures were fixed and imaged by confocal microscopy. Intracellular vacuoles (v) are strongly labeled with GFP-RalA (*arrowheads*) and over time they fuse together to form the lumen compartment (L). Bars equal 20 μm.

during EC lumen formation in vivo in developing quail embryos (Davis et al., 2011). Additional studies using different fluorescent labeling approaches in ECs reach the same conclusion that ECs utilize vacuoles to generate lumens and facilitate the development of the apical surface in vitro and in vivo (Davis et al., 2007; Davis, Speichinger, et al., 2015; Wang et al., 2010; Yu, Castranova, Pham, & Weinstein, 2015).

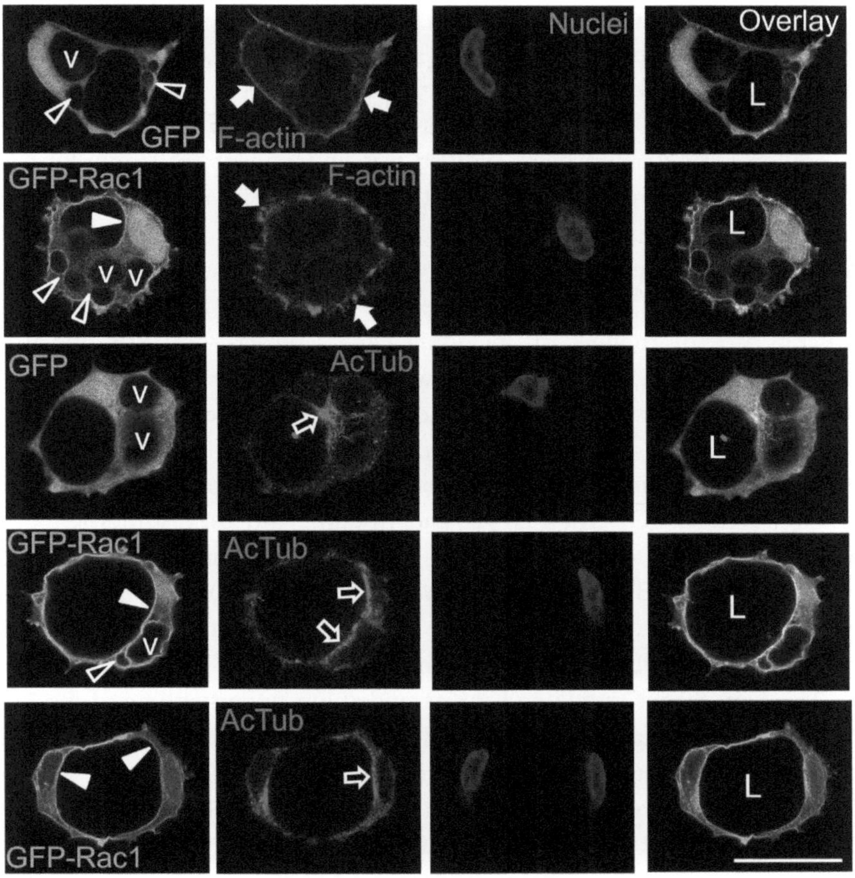

Fig. 6 Intracellular vacuoles are trafficked along the acetylated tubulin cytoskeleton and fuse in a perinuclear, polarized domain to form the EC apical compartment during lumen formation. ECs were induced to express GFP control or GFP-Rac1 and seeded within 3D collagen matrices. After 12 h, cultures were fixed, immunostained with antiacetylated tubulin antibodies, or stained with phalloidin to label F-actin. *Black arrowheads* indicate intracellular vacuoles (v) that form and fuse during EC lumen (L) formation. *White arrowheads* indicate apical targeting of GFP-Rac1. *White arrows* indicate the basal distribution of F-actin, while the *black arrows* indicate the perinuclear, polarized subapical distribution of acetylated tubulin (AcTub) that directs intracellular vacuole trafficking and fusion events. Bar equals 25 μm.

We later extended these studies with siRNA suppression experiments that revealed the same results (Koh, Mahan, & Davis, 2008; Norden et al., 2016). Also, in a recent study, knockout of Cdc42 in mouse ECs leads to marked inhibition of lumen formation in vivo (Barry et al., 2015). In addition, we identified key downstream effectors of Cdc42 and Rac1 including Pak2 (a common effector of Cdc42 and Rac1), Pak4 (selective for Cdc42), and the polarity regulator, Par6b (which associates with Par3) which also control the lumen formation process (Koh, Mahan, & Davis, 2008). Additional work demonstrated a key role for protein kinase C epsilon (PKCε), which we showed was upstream of Src family kinases (Src, Yes), and then Pak kinases which together phosphorylate and activate c-Raf/B-Raf leading to strong Erk activation (Koh et al., 2009) (Fig. 4). All of these kinases appear to be necessary to stimulate and control EC lumen and tube formation, and many of these have been confirmed in vivo to control EC tube morphogenesis including Pak-2 and Pak-4, c-Raf, and Erk kinases. This cascade is activated downstream of integrin-ECM contacts, MT1-MMP-mediated proteolysis and growth factor-dependent signaling (Fig. 4). As we discuss later in this chapter, we have identified five key growth factors which together control human ECs to form tube networks (ie, SCF, IL-3, SDF-1α, FGF-2, and insulin), and they stimulate both EC tubulogenesis, but also establish the conditions for ECs to produce PDGF-BB and HB-EGF (and presumably other factors) to stimulate pericyte recruitment, which is necessary to create EC–pericyte coassembled tubes and the capillary basement membrane matrix.

In a very recent study, we have extended our understanding of the EC lumen formation signaling cascade and have identified three additional small GTPases that control this process including k-Ras, Rac2, and Rap1b (Norden et al., 2016) (Fig. 4). In support of this conclusion was the finding that a Cdc42 and Rac GTPase-activating protein (GAP), Arhgap31, as well as a Ras GAP, Rasa1, were found to inhibit EC lumen formation. siRNA suppression of these two GAPs singly or together markedly stimulates EC tubulogenesis (through Cdc42, Rac, and Ras activation), while suppression of Arhgap29, a RhoGAP, leads to the opposite phenotype and blocks EC tube formation (through Rho activation) (Norden et al., 2016). Interestingly, a series of molecules which interfere with Rho activation are known to positively regulate EC tubulogenesis and these are Pak2, Pak4, Src, Erk1/2, cerebral cavernous malformation (CCM)1 (a Rap1 effector) and CCM2, Rasip1, and Arhgap29 (Kleaveland et al., 2009; Koh, Mahan, & Davis, 2008; Koh et al., 2009; Norden et al., 2016; Whitehead et al., 2009; Xu et al., 2011). Mouse knockout of Rasip1 leads to marked inhibition of EC lumen formation in vivo during vascular development and Rasip1 also interacts directly with Arhgap29 (Xu et al., 2011). siRNA suppression of Rasip1 in vitro recapitulates the phenotype in vivo, in that ECs are able to form cords and networks but they fail to form lumen structures

(Norden et al., 2016). A GFP-Rasip1 construct was shown to target to the developing apical surface during the lumen formation process (Norden et al., 2016). Rasip1 appears to be both a Ras and Rap1 effector (Mitin et al., 2004; Post, Pannekoek, Ponsioen, Vliem, & Bos, 2015; Post et al., 2013) which intersect well with our findings demonstrating a role for both k-Ras and Rap1b in EC lumen formation (Norden et al., 2016). A recent study demonstrates an interaction of Rasip1 with the heart of glass (HEG) protein which also interacts with CCM1 to control vascular and cardiac development (de Kreuk et al., 2016).

In addition, our recent work demonstrates an important role for the small GTPase effectors, IQGAP-1, MRCKβ, beta-Pix, and G protein-coupled receptor kinase-interacting ArfGAP1 (GIT1), in addition to Pak2, Pak4, Rasip1, and Par6b (which interacts with Par3 and thus forms complexes with Par3-associated proteins such as JamB and JamC) (Norden et al., 2016) (Fig. 4). We were able to demonstrate an interaction of Cdc42 and Rac1 with IQGAP1 and Cdc42 with MRCKβ during the EC lumen formation process in 3D collagen matrices. IQGAP1 has important scaffolding properties which promote Raf activation and Erk-dependent signaling and thus would be one molecule which may integrate Cdc42, Rac, and k-Ras (Raf is a key Ras effector) signaling (Norden et al., 2016). Pak2 is known to interact and be regulated by GIT1 which binds to beta-Pix, a Cdc42, and Rac effector (Loo, Ng, Lim, & Manser, 2004). Interestingly, GIT1 and beta-Pix negatively regulate focal adhesion assembly (similarly to Rasip1 in ECs) and may participate in reducing RhoA-dependent cell contractility during the lumen formation process (Kuo, Han, Hsiao, Yates, & Waterman, 2011). Interestingly, MRCKβ been reported to play a functional role in Cdc42-dependent nuclear repositioning during directed cell motility (Gomes, Jani, & Gundersen, 2005) and has been reported to facilitate EC–EC junctional contacts along with other molecules including Rap1 and Rac1 (Ando et al., 2013).

Considerable work remains to elucidate how each of the key GTPases and effectors affect distinct steps during EC lumen and tube formation. Our work over many years suggests that there are a series of definable steps that control the EC lumen formation process. These steps are (i) membrane trafficking events where pinocytic membrane bound vacuoles (enriched in GTPases such as Rac and Ras) are trafficked to a polarized perinuclear region to fuse and develop the apical membrane surface; (ii) pinocytic vacuoles and other intracellular membrane compartments (Weibel–Palade bodies and Golgi-derived membrane vesicles) move along cytoskeletal tracks to fuse in exocytic events with the developing apical membrane; (iii) the establishment of asymmetric cytoskeletal polarity where F-actin is enriched in a basal location, while modified tubulins such as acetylated tubulin are enriched in a subapical location and can be observed to be in contact with vacuoles and vesicles that are being trafficked to the apical domain; and (iv) coordination of signaling, vesicle trafficking, and cytoskeletal polarization with MT1-MMP-mediated proteolysis to develop vascular guidance tunnels in conjunction with EC tube networks to establish a 3D matrix template (ie, 2D-like surfaces in 3D matrices) for tube remodeling, mural cell recruitment, basement membrane assembly, and tube maturation.

7. CYTOSKELETAL CONTROL OF CELL POLARIZATION COUPLED WITH MEMBRANE TRAFFICKING EVENTS TO CREATE A SPECIALIZED APICAL SURFACE DURING EC TUBULOGENESIS

It is clear that the dynamic process of vascular tube morphogenesis requires that EC cytoskeletal elements must be acting in concert to generate the appropriate EC morphologies necessary to form branching tube networks with defined lumens. Importantly, both the actin and microtubule cytoskeletons are required for EC tubulogenesis (Davis & Camarillo, 1996) and it is well known that small GTPases, that control this process, act in part by controlling the coordinated assembly and disassembly of the cytoskeleton which is necessary at different stages to regulate a complex morphogenic process such as tubulogenesis. Our past work revealed that disruption of microtubules in established EC tube networks using chemical agents that depolymerize microtubules (ie, colchicine and vinblastine) cause rapid tube collapse due to RhoA-dependent actomyosin contractility (Bayless & Davis, 2004). Interestingly, an important Rho guanine exchange factor (GEF), GEF-H1 is known to be sequestered and inactivated when associated with polymerized tubulin, but is released and able to cause Rho activation following disruption of microtubules (Krendel, Zenke, & Bokoch, 2002). A key regulator of EC tubulogenesis, Pak4, is able to phosphorylate and inactivate GEF-H1 (Callow, Zozulya, Gishizky, Jallal, & Smeal, 2005).

To investigate how tubulin polymerization might regulate EC tube formation and maintenance, we focused on microtubule tip complex proteins that are known to control how tubulin is polymerized to affect cellular structure and function (Kim, Martinez-Lemus, & Davis, 2013). We identified three important regulators of tubulin assembly, EB1, p150glued, and

Clasp1, which were critical during EC lumen formation. siRNA suppression of each of these molecules markedly inhibited EC lumen and tube formation (Kim et al., 2013). Interestingly, we observed that tubulin modifications such as acetylation and detyrosination were strongly increased as ECs form tubes in 3D matrices. Immunostaining revealed that these modified tubulins could be seen to be distributed in a subapical region supporting the apical membrane surface in both forming and established EC tube networks (Kim et al., 2013) (Fig. 6). This occurred in a manner dependent on these three microtubule tip complex proteins. In contrast, we demonstrated that F-actin, using phalloidin staining, is strongly distributed in a basal location during tube formation and following tube establishment (Fig. 6). Thus, during EC tubulogenesis, the cytoskeleton becomes asymmetrically distributed with modified tubulins in a subapical domain, while F-actin is observed in a basal domain. To further address how tubulin acetylation might affect EC tube formation, we utilized siRNAs to two tubulin deacetylases, histone deacetylase 6 (HDAC6), and sirtuin2, and in both cases, we were able to show that they increased tubulin acetylation and also increased lumen formation (Kim et al., 2013). When the siRNAs were combined, the effects were considerably greater and a very strong increase in EC lumen, and tube formation was observed. In support of these findings, we have also recently shown that the HDAC6 inhibitor, tubacin, markedly increases EC tubulogenesis (Norden et al., 2016). In addition, siRNAs to Arhgap31 (to stimulate Cdc42 and Rac1 activity), Rasa1 (to stimulate k-Ras activity), or the combination of these siRNAs strongly enhance EC lumen formation as well as increase tubulin acetylation and detyrosination directly correlating with the accelerated morphogenic response (Norden et al., 2016). Also, the opposite occurred with Arhgap29, with reduced EC lumen formation, tubulin acetylation and detyrosination compared to control siRNA-treated cultures. Overall, these data suggest that modified tubulins like acetylation and detyrosinated forms are functionally controlling EC tubulogenesis by supporting the developing EC apical membrane surface and regulating membrane trafficking and possibly fusion of vacuoles/vesicles to the apical surface.

It appears clear that a major step in vascular lumen and tube formation is establishment of an asymmetrical cytoskeleton which creates polarity to identify the developing apical domain (Figs. 5 and 6). Our early studies here suggest that this might be necessary to direct membrane vacuoles and vesicles to a polarized perinuclear region where vacuoles and vesicles coalesce and fuse together to create an apical surface (Figs. 5 and 6). Confocal imaging experiments performed over time suggests that this is the case and, also real-time movies of EC lumen formation similarly support this conclusion. Several small GTPases can be shown to target vacuoles and progress over time into the apical domain surface as the lumen is formation. In particular, Rac1, k-Ras, and RalA target particularly well to vacuole membranes (Figs. 5 and 6), while Cdc42 does also, but appears to be more concentrated in a subapical region that supports these vacuoles as they transport and fuse in this apical domain (Bayless & Davis, 2002; Davis et al., 2007; Norden et al., 2016). Other small GTPases including Cdc42, Rac1, Rac2, k-Ras, Rap1b, and RalA can be shown to be polarized and targeted to the EC apical surface and subapical region during lumen formation (Davis, Speichinger, et al., 2015; Norden et al., 2016). We observe these vacuoles/vesicles to be in contact with acetylated tubulin in the perinuclear region where they fuse during EC lumen and tube formation (Davis, Speichinger, et al., 2015; Norden et al., 2016). Interestingly, in this perinuclear region, we have observed strong colocalization of Cdc42 with acetylated tubulin (Norden et al., 2016), suggesting a role for Cdc42 and its effectors in this targeting and vesicle fusion process.

8. GROWTH FACTORS CONTROL VASCULAR TUBE MORPHOGENESIS

Growth factors are known to control many cellular properties including cell proliferation, motility, differentiation, and morphogenesis. An important coregulator of growth factor function is the ECM, which can present or sequester these factors, and can also affect signaling events due to cosignaling between growth factor receptors and ECM receptors such as integrins and syndecans. A fundamental question is what growth factors are necessary to form the blood vs lymphatic vasculatures and are there differences in growth factor requirements for the development and maintenance of the arterial, capillary, and venous blood vessels. Despite considerable efforts to identify growth factors that facilitate the development of these different vascular beds, many questions remain. Although VEGF isoforms such as VEGF-A have been implicated for the creation of the arterial vasculature and VEGF-C for the lymphatic vasculature (Jeltsch et al., 2013; Lohela, Bry, Tammela, & Alitalo, 2009), additional studies need to be performed to determine if these conclusions are oversimplified, and that many more factors are actually necessary to generate these unique vascular beds. The important question is what growth factor combinations directly stimulate the vascular tube morphogenic response. For example, it is possible that the profound effects of VEGF-A and VEGF-C that have been observed during development are due in part to their ability to control the development of the arterial EC vs lymphatic EC lineage in the first place. If the lineages are not generated, then those unique vascular beds would fail to develop, but it remains to be seen if they directly cause the vascular morphogenic responses necessary to create vascular tube networks. In support of the concept that VEGF-A is necessary to generate arterial ECs is that VEGF-A or VEGFR2 mouse knockouts fail to develop hematopoietic stem cells and blood cells in addition to being unable to create the blood vasculature (Shalaby et al., 1997, 1995). Recent studies indicate that ECs within the developing aorta and other vascular beds appear to be the precursors of hematopoietic stem cells which generate all blood cell lineages (Bertrand et al., 2010; Eilken, Nishikawa, & Schroeder, 2009; Hirschi, 2012; Leung et al., 2013; Lo Celso et al., 2009; Zovein et al., 2008). Thus, there appears to be a direct connection between development of blood cell lineages and the developing vasculature which carries those blood cells.

9. HEMATOPOIETIC STEM CELL CYTOKINES (SCF, IL-3, SDF-1α) COMBINED WITH FGF-2 AND INSULIN MARKEDLY STIMULATE EC TUBE MORPHOGENESIS AND EC–PERICYTE TUBE COASSEMBLY

As mentioned earlier, our work investigating how capillary tube networks form shows that VEGF is not able to support EC tube morphogenesis under serum-free defined conditions, and this is the case whether FGF-2 is present or not. After an extensive screening process using our 96-well microassay system of human EC tube morphogenesis in 3D collagen matrices, we identified a combination of five factors that support this process and they are SCF, IL-3, SDF1α, FGF-2, and insulin (Factors) (Stratman et al., 2011) (Fig. 1). The influence of insulin in this "Factor" system (Fig. 7) illustrates an effect that has implications for EC tube morphogenic responses in the context of diabetic patients. To date, no other combination of growth factors has been found to work in our defined serum-free assays (very similar results in either 3D collagen or fibrin matrices) (Figs. 8 and 9). These "Factors" allow for EC tube morphogenesis in assays that mimic vasculogenesis, where individual ECs are randomly suspended in 3D matrices, and in angiogenic sprouting assays where ECs are seeded on a matrix surface and allowed to sprout into the matrix and form tubes. Multiple human EC

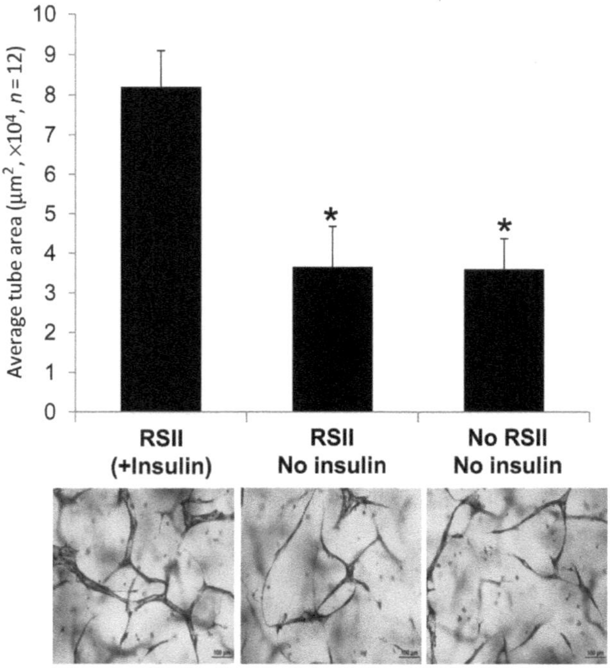

Fig. 7 Insulin signaling is critical for human endothelial cell tubulogenesis. ECs were seeded into 3D collagen matrices containing SCF, IL-3, SDF-1α, and FGF-2. Culture media contained FGF-2 and the indicated RSII supplement with or without insulin, and a third condition without RSII or insulin. The RSII supplement used in our 3D matrix assay system contains sodium selenite, holo-transferrin, fatty acid-free bovine serum albumin, and oleic acid (and we usually add insulin). After 72 h in culture, gels were fixed in 3% glutaraldehyde, stained with 0.1% toluidine blue, and images taken for analysis; representative images are shown here. Data are presented as mean ± standard deviation, $n = 12$ and $^*p < 0.01$. Scale bar is 100 μm.

types respond in a similar manner to these "Factors" including human umbilical vein (HUVEC) or arterial ECs (HUAEC) as well as induced pluripotent stem cell-derived ECs (iPS ECs) (Fig. 10). We also reported that embryonic quail blood vessels are able to sprout and form networks of tubes in response to the "Factors" (Stratman et al., 2011). Consistent with this result, we were able to detect the presence of the "Factor" receptors (c-Kit, IL-3Rα, and CXCR4) in quail embryonic vessels suggesting their potential role in blood vessel morphogenesis in vivo. Furthermore, we demonstrated that addition of chemical inhibitors or blocking antibodies directed to Factor receptors (eg, c-Kit or CXCR4) or the Factors themselves, respectively, to embryonic quail embryos led to disruptions of vessel formation (Stratman et al., 2011).

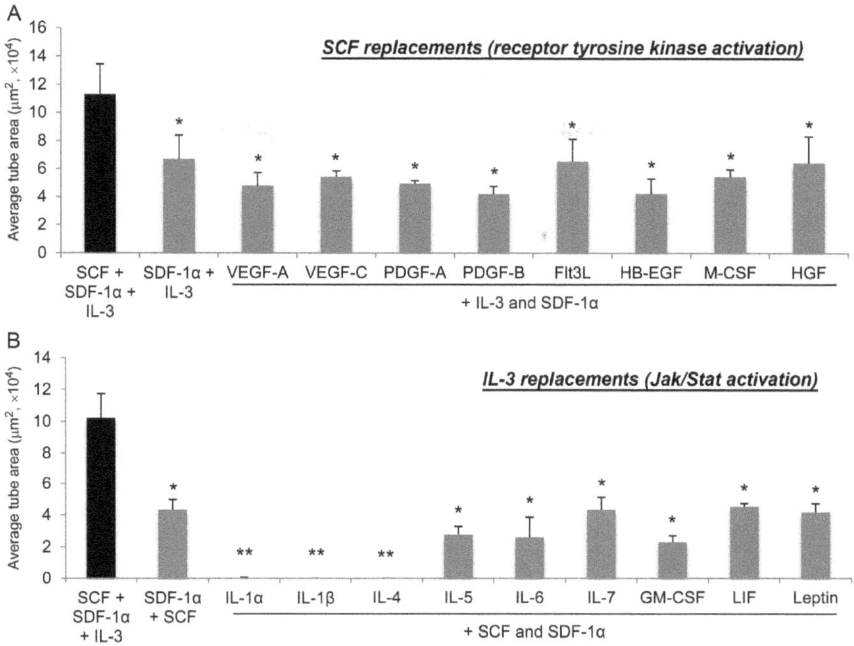

Fig. 8 Hematopoietic stem cell growth factors, SCF, IL-3, and SDF-1α, in conjunction with FGF-2 and insulin, markedly stimulate human EC tube morphogenesis. (A) Growth factors that activate receptor tyrosine kinases were tested in comparison to SCF addition. Each culture contained IL-3, SDF-1α, FGF-2, and insulin, and then we added the indicated growth factor or not. After 72 h, cultures were fixed, stained, and quantitated for EC tube area (*$p < 0.01$, compared to control, *black bar*). (B) Growth factors that activate Jak/Stat signaling were tested in comparison to IL-3. Each culture contained SCF, SDF-1α, FGF-2, and insulin, and then we added the indicated growth factor or not.addition. After 72 h, cultures were fixed, stained, and quantitated for EC tube area (*$p < 0.01$, compared to control, *black bar*). Addition of IL-1α, IL-1β, or IL-4 completely inhibited EC tubulogenesis (**$p < 0.01$ compared to all other conditions).

Another key issue is the critical requirement that the "Factors" allow for EC tube morphogenesis; but also EC tube maturation events whereby pericytes can recruit to tubes. We have demonstrated that EC–pericyte coassembly with vascular basement membrane deposition has been shown to occur with multiple human EC types, including human umbilical vein ECs or arterial ECs, or with different sources of pericytes (human brain-derived or bovine retina-derived) using this defined "Factor"-induced system (Stratman, Malotte, et al., 2009). An important future question is to determine the growth factor requirements for lymphatic vessel formation. We are currently assessing a potential role for the "Factors" in controlling this process.

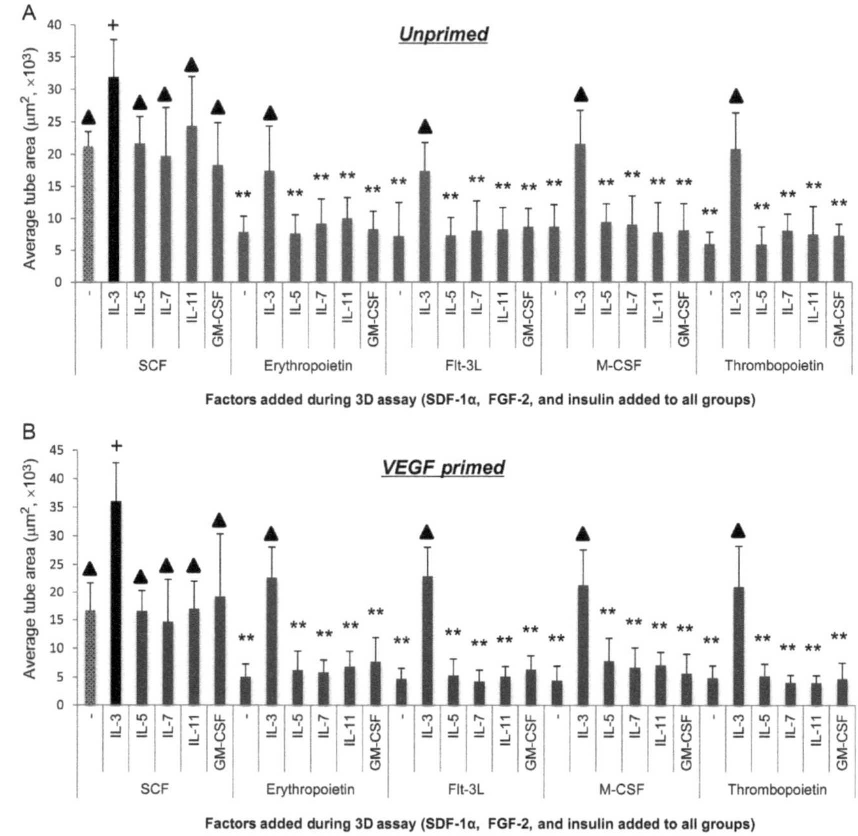

Fig. 9 The unique combination of hematopoietic stem cell cytokines, SCF, IL-3, and SDF-1α, when added with FGF-2 and insulin, optimally stimulates EC tube morphogenesis. To test whether different combinations of hematopoietic stem cell cytokine receptor tyrosine kinase activators with Jak/Stat activators may induce EC tubulogenesis, untreated (A) or VEGF-primed (B) ECs were seeded with the indicated mixtures of growth factors in 3D collagen matrices for 72 h. Cultures were fixed, stained, and quantitated for tube area. Data are presented as mean ± standard deviation, $n = 9$. $^{**}p < 0.01$, indicates a significant decrease vs SCF and SCF + IL-3; $^{+}p < 0.01$, indicates a significant increase vs SCF; and $^{\blacktriangle}p < 0.02$, indicates a significant decrease vs SCF + IL-3.

To further evaluate possible growth factor combinations that might control EC tube morphogenesis, we performed experiments to evaluate possible SCF or IL-3 substitutes in our model (Figs. 8 and 9). Only SCF was able to optimally stimulate tube morphogenesis compared to multiple other receptor tyrosine kinase-activating growth factors (Fig. 8A). Furthermore, only IL-3 was able to optimally stimulate tube morphogenesis compared to other known Jak/Stat-activating growth factors including IL-5 and GM-CSF, which bind to their heterodimeric receptors like IL-3, and each of these

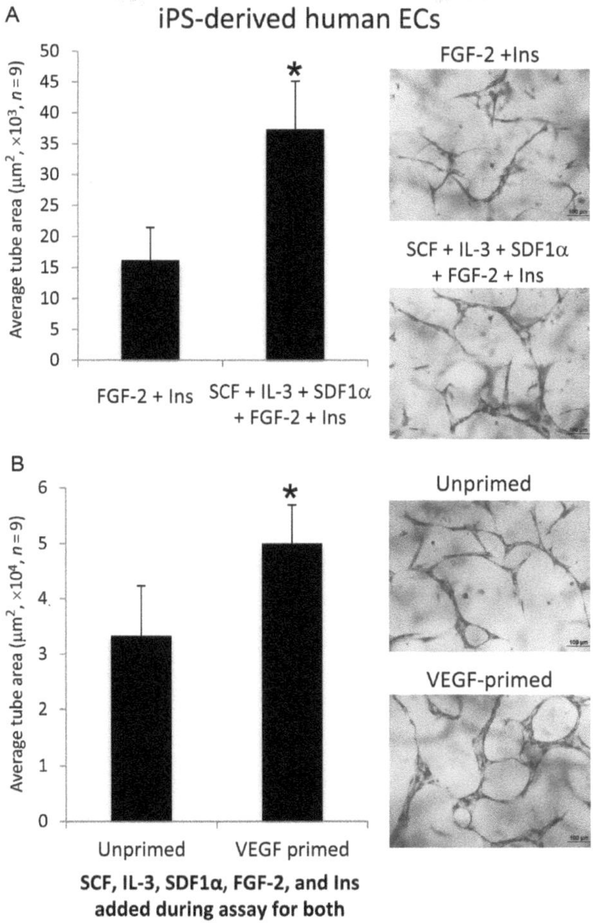

Fig. 10 Human endothelial cells derived from induced pluripotent stem cells form EC-lined tubes in 3D collagen matrices in response to SCF, IL-3, SDF-1α, FGF2, and insulin and are also primed by VEGF. (A) Human iPS-derived ECs (Cellular Dynamics, Madison, WI) were seeded into 3D collagen matrices under the indicated media conditions for 72 h. Cultures were fixed, stained, and quantitated for EC tube area. Data are presented as mean $\pm$ standard deviation, $n=9$ and $*p<0.01$, indicates significance compared to the control FGF-2 + Ins condition. (B) iPS ECs were either left untreated or were primed with VEGF and then seeded into 3D collagen matrices under the indicated media conditions for 72 h. Cultures were fixed, stained, photographed, and quantitated for EC tube area. Data are presented as mean $\pm$ standard deviation, $n=9$ and $*p<0.01$, indicates significance compared to the control unprimed condition. Bar equals 100 μm.

receptors share a common beta chain (Fig. 8B). We also attempted to add other hematopoietic cytokines, which activate receptor tyrosine kinases (ie, Flt3L, M-CSF, erythropoietin, and thrombopoietin), to see if they might be able to combine with alternative Jak/Stat activators different than

IL-3, to stimulate these morphogenic processes (Fig. 9). Finally, we added these other hematopoietic cytokines to the "Factor" combination, and they did not further stimulate EC tube morphogenesis (not shown). Thus, we were unable to identify any other combination than the five "Factors" that we have previously reported to maximally stimulate EC tubulogenesis (Stratman et al., 2011) (Figs. 8 and 9).

It is very intriguing that SCF, IL-3, and SDF-1α have all been reported to play a major role in hematopoietic stem cell (HSC) proliferation and maintenance (which provide hematopoietic cells throughout life). And yet they are central in combination for the ability of ECs to form tubes when added with FGF-2 and insulin. The fact that ECs are precursors for HSCs, an EC type referred to as hemogenic ECs (Hirschi, 2012), appears to be directly related to our findings. It would be make sense if the same "Factors" that stimulate EC tubulogenesis would also initiate HSC generation, as EC-lined tubes are the conduits in which blood cells are propelled throughout tissues. EC tubes would need to be generated first followed by HSCs and differentiated blood cell development. Interestingly, mouse knockout of key angioblast/hemangioblast genes such as the transcription factors, Etv2 and Tal1, which control EC lineage development, also result in the absence of blood cell elements (Lammerts van Bueren & Black, 2012; Porcher et al., 1996).

More work is required to follow through on how to investigate complex growth factor combinations such as the "Factors" that we have identified in an in vivo setting. Previous studies have implicated the involvement of SCF, IL-3, SDF-1α, FGF-2, and insulin as individual factors regulating the embryonic, adult or pathologic vasculature, and hematopoietic stem cells in a variety of contexts (Arai et al., 2005; Brandt, Bhalla, & Hoffman, 1994; Brandt, Briddell, Srour, Leemhuis, & Hoffman, 1992; Brizzi et al., 1994; Broxmeyer, 2008; Bryder & Jacobsen, 2000; Carow, Hangoc, Cooper, Williams, & Broxmeyer, 1991; Dentelli, Rosso, Olgasi, Camussi, & Brizzi, 2011; Dvorak, Seder, Paul, Morgan, & Galli, 1994; He et al., 2010; Heissig, Werb, Rafii, & Hattori, 2003; Jin et al., 2006; Moldenhauer et al., 2015; Suzuki et al., 2014; Tzeng et al., 2011). Considerable work is now demonstrating a fundamental role for SDF-1α in the development of the coronary vasculature and other beds including the gastrointestinal tract in both mouse and zebrafish models (Cavallero et al., 2015; Cha et al., 2012; Harrison et al., 2015; Ivins et al., 2015; Tachibana et al., 1998). One of the major questions in future work is why this particular combination of "Factors" is able to stimulate EC tubulogenesis and sprouting. We hypothesize that coordinated signaling from the five "Factor" receptors in combination with integrin- and MT1-MMP-mediated signals leads to activation of major kinase cascades, small GTPases, and downstream effectors to control EC tubulogenesis (Fig. 4).

10. DEFINING UNIQUE AND SEPARABLE STAGES OF CAPILLARY TUBE FORMATION: VEGF IS AN UPSTREAM PRIMER FOR DOWNSTREAM "FACTOR"-STIMULATED EC TUBULOGENESIS AND EC–PERICYTE TUBE COASSEMBLY

Our longstanding interest in investigating human capillary tube assembly in 3D matrices under serum-free defined conditions, led us to the surprising conclusion that VEGF or the combination of VEGF with FGF-2 was unable to support EC tube morphogenesis. It is generally thought that VEGF is the primary driver of blood vessel formation and yet over many years of investigation, we have been unable to demonstrate this in vitro in a defined 3D system. Instead, our search for the growth factors that stimulate human capillary lumen and tube formation, led us to discover a combination of five growth factors (ie, SCF, IL-3, SDF-1α, FGF-2, and insulin "Factors") that allow for tubes to form in vasculogenic or angiogenic assays, and in both cases, pericytes dramatically recruit to tubes when intermixed with ECs. These data strongly suggest that the "Factors" are the growth factors that stimulate capillary tube networks to form.

An important question arose as to the function of VEGF. VEGF, originally termed vascular permeability factor (Senger, Perruzzi, Feder, & Dvorak, 1986), is typically associated with vascular instability rather than stability in that VEGF treatment leads to increased vascular permeability. In fact, VEGF treatment leads to disassembly of EC–EC junctions (Dejana & Giampietro, 2012; Dejana et al., 2009; Giannotta et al., 2013; Jeltsch et al., 2013), which might be important for initiating vascular morphogenesis, but would not be compatible with tube assembly where EC–EC contacts are necessary for multicellular EC lumens and tubes to form. Thus, we developed the hypothesis that VEGF might be acting in a separate morphogenic step that was upstream of tube morphogenesis. This hypothesis is very consistent with its described role as a vascular permeability factor, and it is quite intriguing that pericyte-derived Ang-1 (an important growth factor for vascular tube stabilization) is known to strongly antagonize this permeability-inducing influence of VEGF (Thurston et al., 2000). It is possible that VEGF-mediated signal transduction, which has been very well studied, is a signaling pathway that controls EC priming, and not tube morphogenesis.

To examine these issues in more detail, we performed experiments by treating ECs with VEGF first and the "Factors" second in the tube morphogenesis assay or vice versa. If the "Factors" were added first during the priming step and then VEGF or VEGF and FGF-2 were added during the tube forming assay, EC tubulogenesis failed to occur over time. However, if VEGF was added as an upstream primer first and then the "Factors" were added during the EC tubulogenesis assay, a marked stimulation of tube formation was observed (Figs. 1 and 11). We observed marked stimulation in the number of EC tip cells as well as the extent of EC tube formation and accelerated EC–pericyte tube coassembly with basement membrane deposition. Thus, VEGF has very strong priming effects that temporally must precede pro-tube morphogenic signals which are supplied

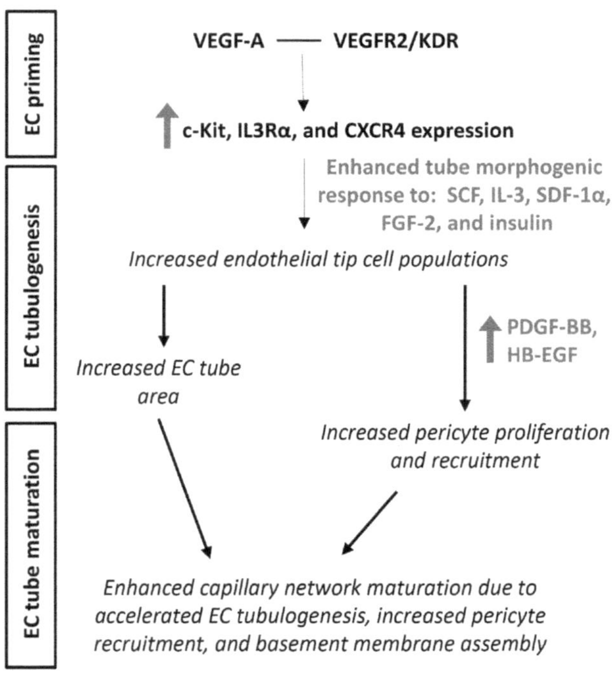

Fig. 11 VEGF is an upstream EC primer that stimulates "Factor"-induced EC tubulogenic and EC–pericyte tube coassembly responses. This schematic diagram illustrates that VEGF is an upstream primer that activates EC tip cells, EC tube formation, and EC–pericyte tube coassembly through upregulation of c-Kit, IL-3Ra, and CXCR4 (to promote responses to the "Factors") and through PDGF-BB and HB-EGF production which stimulates pericyte proliferation, pericyte recruitment, and vascular basement membrane matrix assembly.

by the "Factors." It is clear that considerable more work is necessary to identify the molecular and signaling differences that distinguish EC priming vs EC tube morphogenesis. Overall, our data show that VEGF is an upstream primer for vascular tube morphogenesis but that it does not directly stimulate those events (Fig. 11). In our view, this is a conceptual breakthrough in our understanding of how blood vessels form, and this information may be essential in attempts to improve current therapeutic approaches to appropriately manipulate the vasculature in the context of diseases such as cancer and retinal degeneration.

11. CONCLUSION

Major advances over the last two decades are demonstrating that mechanisms underlying how blood vessels form and function are becoming increasingly complex. Despite great progress, it remains essential to understand the fundamental cell and molecular biology of vascular cells (ECs and mural cells) and how this directly relates to key questions including how ECs are specified; how they form tube networks; how they differentiate into distinct lineages such as arterial, capillary, venous, and lymphatic ECs, determining which growth factors and downstream signaling pathways are necessary to form and maintain different vessel types; how mural cells recruit and communicate with ECs; and how these functions create differentiated and highly stable vascular networks that are critical regulators of tissue development and health. Many disease states arise due to abnormalities in blood vessels that fail to communicate properly with tissue parenchymal cells resulting in serious pathological consequences (Davis, Norden, & Bowers, 2015). Thus, an understanding of the basic biology of the vasculature and how it communicates with specific tissues will be a fundamental aspect of developing new therapeutic strategies to combat particular disease states.

Our laboratory has addressed some of these key questions using predominantly an in vitro strategy which has proved over the years to be a highly effective way to investigate questions underlying how human ECs form lumens and tubes in 3D matrices and how they interact with pericytes to form coassembled capillary networks. An accumulating number of our findings are being recapitulated in vivo through our collaborations and other separate work. It is clear that there is a high degree of complexity to how capillary tubes assemble and mature. This complexity is apparent with regard to the many critical molecules that control tubulogenesis (one of the most important functions of ECs), to demonstrating that ECs and pericytes work together within vascular guidance tunnels to deposit the vascular basement membrane, and defining separable morphogenic steps such as EC priming followed by tube formation that are stimulated by distinct sets of growth factors and molecular signals. Because of this complexity, our belief is that a detailed combined in vitro and in vivo strategy over many years must be employed to ultimately elucidate the molecules and signaling for each of these unique steps and to understand the fundamental biology of the different types of ECs and mural cells in distinct vessel beds and at differing times in health and disease. In our view, many critical diseases have an underlying capillary dysfunction that is central to their pathogenesis, and developing strategies to correct these defects may result in the development of novel therapeutics that might lead to major advancements in the treatment of disease.

CONFLICT OF INTEREST

The authors have no conflicts of interests to declare.

ACKNOWLEDGMENTS

The authors would like to acknowledge the help of Dr. Dae Joong Kim and Annie Smith in the performance and analysis of experiments. This work was supported by NIH Grants HL126518 and HL128584 (to GED).

REFERENCES

Adams, R. H., & Alitalo, K. (2007). Molecular regulation of angiogenesis and lymphangiogenesis. *Nature Reviews. Molecular Cell Biology*, *8*(6), 464–478.

Ando, K., Fukuhara, S., Moriya, T., Obara, Y., Nakahata, N., & Mochizuki, N. (2013). Rap1 potentiates endothelial cell junctions by spatially controlling myosin II activity and actin organization. *The Journal of Cell Biology*, *202*(6), 901–916. http://dx.doi.org/10.1083/jcb.201301115. jcb.201301115 [pii].

Arai, A., Jin, A., Yan, W., Mizuchi, D., Yamamoto, K., Nanki, T., & Miura, O. (2005). SDF-1 synergistically enhances IL-3-induced activation of the Raf-1/MEK/Erk signaling pathway through activation of Rac and its effector Pak kinases to promote hematopoiesis and chemotaxis. *Cellular Signalling*, *17*(4), 497–506.

Armulik, A., Genove, G., & Betsholtz, C. (2011). Pericytes: Developmental, physiological, and pathological perspectives, problems, and promises. *Developmental Cell*, *21*(2), 193–215. http://dx.doi.org/10.1016/j.devcel.2011.07.001. S1534-5807(11)00269-3 [pii].

Arroyo, A. G., & Iruela-Arispe, M. L. (2010). Extracellular matrix, inflammation, and the angiogenic response. *Cardiovascular Research*, *86*(2), 226–235. http://dx.doi.org/10.1093/cvr/cvq049. cvq049 [pii].

Augustin, H. G., Koh, G. Y., Thurston, G., & Alitalo, K. (2009). Control of vascular morphogenesis and homeostasis through the angiopoietin-Tie system. *Nature Reviews. Molecular Cell Biology*, *10*(3), 165–177. http://dx.doi.org/10.1038/nrm2639. nrm2639 [pii].

Barry, D. M., Xu, K., Meadows, S. M., Zheng, Y., Norden, P. R., Davis, G. E., & Cleaver, O. (2015). Cdc42 is required for cytoskeletal support of endothelial cell adhesion during blood vessel formation in mice. *Development, 142*(17), 3058–3070. http://dx.doi.org/10.1242/dev.125260. dev.125260 [pii].

Bayless, K. J., & Davis, G. E. (2002). The Cdc42 and Rac1 GTPases are required for capillary lumen formation in three-dimensional extracellular matrices. *Journal of Cell Science, 115*(Pt. 6), 1123–1136.

Bayless, K. J., & Davis, G. E. (2004). Microtubule depolymerization rapidly collapses capillary tube networks in vitro and angiogenic vessels in vivo through the small GTPase Rho. *The Journal of Biological Chemistry, 279*(12), 11686–11695. http://dx.doi.org/10.1074/jbc.M308373200. M308373200 [pii].

Bayless, K. J., Salazar, R., & Davis, G. E. (2000). RGD-dependent vacuolation and lumen formation observed during endothelial cell morphogenesis in three-dimensional fibrin matrices involves the alpha(v)beta(3) and alpha(5)beta(1) integrins. *The American Journal of Pathology, 156*(5), 1673–1683.

Bertrand, J. Y., Chi, N. C., Santoso, B., Teng, S., Stainier, D. Y., & Traver, D. (2010). Haematopoietic stem cells derive directly from aortic endothelium during development. *Nature, 464*(7285), 108–111. http://dx.doi.org/10.1038/nature08738. nature08738 [pii].

Bowers, S. L., Meng, C. X., Davis, M. T., & Davis, G. E. (2015). Investigating human vascular tube morphogenesis and maturation using endothelial cell-pericyte co-cultures and a doxycycline-inducible genetic system in 3D extracellular matrices. *Methods in Molecular Biology, 1189*, 171–189. http://dx.doi.org/10.1007/978-1-4939-1164-6_12.

Brandt, J. E., Bhalla, K., & Hoffman, R. (1994). Effects of interleukin-3 and c-kit ligand on the survival of various classes of human hematopoietic progenitor cells. *Blood, 83*(6), 1507–1514.

Brandt, J., Briddell, R. A., Srour, E. F., Leemhuis, T. B., & Hoffman, R. (1992). Role of c-kit ligand in the expansion of human hematopoietic progenitor cells. *Blood, 79*(3), 634–641.

Bray, K., Brakebusch, C., & Vargo-Gogola, T. (2011). The Rho GTPase Cdc42 is required for primary mammary epithelial cell morphogenesis in vitro. *Small GTPases, 2*(5), 247–258. http://dx.doi.org/10.4161/sgtp.2.5.18163. 2154-1248-2-5-2 [pii].

Brizzi, M. F., Zini, M. G., Aronica, M. G., Blechman, J. M., Yarden, Y., & Pegoraro, L. (1994). Convergence of signaling by interleukin-3, granulocyte-macrophage colony-stimulating factor, and mast cell growth factor on JAK2 tyrosine kinase. *The Journal of Biological Chemistry, 269*(50), 31680–31684.

Broxmeyer, H. E. (2008). Chemokines in hematopoiesis. *Current Opinion in Hematology, 15*(1), 49–58.

Bryant, D. M., Datta, A., Rodriguez-Fraticelli, A. E., Peranen, J., Martin-Belmonte, F., & Mostov, K. E. (2010). A molecular network for de novo generation of the apical surface and lumen. *Nature Cell Biology, 12*(11), 1035–1045. http://dx.doi.org/10.1038/ncb2106. ncb2106 [pii].

Bryder, D., & Jacobsen, S. E. (2000). Interleukin-3 supports expansion of long-term multilineage repopulating activity after multiple stem cell divisions in vitro. *Blood, 96*(5), 1748–1755.

Callow, M. G., Zozulya, S., Gishizky, M. L., Jallal, B., & Smeal, T. (2005). PAK4 mediates morphological changes through the regulation of GEF-H1. *Journal of Cell Science, 118*(Pt. 9), 1861–1872. http://dx.doi.org/10.1242/jcs.02313. jcs.02313 [pii].

Carmeliet, P., & Jain, R. K. (2011). Molecular mechanisms and clinical applications of angiogenesis. *Nature, 473*(7347), 298–307. http://dx.doi.org/10.1038/nature10144. nature10144 [pii].

Carow, C. E., Hangoc, G., Cooper, S. H., Williams, D. E., & Broxmeyer, H. E. (1991). Mast cell growth factor (c-kit ligand) supports the growth of human multipotential progenitor cells with a high replating potential. *Blood, 78*(9), 2216–2221.

Cascone, I., Napione, L., Maniero, F., Serini, G., & Bussolino, F. (2005). Stable interaction between alpha5beta1 integrin and Tie2 tyrosine kinase receptor regulates endothelial cell response to Ang-1. *The Journal of Cell Biology*, *170*(6), 993–1004. http://dx.doi.org/10.1083/jcb.200507082. jcb.200507082 [pii].

Cavallero, S., Shen, H., Yi, C., Lien, C. L., Kumar, S. R., & Sucov, H. M. (2015). CXCL12 signaling is essential for maturation of the ventricular coronary endothelial plexus and establishment of functional coronary circulation. *Developmental Cell*, *33*(4), 469–477. http://dx.doi.org/10.1016/j.devcel.2015.03.018. S1534-5807(15)00214-2 [pii].

Cha, Y. R., Fujita, M., Butler, M., Isogai, S., Kochhan, E., Siekmann, A. F., & Weinstein, B. M. (2012). Chemokine signaling directs trunk lymphatic network formation along the preexisting blood vasculature. *Developmental Cell*, *22*(4), 824–836. http://dx.doi.org/10.1016/j.devcel.2012.01.011. S1534-5807(12)00045-7 [pii].

Davis, G. E. (2012). Molecular regulation of vasculogenesis and angiogenesis: Recent advances and future directions. In J. W. Homeister & M. S. Willis (Eds.), *Molecular and translational vascular medicine* (pp. 169–206). New York: Springer.

Davis, G. E., Bayless, K. J., Davis, M. J., & Meininger, G. A. (2000). Regulation of tissue injury responses by the exposure of matricryptic sites within extracellular matrix molecules. *The American Journal of Pathology*, *156*(5), 1489–1498. http://dx.doi.org/10.1016/S0002-9440(10)65020-1. S0002-9440(10)65020-1 [pii].

Davis, G. E., Bayless, K. J., & Mavila, A. (2002). Molecular basis of endothelial cell morphogenesis in three-dimensional extracellular matrices. *The Anatomical Record*, *268*(3), 252–275.

Davis, G. E., & Camarillo, C. W. (1996). An alpha 2 beta 1 integrin-dependent pinocytic mechanism involving intracellular vacuole formation and coalescence regulates capillary lumen and tube formation in three-dimensional collagen matrix. *Experimental Cell Research*, *224*(1), 39–51.

Davis, G. E., Kim, D. J., Meng, C. X., Norden, P. R., Speichinger, K. R., Davis, M. T., … Stratman, A. N. (2013). Control of vascular tube morphogenesis and maturation in 3D extracellular matrices by endothelial cells and pericytes. *Methods in Molecular Biology*, *1066*, 17–28. http://dx.doi.org/10.1007/978-1-62703-604-7_2.

Davis, G. E., Koh, W., & Stratman, A. N. (2007). Mechanisms controlling human endothelial lumen formation and tube assembly in three-dimensional extracellular matrices. *Birth Defects Research. Part C, Embryo Today: Reviews*, *81*(4), 270–285.

Davis, G. E., Norden, P. R., & Bowers, S. L. (2015). Molecular control of capillary morphogenesis and maturation by recognition and remodeling of the extracellular matrix: Functional roles of endothelial cells and pericytes in health and disease. *Connective Tissue Research*, *56*(5), 392–402. http://dx.doi.org/10.3109/03008207.2015.1066781.

Davis, G. E., & Senger, D. R. (2005). Endothelial extracellular matrix: Biosynthesis, remodeling, and functions during vascular morphogenesis and neovessel stabilization. *Circulation Research*, *97*(11), 1093–1107.

Davis, G. E., Speichinger, K. R., Norden, P. R., Kim, D. J., & Bowers, S. L. K. (2015). Endothelial cell polarization during lumen formation, tubulogenesis and vessel maturation in 3D extracellular matrices. In K. Ebnet (Ed.), *Cell polarity: Biological role and basic mechanisms* (pp. 205–220). New York: Springer.

Davis, G. E., Stratman, A. N., Sacharidou, A., & Koh, W. (2011). Molecular basis for endothelial lumen formation and tubulogenesis during vasculogenesis and angiogenic sprouting. *International Review of Cell and Molecular Biology*, *288*, 101–165.

de Kreuk, B. J., Gingras, A. R., Knight, J. D., Liu, J. J., Gingras, A. C., & Ginsberg, M. H. (2016). Heart of glass anchors Rasip1 at endothelial cell-cell junctions to support vascular integrity. *eLife*, *5*. http://dx.doi.org/10.7554/eLife.11394. e11394 [pii].

Dejana, E., & Giampietro, C. (2012). Vascular endothelial-cadherin and vascular stability. *Current Opinion in Hematology, 19*(3), 218–223. http://dx.doi.org/10.1097/MOH.0b013e3283523e1c.

Dejana, E., Tournier-Lasserve, E., & Weinstein, B. M. (2009). The control of vascular integrity by endothelial cell junctions: Molecular basis and pathological implications. *Developmental Cell, 16*(2), 209–221. http://dx.doi.org/10.1016/j.devcel.2009.01.004. S1534-5807(09)00032-X [pii].

Dentelli, P., Rosso, A., Olgasi, C., Camussi, G., & Brizzi, M. F. (2011). IL-3 is a novel target to interfere with tumor vasculature. *Oncogene, 30*(50), 4930–4940. http://dx.doi.org/10.1038/onc.2011.204. onc2011204 [pii].

Dvorak, A. M., Seder, R. A., Paul, W. E., Morgan, E. S., & Galli, S. J. (1994). Effects of interleukin-3 with or without the c-kit ligand, stem cell factor, on the survival and cytoplasmic granule formation of mouse basophils and mast cells in vitro. *The American Journal of Pathology, 144*(1), 160–170.

Eelen, G., de Zeeuw, P., Simons, M., & Carmeliet, P. (2015). Endothelial cell metabolism in normal and diseased vasculature. *Circulation Research, 116*(7), 1231–1244. http://dx.doi.org/10.1161/CIRCRESAHA.116.302855. CIRCRESAHA.116.302855 [pii].

Eichmann, A., & Simons, M. (2012). VEGF signaling inside vascular endothelial cells and beyond. *Current Opinion in Cell Biology, 24*(2), 188–193. http://dx.doi.org/10.1016/j.ceb.2012.02.002. S0955-0674(12)00015-4 [pii].

Eichmann, A., & Thomas, J. L. (2013). Molecular parallels between neural and vascular development. *Cold Spring Harbor Perspectives in Medicine, 3*(1), a006551. http://dx.doi.org/10.1101/cshperspect.a006551. cshperspect.a006551 [pii].

Eilken, H. M., Nishikawa, S., & Schroeder, T. (2009). Continuous single-cell imaging of blood generation from haemogenic endothelium. *Nature, 457*(7231), 896–900. http://dx.doi.org/10.1038/nature07760. nature07760 [pii].

Fish, J. E., & Wythe, J. D. (2015). The molecular regulation of arteriovenous specification and maintenance. *Developmental Dynamics, 244*(3), 391–409. http://dx.doi.org/10.1002/dvdy.24252.

Giannotta, M., Trani, M., & Dejana, E. (2013). VE-cadherin and endothelial adherens junctions: Active guardians of vascular integrity. *Developmental Cell, 26*(5), 441–454. http://dx.doi.org/10.1016/j.devcel.2013.08.020. S1534-5807(13)00507-8 [pii].

Gomes, E. R., Jani, S., & Gundersen, G. G. (2005). Nuclear movement regulated by Cdc42, MRCK, myosin, and actin flow establishes MTOC polarization in migrating cells. *Cell, 121*(3), 451–463. http://dx.doi.org/10.1016/j.cell.2005.02.022. S0092-8674(05)00188-1 [pii].

Harrison, M. R., Bussmann, J., Huang, Y., Zhao, L., Osorio, A., Burns, C. G., … Lien, C. L. (2015). Chemokine-guided angiogenesis directs coronary vasculature formation in zebrafish. *Developmental Cell, 33*(4), 442–454. http://dx.doi.org/10.1016/j.devcel.2015.04.001. S1534-5807(15)00243-9 [pii].

He, W. Y., Lan, Y., Yao, H. Y., Li, Z., Wang, X. Y., Li, X. S., … Mao, N. (2010). Interleukin-3 promotes hemangioblast development in mouse aorta-gonad-mesonephros region. *Haematologica, 95*(6), 875–883. http://dx.doi.org/10.3324/haematol.2009.014241. haematol.2009.014241 [pii].

Heissig, B., Werb, Z., Rafii, S., & Hattori, K. (2003). Role of c-kit/Kit ligand signaling in regulating vasculogenesis. *Thrombosis and Haemostasis, 90*(4), 570–576.

Helker, C. S., Schuermann, A., Pollmann, C., Chng, S. C., Kiefer, F., Reversade, B., & Herzog, W. (2015). The hormonal peptide Elabela guides angioblasts to the midline during vasculogenesis. *eLife, 4*. http://dx.doi.org/10.7554/eLife.06726.

Hellstrom, M., Gerhardt, H., Kalen, M., Li, X., Eriksson, U., Wolburg, H., & Betsholtz, C. (2001). Lack of pericytes leads to endothelial hyperplasia and abnormal vascular morphogenesis. *The Journal of Cell Biology, 153*(3), 543–553.

Hellstrom, M., Kalen, M., Lindahl, P., Abramsson, A., & Betsholtz, C. (1999). Role of PDGF-B and PDGFR-beta in recruitment of vascular smooth muscle cells and pericytes during embryonic blood vessel formation in the mouse. *Development*, *126*(14), 3047–3055.

Hirschi, K. K. (2012). Hemogenic endothelium during development and beyond. *Blood*, *119*(21), 4823–4827. http://dx.doi.org/10.1182/blood-2011-12-353466. blood-2011-12-353466 [pii].

Hoang, M. V., Nagy, J. A., & Senger, D. R. (2011). Cdc42-mediated inhibition of GSK-3beta improves angio-architecture and lumen formation during VEGF-driven pathological angiogenesis. *Microvascular Research*, *81*(1), 34–43. http://dx.doi.org/10.1016/j.mvr.2010.09.001. S0026-2862(10)00177-9 [pii].

Hynes, R. O. (2007). Cell-matrix adhesion in vascular development. *Journal of Thrombosis and Haemostasis*, *5*(Suppl. 1), 32–40.

Hynes, R. O. (2009). The extracellular matrix: Not just pretty fibrils. *Science*, *326*(5957), 1216–1219.

Iruela-Arispe, M. L., & Davis, G. E. (2009). Cellular and molecular mechanisms of vascular lumen formation. *Developmental Cell*, *16*(2), 222–231.

Ivins, S., Chappell, J., Vernay, B., Suntharalingham, J., Martineau, A., Mohun, T. J., & Scambler, P. J. (2015). The CXCL12/CXCR4 axis plays a critical role in coronary artery development. *Developmental Cell*, *33*(4), 455–468. http://dx.doi.org/10.1016/j.devcel.2015.03.026. S1534-5807(15)00242-7 [pii].

Jeltsch, M., Leppanen, V. M., Saharinen, P., & Alitalo, K. (2013). Receptor tyrosine kinase-mediated angiogenesis. *Cold Spring Harbor Perspectives in Biology*, *5*(9), a009183. http://dx.doi.org/10.1101/cshperspect.a009183. 5/9/a009183 [pii].

Jin, D. K., Shido, K., Kopp, H. G., Petit, I., Shmelkov, S. V., Young, L. M., … Rafii, S. (2006). Cytokine-mediated deployment of SDF-1 induces revascularization through recruitment of CXCR4+ hemangiocytes. *Nature Medicine*, *12*(5), 557–567.

Kamei, M., Saunders, W. B., Bayless, K. J., Dye, L., Davis, G. E., & Weinstein, B. M. (2006). Endothelial tubes assemble from intracellular vacuoles in vivo. *Nature*, *442*(7101), 453–456. http://dx.doi.org/10.1038/nature04923. nature04923 [pii].

Kesavan, G., Sand, F. W., Greiner, T. U., Johansson, J. K., Kobberup, S., Wu, X., … Semb, H. (2009). Cdc42-mediated tubulogenesis controls cell specification. *Cell*, *139*(4), 791–801. http://dx.doi.org/10.1016/j.cell.2009.08.049. S0092-8674(09)01189-1 [pii].

Kim, D. J., Martinez-Lemus, L. A., & Davis, G. E. (2013). EB1, p150Glued, and Clasp1 control endothelial tubulogenesis through microtubule assembly, acetylation, and apical polarization. *Blood*, *121*(17), 3521–3530. http://dx.doi.org/10.1182/blood-2012-11-470179. blood-2012-11-470179 [pii].

Kleaveland, B., Zheng, X., Liu, J. J., Blum, Y., Tung, J. J., Zou, Z., … Kahn, M. L. (2009). Regulation of cardiovascular development and integrity by the heart of glass-cerebral-cavernous malformation protein pathway. *Nature Medicine*, *15*(2), 169–176. http://dx.doi.org/10.1038/nm.1918. nm.1918 [pii].

Klein, K. R., & Caron, K. M. (2015). Adrenomedullin in lymphangiogenesis: From development to disease. *Cellular and Molecular Life Sciences*, *72*(16), 3115–3126. http://dx.doi.org/10.1007/s00018-015-1921-3.

Koh, W., Mahan, R. D., & Davis, G. E. (2008). Cdc42- and Rac1-mediated endothelial lumen formation requires Pak2, Pak4 and Par3, and PKC-dependent signaling. *Journal of Cell Science*, *121*(Pt. 7), 989–1001.

Koh, W., Sachidanandam, K., Stratman, A. N., Sacharidou, A., Mayo, A. M., Murphy, E. A., … Davis, G. E. (2009). Formation of endothelial lumens requires a coordinated PKC{epsilon}-, Src-, Pak- and Raf-kinase-dependent signaling cascade downstream of Cdc42 activation. *Journal of Cell Science*, *122*(Pt. 11), 1812–1822.

Koh, W., Stratman, A. N., Sacharidou, A., & Davis, G. E. (2008). In vitro three dimensional collagen matrix models of endothelial lumen formation during vasculogenesis and angiogenesis. *Methods in Enzymology*, *443*, 83–101.

Krendel, M., Zenke, F. T., & Bokoch, G. M. (2002). Nucleotide exchange factor GEF-H1 mediates cross-talk between microtubules and the actin cytoskeleton. *Nature Cell Biology*, *4*(4), 294–301. http://dx.doi.org/10.1038/ncb773. ncb773 [pii].

Kuo, J. C., Han, X., Hsiao, C. T., Yates, J. R., 3rd, & Waterman, C. M. (2011). Analysis of the myosin-II-responsive focal adhesion proteome reveals a role for beta-Pix in negative regulation of focal adhesion maturation. *Nature Cell Biology*, *13*(4), 383–393. http://dx.doi.org/10.1038/ncb2216. ncb2216 [pii].

Lammerts van Bueren, K., & Black, B. L. (2012). Regulation of endothelial and hematopoietic development by the ETS transcription factor Etv2. *Current Opinion in Hematology*, *19*(3), 199–205. http://dx.doi.org/10.1097/MOH.0b013e3283523e07.

Leung, A., Ciau-Uitz, A., Pinheiro, P., Monteiro, R., Zuo, J., Vyas, P., ... Porcher, C. (2013). Uncoupling VEGFA functions in arteriogenesis and hematopoietic stem cell specification. *Developmental Cell*, *24*(2), 144–158. http://dx.doi.org/10.1016/j.devcel.2012.12.004. S1534-5807(12)00577-1 [pii].

Lo Celso, C., Fleming, H. E., Wu, J. W., Zhao, C. X., Miake-Lye, S., Fujisaki, J., ... Scadden, D. T. (2009). Live-animal tracking of individual haematopoietic stem/progenitor cells in their niche. *Nature*, *457*(7225), 92–96.

Lohela, M., Bry, M., Tammela, T., & Alitalo, K. (2009). VEGFs and receptors involved in angiogenesis versus lymphangiogenesis. *Current Opinion in Cell Biology*, *21*(2), 154–165. http://dx.doi.org/10.1016/j.ceb.2008.12.012. S0955-0674(09)00013-1 [pii].

Loo, T. H., Ng, Y. W., Lim, L., & Manser, E. (2004). GIT1 activates p21-activated kinase through a mechanism independent of p21 binding. *Molecular and Cellular Biology*, *24*(9), 3849–3859.

Marcelo, K. L., Goldie, L. C., & Hirschi, K. K. (2013). Regulation of endothelial cell differentiation and specification. *Circulation Research*, *112*(9), 1272–1287. http://dx.doi.org/10.1161/CIRCRESAHA.113.300506. 112/9/1272 [pii].

Martin-Belmonte, F., Gassama, A., Datta, A., Yu, W., Rescher, U., Gerke, V., & Mostov, K. (2007). PTEN-mediated apical segregation of phosphoinositides controls epithelial morphogenesis through Cdc42. *Cell*, *128*(2), 383–397. http://dx.doi.org/10.1016/j.cell.2006.11.051. S0092-8674(07)00005-0 [pii].

Mitin, N. Y., Ramocki, M. B., Zullo, A. J., Der, C. J., Konieczny, S. F., & Taparowsky, E. J. (2004). Identification and characterization of rain, a novel Ras-interacting protein with a unique subcellular localization. *The Journal of Biological Chemistry*, *279*(21), 22353–22361. http://dx.doi.org/10.1074/jbc.M312867200. M312867200 [pii].

Moldenhauer, L. M., Cockshell, M. P., Frost, L., Parham, K. A., Tvorogov, D., Tan, L. Y., ... Bonder, C. S. (2015). Interleukin-3 greatly expands non-adherent endothelial forming cells with pro-angiogenic properties. *Stem Cell Research*, *14*(3), 380–395. http://dx.doi.org/10.1016/j.scr.2015.04.002. S1873-5061(15)00044-6 [pii].

Newman, A. C., Nakatsu, M. N., Chou, W., Gershon, P. D., & Hughes, C. C. (2011). The requirement for fibroblasts in angiogenesis: Fibroblast-derived matrix proteins are essential for endothelial cell lumen formation. *Molecular Biology of the Cell*, *22*(20), 3791–3800. http://dx.doi.org/10.1091/mbc.E11-05-0393. mbc.E11-05-0393 [pii].

Norden, P. R., Kim, D. J., Barry, D. M., Cleaver, O. B., & Davis, G. E. (2016). Cdc42 and k-Ras control endothelial tubulogenesis through apical membrane and cytoskeletal polarization: Novel stimulatory roles for GTPase effectors, the small GTPases, Rac2 and Rap1b, and inhibitory influence of Arhgap31 and Rasa1. *PLoS One*, *11*(1), e0147758. http://dx.doi.org/10.1371/journal.pone.0147758. PONE-D-15-42307 [pii].

Porcher, C., Swat, W., Rockwell, K., Fujiwara, Y., Alt, F. W., & Orkin, S. H. (1996). The T cell leukemia oncoprotein SCL/tal-1 is essential for development of all hematopoietic lineages. *Cell*, *86*(1), 47–57. http://dx.doi.org/10.1016/S0092-8674(00)80076-8. S0092-8674(00)80076-8 [pii].

Post, A., Pannekoek, W. J., Ponsioen, B., Vliem, M. J., & Bos, J. L. (2015). Rap1 spatially controls ArhGAP29 to inhibit Rho signaling during endothelial barrier regulation. *Molecular and Cellular Biology*, *35*(14), 2495–2502. http://dx.doi.org/10.1128/MCB.01453-14. MCB.01453-14 [pii].

Post, A., Pannekoek, W. J., Ross, S. H., Verlaan, I., Brouwer, P. M., & Bos, J. L. (2013). Rasip1 mediates Rap1 regulation of Rho in endothelial barrier function through ArhGAP29. *Proceedings of the National Academy of Sciences of the United States of America*, *110*(28), 11427–11432. http://dx.doi.org/10.1073/pnas.1306595110. 1306595110 [pii].

Ramasamy, S. K., Kusumbe, A. P., & Adams, R. H. (2015). Regulation of tissue morphogenesis by endothelial cell-derived signals. *Trends in Cell Biology*, *25*(3), 148–157. http://dx.doi.org/10.1016/j.tcb.2014.11.007. S0962-8924(14)00210-4 [pii].

Rocha, S. F., & Adams, R. H. (2009). Molecular differentiation and specialization of vascular beds. *Angiogenesis*, *12*(2), 139–147. http://dx.doi.org/10.1007/s10456-009-9132-x.

Ruehl, M., Somasundaram, R., Schoenfelder, I., Farndale, R. W., Knight, C. G., Schmid, M., … Schuppan, D. (2002). The epithelial mitogen keratinocyte growth factor binds to collagens via the consensus sequence glycine-proline-hydroxyproline. *The Journal of Biological Chemistry*, *277*(30), 26872–26878. http://dx.doi.org/10.1074/jbc.M202335200. M202335200 [pii].

Sacharidou, A., Stratman, A. N., & Davis, G. E. (2012). Molecular mechanisms controlling vascular lumen formation in three-dimensional extracellular matrices. *Cells, Tissues, Organs*, *195*(1–2), 122–143. http://dx.doi.org/10.1159/000331410. 000331410 [pii].

Saunders, W. B., Bohnsack, B. L., Faske, J. B., Anthis, N. J., Bayless, K. J., Hirschi, K. K., & Davis, G. E. (2006). Coregulation of vascular tube stabilization by endothelial cell TIMP-2 and pericyte TIMP-3. *The Journal of Cell Biology*, *175*(1), 179–191.

Senger, D. R., & Davis, G. E. (2011). Angiogenesis. *Cold Spring Harbor Perspectives in Biology*, *3*(8), a005090. http://dx.doi.org/10.1101/cshperspect.a005090. cshperspect.a005090 [pii].

Senger, D. R., Perruzzi, C. A., Feder, J., & Dvorak, H. F. (1986). A highly conserved vascular permeability factor secreted by a variety of human and rodent tumor cell lines. *Cancer Research*, *46*(11), 5629–5632.

Shalaby, F., Ho, J., Stanford, W. L., Fischer, K. D., Schuh, A. C., Schwartz, L., … Rossant, J. (1997). A requirement for Flk1 in primitive and definitive hematopoiesis and vasculogenesis. *Cell*, *89*(6), 981–990. http://dx.doi.org/10.1016/S0092-8674(00)80283-4. S0092-8674(00)80283-4 [pii].

Shalaby, F., Rossant, J., Yamaguchi, T. P., Gertsenstein, M., Wu, X. F., Breitman, M. L., & Schuh, A. C. (1995). Failure of blood-island formation and vasculogenesis in Flk-1-deficient mice. *Nature*, *376*(6535), 62–66. http://dx.doi.org/10.1038/376062a0.

Simons, M., Alitalo, K., Annex, B. H., Augustin, H. G., Beam, C., Berk, B. C., … Dimmeler, S. (2015). State-of-the-art methods for evaluation of angiogenesis and tissue vascularization: A scientific statement from the American heart association. *Circulation Research*, *116*(11), e99–e132. http://dx.doi.org/10.1161/RES.0000000000000054. RES.0000000000000054 [pii].

Simons, M., & Eichmann, A. (2015). Molecular controls of arterial morphogenesis. *Circulation Research*, *116*(10), 1712–1724. http://dx.doi.org/10.1161/CIRCRESAHA.116.302953. CIRCRESAHA.116.302953 [pii].

Smith, A. O., Bowers, S. L., Stratman, A. N., & Davis, G. E. (2013). Hematopoietic stem cell cytokines and fibroblast growth factor-2 stimulate human endothelial cell-pericyte tube co-assembly in 3D fibrin matrices under serum-free defined conditions. *PLoS One*, *8*(12), e85147. http://dx.doi.org/10.1371/journal.pone.0085147. PONE-D-13-31310 [pii].

Somanath, P. R., Ciocea, A., & Byzova, T. V. (2009). Integrin and growth factor receptor alliance in angiogenesis. *Cell Biochemistry and Biophysics*, *53*(2), 53–64. http://dx.doi.org/10.1007/s12013-008-9040-5.

Stratman, A. N., & Davis, G. E. (2012). Endothelial cell-pericyte interactions stimulate basement membrane matrix assembly: Influence on vascular tube remodeling, maturation, and stabilization. *Microscopy and Microanalysis*, *18*(1), 68–80. http://dx.doi.org/10.1017/S1431927611012402. S1431927611012402 [pii].
Stratman, A. N., Davis, M. J., & Davis, G. E. (2011). VEGF and FGF prime vascular tube morphogenesis and sprouting directed by hematopoietic stem cell cytokines. *Blood*, *117*(14), 3709–3719.
Stratman, A. N., Malotte, K. M., Mahan, R. D., Davis, M. J., & Davis, G. E. (2009). Pericyte recruitment during vasculogenic tube assembly stimulates endothelial basement membrane matrix formation. *Blood*, *114*(24), 5091–5101.
Stratman, A. N., Saunders, W. B., Sacharidou, A., Koh, W., Fisher, K. E., Zawieja, D. C., … Davis, G. E. (2009). Endothelial cell lumen and vascular guidance tunnel formation requires MT1-MMP-dependent proteolysis in 3-dimensional collagen matrices. *Blood*, *114*(2), 237–247.
Stratman, A. N., Schwindt, A. E., Malotte, K. M., & Davis, G. E. (2010). Endothelial-derived PDGF-BB and HB-EGF coordinately regulate pericyte recruitment during vasculogenic tube assembly and stabilization. *Blood*, *116*(22), 4720–4730.
Suzuki, T., Suzuki, S., Fujino, N., Ota, C., Yamada, M., Yamaya, M., … Kubo, H. (2014). c-Kit immunoexpression delineates a putative endothelial progenitor cell population in developing human lungs. *American Journal of Physiology. Lung Cellular and Molecular Physiology*, *306*(9), L855–L865. http://dx.doi.org/10.1152/ajplung.00211.2013. ajplung.00211.2013 [pii].
Swift, M. R., & Weinstein, B. M. (2009). Arterial-venous specification during development. *Circulation Research*, *104*(5), 576–588. http://dx.doi.org/10.1161/CIRCRESAHA.108.188805. 104/5/576 [pii].
Tachibana, K., Hirota, S., Iizasa, H., Yoshida, H., Kawabata, K., Kataoka, Y., … Nagasawa, T. (1998). The chemokine receptor CXCR4 is essential for vascularization of the gastrointestinal tract. *Nature*, *393*(6685), 591–594. http://dx.doi.org/10.1038/31261.
Thurston, G., Rudge, J. S., Ioffe, E., Zhou, H., Ross, L., Croll, S. D., … Yancopoulos, G. D. (2000). Angiopoietin-1 protects the adult vasculature against plasma leakage. *Nature Medicine*, *6*(4), 460–463. http://dx.doi.org/10.1038/74725.
Tirziu, D., & Simons, M. (2009). Endothelium as master regulator of organ development and growth. *Vascular Pharmacology*, *50*(1–2), 1–7. http://dx.doi.org/10.1016/j.vph.2008.08.003. S1537-1891(08)00113-4 [pii].
Tzeng, Y. S., Li, H., Kang, Y. L., Chen, W. C., Cheng, W. C., & Lai, D. M. (2011). Loss of Cxcl12/Sdf-1 in adult mice decreases the quiescent state of hematopoietic stem/progenitor cells and alters the pattern of hematopoietic regeneration after myelosuppression. *Blood*, *117*(2), 429–439. http://dx.doi.org/10.1182/blood-2010-01-266833. blood-2010-01-266833 [pii].
van Hinsbergh, V. W. (2012). Endothelium—Role in regulation of coagulation and inflammation. *Seminars in Immunopathology*, *34*(1), 93–106. http://dx.doi.org/10.1007/s00281-011-0285-5.
Wan, H., Liu, C., Wert, S. E., Xu, W., Liao, Y., Zheng, Y., & Whitsett, J. A. (2013). CDC42 is required for structural patterning of the lung during development. *Developmental Biology*, *374*(1), 46–57. http://dx.doi.org/10.1016/j.ydbio.2012.11.030. S0012-1606(12)00649-5 [pii].
Wang, X., Harris, R. E., Bayston, L. J., & Ashe, H. L. (2008). Type IV collagens regulate BMP signalling in *Drosophila*. *Nature*, *455*(7209), 72–77. http://dx.doi.org/10.1038/nature07214. nature07214 [pii].
Wang, Y., Kaiser, M. S., Larson, J. D., Nasevicius, A., Clark, K. J., Wadman, S. A., … Essner, J. J. (2010). Moesin1 and Ve-cadherin are required in endothelial cells during in vivo tubulogenesis. *Development*, *137*(18), 3119–3128. http://dx.doi.org/10.1242/dev.048785. 137/18/3119 [pii].

Whitehead, K. J., Chan, A. C., Navankasattusas, S., Koh, W., London, N. R., Ling, J., ... Li, D. Y. (2009). The cerebral cavernous malformation signaling pathway promotes vascular integrity via Rho GTPases. *Nature Medicine*, *15*(2), 177–184. http://dx.doi.org/10.1038/nm.1911. nm.1911 [pii].

Wijelath, E. S., Rahman, S., Namekata, M., Murray, J., Nishimura, T., Mostafavi-Pour, Z., ... Sobel, M. (2006). Heparin-II domain of fibronectin is a vascular endothelial growth factor-binding domain: Enhancement of VEGF biological activity by a singular growth factor/matrix protein synergism. *Circulation Research*, *99*(8), 853–860. http://dx.doi.org/10.1161/01.RES.0000246849.17887.66. 01.RES.0000246849.17887.66 [pii].

Xu, K., & Cleaver, O. (2011). Tubulogenesis during blood vessel formation. *Seminars in Cell & Developmental Biology*, *22*(9), 993–1004. http://dx.doi.org/10.1016/j.semcdb.2011.05.001. S1084-9521(11)00076-0 [pii].

Xu, K., Sacharidou, A., Fu, S., Chong, D. C., Skaug, B., Chen, Z. J., ... Cleaver, O. (2011). Blood vessel tubulogenesis requires Rasip1 regulation of GTPase signaling. *Developmental Cell*, *20*, 1–14.

Yu, J. A., Castranova, D., Pham, V. N., & Weinstein, B. M. (2015). Single-cell analysis of endothelial morphogenesis in vivo. *Development*, *142*(17), 2951–2961. http://dx.doi.org/10.1242/dev.123174. dev.123174 [pii].

Zhang, W., Wang, L. J., Xiao, F., Wei, Y., Ke, W., & Xin, H. B. (2012). Intermedin: A novel regulator for vascular remodeling and tumor vessel normalization by regulating vascular endothelial-cadherin and extracellular signal-regulated kinase. *Arteriosclerosis, Thrombosis, and Vascular Biology*, *32*(11), 2721–2732. http://dx.doi.org/10.1161/ATVBAHA.112.300185. ATVBAHA.112.300185 [pii].

Zhong, C., Chrzanowska-Wodnicka, M., Brown, J., Shaub, A., Belkin, A. M., & Burridge, K. (1998). Rho-mediated contractility exposes a cryptic site in fibronectin and induces fibronectin matrix assembly. *The Journal of Cell Biology*, *141*(2), 539–551.

Zovein, A. C., Hofmann, J. J., Lynch, M., French, W. J., Turlo, K. A., Yang, Y., ... Iruela-Arispe, M. L. (2008). Fate tracing reveals the endothelial origin of hematopoietic stem cells. *Cell Stem Cell*, *3*(6), 625–636. http://dx.doi.org/10.1016/j.stem.2008.09.018. S1934-5909(08)00523-7 [pii].

CHAPTER NINE

Stress-Induced Premature Senescence of Endothelial and Endothelial Progenitor Cells

M.S. Goligorsky[*,†,1], **K. Hirschi**[*,†,1]
[*]New York Medical College at Touro University, Valhalla, NY, United States
[†]Yale University, New Haven, CT, United States
[1]Corresponding authors: e-mail address: michael_goligorsky@nymc.edu; Karen.Hirschi@yale.edu

Contents

Abstract

This brief overview of premature senescence of dysfunctional endothelial and endothelial progenitor cells provides information on endothelial cell differentiation and specialization, their ontogeny, and controversies related to endothelial stem and progenitor cells. Stressors responsible for the dysfunction of endothelial and endothelial progenitor cells, as well as cellular mechanisms and consequences of endothelial cell dysfunction are presented. Metabolic signatures of dysfunctional endothelial cells and senescence pathways are described. Emerging strategies to rejuvenate endothelial and endothelial progenitor cells conclude the review.

Advances in Pharmacology, Volume 77
ISSN 1054-3589
http://dx.doi.org/10.1016/bs.apha.2016.04.007

1. INTRODUCTION

A balance between microvascular density and the surrounding tissues in the basin of vascular supply is one of the highly regulated parameters. It is kept constant to insure that the diffusional distances for oxygen supply of the cells do not exceed 500 μm. This constancy of relationship between vascular supply and tissue volume is a cornerstone of the concept developed by J. Folkman attributing tumor shrinkage to cutting the vascular endowment, but it has a much broader applicability to virtually all blood-supplied organs. Proper vascular density is most probably maintained by the long-lived endothelial cells, endothelial progenitor cells, and vascular endothelium stem cells, on the one hand, and proliferative capacity of cells within the basin of such vessels. And yet, despite this tight regulation, microvascular rarefaction is a constant companion of diverse chronic pathological states, thus leading to the loss of differentiated cells, their substitution with myofibroblasts, fibrosis, and organ failure. Understanding the causes and mechanisms of vascular drop-out in chronic diseases is paramount for reducing their deleterious consequences.

2. VASCULAR MORPHOGENESIS

Blood vessels are predominantly composed of endothelial cells that form the inner, luminal layer, and smooth muscle cells that form the surrounding vessel wall. During blood vessel development, endothelial cells are formed first, and undergo rapid expansion and coalescence into capillary plexi that are then remodeled into arterial–venous networks capable of sustaining systemic circulation. Vascular remodeling and maturation involves coordinated migration, cell cycle inhibition, and specification of endothelial subtypes (arterial, venous), as well as smooth muscle cell recruitment. Later in development, a subset of venous endothelial cells bud off to form the lymphatic vasculature.

2.1 Endothelial Cell Differentiation

The de novo emergence of primordial, unspecialized endothelial cells is referred to as vasculogenesis, which begins in the mammal within the extraembryonic yolk sac shortly after gastrulation. Herein, endothelial cells are formed from newly generated mesodermal progenitors, in response to signals from the adjacent visceral endoderm (Belaoussoff, Farrington, & Baron,

1998; Vokes & Krieg, 2002). Later stages of vasculogenesis include the formation of vascular channels and plexi that are remodeled into circulatory networks via the process of angiogenesis.

The signaling pathways that direct the differentiation of endothelial cells from mesodermal progenitors are still not entirely clear and under intense investigation. Murine gene deletion studies revealed that fibroblast growth factor 2 (FGF2 or bFGF) and bone morphogenetic protein 4 (BMP4) are not only critical for mesoderm formation, but also play an important role in endothelial cell specification there from (Marom, Levy, Pillemer, & Fainsod, 2005; Winnier, Blessing, Labosky, & Hogan, 1995; Yamaguchi, Harpal, Henkemeyer, & Rossant, 1994). Mouse embryonic stem (mES) cell differentiation studies suggest that BMP4 promotes mesoderm formation and initiates a program requiring FGF2 to promote the specification of angioblasts, or endothelial progenitors (Park et al., 2004; Pearson, Sroczynska, Lacaud, & Kouskoff, 2008).

Further commitment to an endothelial cell lineage is promoted by signals from the adjacent visceral endoderm, including Indian hedgehog (IHH), which is sufficient to induce the formation of endothelial cells in mouse embryo explants that lack endoderm (Byrd et al., 2002; Vokes & Krieg, 2002). Such effects by IHH may also be mediated by BMP4, as they are during the differentiation of endothelial cells from human ES cells (Kelly & Hirschi, 2009). Vascular endothelial growth factor (VEGF), also produced by the visceral endoderm early in vascular development, is another key regulator of vasculogenesis (Carmeliet et al., 1996; Ferrara et al., 1996; Miquerol, Langille, & Nagy, 2000). VEGF-A predominantly signals through two receptors, VEGFR1 (Flt-1) and VEGFR2 (Flk-1 or Kdr), and mice lacking Flk-1 are embryonic lethal and lack vascular plexus development, despite normal formation of angioblasts (Schuh, Faloon, Hu, Bhimani, & Choi, 1999; Shalaby et al., 1995). Consistent with this, Flk-1 −/− mES cells generate endothelial cells, but they fail to propagate in vitro (Schuh et al., 1999); thus, VEGF-A may regulate the survival and/or propagation of endothelial cells, not necessarily their differentiation.

Transcriptional regulators in the ETS family are also known to play important roles in endothelial cell development. There are ~30 mammalian ETS factors and several, such as *Ets1*, *Erg*, *Fli-1*, and *Etv2*, have been demonstrated to play important roles in vascular development (De Val & Black, 2009; Dejana, Taddei, & Randi, 2007). In fact, regulatory regions for almost all endothelial genes contain ETS binding sites, suggesting that ETS factors could potentially regulate most, if not all, endothelial-specific gene expression

(De Val & Black, 2009; Dejana et al., 2007). *Etv2* (*ER71/etsrp*), or ETS variant 2, appears to play a critical role in the differentiation of mesodermal progenitors toward an endothelial cell fate. *Etv2* is initially broadly expressed within the primitive streak mesoderm, but then restricted to developing vascular endothelial cells (Kataoka et al., 2011; Lee et al., 2008; Salanga, Meadows, Myers, & Krieg, 2010). Mice deficient for *Etv2* are embryonic lethal and lack a yolk sac primitive vascular plexus, dorsal aortae, and endocardium, despite normal mesoderm formation (De Val et al., 2008). Overexpression of *Etv2* leads to ectopic expression of endothelial markers in multiple cell types in vivo, suggesting *Etv2* is necessary and sufficient for vasculogenesis. *Etv2* expression is restricted to Flk-1 + mesoderm in mES cells and its induction within embryoid bodies results in de novo production of Flk-1 + cells (De Val et al., 2008). Thus, *Etv2* appears to regulate expression of VEGF receptor *Flk-1*, which is associated with the specification of an endothelial cell fate. Much remains to be learned about the regulation of this important process.

2.2 Endothelial Cell Specialization

Once primordial endothelial cells are formed, they must be further specialized to perform unique functions. In a properly remodeled circulatory system, the blood vasculature is subdivided into two interconnected yet structurally and functionally distinct networks of arteries and veins. Unlike veins, arteries withstand high blood pressure and are surrounded by several layers of smooth muscle cells and connective tissues. Endothelial cell identity is known to be modulated by hemodynamic flow forces (le Noble et al., 2004); however, the downstream molecular mechanisms that regulate phenotypic changes are not clear.

Nonetheless, signaling pathways that have been implicated in earlier aspects of vascular development are thought to also play significant roles in arterial–venous specification. For example, VEGF-A binds to Flk-1 and coreceptor neuropilin-1 (NP-1) (Gu et al., 2003), activating Notch signaling (Lawson et al., 2001; Lawson, Vogel, & Weinstein, 2002) and promoting *EphrinB2* expression, while suppressing *EphB4*; thus, establishing an arterial identity while suppressing a venous fate in endothelial cells (Wang, Chen, & Anderson, 1998). Studies in mouse ES cell also demonstrate that high VEGF-A levels induce arterial endothelial cells whereas low to intermediate concentrations induce venous differentiation, and inhibition of Notch signaling results in an arterial to venous fate switch (Lanner,

Sohl, & Farnebo, 2007). Wnt signaling is also involved in the specification of an arterial fate; β-catenin (β-cat), a transcriptional coactivator of Wnt signaling pathway, upregulates Notch ligand *Dll4* and promotes arterial specification (Corada et al., 2010; Yamamizu et al., 2010). In addition, Hh signaling also acts upstream of VEGF-A via smoothened receptor to promote arterial endothelial cell specification and repress venous fates (Wilkinson et al., 2012; Williams et al., 2010).

Venous endothelial cell specification is thought to be mediated by COUP-TFII (chicken ovalbumin upstream promoter-transcription factor II), a member of the orphan nuclear receptor superfamily. Endothelial-specific deletion of *COUP-TFII* leads to arterialization of veins, whereas ectopic expression results in fusion of veins and arteries (You et al., 2005). COUP-TFII is thought to suppress NP-1 function and Notch signaling to maintain vein identity.

The lymphatic vasculature is thought to develop primarily from the venous system. It is essential for returning extravasated proteins and cells to the blood stream and for proper immune system surveillance. These roles are clearly distinct from those of blood vessels; thus, generation of lymphatic endothelial cells requires the further specification of a subpopulation of the venous endothelial cells via distinct regulatory pathways. Lymphatic endothelial cell development is thought to be induced by expression of Prospero homeodomain transcription factor (*Prox1*), which is expressed by the subpopulation of venous endothelial cells that sprout, migrate, and form the lymphatic system (Srinivasan et al., 2007; Wigle et al., 2002; Wigle & Oliver, 1999). Genetic ablation of *Prox1* leads to a general failure of lymphatic endothelial cell specification from venous endothelium (Francois et al., 2012; Johnson et al., 2008). In addition, lymphatic endothelial cells were shown to dedifferentiate into blood vessel endothelial cells upon inactivation of *Prox1* during postnatal stages (Johnson et al., 2008). *Prox1* is therefore not only necessary for specification of lymphatic lineage, but also for maintenance of lymphatic endothelial cell identity. Both the SRY-related gene *Sox18* and *COUP-TFII*, which is necessary for specifying venous endothelial cell fate, were shown to regulate *Prox1* expression (Francois et al., 2008; Lee et al., 2009; Lin et al., 2010; Pennisi et al., 2000; Srinivasan et al., 2010). *Sox18* is expressed by cardinal vein endothelial cells that eventually express *Prox1* and form the lymphatic vasculature. Ectopic *Sox18* expression in blood vessel endothelial cells induces *Prox1* and lymphatic endothelial cell marker expression.

Following specification, the lymphatic endothelial cells migrate out of veins to form lymph sacs and then lymphatic vessels (Francois et al., 2012). VEGF-C signaling, via *Flt4* receptor (VEGFR3), is necessary for lymphangiogenic development. Homozygous and heterozygous deletion of VEGF-C in mouse embryos results in a failure of *Prox1*-positive lymphatic endothelial cell progenitors to sprout (Karkkainen et al., 2004), and inhibition of Flt4 function leads to impaired development of the lymphatic system (Makinen et al., 2001). Thus, analogous to the role VEGF-A plays in primordial endothelial cell development, VEGF-C signaling appears to regulate the survival and propagation of lymphatic endothelial cells, and not necessarily their specification.

Hemogenic endothelial cells are also specialized from a small subset of the primordial endothelium within the yolk sac and AGM during embryonic development. These cells acquire hematopoietic potential and give rise to hematopoietic stem and progenitor cells, and their specification and function has recently been reviewed elsewhere (Gritz & Hirschi, 2016).

2.3 Controversial EPC Identity

Although it is not difficult to identify cells that function as endothelial progenitor cells (angioblasts, primordial endothelial cells) based on origin, location, and marker expression, or lack thereof, during embryonic development, their identification in postnatal tissues is far less clear. Cells that can be isolated from blood, bone marrow, and other adult tissues, have been shown to participate in vessel regeneration in injured tissues; however, their origin, phenotype, lineage potential and function in situ, have been heavily debated (as previously reviewed in Hirschi, Ingram, & Yoder, 2008). In addition, not all populations referred to as adult endothelial progenitor cells are the same; thus, leading to confusion and controversy in this field.

3. ONTOGENY OF ENDOTHELIAL AND ENDOTHELIAL PROGENITOR CELLS

Endothelial cell turnover is slow under physiological conditions, averaging from 2 months to 3 years (rev. in Brown, 2010). Damaged endothelial cells are eliminated as a result of patrolling function of the noninvading subset of Ly6C^{low} monocytes. Intravital microscopy studies showed these monocytes crawling on the luminal surface of glomerular and peritubular capillaries and scavenging microparticles (Carlin et al., 2013). In response to nucleic acid "danger" signaling, this subset of cells exhibit longer dwell times in glomerular and peritubular capillaries, attach to the damaged endothelial cells, and recruit neutrophils, which ultimately induce focal necrosis and disposal of cellular debris. Endothelial cells escaping this type of in situ elimination, detach from the basement membrane and appear in the circulation.

The number of colony-forming EPC is declining with age, as determined in healthy aged subjects (Hill et al., 2003), while EPC numbers are inversely correlated with cardiovascular risk (Hill et al., 2003; Werner et al., 2005). The role played by telomere attrition as a driver of replicative cell senescence in aging is well-established. Chen, Brodsky, and Goligorsky (2002) showed that, contrary to the replicative senescence, stress-induced premature senescence (SIPS) of endothelial cells, occurs even in the presence of relatively unaffected telomeres. A diverse group of stress signals, such as pro-oxidants, asymmetric dimethylarginine (ADMA), nonenzymatically glycation-modified proteins induce cell cycle arrest, SIPS and eventual apoptosis in low-passage cultured cells and in young mice. The fact that SIPS can be reversed after the withdrawal of the offending stressor, is of therapeutic importance; however, if the stressor persists, SIPS becomes irreversible, endothelial cells undergo apoptosis, and the process culminates in microvascular rarefaction. What accounts for this point-of-no-return remains unknown.

In aging animals, endothelium-dependent vasorelaxation and angiogenic competence are reduced compared to young ones (Reed, Karres, Eyman, & Vernon, 2007). The same is true for prematurely senescent Klotho-mice (Shimada et al., 2004). Notably, caloric restriction rescues angiogenic competence (Facchetti, Monzani, Cavallini, Bergamini, & La Porta, 2007). Intriguingly, clearance of p16Ink4a-positive senescent cells delays age-associated disorders, but does not lead to extended lifespan (Baker et al., 2011).

One of the downstream targets of caloric restriction, sirtuin-1, is robustly expressed in EPC and EC; however, its expression declines with age and following application of cardiovascular stressors (Chen, Qin, et al., 2012; Chen, Xavier, et al., 2012). Sirtuin-1 deficiency leading to premature senescence of EPC and EC occurs as a result of stress-induced loss of lysosomal membrane integrity and leakage of cathepsin B, which is capable of directly degrading this deacetylase, as shown in in vitro studies. EPC isolated from the bone marrow of mice genetically engineered to lack endothelial sirtuin-1 exhibit higher rate of premature senescence and apoptosis and lower stress tolerance even at a young age (Chen, Qin, et al., 2012; Chen, Xavier, et al., 2012).

4. iPSC TO EPC TRANSITION

The identity of EPC is rather nebulous. The confusion is explainable by the fact that several consensus markers of EPC, either bone marrow-derived, or circulating, or tissue-resident, are also expressed by circulating endothelial cells, endothelial stem cells, or cells of monocytic lineage; these include CD31, CD34, VEGF receptor-2 (Khakoo & Finkel, 2005; Richardson & Yoder, 2011). Two distinct populations of EPC exist in human blood: early and late outgrowth EPC, which are distinguishable on the basis of their proliferative capacity: 20-fold vs 1000-fold expansion, respectively. EPC, and especially the endothelial colony-forming subset, which is enriched in the cord blood (ECFC), have a distinct therapeutic potential (Alev, Li, & Asahara, 2011). Thus, the search for adult autologous sources for such cells is of great value. One of the most remarkable technical accomplishments of the past decade was the establishment and optimization of protocols for generating EPC from iPSC. Yoder and coworkers (Prasain et al., 2014) have successfully achieved this goal by first revisiting the earlier approach by James et al. (2010) based on application of TGF-beta inhibitor to CD144+ cells. However, expression of CD144+ is preceded by the expression of NP-1 in human pluripotent stem cells committed toward endothelial lineage. This earlier finding of Cimato et al. (2009) was explored by exposing pluripotent stem cells to mesodermal differentiation factors with endothelial specification: activin-A, BMP-4, FGF-2, and VEGF (Prasain et al., 2014) which resulted in formation of NP-1+CD31+ cells highly resembling those obtained from the cord blood. It is important to emphasize that NP-1+CD31+ cells did not form teratomas during 6 months of follow-up observation after their inoculation. Remarkably, endothelial cells obtained from patients with peripheral artery disease expressed low levels of NP-1 compared to control cells. Using an NP-1 agonist Fc-NP-1, Prasain et al. (2014) were able to overcome this deficiency and restore proliferative potential while reducing senescence of these cells.

Intriguingly, c-Kit+ cells exhibit somewhat similar robustness in angiogenesis and proliferation. Fang et al. (Fang, Wei, Pentinmikko, Leinonen, & Salven, 2012; Sedwick, 2012) described solitary cells or small clusters of c-Kit+ cells present in all three layers of the vascular wall: adventitial, medial, and intimal. These c-Kit+/VEGFR2+/CD45− cells are clonogenic and differentiate toward EC, SMC, and fibroblasts (Bearzi et al., 2009). Only 0.4% of all vascular wall cells are lin−CD31+CD105+Sca1+CD117/c-Kit+, referred to as vascular endothelial stem cells (VESCs). Adaptive transfer of isolated VESCs demonstrated that a single c-kit+ VESC can generate in vivo functional blood vessels that incorporate into host circulation. VESCs also display long-term self-renewal capacity as shown by repeated rounds of cell isolation and in vivo serial transplantation. Most recent publication on the pulmonary c-Kit+ cells assigns to them a role of an exclusively endothelial precursor (Liu et al., 2015). Genetic lineage tracing studies of c-Kit+ cells revealed that they are present in lung vasculature and coexpress endothelial markers CD31 and VE-cadherin (but not epithelial markers) under basal conditions or during repair after two different lung injury models. It would be necessary in the future to compare VESC (lin−CD31+CD105+Sca1+CD117/c-Kit+) with ECFC (NP-1+CD31+ cells) to address the question: are these overlapping populations or distinct ones? To put this question differently: Is there a single or multiple populations of EPC? Considering the many ways to isolate EPCs, one might deduce that there are multiple populations; however, presently such conclusion has no solid support.

5. METABOLIC SIGNATURES OF EPC, ENDOTHELIAL CELLS, AND THEIR DYSFUNCTIONAL PROFILE

Dysfunctional endothelial cells and their progenitors are a harbinger of diverse cardiovascular and renal diseases. Such cells lose some or all of their highly differentiated functions including production of vasoconstrictors and vasodilators controlling the blood pressure; blood coagulation and fibrinolysis; formation of new blood vessels; regulation of inflammatory processes and the traffic of leukocytes into and out of the vascular lumen; formation of a selective barrier between the blood and interstitium and controlling the transcapillary passage of different molecules; as well as maintenance and regeneration of surrounding tissues. In clinical settings, flow-dependent relaxation of conduit vessels, a function of endothelium-derived relaxing factors, mainly nitric oxide (NO), is suppressed in patients and animals with CKD. Different modifications of testing endothelium-dependent vasorelaxation, together with other surrogate biomarkers such as markers of oxidative stress (8-iso-PGF2α and oxidized LDL), pro-inflammatory markers (high-sensitivity CRP, lipoprotein-associated PLA2, soluble

ICAM-1, IL-6, von Willebrand factor), inhibitor of endothelial nitric oxide synthase (eNOS) ADMA, circulating procoagulants and others have become the standard of diagnosing endothelial cell dysfunction (ECD) (Goligorsky, 2005, 2006; Ludmer et al., 1986). Fig. 1 puts forth diverse traditional and nontraditional risk factors, such as ADMA, advanced glycation end-products, and pro-oxidants, all leading to ECD via its most reliable attribute, uncoupling of eNOS. These metabolic perturbations reviewed previously (Goligorsky, 2005) result in a metabolic shift toward Warburg-type energetics and perturbations in cell cycle. The latter can be reversible or irreversible resulting in apoptosis and vascular rarefaction.

To get broad unbiased insight into endothelial metabolism in health and disease, we isolated renal microvessels from mice with chronically inhibited eNOS screened using 2D electrophoresis, in-gel digestion, and mass-spectrometry analysis revealed at least 13 nonredundant differentially expressed proteins with high level of confidence. Five of those are specific for mitochondria, and two down-regulated proteins are the known components of the Krebs cycle: aconitase-2 and enoyl-coA-hydratase-1 (Addabbo et al., 2009). This deficiency of key enzymes is accompanied by reduced mitochondrial mass, mitochondrial oxidative stress, and switch to the normoxic glycolysis (Warburg type of metabolic hypoxia seen in chronic uncoupling of eNOS) to support energy metabolism. Moreover, by

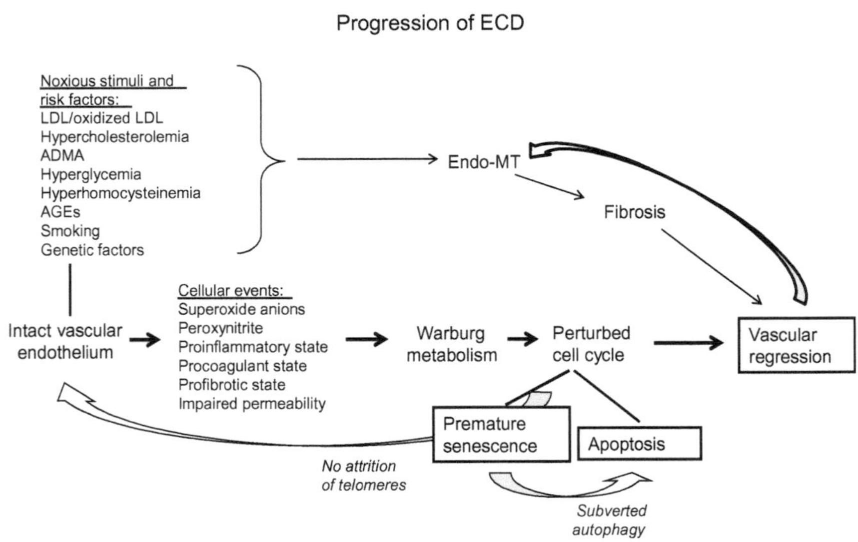

Fig. 1 A schematic overview of progression of endothelial cell dysfunction, instigating noxious, and risk factors leading to altered cell metabolism and cell cycle.

supplying cultured cells with the metabolic intermediate downstream of the deficient aconitase-2—α-ketoglutarate, which bypasses enzymatic bottleneck, we were able to not only restore energy metabolism but also prevent cell death or premature senescence.

It must be interjected here that endothelial cells are thought to heavily rely on normoxic glycolysis, at least in vitro (Verdegem, Moens, Stapor, & Carmeliet, 2014). However, it is proposed that glycolysis is enhanced specifically in "tip" cells. Yet, genetic targeting of cyclophilin D, a component of mitochondrial permeability transition pore, results in enhanced angiogenesis in vivo, suggesting that normally mitochondrial respiration contributes to the phenotype of endothelial cells (Marcu et al., 2015). It is possible that there exists an internal "rheostat" dynamically regulating OXPHOS and glycolytic contribution to energy metabolism and cell phenotype.

Therefore, in our particular case of endothelial dysfunction induced by chronic inhibition of NOS, we next supplemented animals with glutamine. This treatment results in improved vasculopathy (as judged by restored endothelium-dependent vasorelaxation) and decreased proteinuria (Addabbo et al., 2013). In addition, metabolomic studies conducted using liquid chromatography–mass spectrometry analyses disclose multiple metabolite abnormalities developing in ECD and restored by glutamine supplementation. Among those are several lysophospholipids, hippuric acid (all elevated), and glutamine/glutamate itself (reduced levels most probably due to consumption), which become normalized after glutamine supplementation (Addabbo et al., 2013). Hence, metabolic abnormalities affect EC functions and correction of these abnormalities leads to amelioration of ECD and vasculopathy, which contribute to progression of CKD.

Most recently, high-resolution NMR disclosed 44 metabolites associated with stem cell differentiation (Moussaieff et al., 2015). The main metabolic signature of pluripotent embryonic stem cells was found to be represented by normoxic glycolysis and generation of lactate and acetyl-CoA, which supports histone acetylation and blocks its deacetylation. Upon differentiation, cells rapidly lose acetyl-CoA and pluripotency, while showing increased histone deacetylation. Addition of acetate alone can halt differentiation by inducing histone pan-acetylation. Thus, Moussaieff and coauthors conclude that pluripotent state is maintained by normoxic glycolysis and generation of acetyl-CoA as an end-product, whereas metabolic switch depleting acetyl-CoA heralds early differentiation. This remarkable study is somewhat reminiscent of our work in dysfunctional endothelial cells. Analyzing our data in the light of the study by Moussaieff et al. (2015), one wonders whether the dysfunctional state of endothelial cells could be associated with acquisition or readiness to acquire pluripotency ground state. Such a scenario, if confirmed, would create a feedback mechanism for metabolically induced self-renewal.

In addition to that, the above findings of glutamine consumption and depletion in dysfunctional endothelia displaying Warburg-type metabolic profile may shed light on additional biochemical aberrations, some confirmed, others speculative. For instance, impaired expression and activity of sirtuin 1 (SIRT1) and uncoupling of eNOS, lysosomal dysfunction and impaired autophagy, all leading to premature vascular senescence and cardiovascular disease have been previously documented by our laboratory (Brodsky et al., 2004; Chen, Qin, et al., 2012; Chen, Xavier, et al., 2012; Lin et al., 2014; Maizel et al., 2014; Patschan et al., 2008). The hypothetical schema ascribing to Warburg metabolism premature senescence and dysfunction of EPC and EC illustrates its role in reverse Krebs cycle, increased production of acetyl-CoA, and consumption of nicotinamide dinucleotide (NAD) with all the functional sequelae for cells and organisms (Fig. 2).

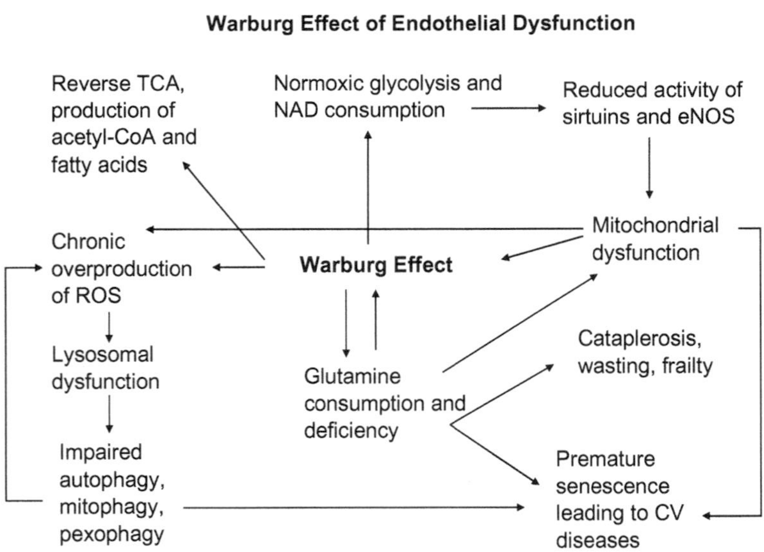

Fig. 2 A schematic depiction of endothelial dysfunction through metabolic aberrations, mitochondrial and lysosomal dysfunction, impaired autophagy, mitophagy and pexophagy leading to normoxic glycolysis (Warburg-type respiration), consumption of glutamine and NAD, and defective SIRT expression/activity. These abnormalities of endothelial metabolism and function result in premature vascular senescence.

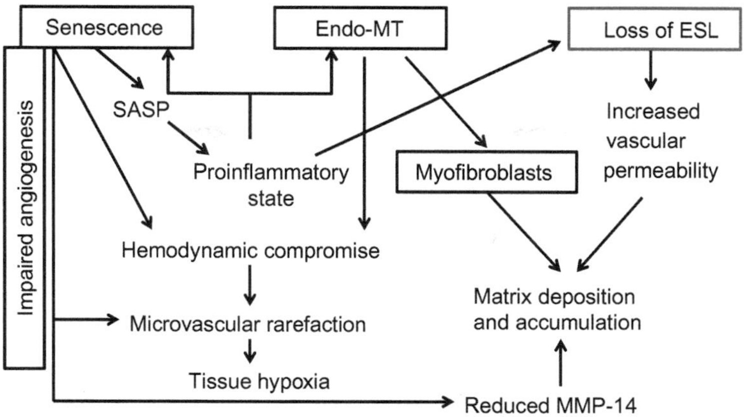

Fig. 3 Endothelium-dependent pathways of fibrosis and progressive organ dysfunction.

Several additional pathways from the dysfunctional endothelium to fibrosis and progressive organ dysfunction are summarized in Fig. 3. It proposes three major processes of premature cell senescence, endothelial–mesenchymal transition (Endo-MT) and the loss of endothelial surface glycocalyx (ESG) as main perpetrators of lost endothelial functions and their consequences leading to accumulation of myofibroblasts and matrix deposition. All three will be discussed later.

Various noxious stimuli lead to SIPS of EPC and vascular endothelium and contribute to development of vasculopathy. We have recently demonstrated that cell stress induces SIPS by depleting SIRT1, a member of the family of NAD-dependent deacetylases, which has been implicated in cellular and organismal longevity. We were able to outline a novel pathway triggered by stress-induced lysosomal membrane permeabilization leading to release of cathepsins and cleavage of SIRT1. As a follow-up, we characterized mice deficient in endothelial SIRT1 and demonstrated that it is dispensable under normal conditions, but leads to exaggerated fibrotic response after injury. MMP-14 deficiency developing in endothelial SIRT1 knockout mice was responsible in part for the observed fibrosis and pharmacological maneuvers restoring MMP-14 prevented it (Chen, Qin, et al., 2012; Chen, Xavier, et al., 2012; Vasko et al., 2014). In fact, Potente et al. (2007) were the first to demonstrate that inhibition of SIRT1 results in modulation of multiple genes, including downregulation of MMP-14, in endothelial cells. This master metalloproteinase, MMP-14, in turn, is crucial for diverse physiological functions such as cell motility, adhesion to extracellular

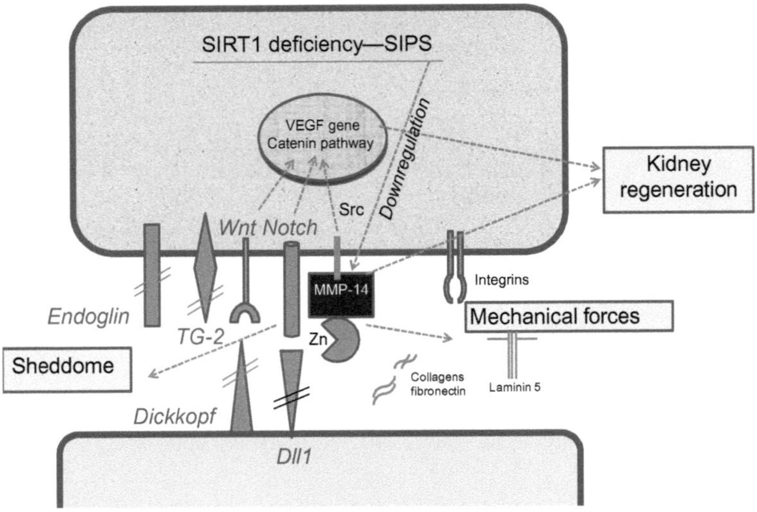

Fig. 4 Cartoon summarizing the distinct potential pathways involved in SIRT1 deficiency- and SIPS-induced down-regulation of MMP-14 consequently leading to perturbations in the sheddome (endoglin, tissue transglutaminase-2 (TG2), Dickkopf, delta-like-1), mechanome (cell and matrix rigidity), and regenerative ability of the kidney (or other organs).

matrix proteins, cell differentiation and signaling. Below, we offer a simplified schema of diverse MMP-14 targets pertinent to functions of EPC and endothelial cells that could be affected by its deficiency (Fig. 4). Among the targets of MMP-14, apart from the matrix proteins, are several epitopes potentially involved in fibrosis, such as endoglin (an extracellular domain of the TGF-β auxiliary receptor richly expressed on endothelial cells), tissue transglutaminase (a multifunctional protein richly expressed on the surface of vascular endothelium and in the extracellular matrix, where it cross-links fibronectin, collagen, vitronectin, and osteopontin), delta-like-1 (a Notch ligand), and Dickkopf-1 (a natural antagonist of Wnt-frizzled interaction through binding to LRP6 leading to the inhibition of Wnt/β-catenin pathway). It is not known whether and how the fragments released by MMP-14 (sheddome) affect signaling.

6. SENESCENCE-ASSOCIATED SECRETORY PRODUCTS

Senescent endothelial cells not only compromise the function of the endothelial lining of the vessels, but also affect the neighboring cells by their secretome referred to as senescence-associated secretory products (SASPs).

SASP is composed of TGF-α, galectin-3, IGFBP-3, -4, and -6, MIC-1 (rev. in Suzuki & Boothman, 2008), among other site-specific components. Dysfunctional senescent endothelial cells produce excessive amounts of collagen XVIII and its C-terminal antiangiogenic fragment, endostatin (O'Riordan, Mendelev, Patschan, Chander, & Goligorsky, 2007). High-resolution mass-spectrometry analysis of the secretome of endothelial progenitor cells disclosed 133 proteins, some membrane-bound, others secretory (Hemmen et al., 2011). Specifically, soluble forms of VEGF receptors, adhesion molecules, semaphorin 3F, and TGF-β, CD109, members of Roundabout family, as well as endothelial markers were detected. Mass spectrometry screen of the secretome of colony-forming units, precursors of mature endothelial cells, identified 272 nonredundant proteins of which 124 were also found in cultured EPC (Pula, Mayr, Evans, et al., 2009). Secretory products included MMP-9, IL-8, MIF, various cathepsins and protease inhibitors, S100 proteins A11, A8, and A4, PAI-2 and apolipoprotein E, as well as a pro-angiogenic and pro-survival factor, thymidine phosphorylase. In many ways, the panoply of functional effects probably afforded by normal and abnormal secretomes remains unresolved.

7. ENDOTHELIAL–MESENCHYMAL TRANSITION

Endo-MT is defined as a process of acquisition by endothelial cells of mesenchymal and stem cell-like phenotypic characteristics. Endo-MT is a normal developmental stage taking place during embryonic formation of heart valves and septa. In adulthood, Endo-MT may turn to be a contributor to vascular drop-out and development of fibrosis, as detected in multiple organs, including heart and kidney (Li, Qu, & Bertram, 2009; O'Riordan et al., 2007; Zeisberg et al., 2007). It has recently been demonstrated that glomerulosclerosis in tensin-2-deficient mice is associated with Endo-MT of glomerular endothelial cells (Kato, Mizuno, & Ito, 2014). Some estimates show that from 30 to 50% of interstitial fibroblasts originate from the endothelium, although other investigations dramatically reduce this figure (LeBlue et al., 2013). Dejana's group has demonstrated Endo-MT in the brain contributing to the onset and progression of cerebral cavernous malformation (Madedaluno et al., 2013). Xu et al. (2015) demonstrated that aberrant Endo-MT causes endocardial fibroelastosis. In *fibrodysplasia ossificans progressiva* Endo-MT is responsible for excessive generation of osteoblasts and chondrocytes from endothelial cells, thus illustrating yet another end-result of this process (Medici et al., 2010). TGF-β is considered to be a major mediator of Endo-MT, whereas BMP-7—a factor counteracting Endo-MT, as demonstrated using endothelial cell fate-tracing technique (rev. in Medici & Kalluri, 2012). Actions of TGF-β are mediated via activin receptor-like kinases 1 and 5 (Alk1 and Alk5). Activation of Alk1 uniquely expressed on endothelial cells results in cell migration, proliferation and angiogenesis, whereas stimulation of Alk5 induces via Smads 2/3 phosphorylation the transcription of SM22α, fibronectin, and PAI-1 which mediate differentiation along the smooth muscle/mesenchymal phenotype and leads to formation of myofibroblasts. Endoglin, another specific endothelial TGF coreceptor, regulates balance between activation of these two Alk pathways. Notably, prolonged activation with TGF-β results in the escape of Alk1 signaling and predominant signaling via Alk 5, thus promoting Endo-MT (rev. in Gaengel, Genove, Armulik, & Betsholtz, 2009). Ubil et al. (2014) described the possibility of reversing cardiac fibrosis by generating endothelial cells from mesenchymal cells.

Gene microarray analysis of cultured endothelial cells treated with an inhibitor of NOS, the most common pathogenic step toward endothelial dysfunction, revealed upregulation of collagen XVIII and its antiangiogenic fragment endostatin, a finding confirmed in vivo in mice chronically treated with NOS inhibitor (O'Riordan et al., 2007). Enhanced generation of endostatin in these animals leads to the development of Endo-MT and eventual rarefaction of renal microvasculature, thus further compounding vascular and parenchymal pathology. In fact, young mice overexpressing endostatin or mice chronically infused with it show signs of premature senescence, microvascular rarefaction, and renal fibrosis (Lin et al., 2014). Endo-MT can be induced not only by TGF-β, but also by activation of Notch which could serve as a regulator of partial or complete Endo-MT (rev. by Welch-Reardon, Wu, & Hughes, 2015). FGF and its downstream target *let*-7 miRNA suppress TGF-β-induced Endo-MT, thus preventing fibrosis (Chen, Qin, et al., 2012; Chen, Xavier, et al., 2012).

An intriguing relationship between ciliated endothelium and development of Endo-MT has recently been recognized. Endothelial cells located at the sites of low or oscillatory blood flow are endowed with primary cilia, which appear to be involved in the induction of shear-sensitive transcription factor Kruppel-like factor 2 (KLF2) (Nauli et al., 2008). In mice lacking endothelial primary cilia KLF2 and KLF4 fail to increase in response to shear stress and are prone to Endo-MT, the situation that can be rescued by overexpression of KLF4 (Egorova et al., 2011).

8. PERTURBATIONS OF ENDOTHELIAL SURFACE GLYCOCALYX

An intact ESG determines most of parameters characterizing normally functioning and dysfunctional endothelial cells, such as regulation of vascular permeability, mediation of blood cell–vessel wall interactions, participation in coagulation and modulation of vasodilation in blood vessels. When it is perturbed, the homeostatic and vasoprotective functions of the glycocalyx become errant and partially contribute to the pathogenesis of several diseases. In particular, diabetes, congestive heart failure, sepsis, and cancer have shown alterations in the endothelial glycocalyx prior to the development of symptoms (Nieuwdorp et al., 2006; Reitsma, Slaaf, Vink, van Zandvoort, & oude Egbrink, 2007; Weinbaum, Tarbell, & Damiano, 2007).

9. MICROVASCULAR ENDOWMENT, MICROVASCULAR PATENCY, AND MICROVASCULAR DROP-OUT

It is conceivable that Endo-MT is at the cross-section of several pathophysiological conditions. Capillary endothelium in the adipose tissue can undergo Endo-MT and give rise to preferential adipogenic differentiation of mesenchymal cells (Gupta et al., 2012; Tran et al., 2012), whereas in patients with *fibrodysplasia ossificans progressiva* generation of osteoblasts and chondrocytes occurs via Endo-MT (Medici et al., 2010). This could account for the expansion of fat cells or chondrocytes and reduction in capillary endowment. It is quite likely that similar processes occur in other tissues. The process of microvascular drop-out has distinct intermediate steps. We surmise that the earlier stage is represented by the loss of vessel patency: endothelial cells are present in usual numbers on immunohistochemical examination of tissues, but the number of actually perfused capillaries is reduced. We observed this dissociation between the vascularity detected through counting CD31-labeled vascular structures and patency, as detected using intravenous injection of tomato lectin specifically labeling endothelia of perfused vessels in murine kidneys undergoing fibrogenesis and in mice exposed to elevated levels of endostatin (Xavier et al., 2015; Lin et al., 2014). It is not excluded that initially microvasculature is nonpatent due to the excessive vasoconstriction, but the total reserve of microvascular profiles remains unchanged, as it occurs in the skin microvasculature even under normal conditions—a large proportion of capillaries in the ambient temperature is nonperfused; however, when the temperature increases, the blood flow across the recruited capillaries jumps nearly 200-fold. Later on, with ensuing stagnation of blood flow, the loss of the shear stress and damage to endothelial cells induce local platelet adhesion, attract macrophages, which in turn recruit other leukocytes, eventually disposing of cell debris (Brown, 2010). Thus, the following stage is likely presented by the obliteration of the capillary lumen, which leaves behind the deendothelialized basement membranes, variably referred to as a "string vessels" or "empty basement membrane tubes", or "empty sleeves" as remnants of preexisting vessels. This remaining scaffold of the basement membrane, rich in endothelial and pericyte growth factors, such as VEGF, bFGF, and PDGF, facilitates the repopulation of string vessels with endothelial cells, leading to deposition of a new layer of the basement membrane material, renewed patency of vessels, and ability to restore blood flow to the tissue. Restoration of endothelial lining of string vessels occurs by proliferation of EC and is reportedly assisted by endothelial progenitor cells (Otani et al., 2002). Sequential rounds of de- and recellularization of string vessels may be responsible for the appearance of thickened basement membranes.

10. ENDOTHELIUM-TARGETING REJUVENATION THERAPEUTICS

Emerging principles of rejuvenation pharmacology are based on interfering with one of three major pathways of cell aging: sirtuins, target of rapamycin (mTOR), and insulin-like growth factor. Pioneering studies of heterochronic parabiosis (sharing circulatory system between young and old mice) have revealed restoration of stem cell competence in aged animals and opened the possibility of rejuvenating therapies (Conboy et al., 2005). A member of the activating/TGF-β superfamily, growth differentiation factor 11 (GDF11) has been identified using modified aptamer-based proteomic technology as one of the factors present in the serum of young mice, reduced in aged animals, and capable of restoring cardiac function upon supplementation of old mice (Loffredo et al., 2013). The issue has been questioned, however, in follow-up studies (Egerman et al., 2015).

Based on the earlier studies of SIRT1 and its activation by resveratrol, a number of small molecule SIRT1 activators has been synthesized and are currently tested (Hubbard & Sinclair, 2014). Sirtuin-activating compounds (STACs) exert their effect by allosteric activation of this deacetylase. Three generations of STACs include, in addition to resveratrol, quercetin, and butein (first generation), SRT 1720, 1460, and 2183 (second generation), and STAC-5, -9, and -10 (third generation), all extending life-span and/or health-span. These compounds are presently undergoing clinical trials. In fact, dietary restriction, which possibly induces SIRT1, acts via mTOR signaling and NAD-dependent pathways accompanied by a shift toward oxidative metabolism (Moroz et al., 2014), both representing novel venues of rejuvenation therapy.

NAD+ is a cofactor necessary for activation of several sirtuins. NAD+ bioavailability is reduced in disease states and aging (Verdin, 2015). A precursor of NAD+, nicotinamide, is paving ways as a therapy to correct NAD+ deficiency.

Another venue for rejuvenation therapy is based on the series of findings implicating mTOR activation and the resulting defect in autophagy in senescence. Application of mTOR inhibitor, rapamycin, has been shown to act as a rejuvenating therapy, even when applied late in life (Harrison et al., 2009).

11. CONCLUSION

This brief overview of the origins of endothelial and endothelial progenitor cells, their function and dysfunction in maintaining (micro)vascular integrity in health and disease, and development of premature senescence of these cells in the course of chronic diseases seeks to paint in broad brushstrokes the picture of abnormal metabolism of dysfunctional endothelium, transition of endothelial cells towards mesenchymal phenotype with vascular drop-out and formation of myofibroblasts and induction of fibrogenesis, culminating in organ fibrosis and functional failure. We also sketch some emerging rejuvenation therapies that may be applicable to the endothelium. However, a long and tortuous path remains before we arrive to a better understanding, prophylaxis, and treatment of ECD.

CONFLICT OF INTEREST

Authors declare no conflicts of interest.

ACKNOWLEDGMENTS

Studies in the authors' labs were supported in part by NIH grants DK54602, DK052783, and DK45462; Westchester Artificial Kidney Foundation/The New York Community Trust (M.S.G.); and NIH grants UH3 EB017103, EB016629, and HL128064 (K.H.).

REFERENCES

Addabbo, F., Chen, Q., Patel, D. P., Rabadi, M., Ratliff, B., Zhang, F., et al. (2013). Glutamine supplementation alleviates vasculopathy and corrects metabolic profile in an in vivo model of endothelial cell dysfunction. *PLoS One*, *8*(6), e65458.

Addabbo, F., Ratliff, B., Park, H. C., Kuo, M. C., Ungvari, Z., Csiszar, A., et al. (2009). The Krebs cycle and mitochondrial mass are early victims of endothelial dysfunction. *American Journal of Pathology*, *174*, 34–43.

Alev, C., Li, M., & Asahara, T. (2011). Endothelial progenitor cells: A novel tool for the therapy of ischemic diseases. *Antioxidants & Redox Signaling*, *15*, 949–965.

Baker, D., Wijshake, J., Tchykonia, T., LeBrasseur, N. K., Childs, B. G., van de Sluis, B., et al. (2011). Clearance of p16Ink4a-positive senescent cells delays ageing-associated disorders. *Nature*, *479*, 232–236.

Bearzi, C., Leri, A., Lo Monaco, F., Rota, M., Gonzalez, A., Hosoda, T., et al. (2009). Identification of coronary vascular progenitor cell in the human heart. *Proceedings of the National Academy of Sciences of the United States of America*, *106*, 15885–15890.

Belaoussoff, M., Farrington, S. M., & Baron, M. H. (1998). Hematopoietic induction and respecification of a-p identity by visceral endoderm signaling in the mouse embryo. *Development (Cambridge, England)*, *125*, 5009–5018.

Brodsky, S. V., Gealekman, O., Chen, J., Zhang, F., Togashi, N., Crabtree, M., et al. (2004). Prevention and reversal of premature endothelial cell senescence and vasculopathy in obesity-induced diabetes by ebselen. *Circulation Research*, *94*(3), 377–384.

Brown, W. (2010). A review of string vessels or collapsed, empty basement membrane tubes. *Journal of Alzheimer's Disease*, *21*, 725–739.

Byrd, N., Becker, S., Maye, P., Narasimhaiah, R., St-Jacques, B., Zhang, X., et al. (2002). Hedgehog is required for murine yolk sac angiogenesis. *Development (Cambridge, England)*, *129*, 361–372.

Carlin, L., Stamatiades, E., Auffray, C., Hanna, R. N., Glover, L., Vizcay-Barrena, G., et al. (2013). Nr4a1-dependent Ly6Clow monocytes monitor endothelial cells and orchestrate their disposal. *Cell*, *153*, 362–375.

Carmeliet, P., Ferreira, V., Breier, G., Pollefeyt, S., Kieckens, L., Gertsenstein, M., et al. (1996). Abnormal blood vessel development and lethality in embryos lacking a single vegf allele. *Nature*, *380*, 435–439.

Chen, J., Brodsky, S. V., & Goligorsky, M. S. (2002). Glycated collagen I induces premature senescence-like phenotypic changes in endothelial cells. *Circulation Research*, *90*, 1290–1298.

Chen, P., Qin, L., Barnes, C., Charisse, K., Zhang, S., Ali, R., et al. (2012). FGF regulates TGF-β signaling and endothelial-to-mesenchymal transition via control of let-7 miRNA expression. *Cell Reports*, *2*, 1684–1696.

Chen, J., Xavier, S., Moskowitz-Kassai, E., Chen, R., Lu, C. Y., Sanduski, K., et al. (2012). Cathepsin cleavage of sirtuin 1 in endothelial progenitor cells mediates stress-induced premature senescence. *American Journal of Pathology*, *180*, 973–983.

Cimato, T., Beers, J., Ding, S., Ma, M., McCoy, J. P., Boehm, M., et al. (2009). Neuropilin-1 identifies endothelial precursors in human and murine embryonic stem cells before CD34 expression. *Circulation*, *119*, 2170–2178.

Conboy, I. M., Conboy, M., Wagers, A., Girma, E., Weissman, I., & Rando, T. (2005). Rejuvenation of aged progenitor cells by exposure to a young systemic environment. *Nature*, *433*, 760–764.

Corada, M., Nyqvist, D., Orsenigo, F., Caprini, A., Giampietro, C., Taketo, M. M., et al. (2010). The wnt/beta-catenin pathway modulates vascular remodeling and specification by upregulating dll4/notch signaling. *Developmental Cell*, *18*, 938–949.

Dejana, E., Taddei, A., & Randi, A. M. (2007). Foxs and Ets in the transcriptional regulation of endothelial cell differentiation and angiogenesis. *Biochimica et Biophysica Acta*, *1775*, 298–312.

De Val, S., & Black, B. L. (2009). Transcriptional control of endothelial cell development. *Developmental Cell*, *16*, 180–195.

De Val, S., Chi, N. C., Meadows, S. M., Minovitsky, S., Anderson, J. P., Harris, I. S., et al. (2008). Combinatorial regulation of endothelial gene expression by ets and forkhead transcription factors. *Cell*, *135*, 1053–1064.

Egerman, M., Cadena, S., Gilbert, J., Meyer, A., Nelson, H., Swalley, S., et al. (2015). GDF11 increases with age and inhibits skeletal muscle regeneration. *Cell Metabolism*, *22*, 164–174.

Egorova, A., Khedoe, P., Goumans, M., Yoder, B., Nauli, S., Dijke, P., et al. (2011). Lack of primary cilia primes shear-induced endothelial-to-mesenchymal transition. *Circulation Research*, *108*, 1093–1101.

Facchetti, F., Monzani, E., Cavallini, G., Bergamini, E., & La Porta, C. A. (2007). Effect of caloric restriction regimen on the angiogenic capacity of aorta and on the expression of endothelin-1 during aging. *Experimental Gerontology*, *42*, 662–667.

Fang, S., Wei, J., Pentinmikko, N., Leinonen, H., & Salven, P. (2012). Generation of functional blood vessels from a single c-kit+ adult vascular endothelial stem cell. *PLoS Biology*, *10*, 1001407.

Ferrara, N., Carver-Moore, K., Chen, H., Dowd, M., Lu, L., O'Shea, K. S., et al. (1996). Heterozygous embryonic lethality induced by targeted inactivation of the vegf gene. *Nature*, *380*, 439–442.

Francois, M., Caprini, A., Hosking, B., Orsenigo, F., Wilhelm, D., Browne, C., et al. (2008). Sox18 induces development of the lymphatic vasculature in mice. *Nature*, *456*, 643–647.

Francois, M., Short, K., Secker, G. A., Combes, A., Schwarz, Q., Davidson, T. L., et al. (2012). Segmental territories along the cardinal veins generate lymph sacs via a ballooning mechanism during embryonic lymphangiogenesis in mice. *Developmental Biology*, *364*, 89–98.

Gaengel, K., Genove, G., Armulik, A., & Betsholtz, C. (2009). Endothelial-mural cell signaling in vascular development and angiogenesis. *Arteriosclerosis, Thrombosis, and Vascular Biology*, *29*, 630–638.

Goligorsky, M. S. (2005). Endothelial cell dysfunction: Can't live with it, how to live without it. *American Journal of Physiology. Renal Physiology*, *288*, F871–F880.

Goligorsky, M. S. (2006). Clinical assessment of endothelial dysfunction: Combine and rule. *Current Opinion in Nephrology and Hypertension*, *15*, 617–624.

Gritz, E., & Hirschi, K. K. (2016). Specification and function of hemogenic endothelium during embryogenesis. *Cellular and Molecular Life Sciences*, *73*(8), 1547–1567.

Gu, C., Rodriguez, E. R., Reimert, D. V., Shu, T., Fritzsch, B., Richards, L. J., et al. (2003). Neuropilin-1 conveys semaphorin and vegf signaling during neural and cardiovascular development. *Developmental Cell*, *5*, 45–57.

Gupta, R., Mepani, R., Kleiner, S., Lo, J. C., Khandekar, M. J., Cohen, P., et al. (2012). Zfp423 expression identifies committed preadipocytes and localizes to adipose endothelial and perivascular cells. *Cell Metabolism*, *15*, 230–239.

Harrison, D., Strong, R., Sharp, Z., Nelson, J., Astle, C., Flurkey, K., et al. (2009). Rapamycin fed late in life extends lifespan in genetically heterogeneous mice. *Nature*, *460*, 392–395.

Hemmen, K., Reinl, T., Buttler, K., Behler, F., Dieken, H., Jänsch, L., et al. (2011). High-resolution mass spectrometric analysis of the secretome from mouse lung endothelial progenitor cells. *Angiogenesis*, *14*, 163–172.

Hill, J., Zalos, G., Halcox, J., Schenke, W., Waclawiw, M., Quyyumi, A., et al. (2003). Circulating endothelial progenitor cells, vascular function, and cardiovascular risk. *New England Journal of Medicine, 348*, 593–600.
Hirschi, K. K., Ingram, D. A., & Yoder, M. C. (2008). Assessing identity, phenotype, and fate of endothelial progenitor cells. *Arteriosclerosis, Thrombosis, and Vascular Biology, 28*, 1584–1595.
Hubbard, B., & Sinclair, D. (2014). Small molecule SIRT1 activators for the treatment of aging and age-related diseases. *Trends in Pharmacological Sciences, 35*, 146–154.
James, D., Nam, H. S., Seandel, M., Nolan, D., Janovitz, T., Tomishima, M., et al. (2010). Expansion and maintenance of human embryonic stem cell-derived endothelial cells by TGFbeta inhibition is Id1 dependent. *Nature Biotechnology, 28*(2), 161–166. http://dx.doi.org/10.1038/nbt.1605. Epub 2010 Jan 17.
Johnson, N. C., Dillard, M. E., Baluk, P., McDonald, D. M., Harvey, N. L., Frase, S. L., et al. (2008). Lymphatic endothelial cell identity is reversible and its maintenance requires prox1 activity. *Genes & Development, 22*, 3282–3291.
Karkkainen, M. J., Haiko, P., Sainio, K., Partanen, J., Taipale, J., Petrova, T. V., et al. (2004). Vascular endothelial growth factor c is required for sprouting of the first lymphatic vessels from embryonic veins. *Nature Immunology, 5*, 74–80.
Kataoka, H., Hayashi, M., Nakagawa, R., Tanaka, Y., Izumi, N., Nishikawa, S., et al. (2011). Etv2/er71 induces vascular mesoderm from flk1+pdgfralpha+ primitive mesoderm. *Blood, 118*, 6975–6986.
Kato, T., Mizuno, S., & Ito, A. (2014). A decrease in glomerular endothelial cells and endothelial-mesenchymal transition during glomerulosclerosis in the tensin2-deficient mice (ICGN strain). *Acta Histochemica et Cytochemica, 47*, 265–271.
Kelly, M. A., & Hirschi, K. K. (2009). Signaling hierarchy regulating human endothelial cell development. *Arteriosclerosis, Thrombosis, and Vascular Biology, 29*, 718–724.
Khakoo, A., & Finkel, T. (2005). Endothelial progenitor cells. *Annual Review of Medicine, 56*, 79–101.
Lanner, F., Sohl, M., & Farnebo, F. (2007). Functional arterial and venous fate is determined by graded vegf signaling and notch status during embryonic stem cell differentiation. *Arteriosclerosis, Thrombosis, and Vascular Biology, 27*, 487–493.
Lawson, N. D., Scheer, N., Pham, V. N., Kim, C. H., Chitnis, A. B., Campos-Ortega, J. A., et al. (2001). Notch signaling is required for arterial-venous differentiation during embryonic vascular development. *Development (Cambridge, England), 128*, 3675–3683.
Lawson, N. D., Vogel, A. M., & Weinstein, B. M. (2002). Sonic hedgehog and vascular endothelial growth factor act upstream of the notch pathway during arterial endothelial differentiation. *Developmental Cell, 3*, 127–136.
LeBlue, V., Taduri, G., O'Connell, J., Teng, Y., Cooke, V., Sugimoto, H., et al. (2013). Origin and function of myofibroblasts in kidney fibrosis. *Nature Medicine, 19*, 1047–1053.
Lee, S., Kang, J., Yoo, J., Ganesan, S. K., Cook, S. C., Aguilar, B., et al. (2009). Prox1 physically and functionally interacts with coup-tfii to specify lymphatic endothelial cell fate. *Blood, 113*, 1856–1859.
Lee, D., Park, C., Lee, H., Lugus, J. J., Kim, S. H., Arentson, E., et al. (2008). Er71 acts downstream of bmp, notch, and wnt signaling in blood and vessel progenitor specification. *Cell Stem Cell, 2*, 497–507.
le Noble, F., Moyon, D., Pardanaud, L., Yuan, L., Djonov, V., Matthijsen, R., et al. (2004). Flow regulates arterial-venous differentiation in the chick embryo yolk sac. *Development (Cambridge, England), 131*, 361–375.
Li, J., Qu, X., & Bertram, J. (2009). Endothelial-myofibroblast transition contributes to the early development of diabetic renal interstitial fibrosis in streptozotocin-induced diabetic mice. *American Journal of Pathology, 175*, 1380–1388.

Lin, F. J., Chen, X., Qin, J., Hong, Y. K., Tsai, M. J., & Tsai, S. Y. (2010). Direct transcriptional regulation of neuropilin-2 by coup-tfii modulates multiple steps in murine lymphatic vessel development. *The Journal of Clinical Investigation*, *120*, 1694–1707.

Lin, C. H., Chen, J., Ziman, B. D., Marshall, S., Maizel, J., & Goligorsky, M. S. (2014). Endostatin and kidney fibrosis in aging: A case for antagonistic pleiotropy? *American Journal of Physiology. Heart and Circulatory Physiology*, *306*(12), H1692–H1699.

Liu, Q., Huang, X., Zhang, H., Tian, X., He, L., Yang, R., et al. (2015). c-Kit+ cells adopt vascular endothelial but not epithelial cell fates during lung maintenance and repair. *Nature Medicine*, *21*, 866–868.

Loffredo, F., Steinhauser, M., Jay, S., Gannon, J., Pancoast, J., Yalamanchi, P., et al. (2013). Growth differentiation factor 11 is a circulating factor that reverses age-related cardiac hypertrophy. *Cell*, *153*, 828–839.

Ludmer, P., Selwyn, A., Shook, T., Wayne, R. R., Mudge, G. H., Alexander, R. W., et al. (1986). Paradoxical vasoconstriction induced by acetylcholine in atherosclerotic coronary arteries. *New England Journal of Medicine*, *315*, 1046–1051.

Madedaluno, L., Rudini, N., Cuttano, R., Bravi, L., Giampietro, C., Corada, M., et al. (2013). EndMT contributes to the onset and progression of cerebral cavernous malformation. *Nature*, *498*, 492–496.

Maizel, J., Xavier, S., Chen, J., Lin, C., Vasko, R., & Goligorsky, M. S. (2014). Sirtuin 1 ablation in endothelial cells is associated with impaired angiogenesis and diastolic dysfunction. *American Journal of Physiology. Heart and Circulatory Physiology*, *307*(12), H1691–H1704.

Makinen, T., Jussila, L., Veikkola, T., Karpanen, T., Kettunen, M. I., Pulkkanen, K. J., et al. (2001). Inhibition of lymphangiogenesis with resulting lymphedema in transgenic mice expressing soluble vegf receptor-3. *Nature Medicine*, *7*, 199–205.

Marcu, R., Kotha, S., Zhi, Z., Qin, W., Neeley, C., Wang, R., et al. (2015). The mitochondrial permeability transition pore regulates endothelial bioenergetics and angiogenesis. *Circulation Research*, *116*, 1336–1345.

Marom, K., Levy, V., Pillemer, G., & Fainsod, A. (2005). Temporal analysis of the early bmp functions identifies distinct anti-organizer and mesoderm patterning phases. *Developmental Biology*, *282*, 442–454.

Medici, D., & Kalluri, R. (2012). Endothelial-mesenchymal transition and its contribution to the emergence of stem cell phenotype. *Seminars in Cancer Biology*, *22*, 379–384.

Medici, D., Shore, E., Lounev, V., Kaplan, S., Kalluri, R., & Olsen, B. (2010). Conversion of vascular endothelial cells into multipotent stem-like cells. *Nature Medicine*, *16*, 1400–1406.

Miquerol, L., Langille, B. L., & Nagy, A. (2000). Embryonic development is disrupted by modest increases in vascular endothelial growth factor gene expression. *Development (Cambridge, England)*, *127*, 3941–3946.

Moroz, N., Carmona, J., Anderson, E., Hart, A., Sinclair, D., & Blackwell, T. (2014). Dietary restriction involves NAD-dependent mechanisms and a shift toward oxidative metabolism. *Aging Cell*, *13*, 1075–1085.

Moussaieff, A., Rouleau, M., Kitsberg, D., Cohen, M., Levy, G., Barasch, D., et al. (2015). Glycolysis-mediated changes in acetyl-CoA and histone acetylation control the early differentiation of embryonic stem cells. *Cell Metabolism*, *21*, 392–402.

Nauli, S., Kawanabe, Y., Kaminski, J., Pearce, W., Ingber, D., & Zhou, J. (2008). Endothelial cilia are fluid shear sensors that regulate calcium signaling and nitric oxide production through polycystin-1. *Circulation*, *117*, 1161–1171.

Nieuwdorp, M., Mooij, H. L., Kroon, J., Atasever, B., Spaan, J. A., Ince, C., et al. (2006). Endothelial glycocalyx damage coincides with microalbuminuria in type 1 diabetes. *Diabetes*, *55*, 1127–1132.

O'Riordan, E., Mendelev, N., Patschan, S., Chander, P., & Goligorsky, M. S. (2007). Chronic NOS inhibition actuates endothelial-mesenchymal transformation. *American Journal of Physiology*, *292*, H285–H294.

Otani, A., Kinder, K., Ewalt, K., Otero, F., Schimmel, P., & Friedlander, M. (2002). Bone marrow-derived stem cells target retinal astrocytes and can promote or inhibit retinal angiogenesis. *Nature Medicine*, *8*, 1004–1010.

Park, C., Afrikanova, I., Chung, Y. S., Zhang, W. J., Arentson, E., Fong Gh, G., et al. (2004). A hierarchical order of factors in the generation of flk1- and scl-expressing hematopoietic and endothelial progenitors from embryonic stem cells. *Development (Cambridge, England)*, *131*, 2749–2762.

Patschan, S., Chen, J., Gealekman, O., Krupincza, K., Wang, M., Shu, L., et al. (2008). Mapping mechanisms and charting the time course of premature cell senescence and apoptosis: Lysosomal dysfunction and ganglioside accumulation in endothelial cells. *American Journal of Physiology. Renal Physiology*, *294*(1), F100–F109.

Pearson, S., Sroczynska, P., Lacaud, G., & Kouskoff, V. (2008). The stepwise specification of embryonic stem cells to hematopoietic fate is driven by sequential exposure to Bmp4, activin A, bFGF and VEGF. *Development (Cambridge, England)*, *135*, 1525–1535.

Pennisi, D., Gardner, J., Chambers, D., Hosking, B., Peters, J., Muscat, G., et al. (2000). Mutations in sox18 underlie cardiovascular and hair follicle defects in ragged mice. *Nature Genetics*, *24*, 434–437.

Potente, M., Ghaeni, L., Baldessari, D., Mostoslavsky, R., Rossig, L., Dequiedt, F., et al. (2007). SIRT1 controls endothelial angiogenic functions during vascular growth. *Genes & Development*, *21*(20), 2644–2658.

Prasain, N., Lee, M., Vemula, S., Meador, J. L., Yoshimoto, M., Ferkowicz, M. J., et al. (2014). Differentiation of human pluripotent stem cells to cells similar to cord-blood endothelial colony-forming cells. *Nature Biotechnology*, *32*, 1151–1157.

Pula, G., Mayr, U., Evans, C., et al. (2009). Proteomics identifies thymidine phosphorylase as a key regulator of the angiogenic potential of colony-forming units and endothelial progenitor cell cultures. *Circulation Research*, *104*, 32–40.

Reed, M., Karres, N., Eyman, D., & Vernon, R. B. (2007). Culture of murine aortic explants in 3-dimensional matrix: A novel, miniaturized assay of angiogenesis in vitro. *Microvascular Research*, *73*, 248–252.

Reitsma, S., Slaaf, D. W., Vink, H., van Zandvoort, M. A., & oude Egbrink, M. G. (2007). The endothelial glycocalyx: Composition, functions, and visualization. *Pflügers Archiv-European Journal of Physiology*, *454*, 345–359.

Richardson, M., & Yoder, M. (2011). Endothelial progenitor cell: Quo Vadis? *Journal of Molecular and Cellular Cardiology*, *50*, 266–272.

Salanga, M. C., Meadows, S. M., Myers, C. T., & Krieg, P. A. (2010). Ets family protein etv2 is required for initiation of the endothelial lineage but not the hematopoietic lineage in the Xenopus embryo. *Developmental Dynamics: An Official Publication of the American Association of Anatomists*, *239*, 1178–1187.

Schuh, A. C., Faloon, P., Hu, Q. L., Bhimani, M., & Choi, K. (1999). In vitro hematopoietic and endothelial potential of flk-1(−/−) embryonic stem cells and embryos. *Proceedings of the National Academy of Sciences of the United States of America*, *96*, 2159–2164.

Sedwick, C. (2012). On the hunt for vascular endothelial stem cells. *PLoS Biology*, *10*(10), e1001408.

Shalaby, F., Rossant, J., Yamaguchi, T. P., Gertsenstein, M., Wu, X. F., Breitman, M. L., et al. (1995). Failure of blood-island formation and vasculogenesis in flk-1-deficient mice. *Nature*, *376*, 62–66.

Shimada, T., Takeshita, T., Murohara, T., Sasaki, K., Egami, K., Shintani, S., et al. (2004). Angiogenesis and vasculogenesis are impaired in the precocious-aging klotho mouse. *Circulation*, *110*, 1148–1155.

Srinivasan, R. S., Dillard, M. E., Lagutin, O. V., Lin, F. J., Tsai, S., Tsai, M. J., et al. (2007). Lineage tracing demonstrates the venous origin of the mammalian lymphatic vasculature. *Genes & Development, 21*, 2422–2432.

Srinivasan, R. S., Geng, X., Yang, Y., Wang, Y., Mukatira, S., Studer, M., et al. (2010). The nuclear hormone receptor coup-tfii is required for the initiation and early maintenance of prox1 expression in lymphatic endothelial cells. *Genes & Development, 24*, 696–707.

Suzuki, M., & Boothman, D. (2008). Stress-induced premature senescence (SIPS)—Influence of SIPS on radiotherapy. *Journal of Radiation Research, 49*, 105–112.

Tran, K.-V., Gealekman, O., Frontini, A., Zingaretti, M. C., Morroni, M., Giordano, A., et al. (2012). The vascular endothelium of the adipose tissue gives rise to both white and brown fat cells. *Cell Metabolism, 15*, 222–229.

Ubil, E., Duan, J., Pillai, I., Rosa-Garrido, M., Wu, Y., Bargiacchi, F., et al. (2014). Mesenchymal-endothelial transition contributes to cardiac neovascularization. *Nature, 514*, 585–590.

Vasko, R., Xavier, S., Chen, J., Lin, C. H., Ratliff, B., Rabadi, M., et al. (2014). Endothelial sirtuin 1 deficiency perpetrates nephrosclerosis through downregulation of matrix metalloproteinase-14: Relevance to fibrosis of vascular senescence. *Journal of the American Society of Nephrology, 25*(2), 276–291.

Verdegem, D., Moens, S., Stapor, P., & Carmeliet, P. (2014). Endothelial cell metabolism: Parallels and divergences with cancer cell metabolism. *Cancer & Metabolism, 2*, 19.

Verdin, E. (2015). NAD+ in aging, metabolism, and neurodegeneration. *Science, 350*, 1208–1213.

Vokes, S. A., & Krieg, P. A. (2002). Endoderm is required for vascular endothelial tube formation, but not for angioblast specification. *Development (Cambridge, England), 129*, 775–785.

Wang, H. U., Chen, Z. F., & Anderson, D. J. (1998). Molecular distinction and angiogenic interaction between embryonic arteries and veins revealed by ephrin-b2 and its receptor eph-b4. *Cell, 93*, 741–753.

Weinbaum, S., Tarbell, J. M., & Damiano, E. R. (2007). The structure and function of the endothelial glycocalyx layer. *Annual Review of Biomedical Engineering, 9*, 121–167.

Welch-Reardon, K., Wu, N., & Hughes, C. (2015). A role for partial endothelial-mesenchymal transitions in angiogenesis? *Arteriosclerosis, Thrombosis, and Vascular Biology, 35*, 303–308.

Werner, N., Kosiol, S., Schiegl, T., Ahlers, P., Walenta, K., Link, A., et al. (2005). Circulating endothelial progenitor cells and cardiovascular outcomes. *New England Journal of Medicine, 353*, 999–1007.

Wigle, J. T., Harvey, N., Detmar, M., Lagutina, I., Grosveld, G., Gunn, M. D., et al. (2002). An essential role for prox1 in the induction of the lymphatic endothelial cell phenotype. *The EMBO Journal, 21*, 1505–1513.

Wigle, J. T., & Oliver, G. (1999). Prox1 function is required for the development of the murine lymphatic system. *Cell, 98*, 769–778.

Wilkinson, R. N., Koudijs, M. J., Patient, R. K., Ingham, P. W., Schulte-Merker, S., & van Eeden, F. J. (2012). Hedgehog signaling via a calcitonin receptor-like receptor can induce arterial differentiation independently of vegf signaling in zebrafish. *Blood, 120*, 477–488.

Williams, C., Kim, S. H., Ni, T. T., Mitchell, L., Ro, H., Penn, J. S., et al. (2010). Hedgehog signaling induces arterial endothelial cell formation by repressing venous cell fate. *Developmental Biology, 341*, 196–204.

Winnier, G., Blessing, M., Labosky, P. A., & Hogan, B. L. (1995). Bone morphogenetic protein-4 is required for mesoderm formation and patterning in the mouse. *Genes & Development, 9*, 2105–2116.

Xavier, S., Vasko, R., Matsumoto, K., Zullo, J., Chen, R., Maizel, J., et al. (2015). Curtailing endothelial TGFb signaling is sufficient to reduce endothelial-mesenchymal transition and fibrosis in CKD. *Journal of the American Society of Nephrology*, *26*(4), 817–829.
Xu, X., Friehs, I., Zhong, H., Melnychenko, I., Tampe, B., Alnour, F., et al. (2015). Endocardial fibroelastosis is caused by aberrant endothelial-to-mesenchymal transition. *Circulation Research*, *116*, 857–866.
Yamaguchi, T. P., Harpal, K., Henkemeyer, M., & Rossant, J. (1994). Fgfr-1 is required for embryonic growth and mesodermal patterning during mouse gastrulation. *Genes & Development*, *8*, 3032–3044.
Yamamizu, K., Matsunaga, T., Uosaki, H., Fukushima, H., Katayama, S., Hiraoka-Kanie, M., et al. (2010). Convergence of notch and beta-catenin signaling induces arterial fate in vascular progenitors. *The Journal of Cell Biology*, *189*, 325–338.
You, L. R., Lin, F. J., Lee, C. T., DeMayo, F. J., Tsai, M. J., & Tsai, S. Y. (2005). Suppression of notch signalling by the coup-tfii transcription factor regulates vein identity. *Nature*, *435*, 98–104.
Zeisberg, E. M., Tarnavski, O., Zeisberg, M., Dorfman, A. L., McMullen, J. R., Gustafsson, E., et al. (2007). Endothelial-to-mesenchymal transition contributes to cardiac fibrosis. *Nature Medicine*, *13*(8), 952–961.

CHAPTER TEN

Estrogens and Coronary Artery Disease: New Clinical Perspectives

M.R. Meyer*[,1], M. Barton[†,1]
*Triemli City Hospital, Zürich, Switzerland
[†]Molecular Internal Medicine, University of Zürich, Zürich, Switzerland
[1]Corresponding authors: e-mail address: matthias.meyer@triemli.zuerich.ch; barton@access.uzh.ch

Contents

Abstract

In premenopausal women, endogenous estrogens are associated with reduced prevalence of arterial hypertension, coronary artery disease, myocardial infarction, and stroke. Clinical trials conducted in the 1990s such as HERS, WHI, and WISDOM have shown that postmenopausal treatment with horse hormone mixtures (so-called conjugated equine estrogens) and synthetic progestins adversely affects female cardiovascular health. Our

Advances in Pharmacology, Volume 77
ISSN 1054-3589
http://dx.doi.org/10.1016/bs.apha.2016.05.003

understanding of rapid (nongenomic) and chronic (genomic) estrogen signaling has since advanced considerably, including identification of a new G protein-coupled estrogen receptor (GPER), which like the "classical" receptors ERα and ERβ is highly abundant in the cardiovascular system. Here, we discuss the role of estrogen receptors in the pathogenesis of coronary artery disease and review natural and synthetic ligands of estrogen receptors as well as their effects in physiology, on cardiovascular risk factors, and atherosclerotic vascular disease. Data from preclinical and clinical studies using nonselective compounds activating GPER, which include selective estrogen receptor modulators such as tamoxifen or raloxifene, selective estrogen receptor downregulators such as Faslodex™ (fulvestrant/ICI 182,780), vitamin B_3 (niacin), green tea catechins, and soy flavonoids such as genistein or resveratrol, strongly suggest that activation of GPER may afford therapeutic benefit for primary and secondary prevention in patients with or at risk for coronary artery disease. Evidence from preclinical studies suggest similar efficacy profiles for selective small molecule GPER agonists such as G-1 which are devoid of uterotrophic activity. Further clinical research in this area is warranted to provide opportunities for future cardiovascular drug development.

ABBREVIATIONS

ADCF adipose-derived contracting factor
cAMP cyclic adenosine monophosphate
CEEs conjugated equine estrogens
DPN diarylpropionitrile
eNOS endothelial nitric oxide synthase
EDCF endothelium-derived contracting factor
ELITE Early versus Late Intervention Trial with Estradiol
ER estrogen receptor
ERα estrogen receptor alpha
ERβ estrogen receptor beta
ERA endothelin receptor antagonist
GPCR G protein-coupled receptor
GPER G protein-coupled estrogen receptor
GPR30 G protein-coupled receptor 30
HERS Heart and Estrogen/progestin Replacement Study
HDL high-density lipoprotein
LDL low-density lipoprotein
MPA medroxyprogesterone acetate
MR mineralocorticoid receptor
NO nitric oxide
PKA protein kinase A
PVAT perivascular adipose tissue
PPT propylpyrazole triol
SERD selective estrogen receptor downregulator
SERM selective estrogen receptor modulator
TXA_2 thromboxane A_2
TNFα tumor necrosis factor α
WHI Women's Health Initiative
WISDOM Women's International Study on long Duration Oestrogen after Menopause

1. INTRODUCTION

Atherosclerosis and its complications such as myocardial infarction and stroke represent the most common causes of death in women (Barrett-Connor, 2013; Lozano et al., 2012; Meyer & Barton, 2009; Turgeon, McDonnell, Martin, & Wise, 2004; Yusuf, Ounpuu, & Anand, 2002), which has important socioeconomical implications (Barton, 2013c; Barton, Husmann, & Meyer, 2016; Heidenreich et al., 2011). Although healthy premenopausal women are largely protected from cardiovascular disease, its prevalence rapidly rises after menopause, consistent with potent salutary vascular effects of endogenous estrogens and particularly its physiologically most relevant form 17β-estradiol (Barrett-Connor, 2013; Schenck-Gustafsson et al., 2011; Fig. 1). These effects have been ascribed, in part, to the potent antiinflammatory properties of estrogens that are lost after menopause (Burke, Farb, Malcom, & Virmani, 2001; Meyer & Barton, 2009; Meyer, Haas, & Barton, 2006; Miller et al., 2007; Novella, Heras, Hermenegildo, & Dantas, 2012; Straub, 2007). Indeed, first manifestations of coronary atherosclerosis are delayed by about 10 years in women compared with men, translating into a considerably lower incidence of coronary artery disease below 64 years of age (Isles, Hole, Hawthorne, & Lever, 1992). These sex differences disappear after natural or surgical menopause (ovariectomy) (McSweeney et al., 2016; Mucowski et al., 2014; Witteman, Grobbee, Kok, Hofman, & Valkenburg, 1989), or under conditions impairing ovarian function and thus estrogen production (Barton, 2013a; Crandall & Barrett-Connor, 2013; McSweeney et al., 2016).

Interestingly, menopause may not alone determine future cardiovascular risk, but risk factors for the development of atherosclerosis such as hypertension, dyslipidemia, obesity, smoking, or hypercholesterolemia may actually cause progressive ischemic damage to the ovaries, and thus determine the age at which menopause occurs (Kok et al., 2006). Not only the presence of risk factors but also the degree of exposure to endogenous estrogens during premenopausal years will contribute to the atherosclerosis burden later in life (Meyer, Haas, & Barton, 2008). In fact, interference with pituitary-controlled release of 17β-estradiol may serve as an independent predictor for coronary atherosclerosis in postmenopausal women (Merz et al., 2006), and low plasma concentrations of 17β-estradiol have been associated with angiographically confirmed coronary artery disease in premenopausal women (Bairey Merz et al., 2003). Healthy women before menopause

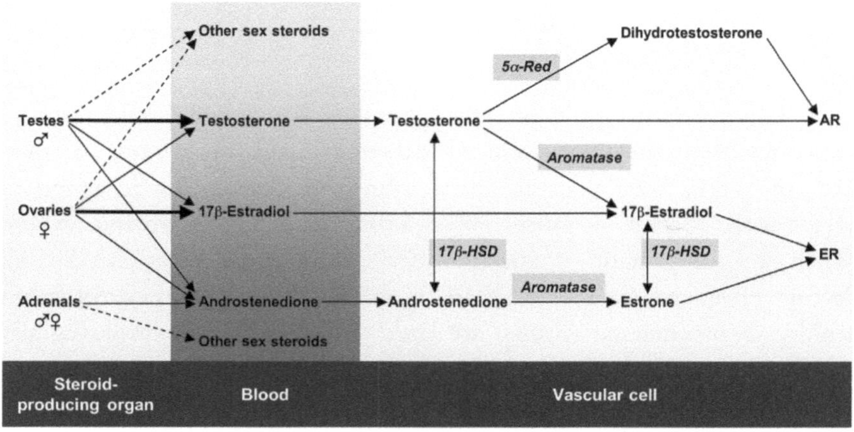

Fig. 1 Cellular sources and vascular production of estrogens and androgens in men and women. In men, testosterone and ~1/5 of 17β-estradiol are produced by the testes, whereas the ovaries produce 17β-estradiol and most of the androgens in women. Androstenedione is formed by the testes, ovaries, and adrenals. The adrenals also synthesize the prohormone dehydroepiandrosterone, which is converted into androgens and/or estrogens in peripheral tissues. Circulating sex hormones bind to sex hormone-binding globulin and, to a lesser extent, to plasma albumin. A small fraction of these sex hormones circulates in the free state. Testosterone enters vascular cells containing the enzyme aromatase which converts testosterone into 17β-estradiol and androstenedione into estrone. Thus, aromatase represents an important local source of estrogens producing ~80% of bioavailable 17β-estradiol in males. Estrone and androstenedione are converted by 17β-hydroxysteroid dehydrogenase (17β-HSD) to more potent sex steroids, 17β-estradiol and testosterone, respectively. Testosterone can then be converted into the most potent androgen, dihydrotestosterone in cells expressing 5α-reductase (5α-Red). Estrogens bind to either nuclear or membrane estrogen receptors ([*ER*]: estrogen receptor α, estrogen receptor β, and GPER), while testosterone or dihydrotestosterone binds to androgen receptors ([*AR*], androgen receptor A or androgen receptor B) eliciting chronic (genomic) and acute (nongenomic) responses in target cells, respectively. *Figure modified from Meyer, M. R., Haas, E., & Barton, M. (2008). Need for research on estrogen receptor function: Importance for postmenopausal hormone therapy and atherosclerosis.* Gender Medicine, 5 *(Suppl. A), S19–S33, with permission from the publisher.*

display greater endothelium-dependent vasodilation (a surrogate of cardiovascular health (Barton, 2010; Feletou & Vanhoutte, 2006; Vanhoutte, Shimokawa, Feletou, & Tang, 2016)) compared to age-matched men (Steinberg et al., 2000). Clinical studies have demonstrated that estrogen therapy improves endothelium-dependent vasodilation in nondiabetic postmenopausal women (Gambacciani et al., 2002; Guzic-Salobir, Keber, Seljeflot, Arnesen, & Vrabic, 2001; Lim et al., 1999). Furthermore, observational studies have demonstrated therapeutic benefits on coronary atherosclerosis in women who opted for hormone therapy within a few years of menopause (Grady et al., 1992; Schierbeck et al., 2012).

In contrast, large-scale, prospective clinical trials such as the HERS (Heart and Estrogen/progestin Replacement Study) (Hulley et al., 1998), WHI (Women's Health Initiative) (Rossouw et al., 2002), and WISDOM (Women's International Study of long Duration Oestrogen after Menopause) (Vickers et al., 2007) showed that treatment of aged postmenopausal women with horse urine-derived hormone mixtures misleadingly dubbed "conjugated equine estrogens" (CEEs, which among other substances contain horse estrogens, progestins, and androgens (Dubey, Imthurn, Barton, & Jackson, 2005)) and medroxyprogesterone acetate (MPA) increases the risk for cardiovascular events (Hulley et al., 1998; Khalil, 2013; Reslan & Khalil, 2012; Rossouw et al., 2002). Although HERS, WHI, and WISDOM have been widely criticized for their design—including the initiation of hormone therapy in women at an advanced age (>65 years) and inclusion in the study despite the presence of several cardiovascular risk factors such as arterial hypertension and obesity, or even established coronary artery disease as well as the use of animal hormone mixtures that do not contain natural human 17β-estradiol (Barton, 2001; Barton & Dubey, 2002; Barton, Meyer, Bhattacharya, & Haas, 2007; Barton, Meyer, & Haas, 2007; Dubey et al., 2005)—hormone therapy is generally not recommended for the primary prevention or treatment of cardiovascular diseases (Miller & Duckles, 2008; Schenck-Gustafsson et al., 2011). Of note, a more recent analysis of the WHI trial suggests that cardiovascular risk reduction may be achieved only after 6 or more years of treatment (Toh, Hernandez-Diaz, Logan, Rossouw, & Hernan, 2010). Indeed, the protective effects of hormone therapy on endothelium-dependent vascular function decrease with age (Barton, Meyer, & Haas, 2007; Sherwood et al., 2007). However, given the potent vasculotropic effects of estrogen and its variation within different stages of reproductive life and the time since menopause, detailed studies on the mechanisms involved in vascular action of 17β-estradiol at different stages of atherosclerosis progression are warranted. Indeed, clinical trials such as the ELITE utilizing 17β-estradiol as bioidentical hormone therapy in early postmenopausal women have shown a significant reduction of atherosclerosis-related cardiovascular events (Hodis et al., 2016; Keaney & Solomon, 2016; Schierbeck et al., 2012), and even in the WHI study younger age at the initiation of "equine estrogen" treatment was associated with a reduction in cardiovascular events after years 7–8 (Harman et al., 2010; McSweeney et al., 2016).

2. ENDOTHELIAL CELL DYSFUNCTION AND ATHEROSCLEROSIS

Functional integrity of endothelial cells is an indicator and a prerequisite for vascular health and counteracts the development of atherosclerosis (Barton, 2013c; Vanhoutte et al., 2016). Endothelial cells, which form the innermost cell monolayer of the arterial wall, are important regulators of vascular homeostasis (Vanhoutte et al., 2016) and critically involved in atherogenesis (Gimbrone & Garcia-Cardena, 2016; Hansson, 2001, 2005; Hansson & Libby, 2006; Libby, Ridker, & Maseri, 2002; Ross, 1999). Proinflammatory conditions such as dyslipidemia, hyperglycemia, hypertension, and smoking trigger an increase in endothelial expression of leukocyte adhesion molecules (Hansson, 2001, 2005; Hansson & Libby, 2006; Libby et al., 2002; Ross, 1999). The resulting recruitment of monocytes and T cells into the subendothelial space initiates a local inflammatory response (Barton, 2013c; Hansson, 2001, 2005; Hansson & Libby, 2006; Libby et al., 2002; Ross, 1999). Inflammatory cytokines subsequently induce monocytes to differentiate into macrophages, which take up modified lipid particles to become foam cells, the prototypical cell of the early atherosclerotic lesion, the "fatty streak" (Hansson, 2001, 2005; Hansson & Libby, 2006; Libby et al., 2002; Ross, 1999). Local production of inflammatory mediators further impairs the numerous protective functions of endothelial cells (Barton, 2013c), favors the perpetuation of inflammation, vasoconstriction/vasospasm, as well as vascular smooth muscle cell growth and migration, and ultimately results in progression of atherosclerosis (Hansson, 2001, 2005; Hansson & Libby, 2006; Libby et al., 2002; Ross, 1999).

In addition to other atheroprotective effects such as salutary changes in lipid profiles, inhibition of vascular smooth muscle cell proliferation, and antioxidant activity (Arnal et al., 2006; Clarkson, 2007; Mendelsohn & Karas, 1999; Meyer et al., 2006, 2008; Miller & Duckles, 2008), 17β-estradiol mediates potent antiinflammatory effects (Arnal et al., 2006; Burke et al., 2001; Miller & Duckles, 2008; Stork, van der Schouw, Grobbee, & Bots, 2004; Straub, 2007) that depend on the presence of mature lymphocytes (Elhage et al., 2000, 2005). This is of particular importance since vascular inflammation is a hallmark not only of advanced, complicated atheroma, but already an essential driver of early, potentially reversible atherosclerotic stages such as abnormalities in endothelial cell function and initiation of fatty streak formation (Hansson, 2001, 2005; Hansson & Libby, 2006;

Libby et al., 2002; Ross, 1999). Given its potent antiinflammatory properties it is not surprising that 17β-estradiol is able to reverse the formation of fatty streaks partly by improving endothelial function and inhibiting the recruitment of inflammatory cells (Arnal et al., 2006; Burke et al., 2001; Clarkson, 2007; Mendelsohn & Karas, 1999; Meyer et al., 2006, 2008; Miller & Duckles, 2008; Pinto et al., 1997; Stork et al., 2004; Straub, 2007; Virdis et al., 2000).

In contrast to early fatty streak lesions, a different pathobiology underlies advanced atherosclerotic lesions that provide an altered substrate, which may become insensitive to salutary effects of estrogen on endothelial cell function but may result in reduced vasodilatory capacity, enhanced inflammatory activity, or even the promotion of plaque instability in response to estrogen (Mendelsohn & Karas, 2005). Though estrogen's ability to counteract early, still reversible atherosclerotic changes appears particularly appealing from a preventive and therapeutic perspective, which is supported by epidemiological and clinical data demonstrating beneficial effects of natural and synthetic estrogens in early but not in late stages of atherosclerosis. Still, surprisingly little information about the targets and signaling pathways involved in estrogen's protective vascular mechanisms during the early stages of the disease.

3. ESTROGEN RECEPTORS AND THE ARTERIAL VASCULAR WALL

Estrogens, discovered 90 years ago by German chemist Adolf Butenandt (Butenandt, 1929, 1931), are sex steroid hormones that are formed in multiple tissues in men and women, including the gonads, adipose, the adrenals, and also in the arterial vascular wall (Barton, 2016; Meyer et al., 2008; Fig. 1). The effects of estrogen are mediated by specific receptors that recognize and bind the hormone transmitting its information to downstream effectors (Matthews & Gustafsson, 2003; Meyer, Haas, Prossnitz, & Barton, 2009). Evidence for the first mammalian estrogen receptor (later termed *esr1* or ERα) was reported as early as 1958 (Jensen, 2014; Jensen & DeSombre, 1973; Jensen, Desombre, Hurst, Kawashima, & Jungblut, 1967; Jensen & Jacobson, 1960; Soloff & Szego, 1969; Talwar, Segal, Evans, & Davidson, 1964). ERα/*esr1* was finally cloned in the mid-1980s (Green, Walter, Greene, et al., 1986; Green, Walter, Kumar, et al., 1986; Greene et al., 1986; Walter et al., 1985), followed by the cloning of a cDNA encoding for a second, highly homologous estrogen receptor termed ERβ in 1996 (Kuiper, Enmark, Pelto-Huikko, Nilsson, & Gustafsson, 1996; Mosselman,

Polman, & Dijkema, 1996). ERα (*esr1*) and ERβ (*esr2*) predominantly function as ligand-activated transcription factors that reside in the cytosol and translocate into the nucleus upon ligand binding to regulate target gene expression (Burns & Korach, 2012; Deroo & Korach, 2006; Hewitt, Winuthayanon, & Korach, 2016). Membrane subpopulations of ERα and ERβ at the plasma membrane also exist that are capable to initiate rapid intracellular signaling (Meyer et al., 2009; Razandi, Pedram, Greene, & Levin, 1999; Razandi, Pedram, Merchenthaler, Greene, & Levin, 2004). Both ERα and ERβ are widely expressed in the cardiovascular system, and their important roles in cardiovascular physiology and disease have been reviewed elsewhere (Burns & Korach, 2012; Deroo & Korach, 2006; Hewitt & Korach, 2002; Hewitt et al., 2016; Mendelsohn & Karas, 1999; Meyer et al., 2006).

Twenty years ago, an orphan 7-transmembrane G protein-coupled receptor (GPCR) was independently cloned by several laboratories (Bonini, Anderson, & Steiner, 1997; Carmeci, Thompson, Ring, Francke, & Weigel, 1997; Feng & Gregor, 1997; Hasbi, O'Dowd, & George, 2005; Kvingedal & Smeland, 1997; McCoy & Perlemutter, 1996; O'Dowd et al., 1998; Owman, Blay, Nilsson, & Lolait, 1996; Takada, Kato, Kondo, Korenaga, & Ando, 1997). This GPCR was initially given the orphan receptor designation G protein-coupled receptor 30 (GPR30) (GPR41 for its rat homologue) as at the time no ligand was known (Barton, 2016). In 2000, Filardo et al. demonstrated that GPR30 mediates rapid, estrogen-dependent effects in human breast cancer cells (Filardo, Quinn, Bland, & Frackelton, 2000; Filardo, Quinn, Frackelton, & Bland, 2002). GPR30 was subsequently shown to bind and induce rapid signaling in response to estrogen (Revankar, Cimino, Sklar, Arterburn, & Prossnitz, 2005; Thomas, Pang, Filardo, & Dong, 2005). The receptor has since become the third established *bona fide* estrogen-binding receptor (Menazza & Murphy, 2016; Prossnitz & Barton, 2011) and has been officially designated "G protein-coupled estrogen receptor-1" (GPER-1) by the *International Union of Basic and Clinical Pharmacology* (Alexander, Mathie, & Peters, 2008).

In teleosts, Lafont et al. recently identified two different isoforms of the *gper* gene that were named *gper-a* and *gper-b*, respectively (Lafont, Rousseau, Tomkiewicz, & Dufour, 2016). The teleost *gper-a*, but not *gper-b*, shares 63% homology with human *gper* and shows high expression in the ovary and pituitary (Lafont et al., 2016). To date only a single mammalian *gper* gene has been identified (Bonini et al., 1997; Carmeci et al., 1997; Feng & Gregor, 1997; Kvingedal & Smeland, 1997; McCoy & Perlemutter, 1996; O'Dowd et al., 1998; Owman et al., 1996; Prossnitz & Barton, 2011; Takada et al., 1997). We will thus use the abbreviation "GPER" in the present work. GPER interacts with other nuclear steroid hormone receptors, including ERα (Barrett Mueller et al., 2014; Clark, Pollard, Rainville, & Vasudevan, 2016; Jia et al., 2016), ERβ (Caroccia et al., 2014), mineralocorticoid receptor (MR), vitamin D receptor, and the glucocorticoid receptor (reviewed in Barton, 2012b; Prossnitz & Barton, 2014). The recently proposed role of GPER to also function as a receptor for aldosterone (Gros, Ding, Liu, Chorazyczewski, & Feldman, 2013; Gros et al., 2011) has not been confirmed as binding of aldosterone to GPER could not be demonstrated (Barton & Meyer, 2015; Cheng et al., 2014; Rigiracciolo et al., 2016; Wendler, Albrecht, & Wehling, 2012); in addition, aldosterone lacks the typical signaling responses induced by GPER-specific ligands (Evans, Bayliss, Reale, & Evans, 2016). However, GPER, through cross talk with ERβ, contributes to the regulation of adrenal aldosterone production (Caroccia et al., 2014). Moreover, even in the absence of a ligand, ERα negatively regulates MR expression, an effect that is further inhibited by adding estrogen as an ERα agonist (Barrett Mueller et al., 2014; Fuller & Young, 2014). Similarly, cross talk between GPER and MR through shared signaling pathways may partly explain that aldosterone-induced responses can be inhibited by the GPER-selective antagonist G15 (Barton & Meyer, 2015).

GPER is expressed in human endothelial (Meyer et al., 2014; Takada et al., 1997) and vascular smooth muscle cells (Haas et al., 2009) as well as intact arteries and veins of patients with atherosclerosis (Haas et al., 2007; Fig. 2). GPER is also expressed in leukocytes of different species, including eosinophils, lymphocytes, monocytes, and macrophages (Barton, 2016; Pelekanou et al., 2016; Prossnitz & Barton, 2009). Acute vasodilation by the GPER-selective agonist G-1 (Broughton, Miller, & Sobey, 2010; Haas et al., 2009; Lindsey, Carver, Prossnitz, & Chappell, 2011; Lindsey, Cohen, Brosnihan, Gallagher, & Chappell, 2009; Meyer, Baretella, Prossnitz, & Barton, 2010) and lack of such responses in isolated arteries from GPER-deficient animals (Haas et al., 2009) confirmed the functional importance of this receptor in the vascular system (Fig. 3). Similar to previous observations in other cell types (Revankar et al., 2005), GPER staining in endothelial and vascular smooth muscle cells shows a predominantly intracellular expression pattern (Fig. 2; Meyer et al., 2014). Specifically, GPER is predominantly expressed at intracellular membranes such as Golgi apparatus and endoplasmic reticulum of endothelial cells (Meyer et al., 2014), although a fraction can also be detected at the plasma membrane (Meyer et al., 2014). Interestingly, at least in human embryonic kidney cells, GPER undergoes constitutive retrograde trafficking from the plasma membrane to the

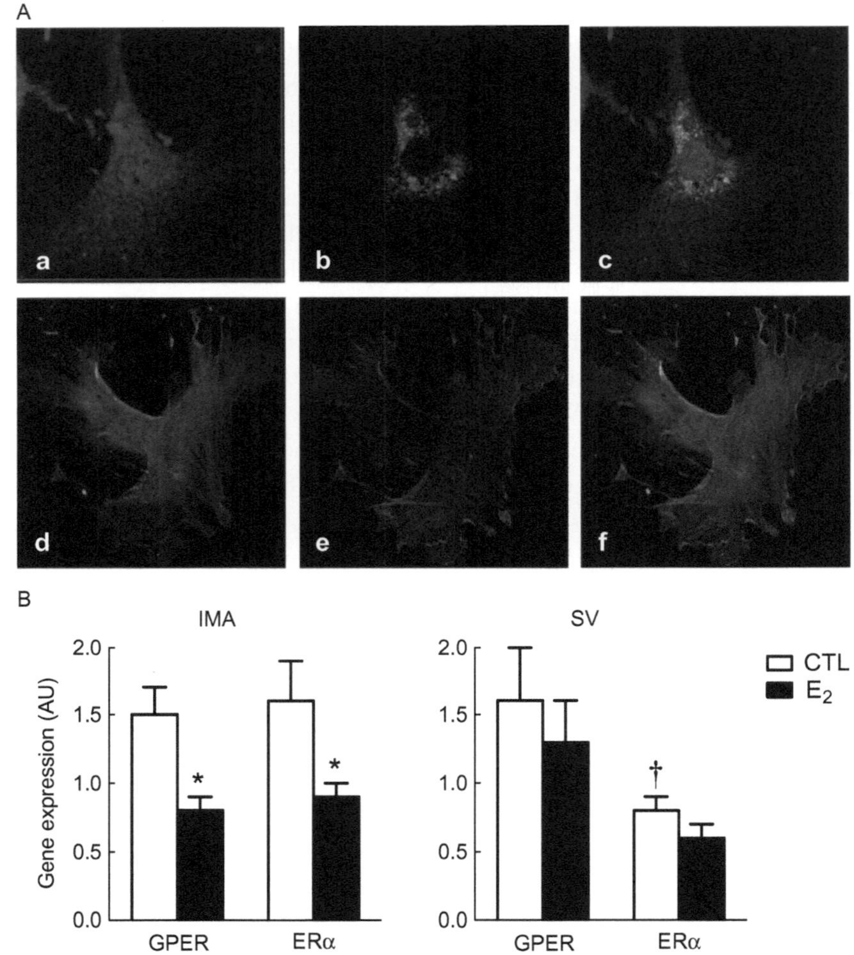

Fig. 2 Expression of GPER in endothelial and vascular smooth muscle cells, arteries, and veins. (A) Human endothelial cells were stained for GPER (*red* [*gray* in the print version], a) or endoplasmic reticulum (*green* [*light gray* in the print version], b). The merged image demonstrates colocalization of both markers (c), suggesting that GPER is expressed predominantly on intracellular membranes. Mouse vascular smooth muscle cells were stained for GPER (*green* [*light gray* in the print version], d) and α-actin (*red* [*gray* in the print version], e). The merged image is shown in (f). The cell nucleus is stained with DAPI (*blue* [*gray* in the print version]). (B) Gene expression of GPER was detected in human internal mammary artery (*IMA*) and saphenous vein (*SV*), and is regulated in response to 17β-estradiol (E_2, 1 μM for 3 h) only in arteries. For comparison, gene expression of ERα is shown. $^*p < 0.05$ vs CTL (control, ethanol 0.2%), $^{\dagger}p < 0.05$ vs IMA. *AU*, arbitrary units. *Panel A is reproduced from Meyer, M. R., Fredette, N. C., Howard, T., Hu, C., Ramesh, C., Daniel, C., …, Prossnitz, E. R. (2014). G protein-coupled estrogen receptor protects from atherosclerosis.* Scientific Reports, *4, 7564. Erratum in:* Scientific Reports, *7565 (2015) 13510. Panel B is adapted from Haas, E., Meyer, M. R., Schurr, U., Bhattacharya, I., Minotti, R., Nguyen, H. H., …, Barton, M. (2007). Differential effects of 17β-estradiol on function and expression of estrogen receptor α, estrogen receptor β, and GPR30 in arteries and veins of patients with atherosclerosis.* Hypertension, 49, *1358–1363, ©2007 by the American Heart Association, with permission from the publisher.*

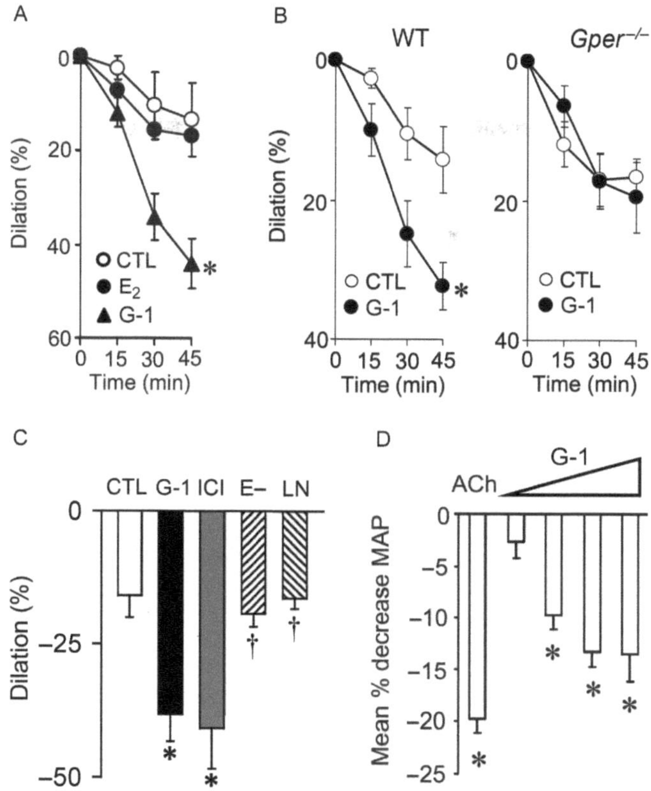

Fig. 3 Activation of GPER induces endothelium-dependent, NO-mediated vasodilation. (A, B) The GPER-selective agonist G-1 (3 μmol/L) induces time-dependent vasodilation in murine carotid arteries, which is more potent than the response to 17β-estradiol (E_2, 3 μmol/L (A)). The vasodilator response to G-1 (3 μmol/L) is absent in *Gper*$^{-/-}$ mice, confirming the selectivity of the GPER agonist (B). (C) In porcine coronary arteries, the GPER agonists G-1 (3 μmol/L) and ICI 182,780 (ICI, 3 μmol/L) induce maximal vasodilation within 40 min. The acute vasodilator response is abolished after removal of the endothelium (E-) or pretreatment with the NO synthase inhibitor L-NAME (LN). (D) Lowering effects on mean arterial blood pressure (MAP) of increasing concentrations of G-1 (41.2 ng/kg, 412 ng/kg, 4.12 μg/kg, 20.6 μg/kg) infused into normotensive male Sprague–Dawley rats. For comparison, the effect of acetylcholine (ACh, 30 ng/kg) is shown. *$p<0.05$ vs CTL (control, ethanol 0.3%), †$p<0.05$ vs G-1. *WT*, wild-type mice. *Panels A, B, and D are reproduced from* Haas, E., Bhattacharya, I., Brailoiu, E., Damjanovic, M., Brailoiu, G. C., Gao, X., Mueller-Guerre, L., Marjon, N. A., Gut, A., Minotti, R., Meyer, M. R., Amann, K., Ammann, E., Perez-Dominguez, A., Genoni, M., Clegg, D. J., Dun, N. J., Resta, T. C., Prossnitz, E. R., & Barton. M. *(2009). Regulatory role of G protein-coupled estrogen receptor for vascular function and obesity.* Circulation Research, 104, *288–291, ©2009 by the American Heart Association, with permission of the publisher. Panel C is adapted from Meyer, M. R., Baretella, O., Prossnitz, E. R., & Barton, M. (2010). Dilation of epicardial coronary arteries by the G protein-coupled estrogen receptor agonists G-1 and ICI 182,780.* Pharmacology, 86, *58–64. Copyright ©2010, Karger Publishers, Basel, Switzerland.*

endoplasmic reticulum, which may facilitate interaction with other signaling proteins (Gaudet, Cheng, Christensen, & Filardo, 2015).

4. ESTROGENS AS MODULATORS OF VASCULAR HOMEOSTASIS

One prototypical vascular action of endogenous estrogens that was first described in healthy men in 1939 (Reynolds & Foster, 1939) is their rapid vasodilator effect (Barton & Meyer, 2009). In premenopausal women, blood pressure is lower than in age-matched men, and inversely correlated with circulating 17β-estradiol levels during the menstrual cycle (Barton & Meyer, 2009; Burt et al., 1995; Minson, Halliwill, Young, & Joyner, 2000). Similarly, elevated estrogen levels during pregnancy contribute to maintaining normotension despite increases in plasma volume and cardiac output (Rang, Wolf, Montfrans, & Karemaker, 2002). In the first years following menopause, blood pressure progressively increases, which results in a higher prevalence of hypertension among women in the seventh decade of life when compared with age-matched men (Burt et al., 1995; Landahl, Bengtsson, Sigurdsson, Svanborg, & Svardsudd, 1986). These findings are compatible with a continuous vasodilator effect mediated by endogenous estrogens, which has been confirmed experimentally in numerous studies (Barton & Meyer, 2009; Meyer, Baretella, Prossnitz, & Barton, 2010). More recently, the GPER-selective agonist G-1 was shown to also induce rapid vasodilation (Fig. 3; Haas et al., 2009), yet the mechanisms involved in GPER-dependent rapid vasodilatory actions elicited by estrogens and related compounds have evolved only recently (Menazza & Murphy, 2016; Meyer et al., 2010; Meyer, Field, Kanagy, Barton, & Prossnitz, 2012; Meyer, Fredette, Barton, & Prossnitz, 2015; Meyer, Prossnitz, & Barton, 2011).

4.1 Endothelium-Dependent Vasodilation and Nitric Oxide

Intact endothelium-dependent vasomotion is a prerequisite of vascular health, and it becomes abnormal and propagates organ injury in diseases such as arterial hypertension, diabetes, chronic kidney disease, and aging (Barton, 2010; Barton, 2013c; Gimbrone & Garcia-Cardena, 2016; Vanhoutte et al., 2016)—the latter showing a pronounced, estrogen-dependent component in postmenopausal women (Barton et al., 2016). Shortly after Furchgott's seminal observation that endothelial cells release a relaxing factor in response to acetylcholine (Furchgott & Zawadzki, 1980) and its subsequent identification as nitric oxide (NO) (Furchgott, 1988; Ignarro, Buga, Wood,

Byrns, & Chaudhuri, 1987), gender differences in endothelium-dependent responses were described (Brandes & Mügge, 1997; Gisclard, Miller, & Vanhoutte, 1988; Maddox, Falcon, Ridinger, Cunard, & Ramwell, 1987; Miller & Vanhoutte, 1990). Accordingly, estrogen treatment of ovariectomized animals for 96 h in vivo enhanced acetylcholine-induced, endothelium-dependent vasodilation in isolated arteries (Gisclard et al., 1988). Similar estrogen-mediated effects potentiating endothelium-dependent relaxation and/or NO formation were observed in human endothelial cells (Caulin-Glaser, Garcia-Cardena, Sarrel, Sessa, & Bender, 1997), in human coronary arteries in vitro (Barton, Cremer, & Mügge, 1998), in human peripheral arteries (Gerhard et al., 1998; Lieberman et al., 1994), in internal mammary arteries (Nechmad et al., 1998), in small resistance arteries (Kublickiene et al., 2008), and in coronary arteries of women (Collins et al., 1995; Reis et al., 1998; Riedel & Mügge, 1993; Rosano, Sarrel, Poole-Wilson, & Collins, 1993) and men in vivo (Karas & Mendelsohn, 1998; Komesaroff et al., 2001; Reis et al., 1998). Consistent with these observations, Rubanyi and associates reported impaired endothelium-dependent vascular function (Sudhir, Chou, Messina, et al., 1997) as well as premature development of coronary atherosclerosis (Sudhir, Chou, Chatterjee, et al., 1997) in a man with a disruptive mutation of ERα/*esr1*. Importantly, estrogen also attenuates constriction to and coronary tissue concentrations of the potent endothelium-derived vasoconstrictor peptide endothelin-1 (Sudhir, Ko, et al., 1997; Webb, Ghatei, McNeill, & Collins, 2000).

In endothelial cells, estrogen rapidly initiates phosphorylation of PI_3 kinase and Akt, leading to activation of endothelial NO synthase (eNOS) and subsequent NO release (Haynes et al., 2000; Simoncini et al., 2000). These effects are in part mediated by subpopulations of ERα and possibly ERβ located to vesicular invaginations of the plasma cell membrane termed caveolae (Chambliss, Yuhanna, Anderson, Mendelsohn, & Shaul, 2002; Chambliss et al., 2000). In intact arteries, activation of ERα may be predominantly involved in rapid, estrogen-induced, NO-mediated endothelium-dependent vasodilation (Barton, Cremer, et al., 1998; Traupe et al., 2007). Interestingly, we reported functional cross talk between vascular ERs already 10 years ago showing that the nonselective ER agonist 17β-estradiol completely abrogated the NO-mediated component of the relaxation response induced by the ERα/GPER agonist propylpyrazole triol (PPT) (Meyer, Prossnitz, et al., 2011; Prossnitz & Arterburn, 2015; Traupe et al., 2007). ICI 182,780 (Faslodex™/fulvestrant), a combined ERα/ERβ downregulator (SERD) but agonist for GPER, also has rapid vasodilator effects (Guo, Razandi, Pedram, Kassab, & Levin, 2005; Meyer et al., 2010; Fig. 3), indicating a novel and additional role of GPER in mediating rapid, endothelium-dependent, NO-mediated vasodilation.

Indeed, recent studies have confirmed a key role of GPER in endothelium-dependent regulation of vascular tone, particularly involving eNOS-derived NO release (Meyer et al., 2014). Among other sources, GPER was originally cloned from cultured human endothelial cells exposed to fluid shear stress, a prototypical stimulus to activate eNOS (Takada et al., 1997). Studies in human and rat endothelial cells stimulated with the GPER-selective agonist G-1 confirmed rapid phosphorylation of eNOS activation sites as well as subsequent NO release (Fig. 4; Altmann et al., 2015; Arefin et al., 2014; Meyer et al., 2014). Recent evidence indicates that eNOS phosphorylation by estrogen involves linkage of calmodulin, an eNOS cofactor, to both ERα and GPER (Tran et al., 2016), again suggesting cross talk between ERs.

NO release induced by 17β-estradiol can be reduced by about 50% in the presence of the GPER-selective antagonist G36, indicating that a significant portion of estrogen-stimulated NO requires GPER activation in human endothelial cells (Fig. 4; Meyer et al., 2014). In intact porcine coronary arteries as well as aorta, carotid, mesenteric and uterine arteries and cerebral arterioles of rodents, G-1 initiates rapid endothelium-dependent vasodilation that is mediated by NO release (Fig. 3; Barton, Meyer, & Prossnitz,

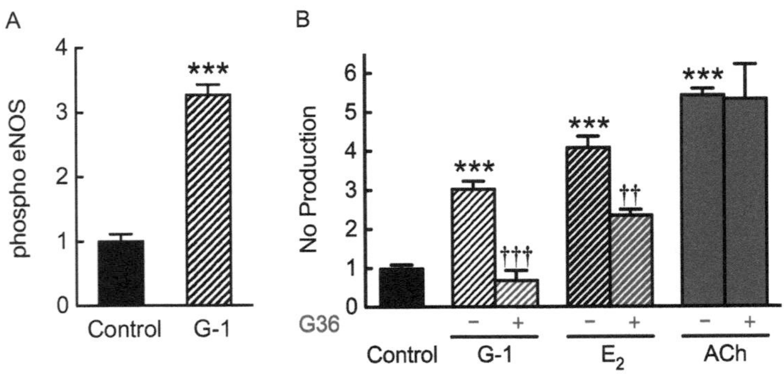

Fig. 4 GPER activates eNOS and NO release in human endothelial cells. (A) The GPER-selective agonist G-1 increases levels of activated eNOS phosphorylated on serine1177. (B) Selective (G-1) and nonselective GPER agonists (E_2, 17β-estradiol) increase endothelial NO release, as does the muscarinic receptor agonist acetylcholine (ACh). G-1-stimulated NO release is completely blocked by the GPER-selective antagonist G36, while responses to E_2 are partially reduced. $^{***}p<0.001$ vs control (DMSO 0.01%); $^{††}p<0.01$, $^{†††}p<0.001$ vs no antagonist. *Figure reproduced from Meyer, M. R., Fredette, N. C., Howard, T. A., Hu, C., Ramesh, C., Daniel, C., Amann, K., Arterburn, J. B., Barton, M., & Prossnitz, E. R. (2014). G protein-coupled estrogen receptor protects from atherosclerosis.* Scientific Reports, 4, *7564. Erratum in:* Scientific Reports, *7565 (2015) 13510.*

2013; Broughton et al., 2010; Jang, Seok, Arterburn, Olatunji, & Kim, 2013; Lindsey, Liu, & Chappell, 2014; Mata et al., 2015; Mazzuca, Mata, Li, Rangan, & Khalil, 2015; Meyer et al., 2010; Murata, Dietrich, Xiang, & Dacey, 2013; Reslan, Yin, do Nascimento, & Khalil, 2013; Tropea et al., 2015). Furthermore, basal NO release in carotid arteries is reduced in GPER-deficient but not wild-type mice fed a proinflammatory diet (Meyer et al., 2014). In contrast, basal and stimulated NO release is preserved in aortas from healthy GPER-deficient mice, but acutely impaired by the GPER-selective antagonist G15 (Meyer et al., 2012). These findings suggest the presence of chronic, GPER-dependent compensatory mechanisms which however may not be sufficient under inflammatory conditions such as atherosclerosis. NO-mediated vasodilation in response to G-1 has also been observed in human resistance arteries isolated from subcutaneous fat biopsies, which surprisingly was found to be maximal in arteries obtained from postmenopausal women (Arefin et al., 2014), suggesting the possibility that GPER-mediated signaling pathways could provide a vascular "back-up mechanism" in arteries that have lost their tone-modulating properties mediated by ovarian estrogens.

4.2 Endothelium-Independent Vasodilation and Cyclic Adenosine Monophosphate

In addition to activating eNOS/increasing NO bioactivity and facilitating endothelium-dependent vasodilation (Barton, Cremer, et al., 1998), 17β-estradiol also directly acts on the vascular smooth muscle. Twenty-five years ago, it was first demonstrated that 17β-estradiol rapidly increases cyclic adenosine monophosphate (cAMP) in human coronary arteries within 30 min of application, an effect that was paralleled by acute vasodilation of precontracted human coronary artery rings (Barton, 1994, 2016; Mügge, Riedel, Barton, Kuhn, & Lichtlen, 1993a, 1993b; Riedel & Mügge, 1993). The acute stimulatory effect on cellular cAMP levels was subsequently confirmed by studies in intact uterine tissue also showing rapid increases in cAMP content within 30 min of 17β-estradiol application (Aronica, Kraus, & Katzenellenbogen, 1994). Comparable observations were later made in pulmonary arterial vascular smooth muscle cells demonstrating increases in cAMP through nongenomic mechanisms (Farhat, Abi-Younes, Dingaan, Vargas, & Ramwell, 1996).

In the arterial vasculature, cAMP mediates activation of protein kinase A (PKA), which inhibits RhoA activity and subsequently activates myosin light chain phosphatase to induce vasodilation (Lindsey et al., 2014; Yu, Li, Klussmann, Stallone, & Han, 2014). Moreover, rapid vasorelaxation in response to 17β-estradiol involves changes in calcium and potassium fluxes (White, 2002) that may also be related to the increases in cyclic nucleotides. GPER is involved in endothelium-independent vasodilation, as shown in porcine coronary and rat mesenteric arteries and cerebral arterioles using its selective agonist G-1 (Lindsey et al., 2014; Murata et al., 2013; Yu et al., 2014, 2011). These responses involve both increases in potassium efflux via large-conductance calcium- and voltage-activated potassium (BK_{Ca}) channels (Yu et al., 2011) as well as cAMP production and activation of PKA (Lindsey et al., 2014; Yu et al., 2014). Prossnitz and associates (Revankar et al., 2005) and Evans et al. (2016) have demonstrated that GPER activation increases cAMP production also in nonvascular cells.

4.3 Endothelium-Dependent Vasoconstriction and Vasoconstrictor Prostanoids

Endothelium-dependent vasodilator tone is counterbalanced by endothelium-dependent vasoconstriction to maintain vascular tone (Barton, 2013c) through activity and/or release of endothelium-derived contracting factor (EDCF; Barton, 2011), which has been identified as cyclooxygenase-derived vasoconstrictor prostanoids (Barton, 2011, 2013c; Barton, Baretella, & Meyer, 2012b; Vanhoutte et al., 2016). The bioactivity of EDCF is increased during ischemia and in almost all disease conditions affecting the cardiovascular system (Barton, 2013c; Barton et al., 2012b; Vanhoutte et al., 2016). It is noteworthy that recent evidence indicates that NO itself may cause endothelium-dependent vasoconstriction (Vanhoutte et al., 2016). Endothelium-dependent modulation of vascular tone by estrogens is not limited to the regulation of NO bioactivity. In the early 1980s it was reported that estradiol stimulates cyclooxygenase and prostacyclin synthetase activity in vascular cells (Chang, Nakao, Orimo, & Murota, 1980) while few years later, Vanhoutte and associates reported effects of in vivo treatment with estrogen for 4 to 14 days on endothelium-dependent and -independent contractile responses in isolated arteries (Gisclard, Flavahan, & Vanhoutte, 1987; Miller & Vanhoutte, 1990). Subsequently, Davidge and coworkers reported that estrogen inhibits the generation of the cyclooxygenase/prostaglandin H synthase products thromboxane A_2 (TXA_2) and prostacyclin, and their stable metabolites TXB_2 and $PGF_1\alpha$ in endothelial cells in vitro consistent with reducing EDCF release (Stewart, Zhang, & Davidge, 1999).

As GPER was originally cloned from shear stress-exposed human endothelial cells (Takada et al., 1997) and shown to increase bioactivity of EDRF/endothelium-derived NO (Altmann et al., 2015; Arefin et al., 2014; Broughton et al., 2010; Jang et al., 2013; Lindsey et al., 2014;

Mata et al., 2015; Meyer, Amann, et al., 2012; Meyer et al., 2010, 2014; Murata et al., 2013; Tropea et al., 2015), it was a logical next step to study whether GPER expression or its activation also affects formation or activity of EDCF. Experiments in young and otherwise healthy male mice deficient in GPER demonstrated that constitutive expression of this receptor is required to suppress endothelium-dependent vasoconstriction (Meyer, Amann, et al., 2012). Consequently, EDCF-mediated contractions are enhanced in GPER-deficient mice, as are those induced by the TXA_2 analog U46619 (Fig. 5; Meyer, Amann, et al., 2012). In subsequent studies we found, studying female mice fed a diet causing generalized vascular inflammation, that the inhibition of increased EDCF-dependent vasoconstriction by endogenous estrogens involves GPER (Meyer et al., 2015). Moreover, we could demonstrate that in human endothelial cells tumor necrosis factor α (TNFα)-stimulated TXA_2 production is prevented by estrogen treatment, an effect that could be blocked by the GPER-selective antagonist G36. These findings are consistent with the notion that GPER is a new inhibitory regulator of EDCF production and activity in vitro and in vivo (Fig. 5; Meyer et al., 2015).

4.4 Adipose-Derived Vasoconstriction and Perivascular Adipose Tissue

Similar to endothelial cells, perivascular adipose tissue (PVAT) is an important source of vasoactive, growth-promoting, and proinflammatory molecules (Barton, 2012a; Barton, Baretella, & Meyer, 2012a), which we have recently collectively termed "*modifiable molecular mediators*" (Barton, 2014; Barton et al., 2016). A role of PVAT as a regulator of vascular tone was first proposed by Soltis in Cassis in 1991 (Löhn et al., 2002; Meyer, Fredette, Barton, & Prossnitz, 2013; Soltis & Cassis, 1991). After the discovery of "*adipose-derived relaxing factor*" (ADRF) by Löhn and coworkers in 2002 (Löhn et al., 2002), research for more than a decade remained focused on PVAT-dependent vasodilation (Gollasch, 2012). Only recently it has become appreciated that PVAT is also an important source of *vasoconstricting* factors that are increasingly formed under disease conditions such as obesity and diabetes (Mendizabal, Llorens, & Nava, 2013; Meyer et al., 2013; Owen et al., 2013).

Adult mice in which GPER has been deleted develop abdominal obesity including increased amounts of PVAT (Davis et al., 2014; Ford et al., 2011; Haas et al., 2009; Meyer et al., 2013; Sharma et al., 2013). While in wild-type mice PVAT had no effect on serotonin-induced contractions, we observed that contractions were markedly enhanced by PVAT in GPER-

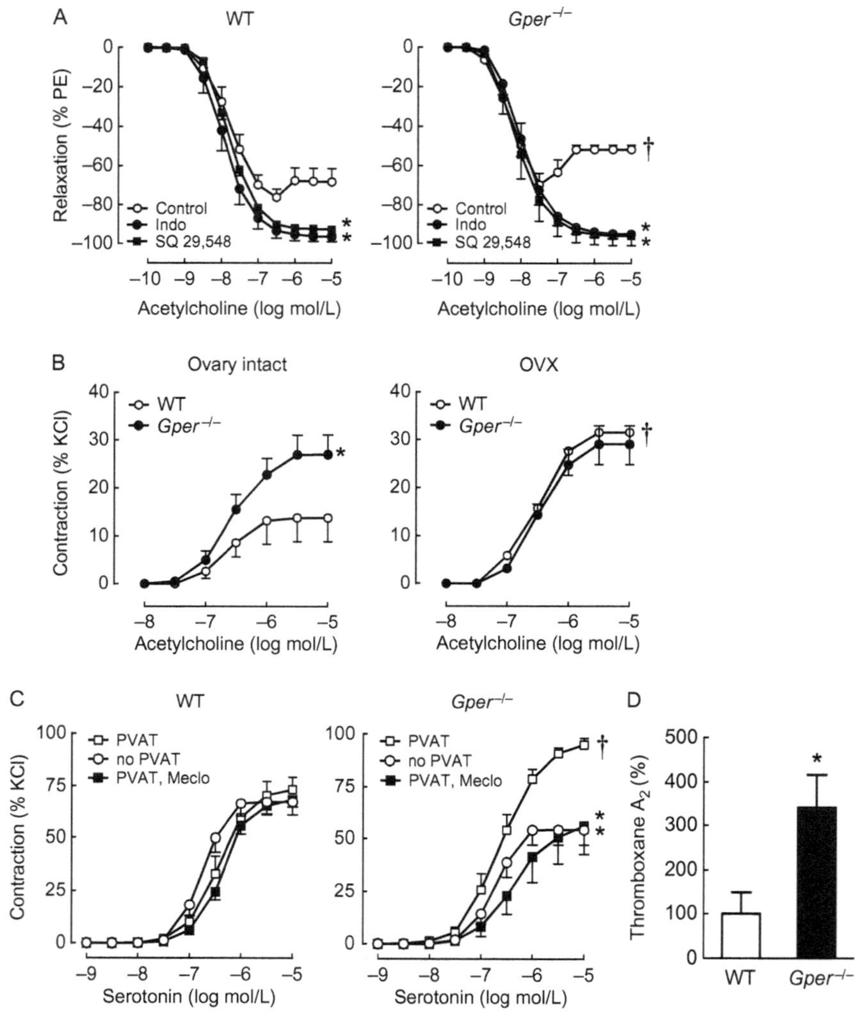

Fig. 5 GPER regulates endothelium-dependent (EDCF) and adipose-dependent (ADCF) vasoconstriction to prostanoids. (A) Responses to "endothelium-derived contracting factor" (EDCF) stimulated by acetylcholine are potentiated in the aorta of *Gper*$^{-/-}$ mice. Contractions are completely blocked by the cyclooxygenase inhibitor indomethacin (Indo) and the thromboxane-prostanoid receptor antagonist SQ 29,548, confirming the identity of EDCF as vasoconstrictor prostanoids. $^*p < 0.05$ vs control, $^{\dagger}p < 0.05$ vs WT (wild-type mice). (B) In ovary-intact mice, acetylcholine-induced, endothelium-dependent, contractions are potentiated in *Gper*-deficient carotid arteries. Removal of endogenous estrogens by ovariectomy (OVX) abolishes their inhibitory effects on vasoconstriction only in WT mice, indicating GPER-dependent signaling. $^*p < 0.05$ vs WT, $^{\dagger}p < 0.05$ vs ovary-intact. (C) Deletion of the *Gper* gene increases serotonin-induced contractions in aorta with perivascular adipose tissue (PVAT) intact. This "adipose-derived contracting factor" (ADCF) is blocked by the cyclooxygenase inhibitor meclofenamate (Meclo), confirming that ADCF

deficient mice (as well as in mice with diet-induced obesity), and that this augmentation was abolished by PVAT removal (Fig. 5; Meyer et al., 2013). Further experiments demonstrated that the factor released by PVAT is an arachidonic acid-derived product of cyclooxygenase and chemically identical with "endothelium-derived contracting factor" (EDCF, ie, prostanoids such as TXA_2) (Meyer et al., 2013; Fig. 5D). Accordingly, we termed this newly discovered contracting factor "*adipose-derived contracting factor*" (ADCF; Meyer et al., 2013). Modulating perivascular fat, which accelerates atherosclerosis (Ohman et al., 2011), as well as the increased adipose-derived vasoconstriction under disease conditions represents an attractive novel target for the treatment of cardiovascular diseases.

4.5 Vascular Smooth Muscle Cell Growth and In Vitro Atherogenesis

Estrogens have long been known to inhibit vascular smooth muscle cell proliferation (Farhat, Lavigne, & Ramwell, 1996; Foegh, Zhao, Farhat, & Ramwell, 1995; Vargas, Wroblewska, Rego, Hatch, & Ramwell, 1993) and collagen formation (Beldekas, Smith, Gerstenfeld, Sonenshein, & Franzblau, 1981), cellular processes that have be ascribed to the modes of action of estrogen-mediated inhibition of atherogenesis (Mendelsohn & Karas, 1999). Of note, the proliferation response after balloon angioplasty is suppressed by estrogen—however, this effect is maintained in mice lacking ERα (Iafrati et al., 1997), ERβ (Karas et al., 1999), or both (Karas et al., 2001) indicating that other additional targets and mechanisms must be involved. It has been 10 years since we detected high expression of GPER in intact human arteries (Fig. 2; Haas et al., 2007; Meyer et al., 2006), and we demonstrated rapid effects of its activation in human vascular smooth muscle

Fig. 5—Cont'd activity is mediated by vasoconstrictor prostanoids. $^{*}p<0.05$ vs PVAT, $^{\dagger}p<0.05$ vs WT. (D) Serotonin-stimulated ADCF release from aortic PVAT of WT and *Gper*$^{-/-}$ mice measured as the stable thromboxane A_2 metabolite, thromboxane A_2. $^{*}p<0.05$ vs WT. *Panel A is reproduced from Meyer, M. R., Amann, K., Field, A. S., Hu, C., Hathaway, H. J., Kanagy, N. L., …, Prossnitz, E. R. (2012). Deletion of G protein-coupled estrogen receptor increases endothelial vasoconstriction.* Hypertension, *59, 507–512, ©2012 by the American Heart Association, with permission from the publisher. Panel B is reproduced from Meyer, M. R., Fredette, N. C., Barton, M., & Prossnitz, E. R. (2015). G protein-coupled estrogen receptor inhibits vascular prostanoid production and activity.* Journal of Endocrinology, *227, 61–69, ©2015 BioScientifica Ltd. Panels C and D are adapted from Meyer, M. R., Fredette, N. C., Barton, M., & Prossnitz, E. R. (2013). Regulation of vascular smooth muscle tone by adipose-derived contracting factor.* PLoS One, *8, e79245.*

cells using selective ligands in vitro (Bhattacharya et al., 2007; Haas et al., 2009, 2007; Meyer et al., 2006).

Given that inhibitory effects of estrogen on vascular smooth muscle proliferation in vivo are maintained in mice deficient either in ERα or ERβ (Iafrati et al., 1997; Karas et al., 1999), we hypothesized that GPER activation might inhibit smooth muscle cell proliferation in vitro (Meyer & Barton, 2009; Meyer et al., 2006, 2008). Indeed, in human arterial vascular smooth muscle cells, activation of GPER by the selective agonist G-1 (Dennis et al., 2009) or the SERD ICI 182,780 (Faslodex™/fulvestrant) was found to potently inhibit serum-stimulated cell proliferation (Haas et al., 2009; Fig. 6). These findings were subsequently confirmed in human and porcine coronary vascular smooth muscle cells by Han and associates (Li et al., 2013). Suppression of vascular cell proliferation in the presence of GPER or upon pharmacological activation of the receptor can also be observed in vivo in ovariectomized female mice with atherosclerosis: atheroma formation is enhanced when GPER is deleted, whereas treatment with the GPER-selective agonist G-1 inhibits it (Barton & Prossnitz, 2015; Meyer et al., 2014; Fig. 7). In line with these findings, inhibition of smooth muscle cell growth can be achieved using green tea catechins (Locher, Emmanuele,

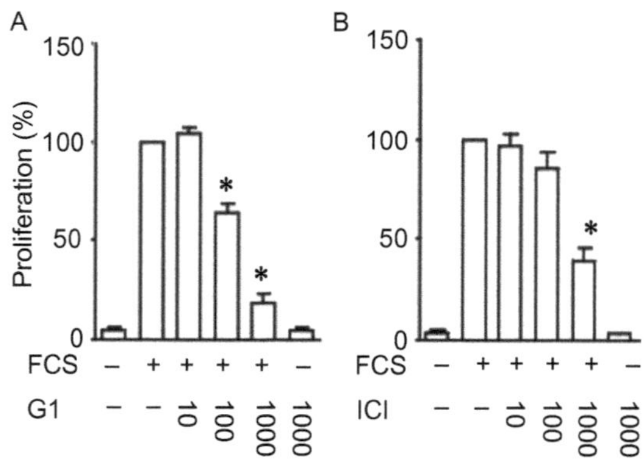

Fig. 6 Effects of GPER activation on human vascular smooth muscle cell growth in an in vitro model of atherogenesis. (A) The GPER-selective agonist G-1 (10, 100, and 1000 nmol/L) inhibits serum-stimulated proliferation of vascular smooth muscle cells. (B) The selective estrogen receptor downregulator ICI 182,780 (Faslodex™/fulvestrant) also inhibits serum-stimulated human vascular smooth muscle cell growth, albeit at higher concentrations. *$p < 0.05$ vs solvent. *Figure ©2009 by the American Heart Association and reproduced from Haas, E., Bhattacharya, I., Brailoiu, E., Damjanovic, M., Brailoiu, G. C., Gao, X., …, Barton, M. (2009). Regulatory role of G protein-coupled estrogen receptor for vascular function and obesity.* Circulation Research, 104, *288–291, ©2009 by the American Heart Association, with permission of the publisher.*

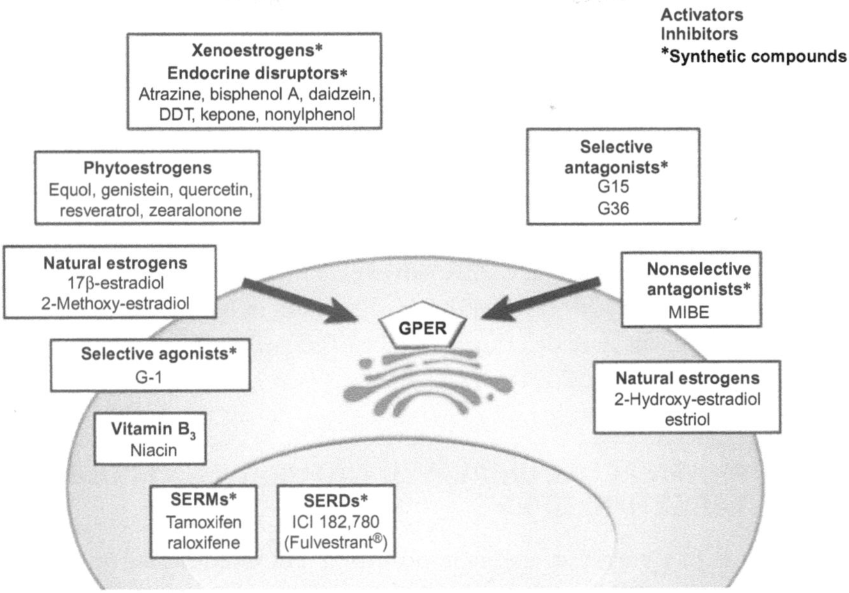

Fig. 7 Selective and nonselective activators and inhibitors of GPER signaling. GPER is activated by natural estrogens including endogenous estrogen (17β-estradiol or 2-methoxyestradiol), dietary isoflavones and phytoestrogens (quercetin, genistein, green tea catechins, resveratrol), as well as highly stable environmental chemicals ("endocrine disruptors" or "environmental pollutants") also known as xenoestrogens (atrazine, bisphenol A, nonylphenol, and the pesticide p,p0-dichlorodiphenyltrichloroethane (DDT)). Zearalenone is plant (fungus)-derived and thus by definition a phytoestrogen as shown; due to its toxicity as an environmental mycotoxin it qualifies as a xenoestrogen. Similarly, daidzein—which like genistein is a soy-derived isoflavone—may also be considered an environmental xenoestrogen. Equol, a metabolite of daidzein, is a phytoestrogen-derivative rather than a pure phytoestrogen. GPER is also activated by niacin (vitamin B_3), selective estrogen receptor modulators (SERMs), and selective estrogen receptor downregulators (SERDs). SERMs and SERDs exhibit inhibitory activity on the "classical" estrogen receptors, ERα and ERβ. GPER is also activated by the selective synthetic small molecule G-1, while GPER-selective small molecules G15 and G36 and the non-ER-selective compound ethyl 3-[5-(2-ethoxycarbonyl-1-methylvinyloxy)-1-methyl-1H-indol-3-yl] but-2-enoate (MIBE) inhibit GPER. The catechol estrogen metabolite 2-hydroxy-17β-estradiol (2-OH-E_2) and the 16-OH-DHEAS metabolite estriol (E_3) also display inhibitor activity toward GPER. *Figure © 2015 and reproduced from Barton, M., & Prossnitz, E. R. (2015). Emerging roles of GPER in diabetes and atherosclerosis.* Trends in Endocrinology & Metabolism, 26, *185–192, with permission of Cell Press.*

Suter, Vetter, & Barton, 2002) which activate ERα, ERβ, and GPER (Moreno-Ulloa et al., 2015), genistein (Fujio, Yamada, Takahashi, & Shibata, 1993; Lee, Wang, Lanzillo, & Fanburg, 1994; Somjen et al., 2002), resveratrol (El-Mowafy & White, 1999; Mizutani, Ikeda, &

Yamori, 2000), niacin (Wulfroth & Grunwald, 1989), and selective estrogen receptor modulators (SERMs) such as tamoxifen (Grainger, Weissberg, & Metcalfe, 1993), raloxifene (Takahashi et al., 2003), and idoxifene (Yue et al., 2000), which are all known to mediate at least part of their effects through pathways involving GPER (Barton, 2016; Prossnitz & Arterburn, 2015). Accordingly, GPER signaling represents an important new mechanism that inhibits vascular smooth muscle cell proliferation and thus atherosclerosis. Finally, activation of GPER can explain why previous studies found that part of the salutary effects of estrogen on smooth muscle cell proliferation are maintained even after deletion of either ERα and/or ERβ (Iafrati et al., 1997; Karas et al., 1999, 2001).

5. PHARMACOLOGICAL AND NATURAL MOLECULES TARGETING GPER

Although a variety of naturally occurring and synthetic substances are capable to activate ERs and initiate vascular responses, it is important to consider the different degree by which each molecule activates or inhibits ERα, ERβ, or GPER (Barton, 2016). 17β-Estradiol, the predominant and physiologically most relevant estrogen in premenopausal women, nonselectively activates all three known vascular estrogen receptors inducing diverse rapid and chronic effects, including acute vasodilation (Barton, 2016). However, 17β-estradiol is not only produced by the ovaries but also synthesized locally in the vascular wall through conversion of androgen precursors by the enzyme aromatase (Fig. 1; Harada et al., 1999; Meyer et al., 2006, 2008). Such localized vascular production of estrogen not only is relevant for the regulation of vascular tone specifically in blood vessels of young men (Lew, Komesaroff, Williams, Dawood, & Sudhir, 2003) but also contributes to inhibition of atherosclerosis (Mukherjee, Dinh, Chaudhuri, & Nathan, 2002). Estrone, the most abundant estrogen in postmenopausal women (Gruber, Tschugguel, Schneeberger, & Huber, 2002), displays lower binding affinities to ERα, ERβ, and GPER, and its vascular actions are less clear. Additional natural and man-made substances targeting GPER with agonistic activity include 2-methoxyestradiol, phytoestrogens such as genistein, daidzein, equol, resveratrol, green tea catechins, niacin, as well as endocrine disruptors/xenoestrogens such as bisphenol A and the pesticide DDT (Barton, 2016; Fig. 7).

To facilitate assigning individual functions to ERα, ERβ, and GPER and to increase the understanding of how these receptors contribute to estrogen biology, ER-selective ligands such as PPT (Stauffer et al., 2000) and diarylpropionitrile (DPN) (Meyers et al., 2001), as well as the GPER-selective agonist G-1 (Bologa et al., 2006) and GPER-selective antagonists G15 (Dennis et al., 2009) and G36 (Dennis et al., 2011) have been developed (Fig. 7), the latter displaying an improved ER counterselectivity. Although originally introduced as selective agonist for ERα, PPT has since been shown to also activate ERβ and GPER (Petrie et al., 2013), which complicates its use as pharmacological tool.

Over the past decade, several compounds previously introduced as therapeutic drug in neoadjuvant cancer therapy have been identified to activate GPER (Filardo et al., 2000; Revankar et al., 2005; Thomas et al., 2005). In particular, these include SERMs, such as hydroxy tamoxifen and raloxifene, as well as SERDs such as Faslodex™/fulvestrant (ICI 182,780; Fig. 7). The discovery that clinically approved drugs are capable of also activating GPER requires us to rethink and extend our therapeutic rationale when applying them to patients ("reverse translational medicine"; Barton, 2016), in particular for the treatment of hormone-sensitive tumors such as breast cancer. This also indicates that older studies investigating vascular function utilizing SERMs and SERDs may in fact point at GPER-dependent effects (Barton, 2012b, 2016). Taken together, the availability of multiple compounds with agonistic or antagonistic activity toward GPER (Fig. 7) has accelerated, yet in some instances also complicated the evaluation of estrogen receptor function in vascular health, disease, and therapy.

6. GPCR-DEPENDENT ESTROGEN ACTIONS: THERAPEUTIC POTENTIAL FOR VASCULAR DISEASE

Previous studies have shown that the "classical" estrogen receptors, ERα and ERβ, do not fully explain the therapeutic efficacy of estrogens to inhibit smooth muscle cell proliferation (Iafrati et al., 1997; Karas et al., 1999), atherogenesis (Hodgin et al., 2001; Villablanca et al., 2009), or vasoconstriction (Hill, Gebre, Schlicker, Jordan, & Necessary, 2010), suggesting the involvement of additional targets and yet unknown mechanisms. As discussed earlier, estrogen signaling via GPCRs (the first one identified being GPER) appears to play a significant role in cardiovascular disease processes. Work in recent years has shown that inflammatory activation can be considered a "common ground" or unifying mechanism for most prevalent noncommunicable cardiovascular, metabolic, and renal diseases as well as aging (Barton et al., 2016). These conditions not only make up for the majority of morbidity and mortality in Western countries but are also responsible for the ever-increasing health costs associated with it (Barton et al., 2016). We will discuss below the evidence for GPCR-dependent estrogen signaling in the most prevalent risk factors and disease conditions affecting the cardiovascular system. The roles of ERα and ERβ in the protection from cardiovascular disease have been reviewed elsewhere (Burns & Korach, 2012; Deroo & Korach, 2006; Hewitt et al., 2016; Orshal & Khalil, 2004).

6.1 Arterial Hypertension

Arterial hypertension, a key risk factor for atherosclerosis and coronary artery disease (Barrett-Connor, 2013; Barton et al., 2016), is a highly prevalent disease that increases with age, obesity, diabetes, renal disease, and after menopause and currently affects over 1 billion patients globally (Barton et al., 2016). Despite intensive efforts on prevention, the prevalence of arterial hypertension and the associated morbidity and mortality continues to increase, partly due to the overall aging of the world population (Barton et al., 2016). Estrogens have been implicated in the lower blood pressure in premenopausal women as compared to men of similar age (Barton & Meyer, 2009), and their blood pressure-lowering effects have been attributed to the potent vasodilator activity which is observed in both men and women (Meyer et al., 2006; Meyer, Prossnitz, et al., 2011; Mügge et al., 1993b). While the roles of ERα and ERβ in hypertension have been well characterized (Barton & Meyer, 2009; Burns & Korach, 2012; Deroo & Korach, 2006; Hewitt et al., 2016; Orshal & Khalil, 2004), particularly work from Lindsey, Chappell, and colleagues has demonstrated that GPER also plays a role in the antihypertensive effects of estrogens, and that selective activation of GPER—which causes rapid vasodilation (Haas et al., 2009)—lowers blood pressure and provides protection from hypertensive organ injury in the kidney and heart (Lindsey et al., 2009; Lindsey, Yamaleyeva, Brosnihan, Gallagher, & Chappell, 2011). Most recent work of Lindsey's group has shown that the GPER-selective agonist G-1 may afford organ protection in experimental hypertension even without reducing systemic blood pressure (Liu et al., 2016) similar to other drugs targeting GPCRs such as endothelin receptor antagonists (ERAs) (Barton, 2008; Barton et al., 1998). Thus, targeting GPER to increase NO bioactivity (a mechanism by which also ERAs exert part of their vasoprotective effects (Barton, Haudenschild, et al., 1998)) appears to be a promising new approach that warrants further studies, possibly also in patients.

6.2 Dyslipidemia

Part of the protective effects of estrogens on atherogenesis in premenopausal women have been associated with their ability to raise high-density lipoprotein (HDL) cholesterol and lower low-density lipoprotein (LDL) cholesterol levels (Barrett-Connor, 2013; Barton, 2013a). While some of the lipid-modulating effects are mediated by either ERα or ERβ (Barton, 2013a), it had been unclear whether and to what extent other targets may also be involved.

GPER is expressed and functional in the liver (Ikeda et al., 2012), suggesting a regulatory role in hepatic lipid metabolism. Indeed, deletion of GPER in male mice results in dyslipidemia, which however becomes apparent only with advanced age (Sharma et al., 2013). Similarly, deletion of GPER in female mice raises total and LDL cholesterol levels, an effect comparable to that of ovariectomy (Meyer et al., 2014). Salutary effects of GPER on dyslipidemia were confirmed and extended by studies by Hussain et al. who found that GPER not only reduces LDL receptor protein expression but also facilitates LDL uptake by hepatocytes (Hussain et al., 2015). Work by Maggiolini and coworkers suggests yet another role of GPER in lipid metabolism as these investigators found that estrogen stimulates fatty acid synthase via GPER activation in cancer cells (Santolla et al., 2012), and that GPER serves as a receptor for niacin (vitamin B_3) (Santolla et al., 2014), which has been used as a clinical drug for the treatment of dyslipidemia for over half a century (Barton, 2016). Whether fatty acid synthase activation by GPER also plays a role in estrogen-dependent protection from vascular disease is currently not known.

6.3 Obesity and Diabetes

Obesity accelerates atherogenesis in young men but not women (McGill et al., 2002), indicating a protective effect of endogenous estrogens. Obesity is a highly prevalent disease globally affecting 640 million patients aged 20 and older (Barton, 2012a, 2013c; NCD Risk Factor Collaboration, 2016; Ng et al., 2014), with numbers expected to increase further due to the growing world population (Barton, 2010; Barton et al., 2016; Heidenreich et al., 2011; Olshansky et al., 2005; Smith, 2016; Stewart, Cutler, & Rosen, 2009). Its prevalence is highest among the elderly (Bramlage et al., 2004; Sakurai et al., 2010) but it increasingly affects children and adolescents (Barton, 2012a). Obesity represents the most prevalent and important preventable cause of arterial hypertension and diabetes and thus also coronary artery disease and stroke (Barton et al., 2016). In addition, obesity adversely affects diverse body functions, leading to fatty liver disease with consecutive abnormalities in metabolic homeostasis, insulin sensitivity, and lipid metabolism, increased vasoconstriction, renal disease, inflammation, and musculoskeletal injury (Barton, 2010; Barton, 2013c; Barton et al., 2016).

Endogenous estrogens in premenopausal women protect from obesity, insulin resistance, and diabetes, evidenced by their increased prevalence after menopause (Barton & Meyer, 2009) and in women with dysregulated ovarian steroid synthesis due to the polycystic ovary syndrome (Barton, 2016; McSweeney et al., 2016). While some of the salutary estrogen-mediated effects on obesity and insulin resistance are mediated by ERα or ERβ (Hevener, Clegg, & Mauvais-Jarvis, 2015; Louet, LeMay, & Mauvais-Jarvis, 2004; Meyer, Clegg, Prossnitz, & Barton, 2011; Nadal et al., 2011; Tiano & Mauvais-Jarvis, 2012), others involve additional estrogen targets (Meyer, Clegg, et al., 2011). We (Haas et al., 2009; Meyer et al., 2013) and others (Davis et al., 2014; Ford et al., 2011; Sharma et al., 2013) found that deletion of GPER results in abdominal obesity, suggesting a protective role of this receptor. Studies by the groups of Nadal (Nadal, Alonso-Magdalena, Soriano, Quesada, & Ropero, 2009; Nadal, Alonso-Magdalena, Soriano, Ropero, & Quesada, 2009), Leeb-Lundberg (Mårtensson et al., 2009), Mauvais-Jarvis (Liu et al., 2013; Mauvais-Jarvis, Clegg, & Hevener, 2013; Tiano et al., 2011; Tiano & Mauvais-Jarvis, 2012), Clegg (Davis et al., 2014; Hevener et al., 2015), and Prossnitz (Sharma et al., 2013; Sharma & Prossnitz, 2011) have now provided compelling evidence for roles of GPER in glucose and insulin metabolism, insulin secretion and sensitivity, and pancreatic beta islet cell survival (Barton & Prossnitz, 2015; Prossnitz & Barton, 2011, 2014).

6.4 Coronary Artery and Cerebrovascular Disease

Administration of the natural estrogen 17β-estradiol acutely improves exercise capacity (Rosano et al., 1993) and increases coronary artery diameter in postmenopausal women with coronary artery disease (Riedel & Mügge, 1993). Similar observatios have been made in primates with coronary atherosclerosis (Williams, Adams, Herrington, & Clarkson, 1992), suggesting that the coronary vasculature retains its sensitivity to the vasodilating effects of estrogen even in the presence of atherosclerotic vascular disease. Indeed, vasodilator effects of 17β-estradiol, a nonselective ER agonist (Prossnitz & Barton, 2011), are preserved in epicardial coronary arteries obtained from male or female patients with coronary artery disease (Karas & Mendelsohn, 1998; Mügge et al., 1993a). PPT, an agonist for ERα (Stauffer et al., 2000) that also activates GPER (Petrie et al., 2013), causes rapid dilation of porcine coronary arteries (Traupe et al., 2007) and small arteries of patients with coronary artery disease (Cruz, Agewall, Schenck-Gustafsson, & Kublickiene, 2008). The dilator response is comparable to that caused by the GPER agonists G-1 and Faslodex™/fulvestrant (Han & White, 2014; Meyer et al., 2010; Yu et al., 2014, 2011). Kublickiene and coworkers also found that the phytoestrogens and GPER agonists genistein and resveratrol induced similar relaxations in arteries of these patients (Cruz et al., 2008).

Inflammation represents one of the main underlying factors promoting atherosclerosis progression and plaque rupture (Libby, 2006, 2013). The protective effects of estrogen on inflammation in atherosclerosis have been recently reviewed (Barton, 2013a; Straub, 2007). Increased inflammation present in advanced atherosclerotic lesions of postmenopausal women can be alleviated by estrogen therapy (Burke et al., 2001), an effect that appears to involve at least in part GPER-dependent mechanisms (Meyer et al., 2014; Pelekanou et al., 2016). Since the protective effects of 17β-estradiol on advanced experimental atherosclerosis are only in part blocked in ovariectomized mice lacking ERα (Hodgin et al., 2001) and since estrogen inhibition of fatty streak formation and recruitment of inflammatory cells is unaffected by deletion of ERα (Villablanca et al., 2009), we had postulated already a decade ago that other estrogen targets must be involved (Meyer & Barton, 2009; Meyer et al., 2006, 2008). Moreover, Oparil and associates recently found age-dependent effects of estrogen on proinflammatory responses in vascular smooth muscle cells and macrophages, and that selective GPER activation was able to recapitulate or even exceed the beneficial effects of estrogen (Bowling et al., 2014; Giordano et al., 2015)

We have recently reproduced the effects of human menopause on atherosclerosis in a model of diet-induced atherogenesis, demonstrating increased plaque formation and inflammation (Meyer et al., 2014). Interestingly, deletion of GPER in premenopausal (ovary-intact) mice had effects comparable to menopause on atherosclerosis progression, inflammation, and dyslipidemia (Fig. 8), which were further aggravated by ovariectomy in GPER-deficient mice (Meyer et al., 2014). This indicates that in females GPER mediates both estrogen-dependent and -independent chronic vascular functions. Notably, treatment with the GPER-selective agonist G-1 markedly reduced atherosclerotic plaque formation and inflammation in

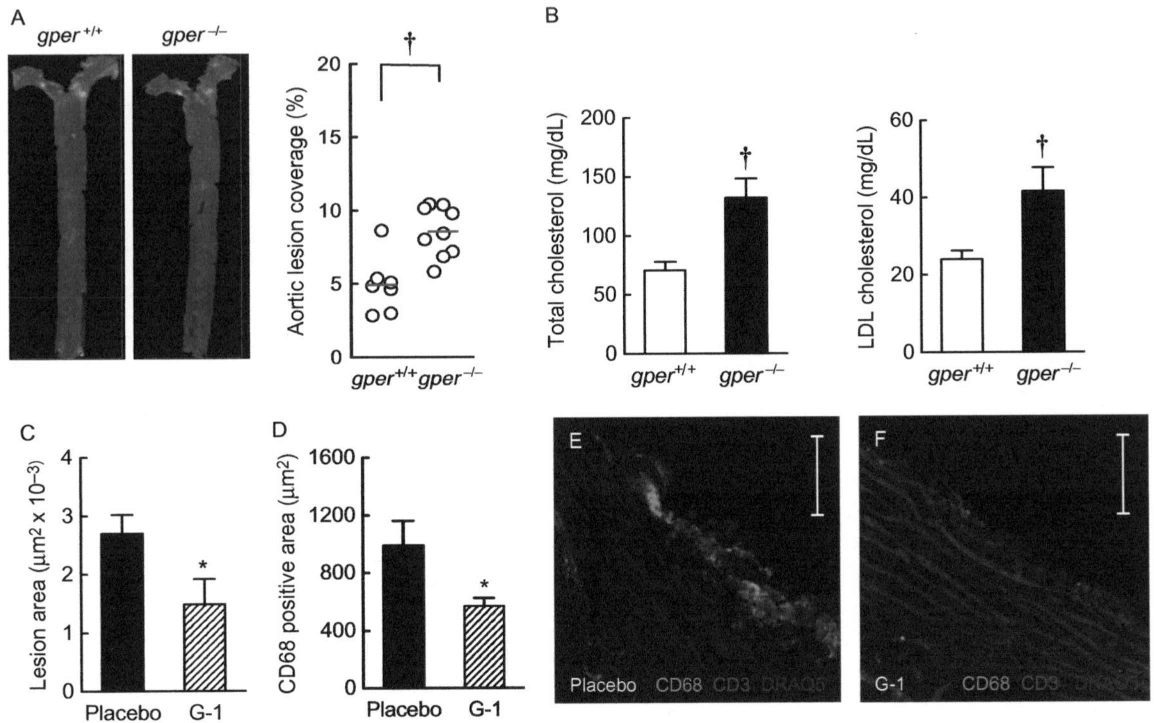

Fig. 8 Protective effects of GPER on atherosclerosis, vascular inflammation, and dyslipidemia. (A) Deletion of GPER (*gper*$^{-/-}$) aggravates diet-induced atherosclerosis in ovary-intact wild-type controls (*gper*$^{+/+}$). (B) Deletion of GPER causes dyslipidemia in ovary-intact wild-type mice with atherosclerosis by increasing total and LDL cholesterol levels. (C, D) Treatment with the GPER-selective agonist G-1 for 12 weeks reduces diet-induced atherosclerotic lesions and macrophage accumulation (stained with CD68 antibody) in the aortic root of surgically postmenopausal mice. (E, F) Staining for macrophages (CD68+) and T cells (CD3+) in surgically postmenopausal mice treated with placebo or G-1. The cell nucleus is stained with DRAQ5 (*blue* (*gray* in the print version)). Scale bar, 100 µm. $^{*}p < 0.05$ vs placebo, $^{\dagger}p < 0.05$ vs *gper*$^{+/+}$. *Panels in this figure are reproduced from Meyer, M. R., Fredette, N. C., Howard, T., Hu, C., Ramesh, C., Daniel, C., …, Prossnitz, E. R. (2014). G protein-coupled estrogen receptor protects from atherosclerosis.* Scientific Reports, 4, *7564. Erratum in:* Scientific Reports, *7565 (2015) 13510.*

surgically postmenopausal mice (Fig. 8) (Meyer et al., 2014), demonstrating that activation of GPER can largely recapitulate the protective effects of endogenous estrogens. Importantly, unlike estrogen in ovary-intact animals, G-1 was devoid of any effects on uterine growth (Meyer et al., 2014), an important unwanted clinical side effect of hormone therapy in postmenopausal women that requires clinical monitoring and progestin therapy.

Cerebrovascular disease is the major cause of stroke and disability and poses a substantial economic burden on health-care systems as well as on patients and their families (Barton et al., 2016). Like coronary artery disease, premenopausal women show a lower prevalence and better outcome after ischemic stroke compared with age-matched men, differences that are abrogated by natural or surgical menopause (Barton & Meyer, 2009; Wise, Dubal, Rau, Brown, & Suzuki, 2005). Natural estrogens such as 17β-estradiol and SERMs exert profound protective effects in the adult and the aging brain (Dhandapani & Brann, 2002; Wise et al., 2005). Activation of GPER has been found to rapidly dilate carotid arteries in a sex-independent fashion (Broughton et al., 2010). In experimental models of stroke and reperfusion, infarction volume is smaller and macrophage and T-cell number in the ischemic hemisphere is lower in ovary-intact females compared with males (Xiong et al., 2015), while expression of GPER after ischemic stroke increases only in males (Broughton et al., 2013). The functional role of GPER in cerebral ischemic injury and whether it depends on estrogen is less clear: while in one study activation of GPER in ovariectomized females (but not ovary-intact females) improved infarct size, G-1 treatment increased infarction zones in males (Broughton et al., 2014); another study found beneficial effects of G-1 treatment on cerebrovascular function after stroke in both males and females, with more pronounced effects in females (Murata et al., 2013). Similar to what can be observed in peripheral arteries (Barton et al., 2013), these data point at complex and sex-dependent roles of GPER and estrogen receptors in brain vascular physiology and stroke pathophysiology that warrant further study, particularly with regard to factors such as vascular autoregulation and oxidative stress that plays important roles in reperfusion injury in the brain.

6.5 Congestive Heart Failure

Coronary artery disease—often as a consequence of arterial hypertension, diabetes, and other risk factors—is one of the most prevalent causes of heart failure (Barton, 2013b; Taylor & Hobbs, 2013), which causes more hospitalizations of older patients than any other disease in developed countries (Heidenreich et al., 2011). Myocardial contractility depends on intact GPCR signaling, with angiotensin and β-adrenergic receptors as most prominent ones that are targeted by standard heart failure therapies (Violin, Soergel, Boerrigter, Burnett, & Lark, 2013). Given that GPER has been found to be highly expressed in the heart already in the initial cloning studies of GPR30 (Feng & Gregor, 1997; Kvingedal & Smeland, 1997; McCoy & Perlemutter, 1996; Owman et al., 1996) or its rat homologue GPR41 (Bonini et al., 1997), it may not be surprising that several groups subsequently demonstrated that targeting GPER may be effective to treat structural and functional changes in heart failure, including hypertrophy, fibrosis, and impaired left ventricular filling in early diastole. In genetically modified hypertensive rats with diastolic heart failure, the GPER-selective agonist G-1 improves diastolic function and attenuates cardiac hypertrophy and fibrosis (De Francesco et al., 2013; Jessup, Lindsey, Wang, Chappell, & Groban, 2010; Wang et al., 2012). Similarly, G-1 treatment reduces cardiac hypertrophy, cardiomyocyte size, and perivascular fibrosis in mice with a heart disease-prone genetic background (Lenhart, Broselid, Barrick, Leeb-Lundberg, & Caron, 2013). G-1 treatment also improves myocardial contractility in rats with heart failure induced by isoproterenol (Kang et al., 2012).

The cardiac L-arginine/NO pathway (Filice et al., 2009) as well as the renin–angiotensin (Jessup et al., 2010; Wang et al., 2012) and endothelin systems (Meyer, Fredette, Sharma, Barton, & Prossnitz, 2016) are all regulated by GPER and play a major role in the pathophysiology of heart failure (Barton, 2014; Lakatta, 2015; Violin et al., 2013). Unraveling GPER-dependent signaling pathways contributing to heart failure is thus important to understand the role of estrogen and GPER in left ventricular remodeling and diastolic dysfunction. Finally, GPER also plays a role in infarction-induced myocardial reperfusion injury and the estrogen-mediated protective effects. Early work by Kimura et al. demonstrated that expression of the rat GPER homologue GPR41 (which was initially designated hypoxia-induced apoptosis receptor, HIA-R) increases strongly in cardiomyocytes after hypoxia, and that a mutant GPR41 form (in which the G protein binding domain was inactivated) inhibits apoptosis of these cells (Kimura et al., 2001). Effects of hypoxia on GPER expression were confirmed by Maggiolini et al. in cardiomyocytes and breast cancer cells (Recchia et al., 2011). Recently, Bopassa and colleagues reported that deletion of GPER, but not that of ERα and ERβ, abrogates the protective effect of estrogen on left ventricular contractility after global ischemia caused by myocardial infarction (Kabir et al., 2015), further supporting a role for GPER in cardiac function and systolic heart failure (Prossnitz & Barton, 2011, 2014). Finally, according to a most recent report estrogen treatment reverses preexisting experimental heart failure induced by pressure overload, and estrogen therapy was also found to reverse myocardial scarring and stimulate angiogenesis (Iorga et al., 2016). The latter seems to be important as endothelial cells constitute by far most of the noncardiomyocyte cells in the heart (Pinto et al., 2016; Zhou & Pu, 2016). The roles of the individual estrogen receptors, particularly that of GPER, contributing to the estrogen-mediated reversal of heart failure (Iorga et al., 2016) and whether this therapeutic benefit can also be achieved clinically in patients has not yet been studied. However, Lee and associates have shown that GPER mediates protective effects by partly preventing ventricular remodeling after myocardial infarction (Lee, Lin, & Chang, 2014).

7. CONCLUSIONS

7.1 New Clinical Perspectives on Currently Approved Drugs Targeting GPER

Several drugs established in clinical use have meanwhile been identified to also target GPER. It is noteworthy that some of these drugs had been patented more than 50 years before their identification as GPER agonists (Barton, 2016), such as the lipid-lowering drug niacin/vitamin B_3 (Gille, Bodor, Ahmed, & Offermanns, 2008; Santolla et al., 2014). Similarly, the SERMs hydroxytamoxifen and raloxifene and the SERD Faslodex™ (fulvestrant, ICI 182,780, Fig. 7; Barton, 2016)—already approved for anti-estrogen therapy of breast cancers or for postmenopausal osteoporosis (Barton, 2016; Prossnitz & Arterburn, 2015)—have now all been identified as GPER agonists (Abdelhamid et al., 2011; Revankar et al., 2005; Thomas et al., 2005). Interestingly, arterial hypotension is a common side effect of both Faslodex™ (Vergote & Abram, 2006) and niacin (Gille et al., 2008), which may point toward concomitant GPER activation causing excessive vasodilation in these patients (Meyer et al., 2010). It is thus important to reassess the clinical data currently available (Barton, 2016) given that the signaling pathways targeted by these drugs will not only include ERα/ERβ or the recently discovered niacin receptors GPR109A (MH74A) and GPR109B (HM74) (Barton, 2016), but likely also GPER.

7.2 Risk Factors, Medical Genetics, and Personalized (Precision) Medicine

As opposed to using applied mathematics of statistical "evidence" as a basis for clinical decision making (so-called "evidence-based medicine"), personalized or "precision" medicine goes back to the individual patient aiming at taking into account clinical information and data of the patient helping the physician to predict whether a preventive or therapeutic measure will beneficially, neutrally, or adversely affect the patient's health. In particular, personalized medicine requires information obtained from medical genetics analyses and involves pharmacogenomics concepts (Sadee, 2005). Susceptibility to low-renin arterial hypertension has been mapped to chromosome 7 (7p22) (Haas et al., 2009; Lafferty et al., 2000) where the human *gper* gene is also located (Haas et al., 2009; Scherer et al., 2003). Consistent with the pressure-lowering and vasodilating effects of GPER-activating drugs (Haas et al., 2009; Meyer, Amann, et al., 2012; Meyer, Prossnitz, et al., 2011), Feldman et al. recently reported increased blood pressure in female carriers of GPER P16L, a hypofunctional genetic variant of the *gper* gene, compared with P16/P16 carriers (Feldman et al., 2014).

Part of the protective effects of estrogens on atherogenesis in premenopausal women have been attributed to their lipid-modifying effect as well as delaying the onset of hypertension, diabetes, and obesity (Barton, 2013a; Barton & Prossnitz, 2015). In premenopausal women, estrogen levels are associated with higher HDL cholesterol and lower LDL cholesterol levels, and dyslipidemia often develops after menopause (Barton, 2013a). Tamoxifen was originally patented in the early 1960s for the treatment of hypercholesterolemia, and related compounds similarly inhibited cholesterol metabolism and biosynthesis (Barton, 2016; Maximov, Lee, & Jordan, 2013). Half a century later, GPER—which is one of the targets of tamoxifen—has been shown to mediate salutary effects on cholesterol metabolism (Barton & Prossnitz, 2015; Hussain et al., 2015; Meyer et al., 2014). Interestingly, other GPER-activating SERMs and SERDs also exert lipid-lowering effects in patients (De Leo et al., 2001; de Valk-de Roo et al., 1999; Yoshii et al., 2015). Moreover, humans carrying the hypofunctional P16L genetic variant of GPER display increased LDL cholesterol levels (Hussain et al., 2015). Similarly, beneficial effects of SERM or flavonoid treatments on metabolic profiles and insulin sensitivity have been reported in patients (de Valk-de Roo et al., 1999; Villa et al., 2009).

Further research in this area is warranted to increase our understanding on how these drugs affect cardiovascular risk factors, to what degree the different ERs mediate their wanted and unwanted effects such as venous thrombosis (Barton & Meyer, 2009) in some but not all patients, and to identify responders vs nonresponders among coronary artery disease patients receiving hormone treatment. Already more than 20 years it was shown that coronary vasodilation in response to sublingual estradiol occurs only in those patients who show an increase in circulating 17β-estradiol levels within 30 min of administration, whereas in some patients 17β-estradiol levels remain unchanged (Riedel & Mügge, 1993), suggesting important individual differences in hormone absorption, pharmacokinetics, and thus local bioavailability. Ultimately, identifying factors of how individual patients respond to hormone therapy will allow to develop and apply theranostic approaches, to provide personalized medicine to the individual patient, and to improve prevention and treatment of cardiovascular disease.

7.3 GPER-Targeting Drugs in Clinical Trials of Coronary Artery Disease

Failures of previous clinical trials investigating the use of CEEs and MPA on cardiovascular outcomes (Hulley et al., 1998; Rossouw et al., 2002) in older postmenopausal women with cardiovascular risk factors or disease are in part drug-related (Barton, 2001, 2011; Dubey et al., 2005; Meyer et al., 2008). However, they may also be explained by the increasing complexity of genomic and rapid vascular estrogen signaling pathways initiated by ERα, ERβ, and GPER, as well as through cross talk between these receptors (Menazza & Murphy, 2016; Meyer & Barton, 2009; Meyer et al., 2008; Turgeon et al., 2004). Importantly, studies utilizing natural (ie, bioidentical) human hormones administered in a cyclic or noncylic fashion to recently postmenopausal and otherwise healthy women reported protective effects of hormone treatment markedly reducing cardiovascular events (Hodis et al., 2016; Keaney & Solomon, 2016; Schierbeck et al., 2012).

Whether separating pharmacological effects to selectively target beneficial vascular signaling pathways by using improved (Madak-Erdogan et al., 2016) and/or low-dose hormone or GPER-activating treatments might translate into cardiovascular clinical benefits while reducing harmful side effects remains to be demonstrated. Some evidence in this regard is already available from clinical studies using SERMs in the treatment of breast cancer and osteoporosis: in younger postmenopausal women, treatment with raloxifene or lasofoxifene reduced the incidence of adverse clinical events related to coronary artery disease (Collins et al., 2009; Cummings et al.,

2010; Ensrud et al., 2010). It is unlikely that the ERα/ERβ blocking effects of SERMs are responsible for the beneficial effects as clinical worsening disease would be expected (Barton, 2016). Thus, GPER may contribute in part to the beneficial cardiovascular effects observed in these studies. In view of the beneficial outcomes previously reported (Collins et al., 2009; Cummings et al., 2010; Ensrud et al., 2010) additional studies to test whether treatment with SERMs in younger and otherwise healthy postmenopausal women has beneficial effects similar to those of bioidentical estrogens (Harman et al., 2010; Hodis et al., 2016; Keaney & Solomon, 2016; McSweeney et al., 2016; Schierbeck et al., 2012) are warranted. In addition, developing novel, highly selective ligands that specifically mimic the beneficial vascular effects of 17β-estradiol might provide new therapeutic approaches (Bologa et al., 2006; Madak-Erdogan et al., 2016). There is now accumulating evidence that GPER signaling contributes to the beneficial cardiovascular effects of premenopausal estrogens (Barton & Prossnitz, 2015; Meyer, Prossnitz, et al., 2011) and that adverse effects of menopause can be alleviated by drugs targeting GPER. The fact that unselective GPER activators are clinically approved for other diseases and readily available (Barton, 2016) and the discovery of novel, small-molecule ligands selective for GPER (Dennis et al., 2009) offers the opportunity for clinical drug development in search for new effective and safe therapeutics for patients with coronary artery disease.

DISCLOSURE STATEMENT

M.R.M. and M.B. are inventors on a US patent application on the use of GPER-targeting compounds. M.B. serves as a consultant for AbbVie, Inc.

ACKNOWLEDGMENTS

This work was supported by Swiss National Science Foundation grants Nr. 135 874 and 141 501 (to. M.R.M.) and Nr. 108 256 and 122 504 (to M.B.).

Role of the funding sources. The funders had no involvement in the drafting, writing, or decision to submit the present article for publication.

REFERENCES

Abdelhamid, R., Luo, J., Vandevrede, L., Kundu, I., Michalsen, B., Litosh, V. A., … Thatcher, G. R. (2011). Benzothiophene selective estrogen receptor modulators provide neuroprotection by a novel GPR30-dependent mechanism. *ACS Chemical Neuroscience*, *2*, 256–268. http://dx.doi.org/10.1021/cn100106a.

Alexander, S. P., Mathie, A., & Peters, J. A. (2008). Guide to receptors and channels (GRAC). *British Journal of Pharmacology*, *153*(Suppl. 2), S1–S209. http://dx.doi.org/10.1038/sj.bjp.0707746. 0707746 [pii] (3rd edition).

Altmann, J. B., Yan, G., Meeks, J. F., Abood, M. E., Brailoiu, E., & Brailoiu, G. C. (2015). G protein-coupled estrogen receptor-mediated effects on cytosolic calcium and nanomechanics in brain microvascular endothelial cells. *Journal of Neurochemistry*, *133*, 629–639. http://dx.doi.org/10.1111/jnc.13066.

Arefin, S., Simoncini, T., Wieland, R., Hammarqvist, F., Spina, S., Goglia, L., & Kublickiene, K. (2014). Vasodilatory effects of the selective GPER agonist G-1 is maximal in arteries of postmenopausal women. *Maturitas*, *78*, 123–130. http://dx.doi.org/10.1016/j.maturitas.2014.04.002.

Arnal, J. F., Douin-Echinard, V., Brouchet, L., Tremollieres, F., Laurell, H., Lenfant, F., … Gourdy, P. (2006). Understanding the oestrogen action in experimental and clinical atherosclerosis. *Fundamental & Clinical Pharmacology*, *20*, 539–548. http://dx.doi.org/10.1111/j.1472-8206.2006.00445.x. FCP445 [pii].

Aronica, S. M., Kraus, W. L., & Katzenellenbogen, B. S. (1994). Estrogen action via the cAMP signaling pathway: Stimulation of adenylate cyclase and cAMP-regulated gene transcription. *Proceedings of the National Academy of Sciences of the United States of America*, *91*, 8517–8521.

Bairey Merz, C. N., Johnson, B. D., Sharaf, B. L., Bittner, V., Berga, S. L., Braunstein, G. D., … Sopko, G. (2003). Hypoestrogenemia of hypothalamic origin and coronary artery disease in premenopausal women: A report from the NHLBI-sponsored WISE study. *Journal of the American College of Cardiology*, *41*, 413–419.

Barrett Mueller, K., Lu, Q., Mohammad, N. N., Luu, V., McCurley, A., Williams, G. H., … Jaffe, I. Z. (2014). Estrogen receptor inhibits mineralocorticoid receptor transcriptional regulatory function. *Endocrinology*, *155*, 4461–4472. http://dx.doi.org/10.1210/en.2014-1270.

Barrett-Connor, E. (2013). Menopause, atherosclerosis, and coronary artery disease. *Current Opinion in Pharmacology*, *13*, 186–191. http://dx.doi.org/10.1016/j.coph.2013.01.005.

Barton, M. (1994). Akute Wirkungen von 17β-Oestradiol auf den Tonus menschlicher Arterien in vitro (pp. 1–79). Hannover Medical School, Hannover, Germany. (Doctor of Medicine (D.M.) dissertation).

Barton, M. (2001). Postmenopausal oestrogen replacement therapy and atherosclerosis: Can current compounds provide cardiovascular protection? *Expert Opinion on Investigational Drugs*, *10*, 789–809. http://dx.doi.org/10.1517/13543784.10.5.789.

Barton, M. (2008). Reversal of proteinuric renal disease and the emerging role of endothelin. *Nature Clinical Practice Nephrology*, *4*, 490–501. http://dx.doi.org/10.1038/ncpneph0891.

Barton, M. (2010). Obesity and aging: Determinants of endothelial cell dysfunction and atherosclerosis. *Pflügers Archiv*, *460*, 825–837. http://dx.doi.org/10.1007/s00424-010-0860-y.

Barton, M. (2011). The discovery of endothelium-dependent contraction: The legacy of Paul M. Vanhoutte. *Pharmacological Research*, *63*, 455–462. http://dx.doi.org/10.1016/j.phrs.2011.02.013.

Barton, M. (2012a). Childhood obesity: A life-long health risk. *Acta Pharmacologica Sinica*, *33*, 189–193. http://dx.doi.org/10.1038/aps.2011.204.

Barton, M. (2012b). Position paper: The membrane estrogen receptor GPER—Clues and questions. *Steroids*, *77*, 935–942. http://dx.doi.org/10.1016/j.steroids.2012.04.001.

Barton, M. (2013a). Cholesterol and atherosclerosis: Modulation by oestrogen. *Current Opinion in Lipidology*, *24*, 214–220. http://dx.doi.org/10.1097/MOL.0b013e3283613a94.

Barton, M. (2013b). Mechanisms and therapy of atherosclerosis and its clinical complications. *Current Opinion in Pharmacology*, *13*, 149–153. http://dx.doi.org/10.1016/j.coph.2013.05.001.

Barton, M. (2013c). Prevention and endothelial therapy of coronary artery disease. *Current Opinion in Pharmacology*, *13*, 226–241. http://dx.doi.org/10.1016/j.coph.2013.05.005.

Barton, M. (2014). Aging and endothelin: Determinants of disease. *Life Sciences*, *118*, 97–109. http://dx.doi.org/10.1016/j.lfs.2014.09.009.

Barton, M. (2016). Not lost in translation: Emerging clinical importance of the G protein-coupled estrogen receptor GPER. *Steroids*. e-pub January 24.

Barton, M., Baretella, O., & Meyer, M. R. (2012a). Obesity and risk of vascular disease: Importance of endothelium-dependent vasoconstriction. *British Journal of Pharmacology*, *165*, 591–602. http://dx.doi.org/10.1111/j.1476-5381.2011.01472.x.

Barton, M., Baretella, O., & Meyer, M. R. (2012b). Obesity and risk of vascular disease: Importance of endothelium-derived vasoconstriction. *British Journal of Pharmacology*, *165*, 591–602. http://dx.doi.org/10.1111/j.1476-5381.2011.01472.x.

Barton, M., Cremer, J., & Mügge, A. (1998). 17β-estradiol acutely improves endothelium-dependent relaxation to bradykinin in isolated human coronary arteries. *European Journal of Pharmacology*, *362*, 73–76.

Barton, M., & Dubey, R. K. (2002). Postmenopausal hormone-replacement therapy. *The New England Journal of Medicine*, *346*, 63–65.

Barton, M., Haudenschild, C. C., d'Uscio, L. V., Shaw, S., Munter, K., & Luscher, T. F. (1998). Endothelin ETA receptor blockade restores NO-mediated endothelial function and inhibits atherosclerosis in apolipoprotein E-deficient mice. *Proceedings of the National Academy of Sciences of the United States of America*, *95*, 14367–14372.

Barton, M., Husmann, M., & Meyer, M. R. (2016). Accelerated vascular aging as a paradigm for hypertensive vascular disease: Prevention and therapy. *The Canadian Journal of Cardiology*. *32*(5). http://dx.doi.org/10.1016/j.cjca.2016.02.062, 680.e4–686.e4.

Barton, M., & Meyer, M. R. (2009). Postmenopausal hypertension: Mechanisms and therapy. *Hypertension*, *54*, 11–18.

Barton, M., & Meyer, M. R. (2015). Nicolaus Copernicus and the rapid vascular responses to aldosterone. *Trends in Endocrinology and Metabolism*, *26*, 396–398. http://dx.doi.org/10.1016/j.tem.2015.05.005.

Barton, M., Meyer, M. R., Bhattacharya, I., & Haas, E. (2007a). WISDOM trial, postmenopausal hormone replacement and cardiovascular events: Why it is still not wise to use equine estrogens. *BMJ*, *335*, 239. http://www.bmj.com/rapid-response/2011/2011/2001/wisdom-trial-postmenopausal-hormone-replacement-and-cardiovascular-events-.

Barton, M., Meyer, M. R., & Haas, E. (2007b). Hormone replacement therapy and atherosclerosis in postmenopausal women: Does aging limit therapeutic benefits? *Arteriosclerosis, Thrombosis, and Vascular Biology*, *27*, 1669–1672.

Barton, M., Meyer, M. R., & Prossnitz, E. R. (2013). Alike but not the same: Anatomic heterogeneity of estrogen receptor-mediated vasodilation. *Journal of Cardiovascular Pharmacology*, *62*, 22–25. http://dx.doi.org/10.1097/FJC.0b013e31829709d4.

Barton, M., & Prossnitz, E. R. (2015). Emerging roles of GPER in diabetes and atherosclerosis. *Trends in Endocrinology and Metabolism*, *26*, 185–192. http://dx.doi.org/10.1016/j.tem.2015.02.003.

Beldekas, J. C., Smith, B., Gerstenfeld, L. C., Sonenshein, G. E., & Franzblau, C. (1981). Effects of 17 beta-estradiol on the biosynthesis of collagen in cultured bovine aortic smooth muscle cells. *Biochemistry*, *20*, 2162–2167.

Bhattacharya, I., Damjanovic, M., Hirsiger, S., Haas, E., Gut, A., Mueller-Guerre, L., … Barton, M. (2007). Selective activation of G-protein coupled estrogen receptor GPR30 inhibits vasoconstriction and induces vascular smooth muscle cell ERK1/2 phosphorylation. *Circulation*, *116*, II74.

Bologa, C. G., Revankar, C. M., Young, S. M., Edwards, B. S., Arterburn, J. B., Kiselyov, A. S., … Prossnitz, E. R. (2006). Virtual and biomolecular screening converge on a selective agonist for GPR30. *Nature Chemical Biology*, *2*, 207–212.

Bonini, J. A., Anderson, S. M., & Steiner, D. F. (1997). Molecular cloning and tissue expression of a novel orphan G protein-coupled receptor from rat lung. *Biochemical and Biophysical Research Communications*, *234*, 190–193. http://dx.doi.org/10.1006/bbrc.1997.6591.

Bowling, M. R., Xing, D., Kapadia, A., Chen, Y. F., Szalai, A. J., Oparil, S., & Hage, F. G. (2014). Estrogen effects on vascular inflammation are age dependent: Role of estrogen receptors. *Arteriosclerosis, Thrombosis, and Vascular Biology*, *34*, 1477–1485. http://dx.doi.org/10.1161/ATVBAHA.114.303629.

Bramlage, P., Wittchen, H. U., Pittrow, D., Kirch, W., Krause, P., Lehnert, H., ... Sharma, A. M. (2004). Recognition and management of overweight and obesity in primary care in Germany. *International Journal of Obesity and Related Metabolic Disorders*, *28*, 1299–1308.

Brandes, R. P., & Mügge, A. (1997). Gender differences in the generation of superoxide anions in the rat aorta. *Life Sciences*, *60*, 391–396.

Broughton, B. R., Brait, V. H., Guida, E., Lee, S., Arumugam, T. V., Gardiner-Mann, C. V., ... Sobey, C. G. (2013). Stroke increases g protein-coupled estrogen receptor expression in the brain of male but not female mice. *Neurosignals*, *21*, 229–239. http://dx.doi.org/10.1159/000338019.

Broughton, B. R., Brait, V. H., Kim, H. A., Lee, S., Chu, H. X., Gardiner-Mann, C. V., ... Sobey, C. G. (2014). Sex-dependent effects of G protein-coupled estrogen receptor activity on outcome after ischemic stroke. *Stroke*, *45*, 835–841. http://dx.doi.org/10.1161/STROKEAHA.113.001499.

Broughton, B. R., Miller, A. A., & Sobey, C. G. (2010). Endothelium-dependent relaxation by G protein-coupled receptor 30 agonists in rat carotid arteries. *American Journal of Physiology. Heart and Circulatory Physiology*, *298*, H1055–H1061.

Burke, A. P., Farb, A., Malcom, G., & Virmani, R. (2001). Effect of menopause on plaque morphologic characteristics in coronary atherosclerosis. *American Heart Journal*, *141*, S58–S62.

Burns, K. A., & Korach, K. S. (2012). Estrogen receptors and human disease: An update. *Archives of Toxicology*, *86*, 1491–1504. http://dx.doi.org/10.1007/s00204-012-0868-5.

Burt, V. L., Whelton, P., Roccella, E. J., Brown, C., Cutler, J. A., Higgins, M., ... Labarthe, D. (1995). Prevalence of hypertension in the US adult population. Results from the Third National Health and Nutrition Examination Survey, 1988–1991. *Hypertension*, *25*, 305–313.

Butenandt, A. (1929). Über "Progynon" ein krystallisiertes weibliches Sexualhormon. *Die Naturwissenschaften*, *17*, 879.

Butenandt, A. (1931). Über die chemische Untersuchung der Sexualhormone. *Zeitschrift für Angewandte Chemie*, *44*, 905–916.

Carmeci, C., Thompson, D. A., Ring, H. Z., Francke, U., & Weigel, R. J. (1997). Identification of a gene (GPR30) with homology to the G-protein-coupled receptor superfamily associated with estrogen receptor expression in breast cancer. *Genomics*, *45*, 607–617.

Caroccia, B., Seccia, T. M., Campos, A. G., Gioco, F., Kuppusamy, M., Ceolotto, G., ... Rossi, G. P. (2014). GPER-1 and estrogen receptor-beta ligands modulate aldosterone synthesis. *Endocrinology*, *155*, 4296–4304. http://dx.doi.org/10.1210/en.2014-1416.

Caulin-Glaser, T., Garcia-Cardena, G., Sarrel, P., Sessa, W. C., & Bender, J. R. (1997). 17 beta-estradiol regulation of human endothelial cell basal nitric oxide release, independent of cytosolic Ca2+ mobilization. *Circulation Research*, *81*, 885–892.

Chambliss, K. L., Yuhanna, I. S., Anderson, R. G., Mendelsohn, M. E., & Shaul, P. W. (2002). ERbeta has nongenomic action in caveolae. *Molecular Endocrinology*, *16*, 938–946.

Chambliss, K. L., Yuhanna, I. S., Mineo, C., Liu, P., German, Z., Sherman, T. S., ... Shaul, P. W. (2000). Estrogen receptor alpha and endothelial nitric oxide synthase are organized into a functional signaling module in caveolae. *Circulation Research*, *87*, E44–E52.

Chang, W. C., Nakao, J., Orimo, H., & Murota, S. I. (1980). Stimulation of prostaglandin cyclooxygenase and prostacyclin synthetase activities by estradiol in rat aortic smooth muscle cells. *Biochimica et Biophysica Acta, 620*, 472–482.

Cheng, S. B., Dong, J., Pang, Y., LaRocca, J., Hixon, M., Thomas, P., & Filardo, E. J. (2014). Anatomical location and redistribution of G protein-coupled estrogen receptor-1 during the estrus cycle in mouse kidney and specific binding to estrogens but not aldosterone. *Molecular and Cellular Endocrinology, 382*, 950–959. http://dx.doi.org/10.1016/j.mce.2013.11.005.

Clark, S., Pollard, K., Rainville, J., & Vasudevan, N. (2016). Detection of the phosphorylation of the estrogen receptor alpha as an outcome of GPR30 activation. *Methods in Molecular Biology, 1366*, 457–470. http://dx.doi.org/10.1007/978-1-4939-3127-9_36.

Clarkson, T. B. (2007). Estrogen effects on arteries vary with stage of reproductive life and extent of subclinical atherosclerosis progression. *Menopause, 14*, 373–384.

Collins, P., Mosca, L., Geiger, M. J., Grady, D., Kornitzer, M., Amewou-Atisso, M. G., ... Wenger, N. K. (2009). Effects of the selective estrogen receptor modulator raloxifene on coronary outcomes in the raloxifene use for the heart trial. Results of subgroup analyses by age and other factors. *Circulation, 119*, 922–930.

Collins, P., Rosano, G. M., Sarrel, P. M., Ulrich, L., Adamopoulos, S., Beale, C. M., ... Poole-Wilson, P. A. (1995). 17 beta-Estradiol attenuates acetylcholine-induced coronary arterial constriction in women but not men with coronary heart disease. *Circulation, 92*, 24–30.

Crandall, C. J., & Barrett-Connor, E. (2013). Endogenous sex steroid levels and cardiovascular disease in relation to the menopause: A systematic review. *Endocrinology and Metabolism Clinics of North America, 42*, 227–253. http://dx.doi.org/10.1016/j.ecl.2013.02.003.

Cruz, M. N., Agewall, S., Schenck-Gustafsson, K., & Kublickiene, K. (2008). Acute dilatation to phytoestrogens and estrogen receptor subtypes expression in small arteries from women with coronary heart disease. *Atherosclerosis, 196*, 49–58. http://dx.doi.org/10.1016/j.atherosclerosis.2007.01.038.

Cummings, S. R., Ensrud, K., Delmas, P. D., LaCroix, A. Z., Vukicevic, S., Reid, D. M., ... Investigators, P. S. (2010). Lasofoxifene in postmenopausal women with osteoporosis. *The New England Journal of Medicine, 362*, 686–696. http://dx.doi.org/10.1056/NEJMoa0808692.

Davis, K. E., Carstens, E. J., Irani, B. G., Gent, L. M., Hahner, L. M., & Clegg, D. J. (2014). Sexually dimorphic role of G protein-coupled estrogen receptor (GPER) in modulating energy homeostasis. *Hormones and Behavior, 66*, 196–207. http://dx.doi.org/10.1016/j.yhbeh.2014.02.004.

De Francesco, E. M., Angelone, T., Pasqua, T., Pupo, M., Cerra, M. C., & Maggiolini, M. (2013). GPER mediates cardiotropic effects in spontaneously hypertensive rat hearts. *PLoS One, 8*, e69322. http://dx.doi.org/10.1371/journal.pone.0069322.

De Leo, V., la Marca, A., Morgante, G., Lanzetta, D., Setacci, C., & Petraglia, F. (2001). Randomized control study of the effects of raloxifene on serum lipids and homocysteine in older women. *American Journal of Obstetrics and Gynecology, 184*, 350–353. http://dx.doi.org/10.1067/mob.2001.111065.

Dennis, M. K., Burai, R., Ramesh, C., Petrie, W. K., Alcon, S. N., Nayak, T. K., ... Prossnitz, E. R. (2009). In vivo effects of a GPR30 antagonist. *Nature Chemical Biology, 5*, 421–427. http://dx.doi.org/10.1038/nchembio.168. nchembio.168 [pii].

Dennis, M. K., Field, A. S., Burai, R., Ramesh, C., Petrie, W. K., Bologa, C. G., ... Prossnitz, E. R. (2011). Identification of a GPER/GPR30 antagonist with improved estrogen receptor counterselectivity. *The Journal of Steroid Biochemistry and Molecular Biology, 127*, 358–366. http://dx.doi.org/10.1016/j.jsbmb.2011.07.002. S0960-0760(11)00147-6 [pii].

Deroo, B. J., & Korach, K. S. (2006). Estrogen receptors and human disease. *The Journal of Clinical Investigation, 116*, 561–570. http://dx.doi.org/10.1172/JCI27987.

de Valk-de Roo, G. W., Stehouwer, C. D., Meijer, P., Mijatovic, V., Kluft, C., Kenemans, P., … Netelenbos, C. (1999). Both raloxifene and estrogen reduce major cardiovascular risk factors in healthy postmenopausal women: A 2-year, placebo-controlled study. *Arteriosclerosis, Thrombosis, and Vascular Biology, 19*, 2993–3000.

Dhandapani, K. M., & Brann, D. W. (2002). Protective effects of estrogen and selective estrogen receptor modulators in the brain. *Biology of Reproduction, 67*, 1379–1385.

Dubey, R. K., Imthurn, B., Barton, M., & Jackson, E. K. (2005). Vascular consequences of menopause and hormone therapy: Importance of timing of treatment and type of estrogen. *Cardiovascular Research, 66*, 295–306. http://dx.doi.org/10.1016/j.cardiores.2004.12.012.

Elhage, R., Clamens, S., Reardon-Alulis, C., Getz, G. S., Fievet, C., Maret, A., … Bayard, F. (2000). Loss of atheroprotective effect of estradiol in immunodeficient mice. *Endocrinology, 141*, 462–465.

Elhage, R., Gourdy, P., Jawien, J., Brouchet, L., Castano, C., Fievet, C., … Bayard, F. (2005). The atheroprotective effect of 17beta-estradiol depends on complex interactions in adaptive immunity. *The American Journal of Pathology, 167*, 267–274. 167/1/267 [pii].

El-Mowafy, A. M., & White, R. E. (1999). Resveratrol inhibits MAPK activity and nuclear translocation in coronary artery smooth muscle: Reversal of endothelin-1 stimulatory effects. *FEBS Letters, 451*, 63–67.

Ensrud, K., LaCroix, A., Thompson, J. R., Thompson, D. D., Eastell, R., Reid, D. M., … Cummings, S. (2010). Lasofoxifene and cardiovascular events in postmenopausal women with osteoporosis: Five-year results from the postmenopausal evaluation and risk reduction with lasofoxifene (PEARL) trial. *Circulation, 122*, 1716–1724. http://dx.doi.org/10.1161/CIRCULATIONAHA.109.924571.

Evans, N. J., Bayliss, A. L., Reale, V., & Evans, P. D. (2016). Characterisation of signalling by the endogenous GPER1 (GPR30) receptor in an embryonic mouse hippocampal cell line (mHippoE-18). *PLoS One, 11*, e0152138. http://dx.doi.org/10.1371/journal.pone.0152138.

Farhat, M. Y., Abi-Younes, S., Dingaan, B., Vargas, R., & Ramwell, P. W. (1996). Estradiol increases cyclic adenosine monophosphate in rat pulmonary vascular smooth muscle cells by a nongenomic mechanism. *The Journal of Pharmacology and Experimental Therapeutics, 276*, 652–657.

Farhat, M. Y., Lavigne, M. C., & Ramwell, P. W. (1996). The vascular protective effects of estrogen. *FASEB Journal, 10*, 615–624.

Feldman, R. D., Gros, R., Ding, Q., Hussain, Y., Ban, M. R., McIntyre, A. D., & Hegele, R. A. (2014). A common hypofunctional genetic variant of GPER is associated with increased blood pressure in women. *British Journal of Clinical Pharmacology, 78*, 1441–1452. http://dx.doi.org/10.1111/bcp.12471.

Feletou, M., & Vanhoutte, P. M. (2006). Endothelial dysfunction: A multifaceted disorder (The Wiggers Award Lecture). *American Journal of Physiology. Heart and Circulatory Physiology, 291*, H985–H1002.

Feng, Y., & Gregor, P. (1997). Cloning of a novel member of the G protein-coupled receptor family related to peptide receptors. *Biochemical and Biophysical Research Communications, 231*, 651–654. http://dx.doi.org/10.1006/bbrc.1997.6161.

Filardo, E. J., Quinn, J. A., Bland, K. I., & Frackelton, A. R., Jr. (2000). Estrogen-induced activation of Erk-1 and Erk-2 requires the G protein-coupled receptor homolog, GPR30, and occurs via trans-activation of the epidermal growth factor receptor through release of HB-EGF. *Molecular Endocrinology, 14*, 1649–1660.

Filardo, E. J., Quinn, J. A., Frackelton, A. R., Jr., & Bland, K. I. (2002). Estrogen action via the G protein-coupled receptor, GPR30: Stimulation of adenylyl cyclase and cAMP-

mediated attenuation of the epidermal growth factor receptor-to-MAPK signaling axis. *Molecular Endocrinology*, *16*, 70–84.
Filice, E., Recchia, A. G., Pellegrino, D., Angelone, T., Maggiolini, M., & Cerra, M. C. (2009). A new membrane G protein-coupled receptor (GPR30) is involved in the cardiac effects of 17beta-estradiol in the male rat. *Journal of Physiology and Pharmacology*, *60*, 3–10.
Foegh, M. L., Zhao, Y., Farhat, M., & Ramwell, P. W. (1995). Oestradiol inhibition of vascular myointimal proliferation following immune, chemical and mechanical injury. *Ciba Foundation Symposium*, *191*, 139–145. discussion 145–139.
Ford, J., Hajibeigi, A., Long, M., Hahner, L., Gore, C., Hsieh, J. T., … Oz, O. K. (2011). GPR30 deficiency causes increased bone mass, mineralization, and growth plate proliferative activity in male mice. *Journal of Bone and Mineral Research*, *26*, 298–307. http://dx.doi.org/10.1002/jbmr.209.
Fujio, Y., Yamada, F., Takahashi, K., & Shibata, N. (1993). Responses of smooth muscle cells to platelet-derived growth factor are inhibited by herbimycin-A tyrosine kinase inhibitor. *Biochemical and Biophysical Research Communications*, *195*, 79–83. http://dx.doi.org/10.1006/bbrc.1993.2012.
Fuller, P. J., & Young, M. J. (2014). Duelling receptors: Estrogen receptor versus mineralocorticoid receptor in the cardiovascular system. *Endocrinology*, *155*, 4117–4119. http://dx.doi.org/10.1210/en.2014-1778.
Furchgott, R. F. (1988). Studies on relaxation of rabbit aorta by sodium nitrite: The basis for the proposal that acid-activatable inhibitory factor from bovine retractor penis is inorganic nitrite and the endothelium-derived relaxing factor is nitric oxide. In P. M. Vanhoutte (Ed.), *Vasodilation: Vascular smooth muscle, peptides, autonomic nerves and endothelium* (pp. 401–414). New York: Raven Press.
Furchgott, R. F., & Zawadzki, J. V. (1980). The obligatory role of endothelial cells in the relaxation of arterial smooth muscle by acetylcholine. *Nature*, *288*, 373–376.
Gambacciani, M., Monteleone, P., Vitale, C., Silvestri, A., Fini, M., Genazzani, A. R., & Rosano, G. M. (2002). Dydrogesterone does not reverse the effects of estradiol on endothelium-dependant vasodilation in postmenopausal women: A randomised clinical trial. *Maturitas*, *43*, 117–123.
Gaudet, H. M., Cheng, S. B., Christensen, E. M., & Filardo, E. J. (2015). The G-protein coupled estrogen receptor, GPER: The inside and inside-out story. *Molecular and Cellular Endocrinology*, *418*(Pt. 3), 207–219. http://dx.doi.org/10.1016/j.mce.2015.07.016.
Gerhard, M., Walsh, B. W., Tawakol, A., Haley, E. A., Creager, S. J., Seely, E. W., … Creager, M. A. (1998). Estradiol therapy combined with progesterone and endothelium-dependent vasodilation in postmenopausal women. *Circulation*, *98*, 1158–1163.
Gille, A., Bodor, E. T., Ahmed, K., & Offermanns, S. (2008). Nicotinic acid: Pharmacological effects and mechanisms of action. *Annual Review of Pharmacology and Toxicology*, *48*, 79–106. http://dx.doi.org/10.1146/annurev.pharmtox.48.113006.094746.
Gimbrone, M. A., Jr., & Garcia-Cardena, G. (2016). Endothelial cell dysfunction and the pathobiology of atherosclerosis. *Circulation Research*, *118*, 620–636. http://dx.doi.org/10.1161/CIRCRESAHA.115.306301.
Giordano, S., Hage, F. G., Xing, D., Chen, Y. F., Allon, S., Chen, C., & Oparil, S. (2015). Estrogen and cardiovascular disease: Is timing everything? *The American Journal of the Medical Sciences*, *350*, 27–35. http://dx.doi.org/10.1097/MAJ.0000000000000512.
Gisclard, V., Flavahan, N. A., & Vanhoutte, P. M. (1987). Alpha adrenergic responses of blood vessels of rabbits after ovariectomy and administration of 17 beta-estradiol. *The Journal of Pharmacology and Experimental Therapeutics*, *240*, 466–470.
Gisclard, V., Miller, V. M., & Vanhoutte, P. M. (1988). Effect of 17 beta-estradiol on endothelium-dependent responses in the rabbit. *The Journal of Pharmacology and Experimental Therapeutics*, *244*, 19–22.

Gollasch, M. (2012). Vasodilator signals from perivascular adipose tissue. *British Journal of Pharmacology*, *165*, 633–642. http://dx.doi.org/10.1111/j.1476-5381.2011.01430.x.
Grady, D., Rubin, S. M., Petitti, D. B., Fox, C. S., Black, D., Ettinger, B., … Cummings, S. R. (1992). Hormone therapy to prevent disease and prolong life in postmenopausal women. *Annals of Internal Medicine*, *117*, 1016–1037.
Grainger, D. J., Weissberg, P. L., & Metcalfe, J. C. (1993). Tamoxifen decreases the rate of proliferation of rat vascular smooth-muscle cells in culture by inducing production of transforming growth factor beta. *The Biochemical Journal*, *294*(Pt. 1), 109–112.
Green, S., Walter, P., Greene, G., Krust, A., Goffin, C., Jensen, E., … Chambon, P. (1986a). Cloning of the human oestrogen receptor cDNA. *Journal of Steroid Biochemistry*, *24*, 77–83.
Green, S., Walter, P., Kumar, V., Krust, A., Bornert, J. M., Argos, P., & Chambon, P. (1986b). Human oestrogen receptor cDNA: Sequence, expression and homology to v-erb-A. *Nature*, *320*, 134–139. http://dx.doi.org/10.1038/320134a0.
Greene, G. L., Gilna, P., Waterfield, M., Baker, A., Hort, Y., & Shine, J. (1986). Sequence and expression of human estrogen receptor complementary DNA. *Science*, *231*, 1150–1154.
Gros, R., Ding, Q., Liu, B., Chorazyczewski, J., & Feldman, R. D. (2013). Aldosterone mediates its rapid effects in vascular endothelial cells through GPER activation. *American Journal of Physiology. Cell Physiology*, *304*, C532–C540. http://dx.doi.org/10.1152/ajpcell.00203.2012.
Gros, R., Ding, Q., Sklar, L. A., Prossnitz, E. E., Arterburn, J. B., Chorazyczewski, J., & Feldman, R. D. (2011). GPR30 expression is required for the mineralocorticoid receptor-independent rapid vascular effects of aldosterone. *Hypertension*, *57*, 442–451. http://dx.doi.org/10.1161/HYPERTENSIONAHA.110.161653. HYPERTENSIONAHA.110.161653 [pii].
Gruber, C. J., Tschugguel, W., Schneeberger, C., & Huber, J. C. (2002). Production and actions of estrogens. *The New England Journal of Medicine*, *346*, 340–352.
Guo, X., Razandi, M., Pedram, A., Kassab, G., & Levin, E. R. (2005). Estrogen induces vascular wall dilation: Mediation through kinase signaling to nitric oxide and estrogen receptors alpha and beta. *The Journal of Biological Chemistry*, *280*, 19704–19710. http://dx.doi.org/10.1074/jbc.M501244200. M501244200 [pii].
Guzic-Salobir, B., Keber, I., Seljeflot, I., Arnesen, H., & Vrabic, L. (2001). Combined hormone replacement therapy improves endothelial function in healthy postmenopausal women. *Journal of Internal Medicine*, *250*, 508–515.
Haas, E., Bhattacharya, I., Brailoiu, E., Damjanovic, M., Brailoiu, G. C., Gao, X., … Barton, M. (2009). Regulatory role of G protein-coupled estrogen receptor for vascular function and obesity. *Circulation Research*, *104*, 288–291. http://dx.doi.org/10.1161/CIRCRESAHA.108.190892. CIRCRESAHA.108.190892 [pii].
Haas, E., Meyer, M. R., Schurr, U., Bhattacharya, I., Minotti, R., Nguyen, H. H., … Barton, M. (2007). Differential effects of 17beta-estradiol on function and expression of estrogen receptor alpha, estrogen receptor beta, and GPR30 in arteries and veins of patients with atherosclerosis. *Hypertension*, *49*, 1358–1363.
Han, G., & White, R. E. (2014). G-protein-coupled estrogen receptor as a new therapeutic target for treating coronary artery disease. *World Journal of Cardiology*, *6*, 367–375. http://dx.doi.org/10.4330/wjc.v6.i6.367.
Hansson, G. K. (2001). Immune mechanisms in atherosclerosis. *Arteriosclerosis, Thrombosis, and Vascular Biology*, *21*, 1876–1890.
Hansson, G. K. (2005). Inflammation, atherosclerosis, and coronary artery disease. *The New England Journal of Medicine*, *352*, 1685–1695.

Hansson, G. K., & Libby, P. (2006). The immune response in atherosclerosis: A double-edged sword. *Nature Reviews Immunology*, *6*, 508–519. http://dx.doi.org/10.1038/nri1882. nri1882 [pii].

Harada, N., Sasano, H., Murakami, H., Ohkuma, T., Nagura, H., & Takagi, Y. (1999). Localized expression of aromatase in human vascular tissues. *Circulation Research*, *84*, 1285–1291.

Harman, S. M., Vittinghoff, E., Brinton, E. A., Budoff, M. J., Cedars, M. I., Lobo, R. A., ... Black, D. M. (2010). Timing and duration of menopausal hormone treatment may affect cardiovascular outcomes. *The American Journal of Medicine*, *124*, 199–205. http://dx.doi.org/10.1016/j.amjmed.2010.09.021.

Hasbi, A., O'Dowd, B. F., & George, S. R. (2005). A G protein-coupled receptor for estrogen: The end of the search? *Molecular Interventions*, *5*, 158–161. http://dx.doi.org/10.1124/mi.5.3.5.

Haynes, M. P., Sinha, D., Russell, K. S., Collinge, M., Fulton, D., Morales-Ruiz, M., ... Bender, J. R. (2000). Membrane estrogen receptor engagement activates endothelial nitric oxide synthase via the PI3-kinase-Akt pathway in human endothelial cells. *Circulation Research*, *87*, 677–682.

Heidenreich, P. A., Trogdon, J. G., Khavjou, O. A., Butler, J., Dracup, K., & Ezekowitz, M. D., ... Interdisciplinary Council on Quality of Care and Outcomes Research. (2011). Forecasting the future of cardiovascular disease in the United States: A policy statement from the American Heart Association. *Circulation*, *123*, 933–944. http://dx.doi.org/10.1161/CIR.0b013e31820a55f5.

Hevener, A. L., Clegg, D. J., & Mauvais-Jarvis, F. (2015). Impaired estrogen receptor action in the pathogenesis of the metabolic syndrome. *Molecular and Cellular Endocrinology*, *418*(Pt 3), 306–321. http://dx.doi.org/10.1016/j.mce.2015.05.020.

Hewitt, S. C., & Korach, K. S. (2002). Estrogen receptors: Structure, mechanisms and function. *Reviews in Endocrine & Metabolic Disorders*, *3*, 193–200.

Hewitt, S. C., Winuthayanon, W., & Korach, K. S. (2016). What's new in estrogen receptor action in the female reproductive tract. *Journal of Molecular Endocrinology*, *56*, R55–R71. http://dx.doi.org/10.1530/JME-15-0254.

Hill, B. J., Gebre, S., Schlicker, B., Jordan, R., & Necessary, S. (2010). Nongenomic inhibition of coronary constriction by 17ss-estradiol, 2-hydroxyestradiol, and 2-methoxyestradiol. *Canadian Journal of Physiology and Pharmacology*, *88*, 147–152. http://dx.doi.org/10.1139/Y09-120.

Hodgin, J. B., Krege, J. H., Reddick, R. L., Korach, K. S., Smithies, O., & Maeda, N. (2001). Estrogen receptor alpha is a major mediator of 17beta-estradiol's atheroprotective effects on lesion size in ApoE-/- mice. *The Journal of Clinical Investigation*, *107*, 333–340.

Hodis, H. N., Mack, W. J., Henderson, V. W., Shoupe, D., Budoff, M. J., & Hwang-Levine, J., ... for the ELITE Research Group (2016). Vascular effects of early versus late postmenopausal treatment with estradiol. *The New England Journal of Medicine*, *374*, 1221–1231. http://dx.doi.org/10.1056/NEJMoa1505241.

Hulley, S., Grady, D., Bush, T., Furberg, C., Herrington, D., Riggs, B., & Vittinghoff, E. (1998). Randomized trial of estrogen plus progestin for secondary prevention of coronary heart disease in postmenopausal women. Heart and Estrogen/progestin Replacement Study (HERS) research group. *JAMA*, *280*, 605–613.

Hussain, Y., Ding, Q., Connelly, P. W., Brunt, J. H., Ban, M. R., McIntyre, A. D., ... Feldman, R. D. (2015). G-protein estrogen receptor as a regulator of low-density lipoprotein cholesterol metabolism: Cellular and population genetic studies. *Arteriosclerosis, Thrombosis, and Vascular Biology*, *35*, 213–221. http://dx.doi.org/10.1161/ATVBAHA.114.304326.

Iafrati, M. D., Karas, R. H., Aronovitz, M., Kim, S., Sullivan, T. R., Jr., Lubahn, D. B., … Mendelsohn, M. E. (1997). Estrogen inhibits the vascular injury response in estrogen receptor alpha-deficient mice. *Nature Medicine*, *3*, 545–548.

Ignarro, L. J., Buga, G. M., Wood, K. S., Byrns, R. E., & Chaudhuri, G. (1987). Endothelium-derived relaxing factor produced and released from artery and vein is nitric oxide. *Proceedings of the National Academy of Sciences of the United States of America*, *84*, 9265–9269.

Ikeda, Y., Tajima, S., Izawa-Ishizawa, Y., Kihira, Y., Ishizawa, K., Tomita, S., … Tamaki, T. (2012). Estrogen regulates hepcidin expression via GPR30-BMP6-dependent signaling in hepatocytes. *PLoS One*, 7, e40465. http://dx.doi.org/10.1371/journal.pone.0040465.

Iorga, A., Li, J., Sharma, S., Umar, S., Bopassa, J. C., Nadadur, R. D., … Eghbali, M. (2016). Rescue of pressure overload-induced heart failure by estrogen therapy. *Journal of the American Heart Association*, 5. http://dx.doi.org/10.1161/JAHA.115.002482.

Isles, C. G., Hole, D. J., Hawthorne, V. M., & Lever, A. F. (1992). Relation between coronary risk and coronary mortality in women of the Renfrew and Paisley survey: Comparison with men. *Lancet*, *339*, 702–706.

Jang, E. J., Seok, Y. M., Arterburn, J. B., Olatunji, L. A., & Kim, I. K. (2013). GPER-1 agonist G1 induces vasorelaxation through activation of epidermal growth factor receptor-dependent signalling pathway. *The Journal of Pharmacy and Pharmacology*, *65*, 1488–1499. http://dx.doi.org/10.1111/jphp.12113.

Jensen, E. V., & DeSombre, E. R. (1973). Estrogen-receptor interaction. *Science*, *182*, 126–134.

Jensen, E. (2014). A conversation with Elwood Jensen. Interview by David D. Moore. *Annual Reviews in Physiology*, *74*, 1–11. http://dx.doi.org/10.1146/annurev-physiol-020911-153327.

Jensen, E. V., & Jacobson, H. I. (1960). Fate of steroid estrogens in target tissues. In G. Pincus & E. Vollmer (Eds.), *Biological activities of steroids in relation to cancer* (pp. 161–174). New York: Academic.

Jensen, E. V., Desombre, E. R., Hurst, D. J., Kawashima, T., & Jungblut, P. W. (1967). Estrogen-receptor interactions in target tissues. *Archives d'Anatomie Microscopique et de Morphologie Expérimentale*, *56*, 547–569.

Jessup, J. A., Lindsey, S. H., Wang, H., Chappell, M. C., & Groban, L. (2010). Attenuation of salt-induced cardiac remodeling and diastolic dysfunction by the GPER agonist G-1 in female mRen2.Lewis rats. *PLoS One*, *5*, e15433. http://dx.doi.org/10.1371/journal.pone.0015433.

Jia, B., Gao, Y., Li, M., Shi, J., Peng, Y., Du, X., … Zhang, J. (2016). GPR30 promotes prostate stromal cell activation via suppression of ERα expression and its downstream signaling pathway. *Endocrinology*. epub May 10. http://dx.doi.org/en20161035.

Kabir, M. E., Singh, H., Lu, R., Olde, B., Leeb-Lundberg, L. M., & Bopassa, J. C. (2015). G Protein-coupled estrogen receptor 1 mediates acute estrogen-induced cardioprotection via MEK/ERK/GSK-3 beta pathway after ischemia/reperfusion. *PLoS One*, *10*, e0135988. http://dx.doi.org/10.1371/journal.pone.0135988.

Kang, S., Liu, Y., Sun, D., Zhou, C., Liu, A., Xu, C., … Sun, H. (2012). Chronic activation of the G protein-coupled receptor 30 with agonist G-1 attenuates heart failure. *PLoS One*, 7, e48185. http://dx.doi.org/10.1371/journal.pone.0048185.

Karas, R. H., Hodgin, J. B., Kwoun, M., Krege, J. H., Aronovitz, M., Mackey, W., … Mendelsohn, M. E. (1999). Estrogen inhibits the vascular injury response in estrogen receptor beta-deficient female mice. *Proceedings of the National Academy of Sciences of the United States of America*, *96*, 15133–15136.

Karas, R. H., & Mendelsohn, M. E. (1998). Rapid vasomotor effects of estrogen: Men are part of the club. *Chest*, *114*, 1508–1509.

Karas, R. H., Schulten, H., Pare, G., Aronovitz, M. J., Ohlsson, C., Gustafsson, J. A., & Mendelsohn, M. E. (2001). Effects of estrogen on the vascular injury response in estrogen receptor alpha, beta (double) knockout mice. *Circulation Research*, *89*, 534–539.

Khalil, R. A. (2013). Estrogen, vascular estrogen receptor and hormone therapy in postmenopausal vascular disease. *Biochemical Pharmacology*, *86*, 1627–1642. http://dx.doi.org/10.1016/j.bcp.2013.09.024.

Keaney, J. F., Jr., & Solomon, C. G. (2016). Postmenopausal hormone therapy and atherosclerosis—Time is of the essence. *The New England Journal of Medicine*, *374*, 1279–1280. http://dx.doi.org/10.1056/NEJMe1602846.

Kimura, M., Mizukami, Y., Miura, T., Fujimoto, K., Kobayashi, S., & Matsuzaki, M. (2001). Orphan G protein-coupled receptor, GPR41, induces apoptosis via a p53/Bax pathway during ischemic hypoxia and reoxygenation. *The Journal of Biological Chemistry*, *276*, 26453–26460. http://dx.doi.org/10.1074/jbc.M101289200.

Kok, H. S., van Asselt, K. M., van der Schouw, Y. T., van der Tweel, I., Peeters, P. H., Wilson, P. W., … Grobbee, D. E. (2006). Heart disease risk determines menopausal age rather than the reverse. *Journal of the American College of Cardiology*, *47*, 1976–1983.

Komesaroff, P. A., Fullerton, M., Esler, M. D., Dart, A., Jennings, G., & Sudhir, K. (2001). Low-dose estrogen supplementation improves vascular function in hypogonadal men. *Hypertension*, *38*, 1011–1016.

Kublickiene, K., Fu, X. D., Svedas, E., Landgren, B. M., Genazzani, A. R., & Simoncini, T. (2008). Effects in postmenopausal women of estradiol and medroxyprogesterone alone and combined on resistance artery function and endothelial morphology and movement. *The Journal of Clinical Endocrinology and Metabolism*, *93*, 1874–1883. http://dx.doi.org/10.1210/jc.2007-2651.

Kuiper, G. G., Enmark, E., Pelto-Huikko, M., Nilsson, S., & Gustafsson, J. Å. (1996). Cloning of a novel receptor expressed in rat prostate and ovary. *Proceedings of the National Academy of Sciences of the United States of America*, *93*, 5925–5930.

Kvingedal, A. M., & Smeland, E. B. (1997). A novel putative G-protein-coupled receptor expressed in lung, heart and lymphoid tissue. *FEBS Letters*, *407*, 59–62. doi:S0014-5793(97)00278-0 [pii].

Lafferty, A. R., Torpy, D. J., Stowasser, M., Taymans, S. E., Lin, J. P., Huggard, P., … Stratakis, C. A. (2000). A novel genetic locus for low renin hypertension: Familial hyperaldosteronism type II maps to chromosome 7 (7p22). *Journal of Medical Genetics*, *37*, 831–835.

Lafont, A. G., Rousseau, K., Tomkiewicz, J., & Dufour, S. (2016). Three nuclear and two membrane estrogen receptors in basal teleosts, *Anguilla* sp.: Identification, evolutionary history and differential expression regulation. *General and Comparative Endocrinology*. http://dx.doi.org/10.1016/j.ygcen.2015.11.021 (in press).

Lakatta, E. G. (2015). So! What's aging? Is cardiovascular aging a disease? *Journal of Molecular and Cellular Cardiology*, *83*, 1–13. http://dx.doi.org/10.1016/j.yjmcc.2015.04.005.

Landahl, S., Bengtsson, C., Sigurdsson, J. A., Svanborg, A., & Svardsudd, K. (1986). Age-related changes in blood pressure. *Hypertension*, *8*, 1044–1049.

Lee, T. M., Lin, S. Z., & Chang, N. C. (2014). Both GPER and membrane oestrogen receptor-alpha activation protect ventricular remodelling in 17beta oestradiol-treated ovariectomized infarcted rats. *Journal of Cellular and Molecular Medicine*, *18*, 2454–2465. http://dx.doi.org/10.1111/jcmm.12430.

Lee, S. L., Wang, W. W., Lanzillo, J. J., & Fanburg, B. L. (1994). Regulation of serotonin-induced DNA synthesis of bovine pulmonary artery smooth muscle cells. *The American Journal of Physiology*, *266*, L53–L60.

Lenhart, P. M., Broselid, S., Barrick, C. J., Leeb-Lundberg, L. M., & Caron, K. M. (2013). G-protein-coupled receptor 30 interacts with receptor activity-modifying protein 3 and confers sex-dependent cardioprotection. *Journal of Molecular Endocrinology, 51*, 191–202. http://dx.doi.org/10.1530/JME-13-0021.

Lew, R., Komesaroff, P., Williams, M., Dawood, T., & Sudhir, K. (2003). Endogenous estrogens influence endothelial function in young men. *Circulation Research, 93*, 1127–1133.

Li, F., Yu, X., Szynkarski, C. K., Meng, C., Zhou, B., Barhoumi, R., … Han, G. (2013). Activation of GPER induces differentiation and inhibition of coronary artery smooth muscle cell proliferation. *PLoS One, 8*, e64771. http://dx.doi.org/10.1371/journal.pone.0064771.

Libby, P. (2006). Inflammation and cardiovascular disease mechanisms. *The American Journal of Clinical Nutrition, 83*, 456S–460S.

Libby, P. (2013). Collagenases and cracks in the plaque. *The Journal of Clinical Investigation, 123*, 3201–3203. http://dx.doi.org/10.1172/JCI67526.

Libby, P., Ridker, P. M., & Maseri, A. (2002). Inflammation and atherosclerosis. *Circulation, 105*, 1135–1143.

Lieberman, E. H., Gerhard, M. D., Uehata, A., Walsh, B. W., Selwyn, A. P., Ganz, P., … Creager, M. A. (1994). Estrogen improves endothelium-dependent, flow-mediated vasodilation in postmenopausal women. *Annals of Internal Medicine, 121*, 936–941.

Lim, S. C., Caballero, A. E., Arora, S., Smakowski, P., Bashoff, E. M., Brown, F. M., … Veves, A. (1999). The effect of hormonal replacement therapy on the vascular reactivity and endothelial function of healthy individuals and individuals with type 2 diabetes. *The Journal of Clinical Endocrinology and Metabolism, 84*, 4159–4164. http://dx.doi.org/10.1210/jcem.84.11.6160.

Lindsey, S. H., Carver, K. A., Prossnitz, E. R., & Chappell, M. C. (2011). Vasodilation in response to the GPR30 agonist G-1 is not different from estradiol in the mRen2.Lewis female rat. *Journal of Cardiovascular Pharmacology*. http://dx.doi.org/10.1097/FJC.0b013e3182135f1c.

Lindsey, S. H., Cohen, J. A., Brosnihan, K. B., Gallagher, P. E., & Chappell, M. C. (2009). Chronic treatment with the G protein-coupled receptor 30 agonist G-1 decreases blood pressure in ovariectomized mRen2.Lewis rats. *Endocrinology, 150*, 3753–3758.

Lindsey, S. H., Liu, L., & Chappell, M. C. (2014). Vasodilation by GPER in mesenteric arteries involves both endothelial nitric oxide and smooth muscle cAMP signaling. *Steroids, 81*, 99–102. http://dx.doi.org/10.1016/j.steroids.2013.10.017.

Lindsey, S. H., Yamaleyeva, L. M., Brosnihan, K. B., Gallagher, P. E., & Chappell, M. C. (2011). Estrogen receptor GPR30 reduces oxidative stress and proteinuria in the salt-sensitive female mRen2.Lewis rat. *Hypertension, 58*, 665–671. http://dx.doi.org/10.1161/HYPERTENSIONAHA.111.175174.

Liu, L., Kashyap, S., Murphy, B., Hutson, D. D., Budish, R. A., Trimmer, E. H., … Lindsey, S. H. (2016). GPER activation ameliorates aortic remodeling induced by salt-sensitive hypertension. *American journal of physiology Heart and Circulatory Physiology, 310*, H953–H961. http://dx.doi.org/10.1152/ajpheart.00631.2015.

Liu, S., Kilic, G., Meyers, M. S., Navarro, G., Wang, Y., Oberholzer, J., & Mauvais-Jarvis, F. (2013). Oestrogens improve human pancreatic islet transplantation in a mouse model of insulin deficient diabetes. *Diabetologia, 56*, 370–381. http://dx.doi.org/10.1007/s00125-012-2764-1.

Locher, R., Emmanuele, L., Suter, P. M., Vetter, W., & Barton, M. (2002). Green tea polyphenols inhibit human vascular smooth muscle cell proliferation stimulated by native low-density lipoprotein. *European Journal of Pharmacology, 434*, 1–7.

Löhn, M., Dubrovska, G., Lauterbach, B., Luft, F. C., Gollasch, M., & Sharma, A. M. (2002). Periadventitial fat releases a vascular relaxing factor. *FASEB Journal, 16*, 1057–1063. http://dx.doi.org/10.1096/fj.02-0024com.

Louet, J. F., LeMay, C., & Mauvais-Jarvis, F. (2004). Antidiabetic actions of estrogen: Insight from human and genetic mouse models. *Current Atherosclerosis Reports, 6*, 180–185.

Lozano, R., Naghavi, M., Foreman, K., Lim, S., Shibuya, K., Aboyans, V., ... Memish, Z. A. (2012). Global and regional mortality from 235 causes of death for 20 age groups in 1990 and 2010: A systematic analysis for the Global Burden of Disease Study 2010. *Lancet, 380*, 2095–2128. http://dx.doi.org/10.1016/S0140-6736(12)61728-0.

Madak-Erdogan, Z., Kim, S. H., Gong, P., Zhao, Y. C., Zhang, H., Chambliss, K. L., ... Katzenellenbogen, B. S. (2016). Design of pathway preferential estrogens that provide beneficial metabolic and vascular effects without stimulating reproductive tissues. *Science Signaling 9*, http://dx.doi.org/10.1126/scisignal.aad8170, ra53.

Maddox, Y. T., Falcon, J. G., Ridinger, M., Cunard, C. M., & Ramwell, P. W. (1987). Endothelium-dependent gender differences in the response of the rat aorta. *The Journal of Pharmacology and Experimental Therapeutics, 240*, 392–395.

Mårtensson, U. E., Salehi, S. A., Windahl, S., Gomez, M. F., Sward, K., Daszkiewicz-Nilsson, J., ... Leeb-Lundberg, L. M. (2009). Deletion of the G protein-coupled receptor 30 impairs glucose tolerance, reduces bone growth, increases blood pressure, and eliminates estradiol-stimulated insulin release in female mice. *Endocrinology, 150*, 687–698. http://dx.doi.org/10.1210/en.2008-0623. en.2008-0623 [pii].

Mata, K. M., Li, W., Reslan, O. M., Siddiqui, W. T., Opsasnick, L. A., & Khalil, R. A. (2015). Adaptive increases in expression and vasodilator activity of estrogen receptor subtypes in a blood vessel-specific pattern during pregnancy. *American Journal of Physiology. Heart and Circulatory Physiology, 309*, H1679–H1696. http://dx.doi.org/10.1152/ajpheart.00532.2015.

Matthews, J., & Gustafsson, J. A. (2003). Estrogen signaling: A subtle balance between ER alpha and ER beta. *Molecular Interventions, 3*, 281–292.

Mauvais-Jarvis, F., Clegg, D. J., & Hevener, A. L. (2013). The role of estrogens in control of energy balance and glucose homeostasis. *Endocrine Reviews, 34*, 309–338. http://dx.doi.org/10.1210/er.2012-1055.

Maximov, P. Y., Lee, T. M., & Jordan, V. C. (2013). The discovery and development of selective estrogen receptor modulators (SERMs) for clinical practice. *Current Clinical Pharmacology, 8*, 135–155.

Mazzuca, M. Q., Mata, K. M., Li, W., Rangan, S. S., & Khalil, R. A. (2015). Estrogen receptor subtypes mediate distinct microvascular dilation and reduction in [Ca2+]I in mesenteric microvessels of female rat. *The Journal of Pharmacology and Experimental Therapeutics, 352*, 291–304. http://dx.doi.org/10.1124/jpet.114.219865.

McCoy, R. L., & Perlemutter, D. H. (1996). *Cloning of novel IL8-related receptors from human hepatic tissue: Human IL8-related receptor (DRY12) mRNA, complete codons*. Submitted June 1996 to the EMBL/GenBank/DDBJ databases. Genbank accession code U58828.

McGill, H. C. J., McMahan, C. A., Herderick, E. E., Zieske, A. W., Malcom, G. T., Tracy, R. E., & Strong, J. P. (2002). Obesity accelerates the progression of coronary atherosclerosis in young men. *Circulation, 105*, 2712–2718.

McSweeney, J. C., Rosenfeld, A. G., Abel, W. M., Braun, L. T., Burke, L. E., Daugherty, S. L. ... on behalf of the American Heart Association Council on Cardiovascular and Stroke Nursing, Council on Clinical Cardiology, Council on Epidemiology and Prevention, Council on Hypertension, Council on Lifestyle and Cardiometabolic Health, and Council on Quality of Care and Outcomes Research. (2016). Preventing and experiencing ischemic heart disease as a woman: State of the science—A scientific statement from the American Heart Association. *Circulation, 133*, 1302–1331. http://dx.doi.org/10.1161/CIR.0000000000000381.

Menazza, S., & Murphy, E. (2016). The expanding complexity of estrogen receptor signaling in the cardiovascular system. *Circulation Research, 118*, 994–1007.

Mendelsohn, M. E., & Karas, R. H. (1999). The protective effects of estrogen on the cardiovascular system. *The New England Journal of Medicine, 340*, 1801–1811.

Mendelsohn, M. E., & Karas, R. H. (2005). Molecular and cellular basis of cardiovascular gender differences. *Science*, *308*, 1583–1587.

Mendizabal, Y., Llorens, S., & Nava, E. (2013). Vasoactive effects of prostaglandins from the perivascular fat of mesenteric resistance arteries in WKY and SHROB rats. *Life Sciences*, *93*, 1023–1032. http://dx.doi.org/10.1016/j.lfs.2013.10.021.

Merz, C. N., Johnson, B. D., Berga, S., Braunstein, G., Reis, S. E., & Bittner, V. (2006). Past oral contraceptive use and angiographic coronary artery disease in postmenopausal women: Data from the National Heart, Lung, and Blood Institute-sponsored Women's Ischemia Syndrome Evaluation. *Fertility and Sterility*, *85*, 1425–1431.

Meyer, M. R., Amann, K., Field, A. S., Hu, C., Hathaway, H. J., Kanagy, N. L., … Prossnitz, E. R. (2012). Deletion of G protein-coupled estrogen receptor increases endothelial vasoconstriction. *Hypertension*, *59*, 507–512. http://dx.doi.org/10.1161/HYPERTENSIONAHA.111.184606.

Meyer, M. R., Baretella, O., Prossnitz, E. R., & Barton, M. (2010). Dilation of epicardial coronary arteries by the G protein-coupled estrogen receptor agonists G-1 and ICI 182,780. *Pharmacology*, *86*, 58–64.

Meyer, M. R., & Barton, M. (2009). ERα, ERβ, and gpER: Novel aspects of oestrogen receptor signalling in atherosclerosis. *Cardiovascular Research*, *83*, 605–610. http://dx.doi.org/10.1093/cvr/cvp187.

Meyer, M. R., Clegg, D. J., Prossnitz, E. R., & Barton, M. (2011). Obesity, insulin resistance and diabetes: Sex differences and role of oestrogen receptors. *Acta Physiologica (Oxford, England)*, *203*, 259–269. http://dx.doi.org/10.1111/j.1748-1716.2010.02237.x.

Meyer, M. R., Field, A. S., Kanagy, N. L., Barton, M., & Prossnitz, E. R. (2012). GPER regulates endothelin-dependent vascular tone and intracellular calcium. *Life Sciences*, *91*, 623–627. http://dx.doi.org/10.1016/j.lfs.2012.01.007.

Meyer, M. R., Fredette, N. C., Barton, M., & Prossnitz, E. R. (2013). Regulation of vascular smooth muscle tone by adipose-derived contracting factor. *PLoS One*, *8*, e79245. http://dx.doi.org/10.1371/journal.pone.0079245.

Meyer, M. R., Fredette, N. C., Barton, M., & Prossnitz, E. R. (2015). G protein-coupled estrogen receptor inhibits vascular prostanoid production and activity. *The Journal of Endocrinology*, *227*, 61–69. http://dx.doi.org/10.1530/JOE-15-0257.

Meyer, M. R., Fredette, N. C., Howard, T., Hu, C., Ramesh, C., Daniel, C., … Prossnitz, E. R. (2014). G protein-coupled estrogen receptor protects from atherosclerosis. *Scientific Reports*, *4*, 7564. Erratum in: Scientific Reports, 7565 (2015) 13510.

Meyer, M. R., Fredette, N. C., Sharma, G., Barton, M., & Prossnitz, E. R. (2016). GPER is required for the age-dependent upregulation of the myocardial endothelin system. *Life Sciences*. http://dx.doi.org/10.1016/j.lfs.2016.02.041.

Meyer, M. R., Haas, E., & Barton, M. (2006). Gender differences of cardiovascular disease: New perspectives for estrogen receptor signaling. *Hypertension*, *47*, 1019–1026.

Meyer, M. R., Haas, E., & Barton, M. (2008). Need for research on estrogen receptor function: Importance for postmenopausal hormone therapy and atherosclerosis. *Gender Medicine*, *5*(Suppl. A), S19–S33.

Meyer, M. R., Haas, E., Prossnitz, E. R., & Barton, M. (2009). Non-genomic regulation of vascular cell function and growth by estrogen. *Molecular and Cellular Endocrinology*, *308*, 9–16.

Meyer, M. R., Prossnitz, E. R., & Barton, M. (2011). The G protein-coupled estrogen receptor GPER/GPR30 as a regulator of cardiovascular function. *Vascular Pharmacology*, *55*, 17–25. http://dx.doi.org/10.1016/j.vph.2011.06.003. S1537-1891(11)00064-4 [pii].

Meyers, M. J., Sun, J., Carlson, K. E., Marriner, G. A., Katzenellenbogen, B. S., & Katzenellenbogen, J. A. (2001). Estrogen receptor-beta potency-selective ligands:

Structure-activity relationship studies of diarylpropionitriles and their acetylene and polar analogues. *Journal of Medicinal Chemistry*, *44*, 4230–4251.
Miller, V. M., & Duckles, S. P. (2008). Vascular actions of estrogens: Functional implications. *Pharmacological Reviews*, *60*, 210–241.
Miller, V. M., & Vanhoutte, P. M. (1990). 17 beta-Estradiol augments endothelium-dependent contractions to arachidonic acid in rabbit aorta. *The American Journal of Physiology*, *258*, R1502–R1507.
Miller, A. P., Xing, D., Feng, W., Fintel, M., Chen, Y. F., & Oparil, S. (2007). Aged rats lose vasoprotective and anti-inflammatory actions of estrogen in injured arteries. *Menopause*, *14*, 251–260.
Minson, C. T., Halliwill, J. R., Young, T. M., & Joyner, M. J. (2000). Influence of the menstrual cycle on sympathetic activity, baroreflex sensitivity, and vascular transduction in young women. *Circulation*, *101*, 862–868.
Mizutani, K., Ikeda, K., & Yamori, Y. (2000). Resveratrol inhibits AGEs-induced proliferation and collagen synthesis activity in vascular smooth muscle cells from stroke-prone spontaneously hypertensive rats. *Biochemical and Biophysical Research Communications*, *274*, 61–67. http://dx.doi.org/10.1006/bbrc.2000.3097.
Moreno-Ulloa, A., Mendez-Luna, D., Beltran-Partida, E., Castillo, C., Guevara, G., Ramirez-Sanchez, I., … Villarreal, F. (2015). The effects of (-)-epicatechin on endothelial cells involve the G protein-coupled estrogen receptor (GPER). *Pharmacological Research*, *100*, 309–320. http://dx.doi.org/10.1016/j.phrs.2015.08.014.
Mosselman, S., Polman, J., & Dijkema, R. (1996). ER beta: Identification and characterization of a novel human estrogen receptor. *FEBS Letters*, *392*, 49–53.
Mucowski, S. J., Mack, W. J., Shoupe, D., Kono, N., Paulson, R., & Hodis, H. N. (2014). Effect of prior oophorectomy on changes in bone mineral density and carotid artery intima-media thickness in postmenopausal women. *Fertility and Sterility*, *101*, 1117–1122. http://dx.doi.org/10.1016/j.fertnstert.2013.12.055.
Mügge, A., Riedel, M., Barton, M., Kuhn, M., & Lichtlen, P. R. (1993a). Endothelium independent relaxation of human coronary arteries by 17 beta-oestradiol in vitro. *Cardiovascular Research*, *27*, 1939–1942.
Mügge, A., Riedel, M., Barton, M., Kuhn, M., & Lichtlen, P. R. (1993b). Estrogen dilates human coronary arteries in vitro by an increase in cyclic GMP and cyclic AMP smooth muscle content. *European Heart Journal*, *14*, 16.
Mukherjee, T. K., Dinh, H., Chaudhuri, G., & Nathan, L. (2002). Testosterone attenuates expression of vascular cell adhesion molecule-1 by conversion to estradiol by aromatase in endothelial cells: Implications in atherosclerosis. *Proceedings of the National Academy of Sciences of the United States of America*, *99*, 4055–4060. http://dx.doi.org/10.1073/pnas.052703199.
Murata, T., Dietrich, H. H., Xiang, C., & Dacey, R. G., Jr. (2013). G protein-coupled estrogen receptor agonist improves cerebral microvascular function after hypoxia/reoxygenation injury in male and female rats. *Stroke*, *44*, 779–785. http://dx.doi.org/10.1161/STROKEAHA.112.678177.
Nadal, A., Alonso-Magdalena, P., Soriano, S., Quesada, I., & Ropero, A. B. (2009a). The pancreatic beta-cell as a target of estrogens and xenoestrogens: Implications for blood glucose homeostasis and diabetes. *Molecular and Cellular Endocrinology*, *304*, 63–68. http://dx.doi.org/10.1016/j.mce.2009.02.016. S0303-7207(09)00150-6 [pii].
Nadal, A., Alonso-Magdalena, P., Soriano, S., Ripoll, C., Fuentes, E., Quesada, I., & Ropero, A. B. (2011). Role of estrogen receptors alpha, beta and GPER1/GPR30 in pancreatic beta-cells. *Frontiers in Bioscience (Landmark edition)*, *16*, 251–260.
Nadal, A., Alonso-Magdalena, P., Soriano, S., Ropero, A. B., & Quesada, I. (2009b). The role of oestrogens in the adaptation of islets to insulin resistance. *The Journal of Physiology*, *587*, 5031–5037. http://dx.doi.org/10.1113/jphysiol.2009.177188.

NCD Risk Factor Collaboration (NCD-RisC) (2016). Trends in adult body-mass index in 200 countries from 1975 to 2014: a pooled analysis of 1698 population-based measurement studies with 19·2 million participants. *Lancet*, *387*, 1377–1396. http://dx.doi.org/10.1016/S0140-6736(16)30054-X. Erratum in: Lancet, 2016, 387, 1998.

Nechmad, A., Merin, G., Schwalb, H., Shimon, D. V., Borman, J. B., Milgalter, E., & Mosseri, M. (1998). Estrogen induces nitric oxide-mediated vasodilation of human mammary arteries in vitro. *Nitric Oxide*, *2*, 460–466. http://dx.doi.org/10.1006/niox.1998.0202.

Ng, M., Fleming, T., Robinson, M., Thomson, B., Graetz, N., Margono, C., … Gakidou, E. (2014). Global, regional, and national prevalence of overweight and obesity in children and adults during 1980-2013: A systematic analysis for the Global Burden of Disease Study 2013. *Lancet*, *384*, 766–781. http://dx.doi.org/10.1016/S0140-6736(14)60460-8.

Novella, S., Heras, M., Hermenegildo, C., & Dantas, A. P. (2012). Effects of estrogen on vascular inflammation: A matter of timing. *Arteriosclerosis, Thrombosis, and Vascular Biology*, *32*, 2035–2042. http://dx.doi.org/10.1161/ATVBAHA.112.250308.

O'Dowd, B. F., Nguyen, T., Marchese, A., Cheng, R., Lynch, K. R., Heng, H. H., … George, S. R. (1998). Discovery of three novel G-protein-coupled receptor genes. *Genomics*, *47*, 310–313. http://dx.doi.org/10.1006/geno.1998.5095. S0888-7543(98)95095-9 [pii].

Ohman, M. K., Luo, W., Wang, H., Guo, C., Abdallah, W., Russo, H. M., & Eitzman, D. T. (2011). Perivascular visceral adipose tissue induces atherosclerosis in apolipoprotein E deficient mice. *Atherosclerosis*, *219*, 33–39. http://dx.doi.org/10.1016/j.atherosclerosis.2011.07.012.

Olshansky, S. J., Passaro, D. J., Hershow, R. C., Layden, J., Carnes, B. A., Brody, J., … Ludwig, D. S. (2005). A potential decline in life expectancy in the United States in the 21st century. *New England Journal of Medicine*, *352*, 1138–1145.

Orshal, J. M., & Khalil, R. A. (2004). Gender, sex hormones, and vascular tone. *American Journal of Physiology Regulatory, Integrative and Comparative Physiology*, *286*, R233–R249. http://dx.doi.org/10.1152/ajpregu.00338.2003.

Owen, M. K., Witzmann, F. A., McKenney, M. L., Lai, X., Berwick, Z. C., Moberly, S. P., … Tune, J. D. (2013). Perivascular adipose tissue potentiates contraction of coronary vascular smooth muscle: Influence of obesity. *Circulation*, *128*, 9–18. http://dx.doi.org/10.1161/CIRCULATIONAHA.112.001238.

Owman, C., Blay, P., Nilsson, C., & Lolait, S. J. (1996). Cloning of human cDNA encoding a novel heptahelix receptor expressed in Burkitt's lymphoma and widely distributed in brain and peripheral tissues. *Biochemical and Biophysical Research Communications*, *228*, 285–292.

Pelekanou, V., Kampa, M., Kiagiadaki, F., Deli, A., Theodoropoulos, P., Agrogiannis, G., … Notas, G. (2016). Estrogen anti-inflammatory activity on human monocytes is mediated through cross-talk between estrogen receptor ERalpha36 and GPR30/GPER1. *Journal of Leukocyte Biology*, *99*, 333–347. http://dx.doi.org/10.1189/jlb.3A0914-430RR.

Petrie, W. K., Dennis, M. K., Hu, C., Dai, D., Arterburn, J. B., Smith, H. O., … Prossnitz, E. R. (2013). G protein-coupled estrogen receptor-selective ligands modulate endometrial tumor growth. *Obstetrics and Gynecology International*, *2013*, 472720. http://dx.doi.org/10.1155/2013/472720.

Pinto, A. R., Ilinykh, A., Ivey, M. J., Kuwabara, J. T., D'Antoni, M. L., Debuque, R., … Tallquist, M. D. (2016). Revisiting cardiac cellular composition. *Circulation Research*, *118*, 400–409. http://dx.doi.org/10.1161/CIRCRESAHA.115.307778.

Pinto, S., Virdis, A., Ghiadoni, L., Bernini, G., Lombardo, M., Petraglia, F., … Salvetti, A. (1997). Endogenous estrogen and acetylcholine-induced vasodilation in normotensive women. *Hypertension*, *29*, 268–273.

Prossnitz, E. R., & Arterburn, J. B. (2015). International Union of Basic and Clinical Pharmacology. XCVII. G protein-coupled estrogen receptor (GPER) and its pharmacologic modulators. *Pharmacological Reviews*, *67*, 505–540.

Prossnitz, E. R., & Barton, M. (2009). Signaling, physiological functions and clinical relevance of the G protein-coupled estrogen receptor GPER. *Prostaglandins & Other Lipid Mediators*, *89*, 89–97.

Prossnitz, E. R., & Barton, M. (2011). The G-protein-coupled estrogen receptor GPER in health and disease. *Nature Reviews Endocrinology*, 7, 715–726. http://dx.doi.org/10.1038/nrendo.2011.122.

Prossnitz, E. R., & Barton, M. (2014). Estrogen biology: New insights into GPER function and clinical opportunities. *Molecular and Cellular Endocrinology*, *389*, 71–83.

Rang, S., Wolf, H., Montfrans, G. A., & Karemaker, J. M. (2002). Non-invasive assessment of autonomic cardiovascular control in normal human pregnancy and pregnancy- associated hypertensive disorders: A review. *Journal of Hypertension*, *20*, 2111–2119.

Razandi, M., Pedram, A., Greene, G. L., & Levin, E. R. (1999). Cell membrane and nuclear estrogen receptors (ERs) originate from a single transcript: Studies of ERalpha and ERbeta expressed in Chinese hamster ovary cells. *Molecular Endocrinology*, *13*, 307–319. http://dx.doi.org/10.1210/mend.13.2.0239.

Razandi, M., Pedram, A., Merchenthaler, I., Greene, G. L., & Levin, E. R. (2004). Plasma membrane estrogen receptors exist and functions as dimers. *Molecular Endocrinology*, *18*, 2854–2865. http://dx.doi.org/10.1210/me.2004-0115.

Recchia, A. G., De Francesco, E. M., Vivacqua, A., Sisci, D., Panno, M. L., Ando, S., & Maggiolini, M. (2011). The G protein-coupled receptor 30 is up-regulated by hypoxia-inducible factor-1alpha (HIF-1alpha) in breast cancer cells and cardiomyocytes. *The Journal of Biological Chemistry*, *286*, 10773–10782. http://dx.doi.org/10.1074/jbc.M110.172247.

Reis, S. E., Holubkov, R., Zell, K. A., Smith, A. J., Cohen, H. A., Feldman, M. D., & Blumenthal, R. S. (1998). Estrogen acutely abolishes abnormal cold-induced coronary constriction in men. *Chest*, *114*, 1556–1561.

Reslan, O. M., & Khalil, R. A. (2012). Vascular effects of estrogenic menopausal hormone therapy. *Reviews on Recent Clinical Trials*, 7, 47–70.

Reslan, O. M., Yin, Z., do Nascimento, G. R., & Khalil, R. A. (2013). Subtype-specific estrogen receptor-mediated vasodilator activity in the cephalic, thoracic, and abdominal vasculature of female rat. *Journal of Cardiovascular Pharmacology*, *62*, 26–40. http://dx.doi.org/10.1097/FJC.0b013e31828bc88a.

Revankar, C. M., Cimino, D. F., Sklar, L. A., Arterburn, J. B., & Prossnitz, E. R. (2005). A transmembrane intracellular estrogen receptor mediates rapid cell signaling. *Science*, *307*, 1625–1630.

Reynolds, S. R. M., & Foster, F. I. (1939). Peripheral vascular action of estrogen in the human male. *The Journal of Clinical Investigation*, *18*, 649–655.

Riedel, M., & Mügge, A. (1993). Oestrogen and protection against myocardial ischaemia. *Lancet*, *342*, 872.

Rigiracciolo, D. C., Scarpelli, A., Lappano, R., Pisano, A., Santolla, M. F., Avino, S., … De Francesco, E. M. (2016). GPER is involved in the stimulatory effects of aldosterone in breast cancer cells and breast tumor-derived endothelial cells. *Oncotarget*, 7, 94–111. http://dx.doi.org/10.18632/oncotarget.6475.

Rosano, G. M., Sarrel, P. M., Poole-Wilson, P. A., & Collins, P. (1993). Beneficial effect of oestrogen on exercise-induced myocardial ischaemia in women with coronary artery disease. *Lancet*, *342*, 133–136.

Ross, R. (1999). Atherosclerosis—An inflammatory disease. *The New England Journal of Medicine*, *340*, 115–126.

Rossouw, J. E., Anderson, G. L., Prentice, R. L., LaCroix, A. Z., Kooperberg, C., Stefanick, M. L., ... Ockene, J. (2002). Risks and benefits of estrogen plus progestin in healthy postmenopausal women: Principal results from the Women's Health Initiative randomized controlled trial. *JAMA*, *288*, 321–333.

Sadee, W. (2005). Pharmacogenomics: Harbinger for the era of personalized medicine? *Molecular Interventions*, *5*, 140–143. http://dx.doi.org/10.1124/mi.5.3.1.

Sakurai, T., Iimuro, S., Araki, A., Umegaki, H., Ohashi, Y., Yokono, K., & Ito, H. (2010). Age-associated increase in abdominal obesity and insulin resistance, and usefulness of AHA/NHLBI definition of metabolic syndrome for predicting cardiovascular disease in Japanese elderly with type 2 diabetes mellitus. *Gerontology*, *56*, 141–149. http://dx.doi.org/10.1159/000246970.

Santolla, M. F., De Francesco, E. M., Lappano, R., Rosano, C., Abonante, S., & Maggiolini, M. (2014). Niacin activates the G protein estrogen receptor (GPER)-mediated signalling. *Cellular Signalling*, *26*, 1466–1475. http://dx.doi.org/10.1016/j.cellsig.2014.03.011.

Santolla, M. F., Lappano, R., De Marco, P., Pupo, M., Vivacqua, A., Sisci, D., ... Maggiolini, M. (2012). G protein-coupled estrogen receptor mediates the up-regulation of fatty acid synthase induced by 17beta-estradiol in cancer cells and cancer-associated fibroblasts. *The Journal of Biological Chemistry*, *287*, 43234–43245. http://dx.doi.org/10.1074/jbc.M112.417303.

Schenck-Gustafsson, K., Brincat, M., Erel, C. T., Gambacciani, M., Lambrinoudaki, I., Moen, M. H., ... EMAS (2011). EMAS position statement: Managing the menopause in the context of coronary heart disease. *Maturitas*, *68*, 94–97. http://dx.doi.org/10.1016/j.maturitas.2010.10.005.

Scherer, S. W., Cheung, J., MacDonald, J. R., Osborne, L. R., Nakabayashi, K., Herbrick, J. A., ... Tsui, L. C. (2003). Human chromosome 7: DNA sequence and biology. *Science*, *300*, 767–772. http://dx.doi.org/10.1126/science.1083423.

Schierbeck, L. L., Rejnmark, L., Tofteng, C. L., Stilgren, L., Eiken, P., Mosekilde, L., ... Jensen, J. E. (2012). Effect of hormone replacement therapy on cardiovascular events in recently postmenopausal women: Randomised trial. *BMJ*, *345*, e6409. http://dx.doi.org/10.1136/bmj.e6409.

Sharma, G., Hu, C., Brigman, J. L., Zhu, G., Hathaway, H. J., & Prossnitz, E. R. (2013). GPER deficiency in male mice results in insulin resistance, dyslipidemia, and a proinflammatory state. *Endocrinology*, *154*, 4136–4145. http://dx.doi.org/10.1210/en.2013-1357.

Sharma, G., & Prossnitz, E. R. (2011). Mechanisms of estradiol-induced insulin secretion by the G protein-coupled estrogen receptor GPR30/GPER in pancreatic beta-cells. *Endocrinology*, *152*, 3030–3039. http://dx.doi.org/10.1210/en.2011-0091.

Sherwood, A., Bower, J. K., McFetridge-Durdle, J., Blumenthal, J. A., Newby, L. K., & Hinderliter, A. L. (2007). Age moderates the short-term effects of transdermal 17beta-estradiol on endothelium-dependent vascular function in postmenopausal women. *Arteriosclerosis, Thrombosis, and Vascular Biology*, *27*, 1782–1787. http://dx.doi.org/10.1161/ATVBAHA.107.145383. ATVBAHA.107.145383 [pii].

Simoncini, T., Hafezi-Moghadam, A., Brazil, D. P., Ley, K., Chin, W. W., & Liao, J. K. (2000). Interaction of oestrogen receptor with the regulatory subunit of phosphatidylinositol-3-OH kinase. *Nature*, *407*, 538–541.

Smith, G. D. (2016). A fatter, healthier but more unequal world. *Lancet*, *387*, 1349–1350. http://dx.doi.org/10.1016/S0140-6736(16)00588-2.

Soloff, M. S., & Szego, C. M. (1969). Purification of estradiol receptor from rat uterus and blockade of its estrogen-binding function by specific antibody. *Biochemical and Biophysical Research Communications*, *34*, 141–147.

Soltis, E. E., & Cassis, L. A. (1991). Influence of perivascular adipose tissue on rat aortic smooth muscle responsiveness. *Clinical and Experimental Hypertension Part A, 13*, 277–296.
Somjen, D., Amir-Zaltsman, Y., Gayer, B., Kulik, T., Knoll, E., Stern, N., ... Kohen, F. (2002). 6-Carboxymethyl genistein: A novel selective oestrogen receptor modulator (SERM) with unique, differential effects on the vasculature, bone and uterus. *The Journal of Endocrinology, 173*, 415–427.
Stauffer, S. R., Coletta, C. J., Tedesco, R., Nishiguchi, G., Carlson, K., Sun, J., ... Katzenellenbogen, J. A. (2000). Pyrazole ligands: Structure-affinity/activity relationships and estrogen receptor-alpha-selective agonists. *Journal of Medicinal Chemistry, 43*, 4934–4947.
Steinberg, H. O., Paradisi, G., Cronin, J., Crowde, K., Hempfling, A., Hook, G., & Baron, A. D. (2000). Type II diabetes abrogates sex differences in endothelial function in premenopausal women. *Circulation, 101*, 2040–2046.
Stewart, K. G., Zhang, Y., & Davidge, S. T. (1999). Estrogen decreases prostaglandin H synthase products from endothelial cells. *Journal of the Society for Gynecologic Investigation, 6*, 322–327.
Stewart, S. T., Cutler, D. M., & Rosen, A. B. (2009). Forecasting the effects of obesity and smoking on U.S. life expectancy. *New England Journal of Medicine, 361*, 2252–2260. http://dx.doi.org/10.1056/NEJMsa0900459.
Stork, S., van der Schouw, Y. T., Grobbee, D. E., & Bots, M. L. (2004). Estrogen, inflammation and cardiovascular risk in women: A critical appraisal. *Trends in Endocrinology and Metabolism, 15*, 66–72. http://dx.doi.org/10.1016/j.tem.2004.01.005. S1043276004000062 [pii].
Straub, R. H. (2007). The complex role of estrogens in inflammation. *Endocrine Reviews, 28*, 521–574. http://dx.doi.org/10.1210/er.2007-0001. er.2007-0001 [pii].
Sudhir, K., Chou, T. M., Chatterjee, K., Smith, E. P., Williams, T. C., Kane, J. P., ... Rubanyi, G. M. (1997a). Premature coronary artery disease associated with a disruptive mutation in the estrogen receptor gene in a man. *Circulation, 96*, 3774–3777.
Sudhir, K., Chou, T. M., Messina, L. M., Hutchison, S. J., Korach, K. S., Chatterjee, K., & Rubanyi, G. M. (1997b). Endothelial dysfunction in a man with disruptive mutation in oestrogen-receptor gene. *Lancet, 349*, 1146–1147. http://dx.doi.org/10.1016/S0140-6736(05)63022-X.
Sudhir, K., Ko, E., Zellner, C., Wong, H. E., Hutchison, S. J., Chou, T. M., & Chatterjee, K. (1997). Physiological concentrations of estradiol attenuate endothelin 1-induced coronary vasoconstriction in vivo. *Circulation, 96*, 3626–3632.
Takada, Y., Kato, C., Kondo, S., Korenaga, R., & Ando, J. (1997). Cloning of cDNAs encoding G protein-coupled receptor expressed in human endothelial cells exposed to fluid shear stress. *Biochemical and Biophysical Research Communications, 240*, 737–741.
Takahashi, K., Ohmichi, M., Yoshida, M., Hisamoto, K., Mabuchi, S., Arimoto-Ishida, E., ... Kurachi, H. (2003). Both estrogen and raloxifene cause G1 arrest of vascular smooth muscle cells. *The Journal of Endocrinology, 178*, 319–329.
Talwar, G. P., Segal, S. J., Evans, A., & Davidson, O. W. (1964). The binding of estradiol in the uterus: a mechanism for depression of RNA synthesis. *Proceedings of the National Academy of Sciences of the United States of America, 52*, 1059–1066.
Taylor, C. J., & Hobbs, F. D. (2013). Heart failure therapy in patients with coronary artery disease. *Current Opinion in Pharmacology, 13*, 205–209. http://dx.doi.org/10.1016/j.coph.2013.01.009.
Thomas, P., Pang, Y., Filardo, E. J., & Dong, J. (2005). Identity of an estrogen membrane receptor coupled to a G protein in human breast cancer cells. *Endocrinology, 146*, 624–632.
Tiano, J. P., Delghingaro-Augusto, V., Le May, C., Liu, S., Kaw, M. K., Khuder, S. S., ... Mauvais-Jarvis, F. (2011). Estrogen receptor activation reduces lipid synthesis in

pancreatic islets and prevents beta cell failure in rodent models of type 2 diabetes. *The Journal of Clinical Investigation, 121*, 3331–3342. http://dx.doi.org/10.1172/JCI44564.

Tiano, J. P., & Mauvais-Jarvis, F. (2012). Importance of oestrogen receptors to preserve functional beta-cell mass in diabetes. *Nature Reviews Endocrinology, 8*, 342–351. http://dx.doi.org/10.1038/nrendo.2011.242.

Toh, S., Hernandez-Diaz, S., Logan, R., Rossouw, J. E., & Hernan, M. A. (2010). Coronary heart disease in postmenopausal recipients of estrogen plus progestin therapy: Does the increased risk ever disappear? A randomized trial. *Annals of Internal Medicine, 152*, 211–217. http://dx.doi.org/10.7326/0003-4819-152-4-201002160-00005.

Tran, Q. K., Firkins, R., Giles, J., Francis, S., Matnishian, V., Tran, P., ... Gebert-Oberle, B. (2016). Estrogen enhances linkage in the vascular endothelial calmodulin network via a feedforward mechanism at the G protein-coupled estrogen receptor 1. *The Journal of Biological Chemistry*. http://dx.doi.org/10.1074/jbc.M115.697334.

Traupe, T., Stettler, C. D., Li, H., Haas, E., Bhattacharya, I., Minotti, R., & Barton, M. (2007). Distinct roles of estrogen receptors alpha and beta mediating acute vasodilation of epicardial coronary arteries. *Hypertension, 49*, 1364–1370.

Tropea, T., De Francesco, E. M., Rigiracciolo, D., Maggiolini, M., Wareing, M., Osol, G., & Mandala, M. (2015). Pregnancy augments G protein estrogen receptor (GPER) induced vasodilation in rat uterine arteries via the nitric oxide—cGMP signaling pathway. *PLoS One, 10*, e0141997. http://dx.doi.org/10.1371/journal.pone.0141997.

Turgeon, J. L., McDonnell, D. P., Martin, K. A., & Wise, P. M. (2004). Hormone therapy: Physiological complexity belies therapeutic simplicity. *Science, 304*, 1269–1273. http://dx.doi.org/10.1126/science.1096725.

Vanhoutte, P. M., Shimokawa, H., Feletou, M., & Tang, E. H. (2016). Endothelial dysfunction and vascular disease—A thirtieth anniversary update. *Acta Physiologica (Oxford, England)*. http://dx.doi.org/10.1111/apha.12646 (in press).

Vargas, R., Wroblewska, B., Rego, A., Hatch, J., & Ramwell, P. W. (1993). Oestradiol inhibits smooth muscle cell proliferation of pig coronary artery. *British Journal of Pharmacology, 109*, 612–617.

Vergote, I., & Abram, P. (2006). Fulvestrant, a new treatment option for advanced breast cancer: Tolerability versus existing agents. *Annals of Oncology, 17*, 200–204. http://dx.doi.org/10.1093/annonc/mdj047.

Vickers, M. R., MacLennan, A. H., Lawton, B., Ford, D., Martin, J., Meredith, S. K., ... WISDOM Group (2007). Main morbidities recorded in the women's international study of long duration oestrogen after menopause (WISDOM): A randomised controlled trial of hormone replacement therapy in postmenopausal women. *BMJ, 335*, 239. http://dx.doi.org/10.1136/bmj.39266.425069.AD.

Villa, P., Costantini, B., Suriano, R., Perri, C., Macri, F., Ricciardi, L., ... Lanzone, A. (2009). The differential effect of the phytoestrogen genistein on cardiovascular risk factors in postmenopausal women: Relationship with the metabolic status. *The Journal of Clinical Endocrinology and Metabolism, 94*, 552–558. http://dx.doi.org/10.1210/jc.2008-0735.

Villablanca, A. C., Tenwolde, A., Lee, M., Huck, M., Mumenthaler, S., & Rutledge, J. C. (2009). 17beta-estradiol prevents early-stage atherosclerosis in estrogen receptor-alpha deficient female mice. *Journal of Cardiovascular Translational Research, 2*, 289–299. http://dx.doi.org/10.1007/s12265-009-9103-z.

Violin, J. D., Soergel, D. G., Boerrigter, G., Burnett, J. C., Jr., & Lark, M. W. (2013). GPCR biased ligands as novel heart failure therapeutics. *Trends in Cardiovascular Medicine, 23*, 242–249. http://dx.doi.org/10.1016/j.tcm.2013.01.002.

Virdis, A., Ghiadoni, L., Pinto, S., Lombardo, M., Petraglia, F., Gennazzani, A., ... Salvetti, A. (2000). Mechanisms responsible for endothelial dysfunction associated with acute estrogen deprivation in normotensive women. *Circulation, 101*, 2258–2263.

Walter, P., Green, S., Greene, G., Krust, A., Bornert, J. M., ... Waterfield, M. (1985). Cloning of the human estrogen receptor cDNA. *Proceedings of the National Academy of Sciences of the United States of America, 82*, 7889–7893.

Wang, H., Jessup, J. A., Lin, M. S., Chagas, C., Lindsey, S. H., & Groban, L. (2012). Activation of GPR30 attenuates diastolic dysfunction and left ventricle remodelling in oophorectomized mRen2.Lewis rats. *Cardiovascular Research, 94*, 96–104. http://dx.doi.org/10.1093/cvr/cvs090.

Webb, C. M., Ghatei, M. A., McNeill, J. G., & Collins, P. (2000). 17beta-estradiol decreases endothelin-1 levels in the coronary circulation of postmenopausal women with coronary artery disease. *Circulation, 102*, 1617–1622.

Wendler, A., Albrecht, C., & Wehling, M. (2012). Nongenomic actions of aldosterone and progesterone revisited. *Steroids, 77*, 1002–1006. http://dx.doi.org/10.1016/j.steroids.2011.12.023.

White, R. E. (2002). Estrogen and vascular function. *Vascular Pharmacology, 38*, 73–80.

Williams, J. K., Adams, M. R., Herrington, D. M., & Clarkson, T. B. (1992). Short-term administration of estrogen and vascular responses of atherosclerotic coronary arteries. *Journal of the American College of Cardiology, 20*, 452–457.

Wise, P. M., Dubal, D. B., Rau, S. W., Brown, C. M., & Suzuki, S. (2005). Are estrogens protective or risk factors in brain injury and neurodegeneration? Reevaluation after the Women's health initiative. *Endocrine Reviews, 26*, 308–312. http://dx.doi.org/10.1210/er.2004-0014.

Witteman, J. C., Grobbee, D. E., Kok, F. J., Hofman, A., & Valkenburg, H. A. (1989). Increased risk of atherosclerosis in women after the menopause. *British Medical Journal, 298*, 642–644.

Wulfroth, P., & Grunwald, J. (1989). Differential effect of nicotinic acid derivatives on smooth muscle and endothelial cell proliferation. *Basic Research in Cardiology, 84*, 291–297.

Xiong, X., Xu, L., Wei, L., White, R. E., Ouyang, Y. B., & Giffard, R. G. (2015). IL-4 is required for sex differences in vulnerability to focal ischemia in mice. *Stroke, 46*, 2271–2276. http://dx.doi.org/10.1161/STROKEAHA.115.008897.

Yoshii, T., Yamada, M., Minami, T., Tsunoda, T., Sasaki, M., Kondo, Y., ... Terauchi,, Y. (2015). The effects of bazedoxifene on bone, glucose, and lipid metabolism in postmenopausal women with type 2 diabetes: An exploratory pilot study. *Journal of Clinical Medicine Research, 7*, 762–769. http://dx.doi.org/10.14740/jocmr2278w.

Yu, X., Li, F., Klussmann, E., Stallone, J. N., & Han, G. (2014). G protein-coupled estrogen receptor 1 mediates relaxation of coronary arteries via cAMP/PKA-dependent activation of MLCP. *American Journal of Physiology Endocrinology and Metabolism, 307*, E398–E407. http://dx.doi.org/10.1152/ajpendo.00534.2013.

Yu, X., Ma, H., Barman, S. A., Liu, A. T., Sellers, M., Stallone, J. N.,Han, G. ... (2011). Activation of G protein-coupled estrogen receptor induces endothelium-independent relaxation of coronary artery smooth muscle. *American Journal of Physiology Endocrinology and Metabolism, 301*, E882–E888. http://dx.doi.org/10.1152/ajpendo.00037.2011. ajpendo.00037.2011 [pii].

Yue, T. L., Vickery-Clark, L., Louden, C. S., Gu, J. L., Ma, X. L., Narayanan, P. K., ... Ohlstein, E. H. (2000). Selective estrogen receptor modulator idoxifene inhibits smooth muscle cell proliferation, enhances reendothelialization, and inhibits neointimal formation in vivo after vascular injury. *Circulation, 102*, III281–III288.

Yusuf, S., Ounpuu, S., & Anand, S. (2002). The global epidemic of atherosclerotic cardiovascular disease. *Medical Principles and Practice, 11*, 3–8.

Zhou, P., & Pu, W. T. (2016). Recounting cardiac cellular composition. *Circulation Research, 118*, 368–370. http://dx.doi.org/10.1161/CIRCRESAHA.116.308139.

CHAPTER ELEVEN

Mechanisms of Endothelial Dysfunction in Hypertensive Pregnancy and Preeclampsia

J.S. Possomato-Vieira, R.A. Khalil[1]

Vascular Surgery Research Laboratories, Brigham and Women's Hospital, Harvard Medical School, Boston, MA, United States

[1]Corresponding author: e-mail address: raouf_khalil@hms.harvard.edu

Contents

Advances in Pharmacology, Volume 77
ISSN 1054-3589
http://dx.doi.org/10.1016/bs.apha.2016.04.008

Abstract

Preeclampsia is a pregnancy-related disorder characterized by hypertension and could lead to maternal and fetal morbidity and mortality. Although the causative factors and pathophysiological mechanisms are unclear, endothelial dysfunction is a major hallmark of preeclampsia. Clinical tests and experimental research have suggested that generalized endotheliosis in the systemic, renal, cerebral, and hepatic circulation could decrease endothelium-derived vasodilators such as nitric oxide, prostacyclin, and hyperpolarization factor and increase vasoconstrictors such as endothelin-1 and thromboxane A2, leading to increased vasoconstriction, hypertension, and other manifestation of preeclampsia. In search for the upstream mechanisms that could cause endothelial dysfunction, certain genetic, demographic, and environmental risk factors have been suggested to cause abnormal expression of uteroplacental integrins, cytokines, and matrix metalloproteinases, leading to decreased maternal tolerance, apoptosis of invasive trophoblast cells, inadequate spiral arteries remodeling, reduced uterine perfusion pressure (RUPP), and placental ischemia/hypoxia. RUPP may cause imbalance between the antiangiogenic factors soluble fms-like tyrosine kinase-1 and soluble endoglin and the proangiogenic factors vascular endothelial growth factor and placental growth factor, or stimulate the release of other circulating bioactive factors such as inflammatory cytokines, hypoxia-inducible factor-1, reactive oxygen species, and angiotensin AT_1 receptor agonistic autoantibodies. These circulating factors could then target endothelial cells and cause generalized endothelial dysfunction. Therapeutic options are currently limited, but understanding the factors involved in endothelial dysfunction could help design new approaches for prediction and management of preeclampsia.

ABBREVIATIONS

ACh acetylcholine
ADMA asymmetric dimethylarginine
AngII angiotensin II
AT_1-AA AngII AT_1R agonistic autoantibodies
AT_1R angiotensin II type 1 receptor
BP blood pressure
$[Ca^{2+}]_i$ intracellular free Ca^{2+} concentration
CEC circulating endothelial cells
cGMP cyclic guanosine monophosphate
COX cyclooxygenase
CSE cystathionine gamma-lyase
EC endothelial cell
EDHF endothelium-derived hyperpolarizing factor
eNOS endothelial nitric oxide synthase
ET-1 endothelin-1
EVT extravillous trophoblast
FMD flow-mediated dilation

GFR glomerular filtration rate
H_2O_2 hydrogen peroxide
H_2S hydrogen sulfide
HELLP hemolysis, elevated liver enzymes, low platelets
HIF hypoxia-inducible factor
HTN hypertension
HTN-Preg hypertension in pregnancy
HUVEC human umbilical vein endothelial cell
ICAM-1 intercellular adhesion molecule-1
IL interleukin
IUGR intrauterine growth restriction
L-NAME N_{ω}-nitro-L-arginine methyl ester
MEGJ myoendothelial gap junction
MMP matrix metalloproteinase
NO nitric oxide
Norm-Preg normal pregnant
$O_2^{\cdot -}$ superoxide anion
PE preeclampsia
PGI_2 prostacyclin
PlGF placental growth factor
ROS reactive oxygen species
RPF renal plasma flow
RUPP reduced uterine perfusion pressure
sEng soluble endoglin
sFlt-1 soluble fms-like tyrosine kinase-1
TC trophoblast cell
TNF-α tumor necrosis factor-α
TXA2 thromboxane A2
VCAM-1 vascular cell adhesion molecule-1
VEGF vascular endothelial growth factor
VSM vascular smooth muscle
VSMC VSM cell

1. INTRODUCTION

Normal pregnancy is associated with several maternal hemodynamic and cardiovascular changes in order to meet the oxygen and nutrient demands of the growing fetus including increased blood volume and cardiac output, systemic vasodilation, decreased vascular resistance, and slight decrease in blood pressure (BP; Poston, McCarthy, & Ritter, 1995; Thornburg, Jacobson, Giraud, & Morton, 2000). These hemodynamic and vascular changes involve increases in various endothelium-derived vasodilator substances and redistribution of blood flow in different maternal tissues and organs (Tanbe & Khalil, 2010; Valdes et al., 2009). In 5–8% of pregnancies, women may have hypertension in pregnancy (HTN-Preg). HTN-Preg could be manifested in one of four forms: chronic HTN that predates pregnancy, preeclampsia (PE)–eclampsia, chronic HTN with superimposed PE, and nonproteinuric gestational HTN (Ali & Khalil, 2015).

PE is diagnosed after the 20th week of pregnancy by new onset HTN (systolic pressure $\geq$140 mmHg and/or diastolic pressure $\geq$90 mmHg), often proteinuria, and may be associated with edema and increased platelet aggregation (Brennan, Morton, & Davidge, 2014). PE may also be a part of HELLP syndrome, manifested as hemolysis, elevated liver enzymes, and low platelet count. If untreated, PE may progress to eclampsia, characterized by severe HTN and convulsions, which could culminate into coma and death, causing an estimated 14% of pregnancy-related maternal deaths (Say et al., 2014). PE may also be associated with intrauterine growth restriction (IUGR) and preterm birth, causing $\sim$13% of premature births in the United States (McBride, Bernstein, Badger, Horbar, & Soll, 2015). The complications associated with PE are particularly important in developing countries where the incidence of HTN-Preg is greater and the rates of maternal mortality and preterm births are higher than those in developed countries (Lain & Roberts, 2002).

Although PE is a major cause of maternal and fetal morbidity and a significant burden on the health care system, its etiology and pathophysiology remain unclear. The ambiguity of the cause of PE has made it difficult to develop efficient approaches for prevention or management, and delivery of the infant and placenta remains the only definitive treatment for PE and prevention of its progression to more serious maternal and fetal complications.

Clinical and experimental studies have suggested endothelial dysfunction as a major mechanism in HTN-Preg (Crews, Herrington, Granger, & Khalil, 2000; Maynard, Epstein, & Karumanchi, 2008; Sanchez-Aranguren, Prada, Riano-Medina, & Lopez, 2014; Tuzcu et al., 2015). PE could develop at early-gestational age <34 weeks or late gestational age $\geq$34 weeks (Jardim et al., 2015). Early PE is generally linked to compromised trophoblast invasion, placental hypoxia, and release of bioactive factors that could target the endothelium, while late PE has been linked to preexisting maternal conditions that could affect endothelial integrity (Brandao, Felix, Patricio Edo, Leite, & Cabral, 2014). Nevertheless, endothelial dysfunction is present in both early and late PE (Brandao et al., 2014). Interestingly, PE remits after delivery, pointing to the placenta as the main culprit. Clinical observations have implicated certain predisposing genetic, demographic, and environmental factors that could cause placental maladaptations. Other studies have shown changes in the levels of various bioactive factors in the placenta and maternal circulation. However, because of the difficulty to perform mechanistic studies in pregnant women, it has been difficult to draw a definitive association between the predisposing factors, placental ischemia/hypoxia, circulating bioactive factors, and endothelial dysfunction in humans. In order to further understand the pathophysiological mechanisms of PE, animal models of HTN-Preg have been developed. Both clinical observations and experimental studies have led to the suggestion that certain genetic and environmental risk factors may trigger reduction in uteroplacental perfusion pressure (RUPP) and the resulting placental ischemia/hypoxia could cause the release of bioactive factors that target endothelial cells (ECs) and cause endothelial dysfunction and HTN (Fig. 1). In support of this paradigm, studies have shown changes in the local and circulating levels of multiple factors including the proangiogenic vascular endothelial growth factor (VEGF) and placental growth factor (PlGF), the antiangiogenic factors soluble fms-like tyrosine kinase-1 (sFlt-1) and soluble endoglin (sEng), cytokines such as tumor necrosis factor-α (TNF-α) and interleukin-6 (IL-6), hypoxia-inducible factor (HIF), reactive oxygen species (ROS), and angiotensin II (AngII) type 1 receptor (AT_1R) agonistic autoantibodies (AT_1-AA). These bioactive factors could in turn cause generalized EC dysfunction and HTN, renal glomerular endotheliosis and increased glomerular permeability leading to proteinuria, and cerebral endotheliosis leading to cerebral edema and seizures (Ali & Khalil, 2015; Shah & Khalil, 2015).

In this chapter, we will discuss evidence of endothelial dysfunction in human PE and animal models of HTN-Preg. We will describe the potential upstream mechanisms that could cause endothelial dysfunction starting with the genetic, demographic, and environmental risk factors, the changes in uteroplacental perfusion and placental ischemia, and the release of bioactive factors that could target ECs in the systemic circulation and various tissues and organs. We will also discuss how the advances in our knowledge of the potential predisposing factors and circulating bioactive factors involved in endothelial dysfunction could help design new biomarkers for diagnosis and novel approaches for management of PE.

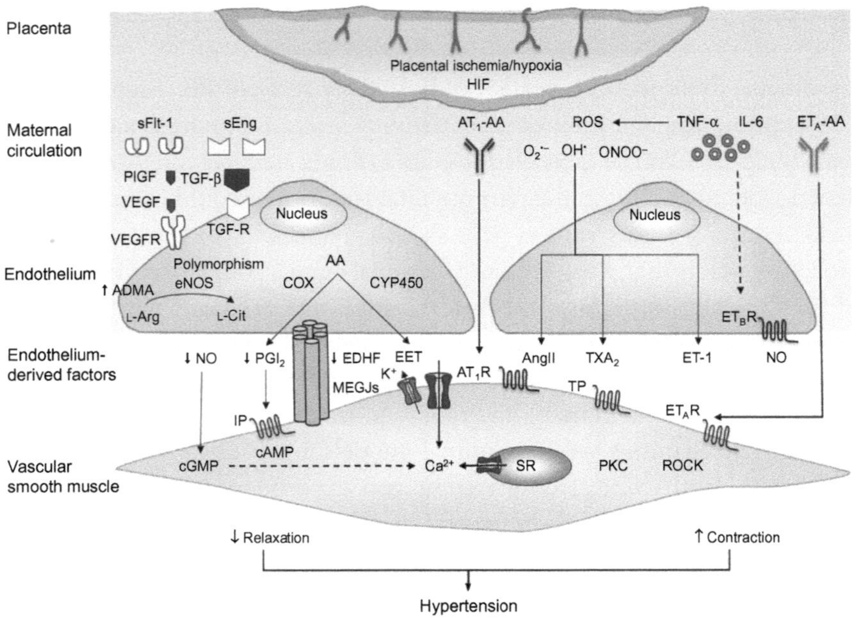

Fig. 1 Circulating bioactive factors cause endothelial dysfunction and hypertension in preeclampsia. Placental ischemia/hypoxia cause an increase in HIF-1 and the release of circulating bioactive factors such as sFlt-1, sEng, AT_1-AA, ROS, TNF-α, IL-6, and ET_A-AA. Excessive sFlt-1 binds VEGF and PlGF and prevents their angiogenic effects mediated by VEGFR binding. sEng binds TGFβ and prevents its angiogenic effects mediated by activation of TGFβ receptor. sFlt-1 and sEng may also lead to endothelial dysfunction and decreased release of the endothelium-derived relaxing factors NO, PGI_2, EDHF, and EET. Polymorphisms in eNOS and elevated ADMA also decrease NO production. AT_1-AA acts on AT_1R and increases the mechanisms of VSM contraction. Cytokines such as TNF-α and IL-6 may increase the production of ROS, which decrease the bioavailability of NO and stimulate the release of vasoconstrictor substances such as AngII, ET-1, and TXA2 which increase VSM $[Ca^{2+}]_i$, PKC, and ROCK and stimulate VSM contraction. Endothelial ET_BR mediates the release of NO, and downregulation of ET_BR in PE decreases ET_BR-mediated relaxation. ET_A-AA stimulates ET_AR and further stimulates VSM contraction. Decreased vascular relaxation and increased contraction cause HTN in PE. *Solid arrows* indicate stimulation. *Dashed arrows* indicate inhibition. *AA*, arachidonic acid; *ADMA*, asymmetric dimethylarginine; *AngII*, angiotensin II; *AT_1-AA*, AngII AT_1R agonistic autoantibodies; *AT_1R*, angiotensin II type 1 receptor; *cAMP*, cyclic adenosine monophosphate; *cGMP*, cyclic guanosine monophosphate; *COX*, cyclooxygenase; *CYP450*, cytochrome 450; *EC*, endothelial cell; *EDHF*, endothelium-derived hyperpolarizing factor; *EET*, epoxyeicosatrienoic acid; *eNOS*, endothelial nitric oxide synthase; *ET-1*, endothelin-1; *ET_AR*, endothelin receptor type A; *ET_A-AA*, ET_AR agonistic autoantibodies; *ET_BR*, endothelin receptor type B; *H_2O_2*, hydrogen peroxide; *HIF*, hypoxia-inducible factor; *IL-6*, interleukin-6; *IP*, PGI_2 receptor; *MEGJs*, myoendothelial gap junctions; *NO*, nitric oxide; *$O_2^{\cdot -}$*, superoxide anion; *OH^-*, hydroxyl ion; *PGI_2*, prostacyclin; *PKC*, protein kinase C; *PlGF*, placental growth factor; *ROCK*, Rho-kinase; *ROS*, reactive oxygen species; *sEng*, soluble endoglin; *sFlt-1*, soluble fms-like tyrosine kinase-1; *SR*, sarcoplasmic reticulum; *TGF-β*, transforming growth factor-β; *TGF-R*, TGF-β receptor; *TNF-α*, tumor necrosis factor-α; *TP*, TXA2 receptor; *TXA2*, thromboxane A2; *VEGF*, vascular endothelial growth factor; *VEGFR*, VEGF receptor; *VSMC*, vascular smooth muscle cell.

2. ENDOTHELIAL DYSFUNCTION IN PREECLAMPSIA

Since the initial observations that ECs play a role in modulation of vascular tonus (Furchgott & Zawadzki, 1980), changes in ECs and the factors released by them have been observed in numerous physiological and pathological conditions. An increase in endothelial function during pregnancy is important for healthy gestation, ensuring a favorable prognosis for the mother and fetus (Brandao et al., 2014). In humans, endothelial function is often assessed using flow-mediated dilation (FMD) of the brachial artery, an ultrasonography test that evaluates EC response to temporary ischemia and reactive hyperemia (Brandao et al., 2014). During normal pregnancy, there is an increase in brachial artery diameter and FMD as gestation progresses (Sierra-Laguado, Garcia, & Lopez-Jaramillo, 2006). Also, endothelium-dependent bradykinin-induced relaxation is increased in human small subcutaneous arteries from pregnant compared with nonpregnant women (Knock & Poston, 1996). Experimental studies corroborated EC adaptations during pregnancy. ATP causes periodic bursts in intracellular free Ca^{2+} concentration ($[Ca^{2+}]_i$) that are more frequent in uterine artery ECs from pregnant compared with nonpregnant ewes (Yi et al., 2010), which could influence the vasodilator response and in turn attenuate the uterine artery myogenic tone and ensure adequate uterine blood flow during pregnancy (Hu et al., 2011).

In contrast with normal pregnancy, women with PE may show systemic EC dysfunction and HTN, glomerular endotheliosis causing renal injury and proteinuria, and cerebral endotheliosis leading to cerebral edema and seizures (Ali & Khalil, 2015). FMD of brachial artery is less in PE than normal pregnant (Norm-Preg) women (Brandao et al., 2014; Guimaraes et al., 2014; Table 1). PE women also show less vasodilation in the radial artery (~7.9%) when compared to Norm-Preg women (~17.4%; Yoshida, Nakao, Kobayashi, & Kobayashi, 2000).

Studies in isolated human vessels have shown decreased bradykinin-induced relaxation in small subcutaneous arteries of PE compared with Norm-Preg women (Knock & Poston, 1996). However, no differences in the magnitude of relaxation or sensitivity to bradykinin were observed in the small myometrial arteries of PE vs Norm-Preg women (Kenny, Baker, Kendall, Randall, & Dunn, 2002). While the endothelial dysfunction in the systemic vessels may explain the elevated BP during PE, the lack of

Table 1 Endothelium-Derived Factors and Markers of Endothelial Function in Human Normal Pregnancy and Preeclampsia

Factor	Specimen (Units)	Normal Pregnancy	Preeclampsia	References
FMD	Brachial artery (mm)	17.55 ± 8.35	5.36 ± 4.61	Guimaraes et al. (2014)
		9.00 ± 5.00	4.00 ± 6.00 early PE	Brandao et al. (2014)
			3.00 ± 3.00 late PE	
Nitrite	Serum (μmol/L)	~40	~48	Noorbakhsh, Kianpour, and Nematbakhsh (2013)
	Serum (μmol/L)	26.45 ± 3.03 first trimester	16.83 ± 0.48 first trimester	Matsubara, Matsubara, Hyodo, Katayama, and Ito (2010)
		25.18 ± 3.92 third trimester	31.28 ± 2.58 third trimester	
	Plasma (μmol/L)	7.26 ± 9.65^{SD}	6.84 ± 7.54^{SD}	Schiessl et al. (2006)
	Plasma (μmol/L)	58.1 ± 40.3^{SD}	36.5 ± 20.9^{SD}	Ehsanipoor et al. (2013)
	Plasma (mM/mg pr)	8.68 ± 2.27^{SD}	5.16 ± 2.66^{SD}	Pimentel et al. (2013)
	Plasma (nM)	177.2 ± 151.3	104.5 ± 84.3	Eleuterio et al. (2013)
	Urine (μmol/L)	361.41 ± 423.12^{SD}	223.88 ± 294.96^{SD}	Schiessl et al. (2006)
	Placental supernatant (μmol/L)	24.9 ± 6.9^{SD}	16.4 ± 5.5^{SD}	Ehsanipoor et al. (2013)
eNOS	Serum (pg/mL)	0.86 ± 0.64	0.67 ± 0.55	Zawiejska, Wender-Ozegowska, Iciek, and Brazert (2014)
	Serum (pg/mL)	1.46 ± 0.31	1.12 ± 0.16 mild PE	Wang et al. (2015)
			0.88 ± 0.23 severe PE	

L-Arginine	Plasma (μM)	62.00 ± 31.06	19.23 ± 10.54	Pimentel et al. (2013)
ADMA	Serum (μM)	0.68 ± 0.20	0.86 ± 0.16	Bian, Shixia, and Duan (2015)
	Plasma (μmol/L)	2.14 ± 0.33	1.88 ± 0.19	Ehsanipoor et al. (2013)
cGMP	Plasma (pmol/mL)	6.82 ± 2.57^{SD}	9.02 ± 4.47^{SD}	Schiessl et al. (2006)
	Urine (nmol/mL)	612.95 ± 440.63^{SD}	518.13 ± 418.68^{SD}	
6-Keto-$PGF_{1\alpha}$	Plasma (pg/mL)	254.51 ± 27.31	158.01 ± 15.20	Lewis, Canzoneri, Gu, Zhao, and Wang (2010)
ET-1	Serum (pg/mL)	17.15 ± 6.14^{SD}	18.48 ± 5.69^{SD}	Celik, Avci, and Isik (2013)
	Serum (ng/mL)	45.9 ± 50.1^{SD}	62.9 ± 67.3^{SD} (HELLP)	Karakus et al. (2016)
	Placental perfusate (pg/mL)			Jain et al. (2014)
	Maternal side	~7	~11	
	Fetal side	~10	~22	
ET_A-AA	Serum	Present	Absent	Velloso et al. (2015)
TXB2	Plasma (pg/mL)	46.36 ± 4.23	41.28 ± 8.42	Lewis et al. (2010)
Endothelial progenitor cells	Whole blood (cells/mL)	1918 ± 563	620 ± 105	Sakashita, Higashi, Soga, Miyoshi, and Kudo (2014)
CEC	Whole blood (cells/mL)	3.0 ± 4.1^{SD}	9.9 ± 7.9^{SD}	Tuzcu et al. (2015)
		5.2 ± 1.4	13.2 ± 5.2	Canbakan et al. (2007)

Continued

Table 1 Endothelium-Derived Factors and Markers of Endothelial Function in Human Normal Pregnancy and Preeclampsia—cont'd

Factor	Specimen (Units)	Normal Pregnancy	Preeclampsia	References
sVCAM-1	Serum (ng/mL)	404 ± 147^{SD}	825 ± 328^{SD}	Tuzcu et al. (2015)
E-Selectin	Serum (ng/mL)	37.5 ± 24.1^{SD}	69.5 ± 38.3^{SD}	Tuzcu et al. (2015)
Endocan	Serum (ng/mL)	15.5 ± 6.19	20.04 ± 12.26	Cakmak et al. (2016)
	Serum (ng/mL)	10.3 ± 3.2	10.7 ± 4.5	Yuksel et al. (2015)
	Plasma (ng/mL)	22.5 (13.8–44.4)	18.2 (10.6–28.0)	Adekola et al. (2015)

Values represent means ± standard error of the mean. *Numbers* in parenthesis indicate range. *6-keto-PGF*$_{1\alpha}$, 6-keto-prostaglandin $F_{1\alpha}$; *ADMA*, asymmetric dimethylarginine; *CEC*, circulating endothelial cells; *cGMP*, cyclic guanosine monophosphate; *eNOS*, endothelial nitric oxide synthase; *ET-1*, endothelin-1; *ET_A-AA*, ET_A receptor agonistic autoantibodies; *FMD*, flow-mediated dilation; *HELLP*, hemolysis, elevated liver enzymes, low platelets; *PE*, preeclampsia; *SD*, standard deviation; *sVCAM-1*, soluble vascular cell adhesion molecule-1; *TXB2*, thromboxane B2.

change in the myometrial vessels may be a compensatory rescue mechanism to maintain blood supply to the fetus.

Circulating endothelial cells (CECs) and other EC factors are often used as markers of endothelial activation/injury. Studies have shown an increase in CEC number and other markers of EC damage such as soluble vascular cell adhesion molecule-1 (sVCAM-1), E-selectin, and endocan in PE compared with Norm-Preg women (Adekola et al., 2015; Cakmak et al., 2016; Canbakan et al., 2007; Tuzcu et al., 2015), although some studies reported no difference in serum endocan levels (Yuksel et al., 2015). Other studies have shown a decrease in circulating endothelial progenitor cells as a marker of endothelial damage in PE women (Sakashita et al., 2014; Table 1).

Studies in animal models of HTN-Preg have supported endothelial dysfunction as an important mechanism in PE. The rat model of surgically induced RUPP in late pregnancy has some of the characteristics of PE including high BP, proteinuria, decreased glomerular filtration rate (GFR) and renal plasma flow (RPF), and IUGR and therefore has been used extensively to understand the molecular mechanisms of PE (Alexander, Kassab, et al., 2001; Amaral et al., 2015; Mazzuca et al., 2014). Acetylcholine (ACh), a known stimulant of ECs, is less potent in inducing relaxation in the aorta and mesenteric microvessels of RUPP than Norm-Preg rats, suggesting endothelial damage in RUPP rats (Crews et al., 2000; Mazzuca et al., 2014). ECs release various factors that promote vascular relaxation such as nitric oxide (NO), prostacyclin (PGI_2), and endothelium-derived hyperpolarizing factor (EDHF) as well as factors that cause vascular contraction such as endothelin-1 (ET-1) and thromboxane A2 (TXA2; Fig. 1). A balance between relaxing and contracting factors is essential to maintain vascular tone, and endothelial dysfunction is associated with an imbalance in the release of these endothelium-derived factors.

3. DECREASED ENDOTHELIUM-DERIVED RELAXING FACTORS IN PREECLAMPSIA

3.1 Nitric Oxide

NO is a potent vasodilator and relaxant of vascular smooth muscle (VSM). NO is a gaseous molecule generated from the conversion of L-arginine to L-citrulline by NO synthase (NOS). NO is produced by endothelial eNOS, inducible iNOS, and neuronal nNOS. NO diffuses into VSM and activates cyclic guanosine monophosphate (cGMP), which promotes Ca^{2+} efflux and thereby decreases VSM $[Ca^{2+}]_i$ and causes VSM relaxation.

Studies using venous occlusion plethysmography showed that infusion of norepinephrine produced a smaller contractile response and that inhibition of NO production using N_{ω}-monomethyl-L-arginine (L-NMMA) caused a greater reduction in hand blood flow in Norm-Preg than nonpregnant women, suggesting increased generation of NO in peripheral vascular beds in healthy pregnancies (Williams, Vallance, Neild, Spencer, & Imms, 1997). Also, nitrites are important metabolites of NO used to indirectly determine NO production, and have been shown to be increased in serum of Norm-Preg compared with nonpregnant women (Shaamash et al., 2000). The plasma concentration and urinary excretion of cGMP, a second messenger of NO, are also increased in normal pregnancy. NOS expression and activity are also increased in human uterine artery (Nelson et al., 2000) and in the placenta along with gestational age (Dotsch et al., 2001), supporting increased NO production during normal pregnancy in humans.

Animal studies support a role of NO during normal pregnancy. Urinary levels of nitrites, eNOS, iNOS, and nNOS mRNA expression and the protein amount of activated phospho-eNOS are increased in Norm-Preg compared with virgin rats (Alexander et al., 1999). Also, BP is increased throughout pregnancy in eNOS knockout mice (Hefler, Reyes, O'Brien, & Gregg, 2001). Other studies have reported an unexpected decrease in BP in eNOS knockout mice during pregnancy (Shesely et al., 2001), suggesting that other NOS isoforms could compensate for the loss of NO production via eNOS, or that NO may not be the only mediator of vascular homeostasis during pregnancy and other mediators could be triggered to maintain vascular function during eNOS deficiency. Of note, the diameter and medial cross-sectional area of uterine artery are reduced, the number of viable pups and pup weight are less, and VSM cell (VSMC) differentiation and proliferation are impaired in pregnant eNOS knockout than Norm-Preg mice (van der Heijden et al., 2005), supporting a role of NO in vascular adaptations during pregnancy.

Endothelial dysfunction is often associated with decreased NO bioavailability due to either decreased synthesis or increased degradation, and changes in NO metabolism could be a factor in PE (Echeverri et al., 2015). Clinical studies have shown conflicting results, reporting increased (Noorbakhsh et al., 2013) or decreased (Ehsanipoor et al., 2013; Eleuterio et al., 2013; Pimentel et al., 2013; Schiessl et al., 2006) serum and plasma nitrite levels in PE compared with Norm-Preg women. Other studies have shown a decrease in serum levels of nitrites in PE during the first trimester that normalizes in the third trimester (Matsubara et al., 2010). Also, urinary nitrite levels may not differ in PE vs Norm-Preg women (Schiessl et al., 2006; Table 1). The discrepancies in nitrite measurement could be related to the difficulty in controlling nitrate intake by diet. However, a study using carefully controlled dietary nitrate/nitrite intake did not show decreased NO production in PE women (Conrad, Kerchner, & Mosher, 1999). Genetic polymorphisms of eNOS could also affect NO production. For instance, Norm-Preg women with the TT phenotype for the T-786C allele have lower plasma nitrite levels than those with the CC phenotype (Sandrim et al., 2010), and the TT phenotype has been proposed as a risk factor for PE in Tunisian women (Ben Ali Gannoun et al., 2015). Lower nitrite levels were also observed in women with the 4a4a vs 4b4b genotype for VNTR polymorphism (Sandrim et al., 2010).

Asymmetric dimethylarginine (ADMA) is a reversible competitive inhibitor of NOS that could also affect NO production (Ehsanipoor et al., 2013; Fig. 1). The observations of decreased ADMA levels in normal pregnancy, and the association of abnormal uterine artery Doppler waveforms with high ADMA concentrations, suggest a role of endogenous NOS inhibitors in adversely affecting maternal vasodilation and BP (Myatt & Webster, 2009). Elevated circulating levels of ADMA are observed in PE and in women presenting with IUGR without PE (Bian et al., 2015; Myatt & Webster, 2009). However, other studies showed a decrease in plasma ADMA levels in PE vs Norm-Preg women (Ehsanipoor et al., 2013). NO production may also be affected by the availability of the NOS substrate L-arginine. While the plasma L-arginine levels do not differ in PE vs Norm-Preg women (Table 1), the transport of L-arginine to the platelets is reduced in PE and could in turn affect NO production in PE women (Pimentel et al., 2013).

The lack of change in whole-body NO despite the increase in BP and the renal damage in PE suggests tissue-specific changes in NOS expression and NO bioavailability such that whole-body NO may not accurately reflect NO activity in the vasculature or the kidneys (Ali & Khalil, 2015). Studies have shown a decrease in nitrites in placentae from PE women (Ehsanipoor et al., 2013). Also, eNOS expression is decreased in umbilical cord of PE compared with Norm-Preg women (Bhavina, Radhika, & Pandian, 2014), and the decrease is greater in women with severe PE (Wang et al., 2015; Zawiejska et al., 2014). However, some studies showed an increase in eNOS mRNA expression in placenta of PE women (Smith-Jackson et al., 2015). Also, while the levels of cGMP are increased during normal pregnancy, the plasma and urinary cGMP levels are not different in PE vs Norm-Preg women (Schiessl et al., 2006; Table 1).

The role of NO has also been examined in animal models of HTN-Preg. In pregnant rats during mid to late gestation NOS blockade with N_{ω}-nitro-L-arginine methyl ester (L-NAME) causes some of the changes observed in human PE including increases in BP and renal vasoconstriction, proteinuria, thrombocytopenia, and IUGR (Kemse, Kale, & Joshi, 2014; Khalil, Crews, Novak, Kassab, & Granger, 1998; Ramesar, Mackraj, Gathiram, & Moodley, 2011), supporting a role of increased NO production during gestation. Similar to the observation in humans, measurements of NO production in HTN-Preg animals have not produced consistent results. Studies showed no difference in nitrite levels in rats treated with L-NAME compared with Norm-Preg (Ramesar et al., 2011). Also, while plasma nitrite levels were lower in RUPP than Norm-Preg rats (Amaral et al., 2015), no difference was observed in urinary nitrite levels (Alexander, Kassab, et al., 2001; Javadian et al., 2013; Table 2). Also, consistent with the studies in human, no changes were observed in the NOS substrate L-arginine in the circulation of RUPP vs Norm-Preg rats (Alexander et al., 2004; Table 2). Vascular function studies have shown an increase in aortic vascular reactivity to phenylephrine in pregnant rats treated with L-NAME (Khalil et al., 1998). Also, ACh-induced relaxation, eNOS expression, and NO production are reduced in mesenteric artery and aorta of RUPP vs Norm-Preg rats, supporting specific reduction in NO synthesis in the vasculature (Amaral et al., 2015; Crews et al., 2000; Mazzuca et al., 2014; Table 2). In another rat model of HTN-Preg produced by infusion of deoxycorticosterone and replacement of drinking water with a 0.9% saline (DOCA-salt), NO-dependent relaxation was reduced in mesenteric vessels despite elevation of eNOS mRNA expression (Mitchell, Cook, Danchuk, & Puschett, 2007).

Thus while clinical observations and experimental studies point to the importance of NO during gestation and that its deficiency could be a possible mechanism in HTN-Preg, due to the difficulties in assessment of nitrite levels in humans and inconsistent data in animal models of HTN-Preg, whether NO production is decreased in PE needs to be further examined.

Table 2 Endothelium-Derived Factors and Markers of Endothelial Function in Norm-Preg and RUPP Rats

Factor	Specimen (Units)	Norm-Preg	RUPP	References
Nitrite	Plasma (μmol/L)	26.34 ± 3.5	14.58 ± 3.1	Amaral et al. (2015)
	Urine (μmol/24 h)	~40	~46	Javadian et al. (2013)
	Urine (μmol/24 h)	46.4 ± 5.3	49.8 ± 6.4	Alexander, Kassab, et al. (2001)
	Aorta (pmol/mg)	~750	~490	Mazzuca et al. (2014)
eNOS mRNA	Aorta (AU)	0.65 ± 0.11	0.33 ± 0.01	Amaral et al. (2015)
L-arginine	Serum (nmol/mL)	191 ± 29	142 ± 22	Alexander, Llinas, Kruckeberg, and Granger (2004)
PreproET-1	Renal cortex (DU)	24.0 ± 2.6	44.1 ± 6.3	Alexander, Rinewalt, et al. (2001)
	Renal medulla (DU)	29.4 ± 2.8	37.7 ± 2.4	
ET-1	Plasma (pg/mL)	0.46 ± 0.06	0.99 ± 0.25	Johnson et al. (2014)
TXB2	Urine (pg/24 h)	2646 ± 257	3663 ± 488	Llinas et al. (2002)

Values represent means ± standard error of the mean. *AU*, arbitrary unit of band intensity normalized to β-actin in Western blot analysis; *DU*, densitometry unit of band intensity normalized to β-actin in ribonuclease protection assay; *eNOS*, endothelial nitric oxide synthase; *ET-1*, endothelin-1; *TXB2*, thromboxane B2.

3.2 Prostacyclin

PGI_2 is a prostanoid produced from the metabolism of arachidonic acid by cyclooxygenase-1 (COX-1) and COX-2. PGI_2 promotes VSM relaxation and inhibits platelet aggregation acting via IP receptor, a cell surface G-protein-coupled receptor. Stimulation of IP receptor in VSM causes an increase in cyclic adenosine monophosphate (cAMP) and leads to a decrease in $[Ca^{2+}]_i$ and VSM relaxation (Fig. 1). PGI_2 also promotes angiogenesis, decreases hypoxic injury by inducing neovascularization in ischemic tissues, and enhances the biological effects of proangiogenic factors in response to ischemia (Majed & Khalil, 2012).

Endothelium-derived PGI_2 may contribute to the hemodynamic and vascular changes during pregnancy. Plasma and urinary levels of 6-keto-prostaglandin $F_{1\alpha}$ (6-keto-$PGF_{1\alpha}$), a hydration product of PGI_2, increase during pregnancy (Majed & Khalil, 2012). Also, PGI_2 synthase mRNA and protein expression and 6-keto-$PGF_{1\alpha}$ levels are elevated in pig endometrium during normal pregnancy (Morawska, Kaczmarek, & Blitek, 2012). However, vascular function studies showed little contribution of PGI_2 to bradykinin-induced relaxation in small subcutaneous arteries from Norm-Preg women (Luksha, Nisell, & Kublickiene, 2004).

Maternal plasma 6-keto-$PGF_{1\alpha}$ levels are decreased in PE compared with Norm-Preg women (Lewis et al., 2010), and the decrease in PGI_2 production may occur even before the onset of clinical symptoms of PE (Mills et al., 1999). As growth factors stimulate PGI_2 production, changes in growth factors during PE may affect PGI_2 levels. Also, hypoxia is a potential pathogenic mechanism in PE, which could cause downregulation of COX-1 with consequent reduction in endothelium-derived PGI_2 (Shah & Khalil, 2015).

In pregnant rats, treatment with high doses of testosterone produces PE-like features, decreased placental IP receptor expression, and impaired PGI_2-mediated relaxation in uterine artery (Chinnathambi et al., 2014), and the impaired relaxation may contribute to the increased vascular resistance and HTN-Preg. On the other hand, measurement of 6-keto-$PGF_{1\alpha}$ concentrations showed no difference in human umbilical vein endothelial cells (HUVECs) from PE and Norm-Preg women (Parra, Lees, Mann, Pearson, & Nicolaides, 2001). Interestingly, predominant basal directional release of constrictor prostanoids, but not PGI_2, was observed in placental trophoblasts from PE pregnancies (Zhao, Gu, Lewis, & Wang, 2008). Thus, while PGI_2 production may not be altered, the greater production of constrictor prostanoids and consequent imbalance between vasodilator and vasoconstrictor prostanoids may contribute to increased placental vasoconstriction in PE.

3.3 Endothelium-Derived Hyperpolarizing Factor

EDHF is a relaxing factor with particular importance in the control of small resistance vessels, local organ blood flow, peripheral vascular resistance, and BP. EDHF's role in vasodilation increases as arterial diameter decreases, causing greater dilation in the smaller distal than the bigger proximal uterine vessels (Gokina, Kuzina, & Vance, 2010). Although the nature of EDHF is unclear, it often presents as K^+ efflux from ECs through intermediate- and small-conductance Ca^{2+}-activated K^+ channels (IK_{Ca} and SK_{Ca}, respectively) causing hyperpolarization of ECs. EC hyperpolarization then spreads via myoendothelial gap junctions (MEGJs) and connexins to cause VSM hyperpolarization, reduction of Ca^{2+} influx via voltage-gated Ca^{2+} channels, and suppression of the activity of phospholipase C, an enzyme involved in transduction pathways in VSM. The opening of EC IK_{Ca} and SK_{Ca} could also cause some accumulation of K^+ ion in the myoendothelial interface which could induce VSM hyperpolarization by activating the inwardly rectifying K^+ (K_{IR}) channels and the Na^+/K^+-ATPase (Coleman, Tare, & Parkington, 2004). EDHF relaxation may also be caused by diffusible factors released from ECs. EDHF may be a product of cytochrome 450 (CYP450), such as epoxyeicosatrienoic acid (EET), which activates large-conductance K_{Ca} (BK_{Ca}) and causes hyperpolarization in VSM (Fig. 1). In some vessels, hydrogen peroxide (H_2O_2) may mimic EDHF-mediated responses by mechanisms involving K_{Ca} activation (Shimokawa, 2010). Thus multiple EDHFs may exist and the identity of EDHF could vary depending on the vascular bed and animal species studied (Feletou & Vanhoutte, 2009).

In small subcutaneous and myometrial arteries of Norm-Preg women, EDHF is responsible for ~50% of bradykinin-induced relaxation, acting together with NO to maintain proper vascular tonus (Kenny et al., 2002; Luksha et al., 2004). The gap junction proteins connexins 37, 40, and 43 are partly involved in EDHF-mediated vascular response during normal pregnancy (Hammond, Mathewson, Baker, Mayhew, & Dunn, 2011). An increase in EC $[Ca^{2+}]_i$ may activate IK_{Ca} and SK_{Ca} and promote EDHF-mediated dilation in uterine radial arteries of pregnant rats (Gokina et al., 2010). Also, PGI_2 may activate endothelial SK_{Ca} in mesenteric arteries of Norm-Preg rats and cause EDHF-mediated activation of BK_{Ca} in VSM (Mandala, Gokina, Barron, & Osol, 2012; Orie, Fry, & Clapp, 2006). The delayed rectifier type of voltage-sensitive K^+ channels (K_v) may also play a role in EDHF-mediated dilation in uterine artery of pregnant rats (Fulep, Vedernikov, Saade, & Garfield, 2001).

Studies in subcutaneous arteries from Norm-Preg women have shown that MEGJs alone may be the main pathway in EDHF-mediated relaxation, while in women with PE MEGJs alone or in combination with H_2O_2 or CYP450 epoxygenase metabolites of arachidonic acid could mediate EDHF-induced vasodilation. The changes in the role of MEGJs may be caused by morphological changes within the vascular wall during PE (Luksha et al., 2008). Studies in mice have shown pregnancy-associated adaptations in the form of decreased sensitivity to phenylephrine and enhanced bradykinin-induced vasodilation in Norm-Preg wild-type mice but not in knockout mice lacking pregnane X receptor, a nuclear receptor that induces the expression of CYP450. Also, treatment with CYP450 inhibitor changed the vasodilatory response to bradykinin in wild-type but not the knockout mice, supporting that metabolites of CYP450 such as EET may play a role in the vascular adaptations during pregnancy (Hagedorn, Cooke, Falck, Mitchell, & Davidge, 2007). As EET is one of the possible factors involved in EDHF-mediated relaxation, it is plausible to suggest that alterations in EDHF may lead to impaired vascular function and HTN-Preg. Although studies in mesenteric microvessels have suggested that the EDHF relaxation may not be compromised in RUPP vs Norm-Preg rats (Mazzuca et al., 2014), decreased EDHF-mediated relaxation is thought to contribute to the vasoconstriction observed in HTN and diabetes, and its role in PE needs further investigation.

4. INCREASED ENDOTHELIUM-DERIVED CONTRACTING FACTORS IN PREECLAMPSIA

4.1 Endothelin-1

ET-1 is a major endothelium-derived vasoconstrictor that could play a role in PE (George & Granger, 2011). ET-1 synthesis is initiated by the production of the long 203-amino-acid preproET, which is cleaved by furin-like protease to biologically inactive 37- to 41-amino-acid big-ET. Big-ET is cleaved by ET-converting enzymes, members of the metalloprotease family, to produce active 21-amino-acid ET-1.

Some of the circulating factors in PE including cytokines, hypoxia, and AT_1-AA may stimulate ECs to produce ET-1 (George & Granger, 2011). This is supported by reports that serum from PE women causes HUVECs to produce greater amounts of ET-1 than Norm-Preg serum (Scalera, Schlembach, & Beinder, 2001). Some studies suggest that plasma ET-1 levels are elevated in PE (Roberts, 1998). ET-1 levels are higher during later stages of PE and return to normal levels within 48 h after delivery (Taylor, Varma, Teng, & Roberts, 1990), suggesting that ET-1 may be involved in the progression rather than the initiation of PE. However, in most studies serum ET-1 levels do not differ in PE vs Norm-Preg women, and higher levels of ET-1 are observed mainly in patients with HELLP syndrome

(Celik et al., 2013; Karakus et al., 2016; Naiker et al., 2001). Of note, ET-1 is released in a paracrine fashion from EC directly on VSMCs, and the increases in ET-1 levels in PE may be localized in tissues. Studies have shown a four- to eightfold increase in ET-1 levels in umbilical cord cells (Kourembanas, McQuillan, Leung, & Faller, 1993) and in renal tissues during later stages of PE (Taylor et al., 1990). In perfused placentas under hypoxia, both the maternal and fetal side produced increased levels of ET-1 (Jain et al., 2014). In RUPP rats, preproET levels show a 45% increase in renal cortex and 22% increase in the medulla (Alexander, Rinewalt, et al., 2001). Thus, measurements of circulating ET-1 may not accurately reflect what happens locally in tissues. It is possible that in severe PE and in HELLP syndrome the production of ET-1 is so augmented that ET-1 release "loses" its paracrine directionality, and consequently an increase in circulating ET-1 levels is detected. This is supported by reports that rat models that mimic severe PE and HELPP syndrome have increased plasma ET-1 levels (Johnson et al., 2014; Morris et al., 2015; Table 2).

ET-1 may play a role in the pathogenesis of PE by inducing apoptosis of trophoblast cells (TCs) and increasing oxidant and antiangiogenic substances (Fiore et al., 2005; George & Granger, 2011), but most studies have focused on the effects of ET-1 on the vasculature. ET-1 activates endothelin receptor type A (ET_AR) and endothelin receptor type B (ET_BR; Mazzuca & Khalil, 2012). ET-1 activation of VSM ET_AR stimulates Ca^{2+} release from the intracellular stores and Ca^{2+} entry through Ca^{2+} channels and causes protein kinase C (PKC)-dependent inhibition of K^+ channels leading to increased $[Ca^{2+}]_i$ and VSM contraction (Mazzuca & Khalil, 2012; Fig. 1). Of note, during hypoxia, up to 80% of ET-1-induced contraction is independent of $[Ca^{2+}]_i$ or PKC (Weigand, Sylvester, & Shimoda, 2006). ET-1-induced vasoconstriction is reduced in mesenteric vessels of Norm-Preg compared with nonpregnant rats (Mazzuca, Dang, & Khalil, 2013), and VSM ET_AR and ET_BR are reduced in aortic media and VSMCs of late-pregnant rat (Ou, Dang, Mazzuca, Basile, & Khalil, 2014). Interestingly, among Brazilian women with PE, 52% of the patients diagnosed with severe PE exhibited increases in ET_AR agonistic autoantibodies (ET_A-AA) which could target ET_AR and increase vasoconstriction (Velloso et al., 2015). Experimental studies have also suggested increases in ET_AR activity in PE and have shown that treatment with ET_AR antagonist reduces BP in RUPP and other models of HTN-Preg (Alexander, Rinewalt, et al., 2001; LaMarca, Cockrell, Sullivan, Bennett, & Granger, 2005; Murphy, LaMarca, Cockrell, & Granger, 2010).

ET-1 also activates ET_BR in ECs and stimulates the release of NO, PGI_2, and EDHF which in turn reduce myogenic vascular tone and promote vasodilation of renal arteries and hyperfiltration in pregnant rats (Conrad, Gandley, Ogawa, Nakanishi, & Danielson, 1999; Mazzuca & Khalil, 2012). ET_BR may be reduced in ECs and renal cells of RUPP rats, and downregulation of ET_BR may impair trophoblast invasion in PE, and decrease microvascular vasodilator activity in RUPP rats. Also, ET_BR-mediated NO production is less in the aorta and mesenteric artery of RUPP than Norm-Preg rats, suggesting that downregulation of endothelial ET_BR could play a role in HTN-Preg (Mazzuca et al., 2014).

4.2 Thromboxane A2

TXA2 is one of the COX-1 prostanoid products in the platelets. TXA2 stimulates platelet aggregation, and VSMC proliferation and mitogenesis (Schramm & Clowse, 2014). TXA2 also causes vasoconstriction by activating prostanoid receptors, $[Ca^{2+}]_i$ and protein kinases such as PKC, mitogen-activated protein kinase (MAPK), and Rho-kinase (ROCK) in VSM (Goulopoulou, Hannan, Matsumoto, & Webb, 2012; Fig. 1). Studies have shown decreased effects of TXA2 analogs in uterine artery of Norm-Preg guinea pigs (Weiner, Thompson, Liu, & Herrig, 1992). Other studies have shown that the overall response to TXA2 analogs is not altered during pregnancy in rats, and any changes in vasoconstriction may be further down through PKC and ROCK signaling pathways (Goulopoulou et al., 2012).

Increased lipid peroxides may activate COX-1 and in turn increase TXA2 synthesis by the platelets in PE (Walsh, 2004). Some studies suggest that TXB2, a stable metabolite of TXA2, is increased in placentas of PE compared with Norm-Preg women (Wang, Walsh, & Kay, 1992). Other studies have shown no difference in plasma TXB2 levels, but rather a 25% reduction in 6-keto-$PGF_{1\alpha}$/TXA2 ratio in PE vs Norm-Preg women (Lewis et al., 2010). Also, TCs from PE women produce more TXA2 than Norm-Preg TCs (Zhao et al., 2008). A decrease in production of PGI_2 and an increase in TXA2 may contribute to the vascular dysfunction in PE (Walsh, 2004).

In experimental studies, measurements of TXB2/6-keto-$PGF_{1\alpha}$ ratio have shown greater levels in a transgenic-activated renin–angiotensin system rat model of HTN-Preg compared with Norm-Preg rats (Verlohren et al., 2008). Also, uterine artery rings from the transgenic HTN-Preg rats showed relaxation at low ACh doses, but increased contraction at higher doses. This endothelium-dependent contraction was blocked by indomethacin or a TXA2 receptor blocker, suggesting that high ACh doses may facilitate the release of endothelial vasoconstrictor prostanoids, and this mechanism may contribute to the HTN during PE. Studies have shown that infusion of TXA2 analogs causes HTN in pregnant rats (Kriston, Venuto, Baylis, & Losonczy, 1999; Losonczy, Singh, Schoenl, Mucha, & Venuto, 1995). While some studies showed that urinary excretion of TXB2 was higher in RUPP than in Norm-Preg rats, infusion of a TXA2 receptor antagonist did not reduce BP or change GFR or effective RPF in RUPP rats, suggesting that increased TXA2 production may not play a major role in mediating HTN and renal vasoconstriction in RUPP rats (Llinas et al., 2002; Table 2). These conflicting reports make it important to further test the role of TXA2 in HTN-Preg and PE.

5. RISK FACTORS FOR ENDOTHELIAL DYSFUNCTION IN PREECLAMPSIA

While several studies support a role of endothelial dysfunction in PE, the factors leading to endothelial damage remain the subject of extensive research. Several predisposing risk factors including genetic, demographic, and environmental factors have been suggested to cause localized abnormalities in placental development, and these abnormalities would in turn lead to RUPP, placental ischemia/hypoxia, release of circulating bioactive factors, and consequently endothelial dysfunction and HTN-Preg.

5.1 Genetic Risk Factors for Preeclampsia

Placental gene mutations are associated with the incidence of PE, and 31 of 36 placental genes are downregulated in PE (Founds et al., 2009). Mutations in placental mitochondrial genes could cause incomplete reduction of O_2 and in turn increase production of ROS and lead to oxidative stress in the placenta and maternal vessels (Trifonova et al., 2014). *ACVR2A* on chromosome 2q22 and *STOX1* on chromosome 10q22 are two of the first PE susceptibility genes identified and both involve normal variations single-nucleotide polymorphism. *STOX1* Y153H common polymorphism was observed in families where several generations of women exhibited severe early PE, and has been linked to trophoblast dysfunction and IUGR (van Dijk & Oudejans, 2011). Also, wild-type female mice crossed with transgenic male mice overexpressing human *STOX1* gene show characteristics of PE including HTN and proteinuria (Doridot et al., 2013). Other genes such as *FOXP3* are important in activation of regulatory T cell (Tregs) and the control of immune response and maternal tolerance during normal pregnancy, and downregulation of *FOXP3* and its polymorphism may predispose women to PE (Rahimzadeh, Norouzian, Arabpour, & Naderi, 2016; Sasaki et al., 2007). Polymorphisms of *eNOS* gene could also be a risk factor for PE. The VNTRa and 894T alleles of *eNOS* gene are associated with early and late severe PE, respectively. For the *eNOS* VNTRb/a polymorphism, plasma NO metabolites are lower in subjects homozygous for the "a" allele. Also, the eNOS 894T allele is subject to selective proteolytic cleavage in ECs and vascular tissues, and this could account for the reduced vascular NO generation in homozygous subjects for this variant (Alpoim et al., 2014). The T−786C allele is also increased in PE compared with Norm-Preg women (Ben Ali Gannoun et al., 2015; Leonardo et al., 2015). Paternal genes may also play a role, and a paternal history of PE was associated with a 2.7% risk of PE when compared to men whose mothers had normal pregnancy (Esplin et al., 2001). However, other studies suggest a limited role of paternal genes in the development of PE (Boyd, Tahir, Wohlfahrt, & Melbye, 2013).

5.2 Demographic, Environmental, and Other Risk Factors for Preeclampsia

Ethnic background, extreme maternal age <16 or >40 years old, personal and family history of PE, primiparity, multiple pregnancy, and environmental factors have been suggested as risk factors for PE (Jardim et al., 2015; Shah & Khalil, 2015; Tanbe & Khalil, 2010).

Ethnic background may influence the incidence of PE, with African-American women having the highest rate (5.2%), while Asian women have the lowest rate (3.5%; Rosenberg, Garbers, Lipkind, & Chiasson, 2005). Age could also be a factor, and studies from Finland and India showed that women with advanced age may be at higher risk of developing PE than young women (Kanagal et al., 2014; Lamminpaa, Vehvilainen-Julkunen, Gissler, & Heinonen, 2012). The maternal lifestyle, diet, and prepregnancy overweight or obesity could increase the risk of PE (Wei et al., 2015). While the incidence of PE is ~3% in normal weight women (body mass index, BMI = 18.5–24.9), the incidence increases to 7% in women with class I obesity (BMI = 30–34.9) and to 13% in super-obese women (BMI = 50) (Spradley, Palei, & Granger, 2015). Preexisting medical conditions such as diabetes, renal or cardiac disease, systemic lupus erythematosus, mental stress, previous neonatal macrosomia, history of reproductive tract surgery and antepartum hemorrhage, and chronic respiratory conditions may also be associated with PE (Tanbe & Khalil, 2010). Of note, some risk factors such as advanced age and obesity or medical conditions such as diabetes are often associated with endothelial damage and the preexisting endothelial dysfunction could be an important mechanism in the development of PE.

6. PLACENTAL ISCHEMIA AS AN INITIATING EVENT IN PREECLAMPSIA

The placenta is a maternal–fetal interface developed in early pregnancy. Vasculogenesis, angiogenesis, and trophoblast-mediated remodeling are required to ensure proper placental and vascular development. Vasculogenesis is the development of de novo vessels from pluripotent mesenchymal stem cells ~18–35 days after conception in humans. Angiogenesis is the formation of new blood vessels by branching from preexisting ones and is regulated by the actions of proangiogenic factors and the invasive capacity of TCs (Pereira et al., 2015).

Adequate placental vascularization is essential for healthy pregnancy. During the first trimester, placental extravillous trophoblasts (EVTs) invade the maternal decidua up to one-third of the myometrium and change the spiral arteries from low-capacity high-resistance into high-capacity low-resistance vessels, thus ensuring sufficient blood supply to the growing fetus (Fig. 2). This also involves substitution of arterial VSM and elastic tissue with fibrinoid material (VanWijk, Kublickiene, Boer, & VanBavel, 2000).

The remittance of symptoms of PE after delivery has pointed to the placenta as a central culprit in this disorder. Abnormal placental development, RUPP, and consequent placental ischemia/hypoxia are important initial events in PE (Alexander, Kassab, et al., 2001; Gilbert, Babcock, & Granger, 2007; Khalil & Granger, 2002). Inadequate placentation could be initiated by abnormal expression of cytokines, major histocompatibility complex (MHC) molecules, natural killer (NK) cells, and macrophages leading to abnormal inflammatory and immune responses and apoptosis of TCs. Also, abnormal expression of integrins and matrix metalloproteinases (MMPs) could lead to decreased extracellular matrix (ECM) remodeling, shallow trophoblastic invasion, and poor remodeling of spiral arteries (Fig. 2).

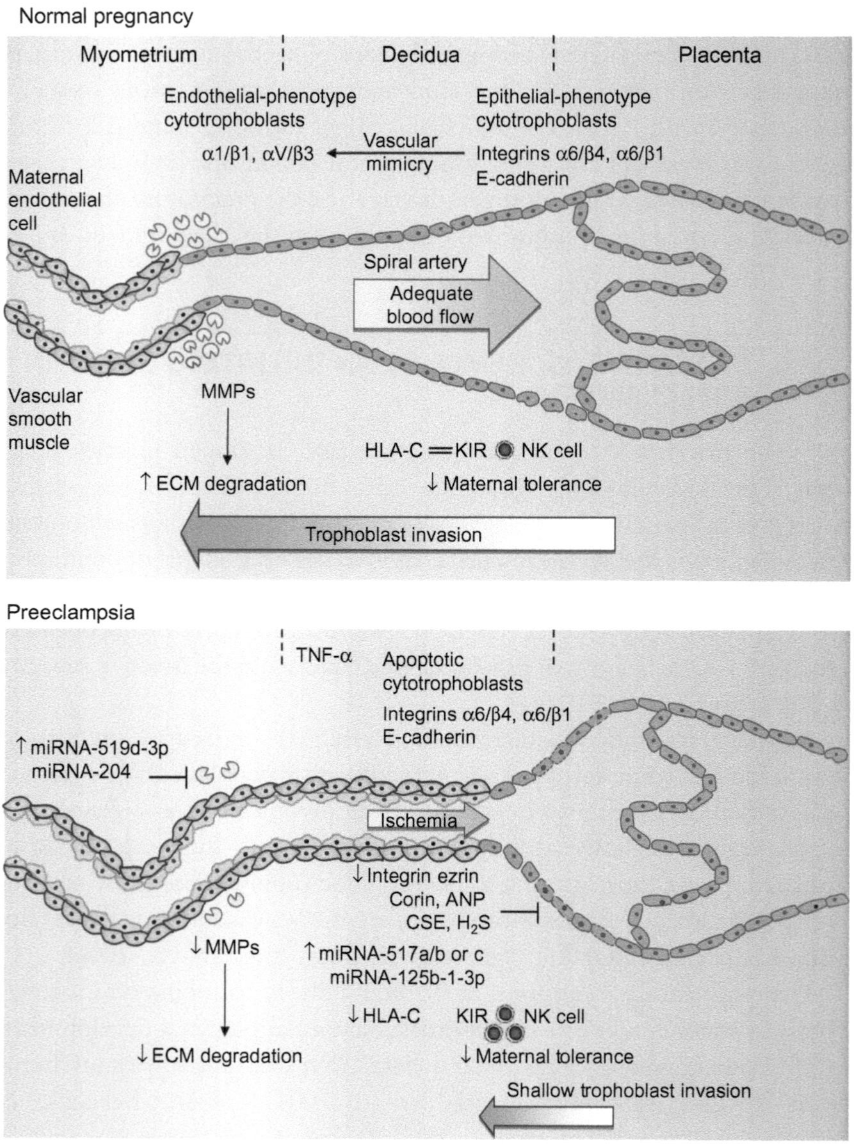

Fig. 2 Deficient placentation in preeclampsia. During normal pregnancy, cytotrophoblasts initially express adhesion molecules characteristic of epithelial cells such as integrins α6/β4 and α6/β1, and E-cadherin. As cytotrophoblasts take the invasive pathway, they express endothelial-phenotype integrins α1/β1 and αV/β3 (vascular mimicry). MMPs also cause degradation of ECM and allow vascular remodeling. Cytotrophoblasts overexpress HLA-C that interacts with the inhibitory KIR receptor and decreases NK cells, leading to maternal tolerance. As a result, cytotrophoblasts invade the decidua to one-third of the myometrium, causing remodeling of the spiral arteries

6.1 Immune Responses and Inadequate Placentation in Preeclampsia

Human pregnancy poses a challenge to maternal immune tolerance. Mother must tolerate the semiallogenic fetus throughout gestation and the fetus must be protected from rejection by the maternal immune system (Zarate, Saucedo, Valencia, Manuel, & Hernandez, 2014). In PE, proinflammatory cytokines such as TNF-α and IL-6 are increased and the inflammatory responses are enhanced. Interestingly, in conditions associated with suppressed immune response such as HIV, the incidence of HTN disorders and PE seems to be lower (Hall, Gebhardt, Theron, & Grove, 2014).

In normal pregnancy, TCs overexpress the MHC-I molecules HLA-C, HLA-E, and HLA-G which interact, respectively, with the inhibitory KIR, CD 94/NKGs, and ILT-2 receptors on NK cells and thereby prevent maternal NK cells from killing placental and fetal tissues (Trowsdale & Moffett, 2008). Decreased HLA-C/KIR interaction increases the risk of PE (Hiby et al., 2004). The decidual NK (dNK) cells are the most abundant immune cells at the maternal–fetal interface constituting 50–70% of the host defense cells. dNK cells may be stimulated by IL-15 to release interferon-γ (INF-γ) which is important for remodeling of the spiral arteries, and a decrease in placental IL-15 mRNA expression and protein levels may be associated with PE (Agarwal, Loganath, Roy, Wong, & Ng, 2001), although other studies link an increase in IL-15 and release of INF-γ to higher incidence of miscarriages and PE (Sones, Lob, Isroff, & Davisson, 2014).

Activation of complement system is important to the development of healthy pregnancy; however, excess activation of complement can lead to PE. In PE, increased levels of complement activation products Bb, C3a, and C5a are observed (Lillegard et al., 2013). Also, studies have found more

---

from small-caliber resistance to high-caliber capacitance vessels, and adequate placental blood flow. In PE, increased immune response causes release of cytokines such as TNF-α, apoptosis of cytotrophoblasts, and continued expression of integrins α6/β4 and α6/β1, and E-cadherin. Decreased expression of the integrin ezrin, corin/ANP, and CSE/H_2S pathway and increased miRNA-517a/b, -517c, and -125b-1-3p prevent cytotrophoblasts from changing into an invasive endothelial phenotype. Increased miRNA-519d-3p and -204 also decrease MMPs leading to decreased ECM degradation and vascular remodeling. Decreased HLA-C interaction with the inhibitory KIR receptor increases NK cells and further decreases maternal tolerance. The decrease in trophoblast invasion of spiral arteries and vascular remodeling leads to shallow placentation to only superficial layers of the decidua, leading to decreased blood flow and placental ischemia. *ANP*, atrial natriuretic peptide; *CSE*, cystathionine gamma-lyase; *ECM*, extracellular matrix; *H_2S*, hydrogen sulfide; *MMP*, matrix metalloproteinase; *NK*, natural killer cell.

neutrophils adhered to the endothelium of resistance-sized vessels in the subcutaneous fat of PE compared with Norm-Preg women. As neutrophils may produce toxic substances, the higher neutrophil infiltration may explain the endothelial dysfunction in PE (Leik & Walsh, 2004). In RUPP rats, inhibition of complement activation causes a decrease in BP independent of alterations in proangiogenic levels, suggesting complement activation in HTN-Preg (Lillegard et al., 2013). Also, depletion of neutrophils causes a decrease in BP in RUPP rats, supporting a role for innate immune response in HTN-Preg (Regal et al., 2015).

6.2 Integrins and Reduced TC Invasion and Spiral Arteries Remodeling in Preeclampsia

Integrins, adhesion molecules, and other factors could affect trophoblast invasion and remodeling of the spiral arteries. Cytotrophoblasts initially express adhesion molecules characteristic of epithelial cells such as integrins $\alpha 6/\beta 4$ and $\alpha 6/\beta 1$, and E-cadherin. In normal pregnancy as cytotrophoblasts take the invasive pathway, epithelial cell-like adhesion molecules cease to be expressed and expression of endothelial-phenotype integrins $\alpha 1/\beta 1$ and $\alpha V/\beta 3$ is observed, a process referred to as vascular mimicry or pseudovasculogenesis (McMaster, Zhou, & Fisher, 2004; Fig. 2). During hypoxia, expression of fibronectin and integrin $\alpha 5$ are high, while expression of integrin $\alpha 1$ is low, suggesting that decreased oxygenation could alter the synthesis of integrins and fibronectin in the placenta (Iwaki et al., 2004). In PE, abnormal expression of epithelial cell-like adhesion molecules and apoptosis of TCs lead to limited invasion of spiral arteries, RUPP, and placental ischemia (McMaster et al., 2004; Roberts & Escudero, 2012; van Dijk & Oudejans, 2013). Ezrin, an integrin involved in cell surface adhesion, migration, and organization, is downregulated in PE syncytiotrophoblast microvesicles, leading to shallow trophoblast invasion and defective placental vascularization (Baig et al., 2014). Also, retention of smooth muscle in the spiral arteries increases the vessels constriction (Perez-Sepulveda, Torres, Khoury, & Illanes, 2014), further decreasing uteroplacental blood flow and promoting placental ischemia (Fig. 2).

Other molecules could influence trophoblast invasion and vascular remodeling during pregnancy. During normal pregnancy, downregulation of EC adhesion molecules minimizes leukocyte adhesion to ECs and thereby ensures patency and maintained blood flow in the spiral arteries. In contrast, in PE, plasma levels of soluble intracellular adhesion molecule-1 (sICAM-1) and sVCAM-1 are increased, which may facilitate leukocyte adhesion and decrease blood flow in spiral arteries (Fei, Hongxiang, Qi, & Daozhen, 2012; Rios et al., 2015).

Corin is a membrane-bound protease that activates atrial natriuretic peptide (ANP), an important factor for trophoblast invasion. Corin expression is decreased in uterus of PE vs Norm-Preg women. Also, corin knockout mice have high BP and proteinuria in late pregnancy and show impaired trophoblast invasion and spiral artery remodeling (Cui et al., 2012).

Hydrogen sulfide (H_2S), a gaseous molecule endogenously produced by the enzyme cystathionine gamma-lyase (CSE), may also play a role in PE. Inhibition of CSE abolishes the invasion of first-trimester EVTs, suggesting that dysregulation of the CSE/H_2S pathway may dysregulate maternal spiral artery remodeling and placental development. In support, low levels of H_2S were reported in PE plasma (Wang et al., 2013).

MicroRNA (miRNA) may also play a role in impaired TC invasion in PE. miRNA-125b-1-3p is overexpressed in PE placentas and inhibits the expression of S1PR1, a G-coupled receptor in human EVTs that facilitates TC invasion, and the decrease in S1PR1 expression may decrease TC invasion (Li, Pan, et al., 2014). Also, miRNA-517a/b and miRNA-517c are overexpressed in PE placentas, and in vitro studies of EVTs have shown that under hypoxic conditions these miRNAs are expressed and cause a decrease in TC invasion (Anton, Olarerin-George, Hogenesch, & Elovitz, 2015).

Due to defective trophoblast invasion, arterial blood flow becomes intermittent, resulting in periods of ischemia/reperfusion and a hypoxic environment which favors oxidative stress, oxidative damage, and inflammation, and leads to endothelial dysfunction in PE (Sanchez-Aranguren et al., 2014).

6.3 Role of MMPs in Abnormal Placentation

MMPs are zinc-dependent degradative enzymes which include collagenases, gelatinases, stromelysins, and other subtypes, and are involved in tissue remodeling and other cellular functions (Li, Mata, Mazzuca, & Khalil, 2014; Montagnana et al., 2009). MMPs are produced as pro-MMPs which are cleaved into active MMPs and play a role in endometrial tissue remodeling during the estrous cycle, menstrual cycle, and pregnancy in human and animals (Li, Mata, et al., 2014).

MMP-2 and -9 may be involved in ECM remodeling and trophoblast invasion of the spiral arteries during pregnancy. MMP-2 is the main collagenolytic enzyme in umbilical cord artery (Ali & Khalil, 2015) and serum MMP-9 levels are elevated in Norm-Preg women (Montagnana et al., 2009; Table 3). MMP-2 and -9 expression and gelatinase activity are upregulated

Table 3 Bioactive Factors in Human Normal Pregnancy and Preeclampsia

Factor	Specimen (Units)	Normal Pregnancy	Preeclampsia	References
VEGF	Serum (ng/mL)	13.9	51.7	Hunter et al. (2000)
	Serum (pg/mL)	14.20 ± 14.54^{SD}	314.45 ± 260.74^{SD}	Celik et al. (2013)
	Serum (pg/mL)	90.55 (90–521)	90 (90–211)	Masoura et al. (2014)
	Plasma (ng/mL)	6.83 (5.32–8.94)	25.24 (21.22–42.92)	Tsatsaris et al. (2003)
	Villous explants (ng/mL/mg protein)	55 ± 64	117 ± 3.7	Ahmad and Ahmed (2004)
PlGF	Serum (pg/mL)	432.9 (359.6–815.2)	65.5 (33.0–101.2)	Ramma et al. (2012)
	Serum (pg/mL)	217.30 ± 74.48	115.72 ± 32.55	Bian et al. (2015)
	Serum (pg/mL)	183 (126–307)	98.0 (63.7–146)	Molvarec, Czegle, Szijarto, and Rigo (2015)
	Plasma (pg/mL)	585.9 (356–1101)	67.41 (57.1–118.5)	Tsatsaris et al. (2003)
	Plasma (pg/mL)	324.6 (101.8–881.4) $\leq$34 weeks	18.3 (9.9–22.5) early PE	March et al. (2015)
		206.8 (60.9–395.7) >34 weeks	82.6 (57.1–215.6) late PE	
sFlt-1	Serum (ng/mL)	1.5 (1.1–1.8)	21.5 (15.2–30.5)	Ramma et al. (2012)
	Serum (ng/mL)	1.8 ± 1.6^{SD}	6.6 ± 5.5^{SD}	Tuzcu et al. (2015)
	Serum (ng/mL)	0.308 ± 0.019	0.321 ± 0.023	Bian et al. (2015)
	Serum (pg/mL)	3252 (2509–4751)	6814 (3736–12.720)	Molvarec et al. (2015)

	Plasma (ng/mL)	0.12 (0–0.29)	2.69 (2.31–2.97)	Tsatsaris et al. (2003)
	Plasma (pg/mL)	3391 (2412–4918) ≤34 weeks	12,895 (8303–17417) early PE	March et al. (2015)
		4378 (2618–5731) >34 weeks	6304 (3127–10638) late PE	
	Villous explants (ng/mL/mg protein)	28 ± 1.7	128 ± 9.8	Ahmad and Ahmed (2004)
sFlt-1 e15a	Serum (pg/mL)	~10,000	~80,000	Palmer et al. (2015)
sFlt-1/PlGF ratio	Serum	15.6 (8.52–36.6)	70.5 (31.8–144)	Molvarec et al. (2015)
	Plasma	9.6 (3.5–58.6) ≤34 weeks	703.1 (146.6–1614.9) early PE	March et al. (2015)
		22.4 (10.2–58.7) >34 weeks	77.0 (18.3–145.1) late PE	
sEng	Serum (ng/mL)	~10	~38 mild PE	Venkatesha et al. (2006)
			~50 severe PE	
			~100 HELLP	
	Serum (ng/mL)	4.3 (3.5–6.1)	70.1 (41.3–109.4)	Ramma et al. (2012)
	Serum (ng/mL)	9.8	46.4 early PE	Levine et al. (2006)
	Serum (ng/mL)	13.3	31.0 late PE	Levine et al. (2006)

Continued

Table 3 Bioactive Factors in Human Normal Pregnancy and Preeclampsia—cont'd

Factor	Specimen (Units)	Normal Pregnancy	Preeclampsia	References
TNF-α	Serum (pg/mL)	14.62 ± 5.61	26.49 ± 12.14	Cakmak et al. (2016)
	Monocytes supernatant (pg/mL)	~25	~130	Matias et al. (2015)
LIGHT	Plasma (pg/mL)	~2	~46	Wang et al. (2014)
IL-6	Serum (pg/mL)	0.6 (0.4–1.0)	1.1 (0.6–7.9)	Ramma et al. (2012)
IL-1β	Serum (pg/mL)	0.10	0.16	Siljee et al. (2013)
	Monocytes supernatant (pg/mL)	~260	~600	Matias et al. (2015)
IL-17	Serum (pg/mL)	0	0.47 (0–0.53)	Molvarec et al. (2015)
IL-18	Monocytes supernatant (pg/mL)	~14	~22	Matias et al. (2015)
HIF-1α	Blood (μL)	~0.8	~2.4	Akhilesh et al. (2013)
$O_2^{\cdot -}$	Neutrophills	3.63 ± 0.91 nmol/10^6 cells^{SD}	6.20 ± 0.92 nmol/10^6 cells^{SD}	Tsukimori, Fukushima, Tsushima, and Nakano (2005)
	Monocytes	~1.7 nmol/10^5 cells	~2.2 nmol/10^5 cells	Peracoli et al. (2011)
H_2O_2	Monocytes	~1.4 nmol/10^5 cells	~1.7 nmol/10^5 cells	Peracoli et al. (2011)
HO-1 mRNA	Blood (RC)	9.87 (8.61–10.53)	9.13 (5.42–10.19)	Nakamura et al. (2009)

HO-2 mRNA	Blood (RC)	7.05 (3.19–7.47)	6.81 (4.73–7.34)	Nakamura et al. (2009)
SOD mRNA	Blood (RC)	5.91 (4.95–6.44)	5.40 (3.90–6.23)	Nakamura et al. (2009)
GPx mRNA	Blood (RC)	7.56 (7.03–8.10)	6.90 (4.54–7.52)	Nakamura et al. (2009)
CAT mRNA	Blood (RC)	7.38 (4.39–7.77)	7.07 (4.90–7.63)	Nakamura et al. (2009)
TAOC	Serum (mmol/L)	1.1 (1.0–1.2)	0.5 (0.2–0.6)	Turpin, Sakyi, Owiredu, Ephraim, and Anto (2015)
AngII	Chorionic villi	15 ± 2 fmol/mg protein	26 ± 6 fmol/mg protein	Anton et al. (2008)
AT_1R mRNA	Chorionic villi (RGE)	1.0 ± 0.1	3.0 ± 0.7	Anton et al. (2008)
AT_1-AA	Serum (OD)	0.27 ± 0.12	0.54 ± 0.13 OD	Bai et al. (2013)
	Serum (OD)	0.315 ± 0.093	0.703 ± 0.132 early PE	Yang et al. (2015)
			0.567 ± 0.111 late PE	
	Serum (RLU)	~12	~75	Siddiqui et al. (2013)
	% Over baseline			

Continued

Table 3 Bioactive Factors in Human Normal Pregnancy and Preeclampsia—cont'd

Factor	Specimen (Units)	Normal Pregnancy	Preeclampsia	References
MMP-1	Umbilical serum (pg/mL)	294.33 ± 11.53	177.67 ± 12.63	Deng, Ling, Liu, Zhao, and Lv (2015)
MMP-2	Serum (ng/mL)	669 (560–760)	834 (656–1002)	Montagnana et al. (2009)
	Plasma (ng/mL)	241.1 ± 35.3^{SD}	290.5 ± 48.4^{SD}	Eleuterio et al. (2015)
MMP-9	Serum (ng/mL)	390 (277–569)	290 (280–470)	Montagnana et al. (2009)
	Plasma (ng/mL)	240.0 ± 197.7^{SD}	262.4 ± 153.8^{SD}	Eleuterio et al. (2015)
TIMP-1	Serum (ng/mL)	148 (121–188)	213 (212–220)	Montagnana et al. (2009)
	Plasma (ng/mL)	142.8 ± 39.2^{SD}	187.1 ± 35.4^{SD}	Eleuterio et al. (2015)
	Umbilical serum (pg/mL)	1304.20 ± 69.66	1363.00 ± 71.50	Deng et al. (2015)
TIMP-2	Serum (ng/mL)	228 (207–267)	232 (225–245)	Montagnana et al. (2009)
	Plasma (ng/mL)	158.3 ± 32.3^{SD}	194.3 ± 49.3^{SD}	Eleuterio et al. (2015)
Uric acid	Serum (μM)	185 (28.7–642) first trimester 217 (102–428) second trimester 278 (68.4–535) third trimester	191 (120–457) first trimester 215(130–248) second trimester 350 (157–720) third trimester	Chen et al. (2016)

Values represent means ± standard error of the mean. ^{SD}indicates standard deviation. *Numbers* in parenthesis indicate range. *AngII*, angiotensin II; *AT1-AA*, AngII AT_1R agonistic autoantibodies; *AT_1R*, angiotensin II type 1 receptor; *CAT*, catalase; *GPx*, glutathione peroxidase; *H_2O_2*, hydrogen peroxide; *HIF-1α*, hypoxia-inducible factor-1α; *HO*, hemeoxygenase; *IL*, interleukin; *MMP*, matrix metalloproteinase; $O_2^{\cdot -}$ superoxide anion; *PE*, preeclampsia; *PlGF*, placental growth factor; *sEng*, soluble endoglin; *sFlt-1*, soluble fms-like tyrosine kinase-1; *SOD*, superoxide dismutase; *TAOC*, total antioxidant capacity; *TIMP*, tissue inhibitor of metalloproteinase; *TNF-α*, tumor necrosis factor-α; *VEGF*, vascular endothelial growth factor; *OD*, optical density measured by spectrophotometry; *RC*, relative concentration to total mRNA extracted from Norm-Preg placenta; *RGE*, relative gene expression of target gene compared to expression in Norm-Preg; *RLU*, relative light unit from CHO.AT_1A cells encoding rat AT_1R and luciferase reporter.

in the myometrium and aorta of Norm-Preg rats (Li, Mata, et al., 2014). Genetic polymorphisms may alter MMP-2 and -9 transcription in PE (Palei, Granger, & Tanus-Santos, 2013). Also, miRNA-519d-3p and miRNA-204 are overexpressed in PE patients (Choi et al., 2013; Li, Ge, Guo, & Lu, 2013) and could target MMP-2 and -9, respectively, and in turn decrease trophoblast invasiveness (Yu, Wang, Liu, & Guan, 2015; Fig. 2). Some studies showed an increase in circulating MMP-2 and -9 in PE vs Norm-Preg women (Eleuterio et al., 2015; Table 3). Other studies showed a decrease in serum MMP-9 in PE (Montagnana et al., 2009). MMP-2 and -9 levels are also reduced in uterus, placenta, and aorta of RUPP rats (Table 4), and low MMP levels may cause excessive collagen deposition, affect smooth muscle growth, and decrease spiral arteries remodeling (Li, Mata, et al., 2014). Also, in an in vitro model of first-trimester trophoblasts, suppression of MMP-9 expression inhibited TC invasive capability, supporting a role of MMP-9 in modulating trophoblast invasion (Yu et al., 2015). MMP-1 is also expressed in cytotrophoblasts and syncytiotrophoblasts of the placenta and decidua and may play a role in trophoblast invasion. MMP-1 levels in the umbilical cord blood, placenta, and decidua are lower in PE than Norm-Preg women, and the low levels of MMP-1 are in a positive correlation with the severity of the disease (Deng et al., 2015).

In addition to their proteolytic activities, MMPs may increase cytokines and ROS in PE (Palei, Spradley, Warrington, George, & Granger, 2013), and MMPs may also modulate vascular tonus. MMP-2 induces vasodilation in rat vena cava via hyperpolarization and activation of K^+ channels (Raffetto, Ross, & Khalil, 2007). Other studies have shown that MMP-2 and -9 may increase the production of ET-1 and related peptides and decrease vasodilator peptides such as adrenomedullin (Fernandez-Patron et al., 2001; Nascimento et al., 2015), leading to an imbalance between dilator and constrictor factors and causing endothelial dysfunction. MMP-2 may also enhance big-ET-1-induced constriction in mesenteric vessels of RUPP rats (Abdalvand, Morton, Bourque, Quon, & Davidge, 2013). Also, in omental vessels of Norm-Preg women, MMP-1 causes vasoconstriction and enhances reactivity to AngII via an endothelium-dependent protease-activated receptor and ET-1 pathway (Nugent, Mishra, Strauss, & Walsh, 2016).

MMP activity is modulated by tissue inhibitors of metalloproteinases (TIMPs; Montagnana et al., 2009). Measurements of TIMP-1 and -2 in the circulation and umbilical serum have shown increases or no change in PE vs Norm-Preg women (Deng et al., 2015; Eleuterio et al., 2015;

Table 4 Bioactive Factors in Normal Pregnant and RUPP Rats

Factor	Specimen (Units)	Norm-Preg	RUPP	References
VEGF	Plasma (pg/mL)	1017 ± 95	670 ± 68	George, Palei, Dent, and Granger (2013)
	Plasma (pg/mL)	830 ± 33	594 ± 34	Gilbert et al. (2007)
	Placenta (pg/mg)	30 ± 2	43 ± 5	George et al. (2013)
PlGF	Plasma (pg/mL)	1.7 ± 0.5	0.28 ± 0.05	Gilbert et al. (2007)
TGF-β	Serum (pg/mL)	1036 ± 82 pg/mL	567 ± 88 pg/mL	Cornelius et al. (2015)
sFlt-1	Plasma (pg/mL)	82 ± 26	660 ± 270	Gilbert et al. (2007)
	Plasma (pg/mL)	1432 ± 255	3431 ± 454	Murphy and Cockrell (2015)
	Placenta (pg/mg)	643 ± 44	809 ± 81	George et al. (2013)
	$CD4^+$ T supernatant (pg/mL)	1046 ± 280	2500 ± 650	Wallace et al. (2011)
sFlt-1/PlGF ratio	Plasma	8.9 ± 1.6	37.2 ± 7.8	Gilbert et al. (2007)
sEng	Serum (APU)	~0.05	~0.10	Gilbert, Gilbert, Arany, and Granger (2009)
	Placenta (APU)	~1.5	~4.8	Gilbert et al. (2009)
TNF-α	Plasma (pg/mL)	16.0 ± 6.4	61.4 ± 12.2	Cornelius et al. (2015)
	$CD4^+$ T (pg/mL)	133 ± 23	250 ± 50	Wallace et al. (2011)
IL-6	Plasma (pg/mL)	30 ± 7	74 ± 15	Cornelius et al. (2015)
	$CD4^+$ T supernatant (pg/mL)	287 ± 12	778 ± 29	Wallace et al. (2011)

IL-10	Plasma (pg/mL)	77.3 ± 22.2	19.6 ± 4.8	Cornelius et al. (2015)
HIF-1α	Placenta (APU)	0.68 ± 0.09	1.42 ± 0.25	Gilbert et al. (2009)
Total ROS	Placenta (RLU)	240.9 ± 24.1	339.3 ± 58.7	Cornelius et al. (2015)
	Aorta (AU)	5.2 ± 0.4	7.7 ± 0.3	Amaral et al. (2013)
HO-1	Placenta (APU)	2.5 ± 0.1	1.4 ± 0.3	Gilbert et al. (2009)
8-Isoprostane	Plasma (ng/mL)	305 ± 85	689 ± 8	Amaral et al. (2013)
AT_1-AA	Serum (bpm)	0.6 ± 0.3	15.3 ± 1.6	LaMarca, Wallukat, et al. (2008)
	Plasma (bpm)	0.08 ± 0.25	17.81 ± 1.1	Cornelius et al. (2015)
	Serum (bpm)	1.1	14.8	Novotny et al. (2012)
MMP-2	Uterus (OD)	~1.0	~0.7	Li, Mata, et al. (2014)
	Placenta (OD)	~0.9	~0.5	
	Aorta (OD)	~0.9	~0.6	
MMP-9	Uterus (OD)	~0.4	~0.2	Li, Mata, et al. (2014)
	Placenta (OD)	~0.3	~0.2	
	Aorta (OD)	~0.4	~0.2	

Values represent means ± standard error of the mean. *AT_1AA*, angiotensin II type 1 receptor agonistic autoantibodies; *HIF-1α*, hypoxia-inducible factor-1α; *HO*, hemeoxygenase; *IL*, interleukin; *MMP*, matrix metalloproteinase; *PlGF*, placental growth factor; *ROS*, reactive oxygen species; *sEng*, soluble endoglin; *sFlt-1*, soluble fms-like tyrosine kinase-1; *TGF-β*, transforming growth factor-β; *TNF-α*, tumor necrosis factor-α; *VEGF*, vascular endothelial growth factor. *APU*, arbitrary pixel unit relative to the 67-kDa band corresponding to albumin in Ponceau-stained membranes; *AU*, arbitrary unit of fluorescence generated by dihydroethidium oxidation; bpm, beats/min measured in spontaneously beating neonatal rat cardiomyocytes exposed to sera containing AT1-AA and antagonized specifically by AT_1R antagonist; *OD*, optical densitometry of Western blot bands normalized to β-actin; *RLU*, relative light unit measured by lucigenin luminescence.

Montagnana et al., 2009), while other studies have shown increases in TIMP-1 and -3 in PE patients (Zhu, Zhong, Pang, & Yu, 2014). The role of MMPs and TIMPs in modulating vascular function and the endothelial dysfunction associated with PE should be further examined in future studies.

7. CIRCULATING BIOACTIVE FACTORS IN PREECLAMPSIA

Placental hypoxia/ischemia is believed to trigger the release of several bioactive factors including the antiangiogenic factors sFlt-1 and sEng, proinflammatory cytokines such as TNF-α and IL-6, HIF, ROS, and AT_1-AA (Fig. 3). These factors could cause EC dysfunction, severe vasoconstriction, and the increases in BP observed in PE women and in animal models of HTN-Preg (Jardim et al., 2015; Shah & Khalil, 2015).

7.1 Proangiogenic and Antiangiogenic Factors in Preeclampsia

7.1.1 Vascular Endothelial Growth Factor

VEGF gene is located on chromosome 6 (6p21.3), which consists of eight exons involved in the expression of a family of proteins including VEGF-A, VEGF-B, VEGF-C, VEGF-D, and PlGF (Jardim et al., 2015). VEGF-A, VEGF-B, and PlGF bind to tyrosine kinase receptor flt-1 (Flt-1 or VEGFR-1) and only VEGF-A binds to VEGFR-2 (Flk-1 or KDR) to induce the development of the placental vasculature (Jardim et al., 2015). VEGF regulates EC proliferation, angiogenesis, and vascular permeability (Jardim et al., 2015; Shah & Khalil, 2015). In ECs, VEGF increases $[Ca^{2+}]_i$, Ca^{2+}/calmodulin, eNOS activity, and PGI_2 leading to decreased vessel tonicity and BP (Cindrova-Davies, Sanders, Burton, & Charnock-Jones, 2011; He et al., 1999; Shen, Lee, & Zioncheck, 1999). VEGF could also induce Akt activation and eNOS Ser^{1177} phosphorylation in ECs, leading to Ca^{2+}-independent generation of NO. In HUVECs, blockade of VEGFR leads to decreased Akt activity and eNOS phosphorylation and impaired endothelial function, supporting a link between changes in VEGF activity and endothelial damage in PE (Cindrova-Davies et al., 2011).

Circulating levels of VEGF vary in PE, with some studies showing an increase in circulating VEGF (Celik et al., 2013; Hunter et al., 2000; Tsatsaris et al., 2003), while other reports show decreased or unchanged levels of serum VEGF (Masoura et al., 2014; Maynard et al., 2003). VEGF production is higher in villous explants from PE than Norm-Preg women (Ahmad & Ahmed, 2004). It is likely that during the vasoconstrictive and

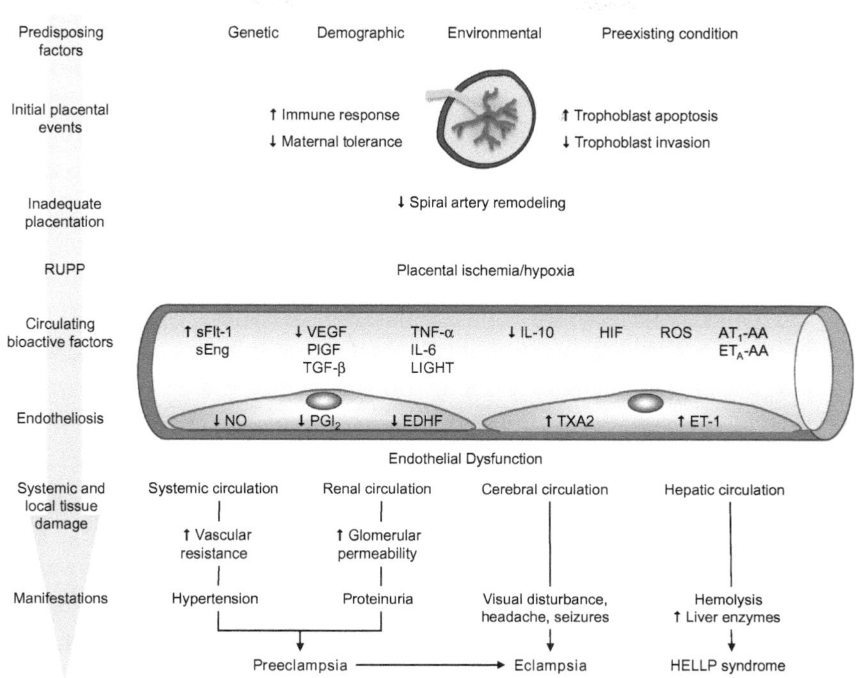

Fig. 3 Predisposing risk factors, intermediary bioactive factors, and endothelial damage in preeclampsia. Genetic, demographic, environmental, and other risk factors cause abnormal placentation. Increased immune response, trophoblast cell apoptosis, and decreased trophoblast invasion cause poor remodeling of spiral arteries and reduced uteroplacental perfusion pressure (RUPP). RUPP triggers the release of several circulating bioactive factors. Bioactive factors target endothelial cells in the systemic circulation causing generalized vasoconstriction, increased vascular resistance and HTN, renal circulation causing increased glomerular permeability and proteinuria, cerebral circulation causing visual disturbance, headaches, seizures, and eclampsia, and the hepatic circulation causing hemolysis, elevated liver enzymes, and low platelets (HELLP syndrome). *AT_1-AA*, angiotensin II type 1 receptor agonistic autoantibodies; *EDHF*, endothelium-derived hyperpolarizing factor; *ET-1*, endothelin-1; *ET_A-AA*, endothelin receptor type A agonistic autoantibodies; *HELLP*, hemolysis, elevated liver enzymes, low platelets; *HIF*, hypoxia-inducible factor; *IL*, interleukin; *NO*, nitric oxide; *PGI_2*, prostacyclin; *PlGF*, placental growth factor; *ROS*, reactive oxygen species; *sEng*, soluble endoglin; *sFlt-1*, soluble fms-like tyrosine kinase-1; *TGF-β*, transforming growth factor-β; *TNF-α*, tumor necrosis factor-α; *TXA2*, thromboxane A2; *VEGF*, vascular endothelial growth factor.

latent phase of PE the increase in vascular shear stress could increase circulating VEGF (Tanbe & Khalil, 2010). Also, women with the T allele of VEGF 936C/T produce lower levels of VEGF and have a higher risk of developing PE than women with VEGF 936C/C (Papazoglou, Galazios, Koukourakis, Kontomanolis, & Maltezos, 2004). While VEGF plasma levels are decreased in RUPP rats (Gilbert et al., 2007), in agreement with the findings in human villous explants, placenta from RUPP rats show greater production of VEGF (George et al., 2013; Table 4). The differences in the results may be related to differences in the methods of VEGF measurement (Maynard et al., 2003). In PE, there is an increase in circulating anti-angiogenic factors that may bind to VEGF. Thus, total (bound and unbound) VEGF measured using radioimmunoassay or competitive enzyme immunoassay could be higher, while free VEGF measured using ELISAs may be lower in PE than Norm-Preg women (Bates, 2011).

Changes in VEGF activity could also influence protein excretion, and a decrease in VEGF may play a role in the glomerular endotheliosis in PE. VEGF is synthesized constitutively in the glomerulus by podocytes where it maintains endothelial health and induces the formation of fenestrae. Interestingly, endotheliosis and loss of fenestrae are observed in genetic glomerular VEGF deficiency (Stillman & Karumanchi, 2007). Also, in clinical cancer trials, the use of VEGF-neutralizing antibodies therapy is associated with proteinuria (Zhu, Wu, Dahut, & Parikh, 2007). In nonpregnant mice, downregulation of VEGF by infusion of VEGF antibodies leads to glomerular endotheliosis and proteinuria similar to what is seen in PE (Eremina et al., 2008). Also, mice lacking one VEGF allele in renal podocytes develop the typical renal pathology found in PE, and infusion of VEGF ameliorates the renal lesions, glomerulonephritis, and thrombotic microangiopathy in RUPP rats (Kim et al., 2000; Masuda et al., 2001). Thus, a decrease in VEGF activity in PE could cause reduction in GFR and leads to proteinuria.

7.1.2 Placental Growth Factor

PlGF is a proangiogenic factor that binds to VEGFR-1 and enhances the angiogenic effects of VEGF (Romero et al., 2008). PlGF has only 1/10th the affinity for VEGFR-1 compared to VEGF, but its level is ~40 times higher than VEGF during normal pregnancy. PIGF dilates uterine vessels and promotes EC growth, vasculogenesis, and placental development (Shah & Khalil, 2015).

Plasma PlGF levels are low in nonpregnant women (~44 pg/mL), but are substantially higher in Norm-Preg women (Romero et al., 2008). PlGF levels during 21 and 22 weeks of gestation are ~353 pg/mL, rising steadily during gestation to reach a median level of ~574 pg/mL after 29 and 30 weeks of gestation (Krauss, Pauer, & Augustin, 2004). During PE, circulating PlGF levels decrease (Bian et al., 2015; Molvarec et al., 2015; Ramma et al., 2012; Tsatsaris et al., 2003) and the decrease is more apparent in early than late PE (March et al., 2015). PlGF has four alternatively spliced mRNA species (PIGF 1–4), and its predominant isoform PIGF-1 is downregulated in PE (Bates, 2011). RUPP and DOCA-salt rats also show a decrease in circulating levels of PlGF (Agunanne, Uddin, Horvat, & Puschett, 2010; Gilbert et al., 2007).

In addition to its growth-promoting effects, PlGF could exert vasodilator effects via VEGFR-1 and an EDHF-dependent pathway involving activation of SK_{Ca} (Mandala et al., 2012; Morton & Davidge, 2013). Also, in mesenteric resistance arteries from pregnant rats treated with L-NAME, L-NNA, and indomethacin, a second application of PlGF could produce a greater reduction in VSM $[Ca^{2+}]_i$ and greater vasodilation. Since VEGF and PlGF signaling is thought to involve receptor dimerization, initial exposure to PlGF may facilitate subsequent responses via stimulated formation of receptor homodimers and their associated submembrane signaling pathways leading to increases in both the speed and the amplitude of the vasodilatory response to repeated stimulation (Mandala et al., 2012). As the levels of PlGF are likely reduced in PE, the vasodilatory responses of PlGF are expected to be decreased.

7.1.3 Soluble fms-Like Tyrosine Kinase-1

sFlt-1 (sVEGFR-1) is an antiangiogenic factor expressed as an alternatively spliced variant of VEGFR-1 and lacks both the transmembrane and cytoplasmic domains. sFlt-1 binds to VEGF and PlGF in the circulation and inhibits their action on cell surface receptors. sFlt-1 may also form a heterodimer with the surface receptors and inhibit any signaling action they might have (Charnock-Jones, 2016). TCs express sFlt-1 mRNA, and sFlt-1 level is ~1.5 ng/mL in Norm-Preg compared to ~0.15 ng/mL in nonpregnant women (Shah & Khalil, 2015). sFlt-1 levels are stable in Norm-Preg women, showing an increase after the 36th week of gestation. Throughout the third trimester, the increase in sFlt1 persists and there is a reduction in VEGF and PlGF. In PE, the increase in sFlt-1 and decrease in PlGF occur at more pronounced levels than those seen in Norm-Preg women (Jardim et al., 2015). *sFlt-1* gene is localized on chromosome 13q12, and an extra copy of this gene in women with trisomy 13 may lead to excess circulating sFlt-1, reduced PlGF, and increased risk of PE (Kakigano et al., 2013). Several reports have shown higher circulating levels of sFlt-1 in early and late PE (Bian et al., 2015; March et al., 2015; Molvarec et al., 2015; Ramma et al., 2012; Tsatsaris et al., 2003; Tuzcu et al., 2015). Serum sFlt-1 is also higher in women with previous PE (~0.5 ng/mL) than in women with previous normal pregnancy (~0.3 ng/mL), even 6 months or more after delivery (Tuzcu et al., 2015). sFlt-1 levels are also increased in villous explants from PE vs Norm-Preg women (Ahmad & Ahmed, 2004).

During placental hypoxia, HIF-1 may bind to the promoter region of *flt-1* gene leading to upregulation of sFlt-1 (Ahmad & Ahmed, 2004; Maynard et al., 2003). In vitro studies in EVTs have shown that overexpression of miR-517a/b and miR-517c increases the expression of TNFSF15, a cytokine involved in Flt-1 splicing, leading to increases in sFlt-1 release (Anton et al., 2015). sFlt-1 e15a, a splice variant of sFlt-1, is the most abundant form released by the placenta. sFlt-1 e15a binds and inhibits VEGF and in turn decreases EC migration, invasion, and tube formation. sFlt-1 e15a is expressed in syncytiotrophoblasts and shows a 10-fold increase in serum levels in PE vs Norm-Preg women (Palmer et al., 2015).

In addition to changes in the levels of sFlt-1, a decline in VEGF/sFlt-1 and PlGF/sFlt-1 ratio by 53% and 70%, respectively, is observed in PE placenta (Ahmad & Ahmed, 2004). The sFlt-1/PlGF ratio in the circulation is higher in PE than in Norm-Preg women from second trimester onward and can be a good predictor of the onset of PE (March et al., 2015; Molvarec et al., 2015), although some reports suggest that it is lower in late compared to early PE (March et al., 2015; Noori, Donald, Angelakopoulou, Hingorani, & Williams, 2010). Circulating sFlt-1 levels and sFlt-1/PlGF ratio are higher in twin than in singleton pregnancies possibly due to greater placental mass (Bdolah et al., 2008; Faupel-Badger et al., 2015), supporting that the placenta is a likely source of these factors. Studies have suggested a direct association between the degree of angiogenic imbalance and ET-1 levels. Patients with sFlt-1/PlGF ratio >85 have higher levels of ET-1 than patients with sFlt-1/PlGF ratio <85 (Verdonk et al., 2015). Also, extracorporeal removal of circulating sFlt-1 in PE patients decreases sFlt-1/PlGF ratio and improves symptoms and prolongs pregnancy (Thadhani et al., 2016).

Of note, the decrease in FMD and elevation of BP in women who subsequently develop late PE may occur before the rise in plasma sFlt-1 levels (26–36 weeks). These findings suggest the involvement of other bioactive factors and that the impaired endothelial function may represent a baseline risk factor for PE rather than a consequence of elevated sFlt-1 levels in cases of late PE (Noori et al., 2010).

RUPP rats show increases in plasma and placental levels of sFlt-1 and plasma sFlt/PlGF ratio (George et al., 2013; Gilbert et al., 2007; Murphy & Cockrell, 2015; Table 4). Other animal models of HTN-Preg show increased or unchanged circulating levels of sFlt-1 (Agunanne et al., 2010; Bobek, Surmon, Mirabito, Makris, & Hennessy, 2015; Gillis, Williams, Garrett, Mooney, & Sasser, 2015; Ramesar et al., 2011; Siddiqui et al., 2013; Sunderland et al., 2011). Importantly, pregnant rats treated with exogenous sFlt-1 develop HTN, proteinuria, marked glomerular endotheliosis with occlusion of renal capillaries, and focal fibrin deposition in glomerular cells (Holwerda et al., 2014; Jiang et al., 2015; Maynard et al., 2003). Also, mice treated with sFlt-1 show an increase in the vascular response to ET-1 (Amraoui et al., 2014). Exposure of ECs to PE plasma results in reduced angiogenesis, and removal of sFlt-1 or treatment with VEGF or a sFlt-1 antibody reverses the antiangiogenic effects of sFlt-1 and restores EC function and angiogenesis (Ahmad & Ahmed, 2004).

Of note, VEGF has been shown to stimulate sFlt-1 production in human placental explants through an action on VEGFR-2 (Fan et al., 2014). It appears that VEGF levels are tightly controlled at the maternal–fetal interface, partly through feedback modulation by sFlt-1, which may play a critical local role in preventing any damage to the placenta or fetus that would otherwise be caused by excess VEGF even during normal pregnancy (Fan et al., 2014). Dysregulation of this feedback mechanism may play a role in the pathogenesis of PE.

7.1.4 Soluble Endoglin

Endoglin (Eng) is a coreceptor for transforming growth factor (TGF)-β1 and TGF-β3 that is highly expressed on cell membranes of ECs and syncytiotrophoblasts. TGF-β1 induces the migration and proliferation of ECs by binding to TGF receptors (Jardim et al., 2015). Eng is a proliferation-associated and hypoxia-inducible protein abundantly expressed in angiogenic ECs (Luft, 2006). In contrast, sEng acts as an antiangiogenic protein that inhibits TGF-β1 signaling and TGF-β1-induced eNOS activation and vasodilation (Jardim et al., 2015). Mutations in *Eng* gene result in loss of capillaries, arteriovenous malformations, and hereditary hemorrhagic telangiectasia (Maynard et al., 2008).

sEng is barely detectable in nonpregnant women and has much lower levels in the serum of Norm-Preg women (Venkatesha et al., 2006). sEng levels are 3-, 5-, and 10-fold higher in mild PE, severe PE, and HELLP syndrome, respectively, compared to gestational age-matched preterm controls (Venkatesha et al., 2006). Serum sEng levels may be increased in both early and late PE (Levine et al., 2006; Ramma et al., 2012), although one study showed that sEng levels increase at 10–17 gestational weeks in women who developed early PE, but not in those that developed late PE (Noori et al., 2010). sEng expression also increases in placental extracts exposed to a hypoxic environment (Yinon et al., 2008).

In RUPP rats, the levels of sEng increase in both serum and placenta (Gilbert et al., 2009), while serum TGF-β levels decrease (Cornelius et al., 2015; Table 4). However, no changes in sEng levels were observed in rat models of HTN-Preg produced by DOCA-salt or L-NAME (Agunanne et al., 2010; Ramesar et al., 2011; Table 5). sEng may act in concert with sFlt-1 to induce vascular permeability, proteinuria, IUGR, and severe HTN in pregnant rats (Venkatesha et al., 2006). Pregnant rats infused with both sFlt-1 and sEng via miniosmotic pump show HELLP syndrome-like features (Morris et al., 2015). Also, sEng impairs the formation of endothelial tubes in cultured HUVECs (Venkatesha et al., 2006).

7.2 Cytokines: TNF-α and Interleukins

Placental ischemia may enhance the synthesis of inflammatory cytokines. Immediately after placental reperfusion injury, reestablished blood flow releases cytokines and inflammatory factors like TNF-α and ILs (Zarate et al., 2014). TNF-α is released from the hypoxic placenta and is a primary modifier of the immune reaction. TNF-α increases vascular permeability, fibroblast proliferation, and lymphocyte activation and induces production of IL-6 and IL-8. TNF-α downregulates eNOS and mitochondrial biogenesis, leading to mitochondrial dysfunction, oxidative stress, and increased ROS (Sanchez-Aranguren et al., 2014). TNF-α can also modify the expression of adhesion molecules in placental vessels (LaMarca, Speed, et al. 2008) and induce abnormal production of MMPs in PE (Palei, Granger, et al., 2013).

Circulating TNF-α levels are increased in PE compared with Norm-Preg women (Cakmak et al., 2016; Moreno-Eutimio et al., 2014; Pinheiro et al., 2015), although the placental levels of TNF-α may not be different in PE vs normal pregnancy (Peixoto et al., 2016). LIGHT, a TNF superfamily member, is also increased in PE and may contribute to placental damage and ischemia (Wang et al., 2014). Plasma levels and $CD4^+$ T cell production of TNF-α are increased in RUPP compared with Norm-Preg rats (Cornelius et al., 2015; LaMarca, Speed, et al., 2008; Wallace et al., 2011). Infusion of TNF-α causes HTN and proteinuria during late pregnancy in baboons (Sunderland et al., 2011), rats (Alexander, Cockrell, Massey, Bennett, & Granger, 2002), and mice (Bobek et al., 2015;

Table 5 Characteristics, Specific EC Changes, and Specific Biomarkers in Other Animal Models of HTN-Preg

Model	Procedure	Characteristics	Norm-Preg	HTN-Preg	References
L-NAME	Rats received L-NAME 0.3 g/L in drinking water on gestation days 1–19	= plasma nitrite (μM)	11.20 ± 1.05	9.87 ± 0.59	Ramesar et al. (2011)
		= plasma L-arginine (μM)	10.47 ± 2.31	5.08 ± 0.81	
		= plasma sEng (ng/mL)	178.52 ± 5.33	183.44 ± 8.294	
		↑ plasma sFlt-1 (pg/mL)	774.91 ± 26.81	1228.80 ± 116.29	
	Rats received 50 mg/kg/day of L-NAME by oral gavage on gestation days 14–19	↑ BP (mmHg)	~119	~136	Kemse et al. (2014)
		= litter size (number of pups)	~11	~12	
		Pup weight (g)	~3.5	~3.0	
		Plasma MDA (μM/mL)	~46	~125	
		Placental lysate TNF-α (pg/mL)	~300	~500	
DOCA-salt	Female rats received 12.5 mg of DOCA (i.p.) and 0.9% saline to drink, followed by weekly i.p. injections of DOCA (6.25 mg) during pregnancy	↑ Systolic BP (mmHg)	90 ± 4	138 ± 4	Mitchell et al. (2007)
		Proteinuria (mg/24 h)	5.7 ± 0.6	9.8 ± 0.3	
		↓ Litter weight (g)	73 ± 5	36 ± 6	
		↓ # of pups	16 ± 2	10 ± 1	
		↓ Vascular ACh relaxation (%)	102 ± 4	68 ± 2	
		↑ Aortic superoxide (RLU)	~15,000	~28,000	

Continued

Table 5 Characteristics, Specific EC Changes, and Specific Biomarkers in Other Animal Models of HTN-Preg—cont'd

Model	Procedure	Characteristics	Norm-Preg	HTN-Preg	References
		↑ Aortic Peroxynitrite (RLU)	~20,000	~38,000	
		↑ Aortic eNOS (AU)	~0.6	~1.0	
		↑ BP (mmHg)	102 ± 4	140 ± 8	Agunanne et al. (2010)
		= Plasma sFlt-1 (pg/mL)	2778.8 ± 52.6	2781.8 ± 30.9	
		↓ Plasma VEGF (pg/mL)	29.2 ± 1.8	19.2 ± 1.3	
		↓ Plasma PlGF (pg/mL)	26.5 ± 1.9	18.2 ± 1.6	
		= Plasma sEng (pg/mL)	821.5 ± 22.4	830.7 ± 25.3	
		Plasma sFlt-1/PlGF ratio	105 ± 3	154 ± 13	
Dahl salt-sensitive (superimposed PE)	Dahl S rats are a genetic model of hypertension and kidney disease	↑ BP (mmHg): Early	~105	~158	Gillis et al. (2015)
		Mid	~98	~160	
		Late	~96	~162	
		Proteinuria (mg/24 h): Early	<10	~98	
		Mid	<10	~100	
		Late	<10	~250	
		↑ UARI	~0.5	~0.7	
		↑ Placental HIF-1α (RDU)	1.0 ± 0.1	1.4 ± 0.1	

		↑ Placental TNF-α (pg/g)	92 ± 5	152 ± 8	
		↑ Plasma sFlt-1 (RDU)	1.00 ± 0.09	1.83 ± 0.16	
		↓ Number of pups	~13	~9	
		↓ Pup weight (g)	~2.5	~1.8	
		↓ Pup length (mm)	~34	~27	
Recombinant sFlt-1-Fc	Mice were injected (i.p.) with recombinant mouse sFlt-1-Fc (12 μg sFlt-1 in saline) on gestation days 9–16	↑ Systolic BP (mmHg)	112.2 ± 3.8	160.8 ± 3.3	Jiang et al. (2015)
		↑ Urine albumin/creatinine ratio (μg/mg)	53.4 ± 2.6	332.3 ± 29.7	
		↑ Serum sFlt-1 levels (ng/mL)	0.9 ± 0.03	2.5 ± 0.13	
		= Number of pups	11.4 ± 0.5	10.6 ± 0.9	
		= Pup weight (g)	1.3 ± 0.1	1.2 ± 0.1	
sFlt-1 and sEng infused	Sprague–Dawley rats were infused with sFlt-1 (4.7 μg/kg) and sEng (7 μg/kg) via miniosmotic pumps on gestation days 12–20	↑ BP (mmHg)	99.5 ± 2.32	121.5 ± 2.44	Morris et al. (2015)
		↑ Plasma lactate dehydrogenase (IU/L)	408.5 ± 52.6	1137 ± 124.2	
		↑ Plasma aspartate aminotransferase (IU/L)	89.29 ± 9.03	140.5 ± 12.66	
		↓ Platelets (platelets/μL)	597,250 ± 34,775	347,125 ± 59,821	
		↑ Urinary protein: creatinine (mg/g)	~1.4	~2.0	
		↑ Plasma endothelin-1 (ng/mL)	~110	~175	

Continued

Table 5 Characteristics, Specific EC Changes, and Specific Biomarkers in Other Animal Models of HTN-Preg—cont'd

Model	Procedure	Characteristics	Norm-Preg	HTN-Preg	References
TNF-α Infused	Mice were infused with TNF-α (500 ng/kg/day) via subcutaneous miniosmotic pump on gestation days 13–17	↑ BP (mmHg)	—	—	Bobek et al. (2015)
		Proteinuria (g/L)	~0.2	~0.7	
		= Number of pups	~7	~8	
		=Pup weight (g)	~0.9	~0.9	
		= sFlt-1 serum levels (ng/mL)	99 ± 18	86 ± 25	
		↑ Expression of placental HIF-1α (RE)	~1.0	~1.5	
AT_1-AA treated	Pregnant C57BL/6 mice received on gestation days 13.5 and 14.5 intraorbital injections of IgG obtained from Norm-Preg or PE women	↑ BP (mmHg)	124 ± 2	142 ± 8	Siddiqui et al. (2013)
		Proteinuria (μg albumin/mg creatinine)	132 ± 17	242 ± 27	
		↑ Serum sFlt-1 (ng/mL)	~21	~50	

Values represent means ± standard error of the mean. *ACh*, acetylcholine; *AU*, arbitrary unit of bands intensity in Western blot analysis; *BP*, blood pressure; *DOCA*, deoxycorticosterone acetate; *eNOS*, endothelial nitric oxide synthase; *HIF*, hypoxia-inducible factor; *IU*, international units; *L-NAME*, N_ω-nitro-L-arginine methyl ester; *MDA*, malondialdehyde; *PlGF*, placental growth factor; *RDU*, relative densitometry units of Western blot bands normalized to Norm-Preg bands; *RE*, relative expression normalized to β-actin; *RLU*, relative light units of chemiluminescence from lucigenin-superoxide and luminol-peroxynitrite reaction; *sEng*, soluble endoglin; *sFlt-1*, soluble fms-like tyrosine kinase-1; *TNF-α*, tumor necrosis factor-α; *UARI*, uterine artery resistance index; *VEGF*, vascular endothelial growth factor.

Table 5). Similarly, infusion of LIGHT, increases BP, induces the expression of ET-1 and sFlt-1, and causes proteinuria in pregnant rats (Wang et al., 2014). TNF-α may also work in concert with other inflammatory cytokines to increase ET-1 levels and cause HTN in RUPP rats (LaMarca, Speed, et al., 2008). TNF-α may also act in synergy with sFlt-1 to promote a proinflammatory state and endothelial dysfunction. In HUVECs, treatment with TNF-α and sFlt-1 causes an increase in the adhesion molecules ICAM and VCAM and markers of endothelial dysfunction such as von Willebrand factor and ET-1 (Cindrova-Davies et al., 2011). In support of a role of TNF-α in HTN-Preg, blockade of TNF-α with Etanercept reduced HTN in RUPP rats. Also, in HUVECs, treatment with serum from RUPP rats treated with the TNF-α blocker produced less ET-1 than serum from untreated RUPP rats (LaMarca, Speed, et al., 2008).

IL-6 is a proinflammatory cytokine that may also be elevated in PE (Moreno-Eutimio et al., 2014; Ramma et al., 2012). RUPP rats show increased plasma levels and higher $CD4^+$ T cell production of IL-6 (Cornelius et al., 2015; Wallace et al., 2011), and infusion of IL-6 leads to HTN and proteinuria in pregnant rats (Lamarca, Brewer, & Wallace, 2011). IL-6 may stimulate dimerization of surface receptor GP-130 on ECs leading to abnormal intracellular signaling and vascular dysfunction. IL-6 also disrupts the tight junctions in ECs and increases vascular permeability (Lockwood et al., 2008). Chronic infusion of TNF-α or IL-6 in late-pregnant rats leads to increased BP, enhanced vascular contraction, and reduced endothelium-dependent relaxation via the NO–cGMP pathway (Davis et al., 2002; Orshal & Khalil, 2004). Also, IL-1β is a proinflammatory cytokine that could contribute to the inflammatory response and disrupts ECs in PE. Monocytes of PE women produce more IL-1β than those of Norm-Preg women (Matias et al., 2015).

IL-10 is an antiinflammatory cytokine whose levels are reduced in the plasma and placenta of PE women (Peixoto et al., 2016; Pinheiro et al., 2015) and in plasma of RUPP rats (Cornelius et al., 2015). Also, hypoxic placental trophoblasts show decreased IL-10 and increased proinflammatory cytokines (Royle et al., 2009).

The source of proinflammatory cytokine in PE is mostly in the maternal circulation. Monocytes/macrophages are the main reservoir of proinflammatory cytokines and are the first activated cells in nonspecific immune response (Rahardjo, Widjajanto, Sujuti, & Keman, 2014). Monocytes treated with PE plasma produce more TNF-α and IL-6 than cells treated with Norm-Preg plasma (Rahardjo et al., 2014). This is likely due to the regulatory effects of IL-10 on monocytes regulating the inflammatory response that occurs during normal pregnancy by controlling TNF-α and IL-1β gene expression (Matias et al., 2015), and this IL-10-mediated regulatory activity could be lost in PE. Interestingly, uric acid is an important stimulator of monocytes to release cytokines, and PE patients often have hyperuricemia. Also, monocytes from PE patients exhibiting high levels of uric acid produce more TNF-α and IL-1β than those from Norm-Preg women (Matias et al., 2015).

7.3 Hypoxia-Inducible Factor

HIF is a transcriptional factor that plays a role in the physiologic responses to hypoxia and may be involved in PE. HIF-1 is a heterodimer consisting of an oxygen-regulated HIF-1α and HIF-2α subunits and a constitutively expressed HIF-1β subunit. Of note, de novo synthesis of HIF-1α may largely occur in response to nonhypoxic stimuli such as inflammatory factors, and TNF-α causes upregulation of HIF-1α mRNA (Bobek et al., 2015). More than 100 genes are regulated downstream from HIF-1 including VEGF, leptin, TGF-β3, and NOS. Also, DNA microarrays have shown that more than 2% of all human genes are regulated directly or indirectly by HIF-1 in arterial ECs (Ali & Khalil, 2015).

Circulating HIF-1α levels are increased in PE women (Akhilesh et al., 2013), although the molecular mechanisms underlying the changes in HIF expression and the genes affected are unclear. HIF-1α has been suggested to upregulate sFlt-1 and sEng, bind to ET-1 gene, regulate ET-1 expression, contribute to shallow trophoblast invasion, and induce AngII-converting enzyme (ACE) expression in the lungs and kidney during PE (Ali & Khalil, 2015; Tal, 2012).

Experimental models support a role of HIF-1α in HTN-Preg. RUPP rats show elevated placental levels of HIF-1α (Gilbert et al., 2009). In mice models of HTN-Preg, knockdown of HIF-1α mRNA using specific siRNA attenuates hallmark features of PE such as high BP, proteinuria, renal damage, and elevated sFlt-1 levels (Iriyama et al., 2015). Also, HIF-1α may increase sFlt-1 production in human villous trophoblast (Iriyama et al., 2015).

7.4 Reactive Oxygen Species

ROS such as superoxide anion ($O_2^{\cdot -}$), H_2O_2, and hydroxyl ion (OH^-) contain highly reactive O_2. During normal pregnancy, ROS generation is increased and may be necessary for proper physiology (Sanchez-Aranguren et al., 2014), and therefore pregnancy may represent a state of oxidative stress due to increased maternal metabolism and metabolic activity of the placenta (Myatt & Webster, 2009). However, in normal pregnancy, placental production of ROS is counterbalanced by the presence of abundant antioxidants (Ali & Khalil, 2015).

During PE, the mRNA expression of antioxidants such as hemeoxygenase-1 (HO-1), HO-2, copper/zinc superoxide dismutase (SOD), glutathione peroxidase (GPx), and catalase is decreased and antioxidants may fail to counterbalance the increased ROS production (Nakamura et al., 2009). This leads to lipid peroxidation, increased TXA2, and loss of GPx activity in placenta (Reslan & Khalil, 2010). The mRNA levels of HO-1, HO-2, SOD, GPx, and catalase are decreased in the cellular component of the maternal blood of PE women (Nakamura et al., 2009). Also, the total antioxidant capacity (TAOC) of serum from PE is reduced when compared with that of Norm-Preg women (Turpin et al., 2015). Interestingly, antioxidant levels were found to be decreased in women who were subsequently diagnosed with early PE (Cohen et al., 2015), suggesting a role of oxidative stress in the pathogenesis of early PE. In PE, reduced brachial artery FMD is associated with decreased plasma levels of the antioxidant ascorbate and is ameliorated with ascorbic acid administration (Chambers et al., 2001). Also, placental HO-1 is reduced in RUPP vs Norm-Preg rats (Gilbert et al., 2009).

Neutrophils and monocytes represent an important source of ROS and EC damage in PE. Monocytes from PE women release more H_2O_2 (Peracoli et al., 2011) and stimulated cells from women with PE produce more $O_2^{\cdot-}$ and cause more EC damage compared to those from Norm-Preg women (Peracoli et al., 2011; Tsukimori et al., 2005). Of note, neutrophils also produce NO, which can protect cells from $O_2^{\cdot-}$-induced damage during normal pregnancy. However, in PE, excess $O_2^{\cdot-}$ scavenge the NO produced by neutrophils to form peroxynitrite ($ONOO^-$), effectively reducing NO bioavailability and causing EC injury (Tsukimori et al., 2005). Another source of $O_2^{\cdot-}$ formation is NADPH oxidase, a membrane-bound enzyme complex that catalyzes the one-electron reduction of oxygen to $O_2^{\cdot-}$ via NADPH. NADPH oxidase isoform NOX1 is overexpressed in PE placentas (Cui, Brockman, Campos, & Myatt, 2006). HUVECs treated with serum from PE women showed an increase in mRNA expression of the NADPH oxidase subunit $gp91^{phox}$, leading to high $O_2^{\cdot-}$ production (Matsubara et al., 2010). Also, in HUVECs treatment with PE serum induces overexpression of iNOS (Matsubara et al., 2010), which could produce excessive amounts of NO and in turn increases ROS production and EC injury and leads to HTN. This is supported by the observation that RUPP rats treated with specific inhibitors of iNOS show a decrease in BP, aortic ROS levels, and NADPH-dependent ROS production (Amaral et al., 2013). Biopterin (BH_4) promotes eNOS dimerization and activity. During hypoxia, a reduction in BH_4 may induce an uncoupling reaction leading to ROS production and decreased NO formation (Mitchell et al., 2007). In DOCA-salt rats, supplementation with BH_4 such as sepiapterin increases NO levels and reduces $O_2^{\cdot -}$ and $ONOO^-$ production (Mitchell et al., 2007).

Other oxidative markers such as malondialdehyde, a marker of lipid peroxidation, and prostaglandin $F_{2\alpha}$ in serum from women at 10–14 gestation weeks were found to be increased in PE. This correlates with the gradual oxidative damage of the placenta, even before the onset of clinical symptoms of PE (Genc et al., 2011). Also, plasma 8-isoprostane, marker of oxidative stress, and total aortic and placental ROS are elevated in RUPP vs Norm-Preg rats (Amaral et al., 2013; Cornelius et al., 2015). Oxidative stress can also modulate miRNA expression in placenta and this could be a potential mechanism in the pathogenesis of PE. Studies in first-trimester villous trophoblasts showed that excessive oxidative stress could affect the expression of several miRNAs including those involved in angiogenesis, apoptosis, immune response, and inflammation (Cross, Tolba, Rondelli, Xu, & Abdel-Rahman, 2015).

7.5 AngII and AT_1 Receptor Agonistic Autoantibodies

AngII is an important circulating hormone in the control of BP and electrolyte homeostasis. AngII acts via AT_1R to promote vasoconstriction, vascular growth, and inflammation and increase $[Ca^{2+}]_i$ and Rho/Rho-kinase activity in VSM. AngII can also act via endothelial AT_2R to activate eNOS, and increase production of NO and PGI_2, counteracting AngII-induced vasoconstriction.

It was first demonstrated that the dose of AngII required to elicit a pressor response of 20 mmHg in diastolic BP in 23–26 weeks' Norm-Preg women was lower in women who subsequently developed PE (Gant, Daley, Chand, Whalley, & MacDonald, 1973). This was a significant study as it showed that the increased pressor response to AngII in PE women occurred long before the manifestation of clinical symptoms. Although normal pregnancy is associated with increased plasma levels of renin and AngII, the pressor response to AngII is decreased due to decreased expression of AT_1R or increased expression of AT_2R.

Angiotensinogen levels may be similar in PE and Norm-Preg women, but PE patients have lower levels of the free thiol form of angiotensinogen and higher levels of the oxidized sulfhydryl-bridge form, which preferentially interacts with renin resulting in increased generation of angiotensin (Rahgozar et al., 2015). AngII levels and AT_1R gene expression are increased in chorionic villi and placenta of PE women (Anton et al., 2008; Mistry, Kurlak, & Broughton Pipkin, 2013). Increased plasma hemopexin activity downregulates AT_1R in human monocytes and ECs in pregnant rats, and inhibition of hemopexin during PE enhances AT_1R expression and vasoconstriction (Bakker et al., 2009).

AT_1-AA is a bioactive factor that mediates its responses via AT_1R. AT_1-AA levels are elevated in serum of PE vs Norm-Preg women (Bai et al., 2013; Siddiqui et al., 2013) and further elevated in severe PE and in early than late PE (Yang et al., 2015). AT_1-AA may enhance collagen-induced platelet aggregation, thus contributing to the hypercoagulability observed in PE (Bai et al., 2013). Circulating AT_1-AA is increased in RUPP vs Norm-Preg rats (Cornelius et al., 2015, 2013; LaMarca, Wallukat, et al., 2008; Novotny et al., 2012) and AT_1-AA infusion in pregnant mice induces several features of PE such as elevated BP, proteinuria, and increased sFlt-1 levels (Siddiqui et al., 2013; Table 5). Also, AT_1-AA infusion in pregnant rats causes increases in ET-1 levels, 11-fold in renal cortex, and 4-fold in the placenta (LaMarca et al., 2009). Also, the vasodilatory response to the endothelial-dependent agonist ACh is reduced in renal interlobar arteries of pregnant rats infused with AT1-AA, suggesting renal endothelial damage as an underlying mechanism in HTN-Preg. Interestingly, the impaired vasodilator response to ACh was abolished by ET_AR blockade, suggesting cross talk between the bioactive factors in promoting endothelial dysfunction in HTN-Preg (Parrish et al., 2011). AT_1-AA reduces trophoblast invasion, increases ROS and $[Ca^{2+}]_i$, activates the tissue factor causing thrombosis, increases BP, causes vascular damage to the adrenal glands, and reduces aldosterone secretion in sFlt-1-treated pregnant mice (Siddiqui et al., 2013; Xia, Ramin, & Kellems, 2007). In cultured trophoblasts, stimulation of AT_1R with IgG isolated from PE women causes an increase in sFlt-1 levels (Zhou et al., 2008). Also, HUVECs treated with AT_1-AA isolated from PE patients induces the release of lactate dehydrogenase (LDH; Yang et al., 2014), a marker of both cellular death and necrosis, suggesting that AT_1-AA may directly cause EC necrosis via AT_1R activation. Also in

HUVECs, AT_1-AA induces the activity of caspase-3 and caspase-8, suggesting that it may promote EC apoptosis (Yang et al., 2014). While the mechanisms of the generation of AT_1-AA in PE are not clearly understood, studies have shown that the plasma levels of AT_1-AA are increased in pregnant rats infused with TNF-α, suggesting the involvement of cytokine-induced pathways (LaMarca, Wallukat, et al., 2008).

8. IMPLICATIONS FOR PREDICTION AND MANAGEMENT OF PREECLAMPSIA

Because PE has a relatively long preclinical phase before clinically manifesting in late gestation, the identification of women at risk, early diagnosis using various biomarkers, and prompt management could improve the maternal and perinatal outcome (Ali & Khalil, 2015).

A decrease in FMD could be an early indicator of compromised EC function between the 24th and 28th gestational weeks and before the clinical diagnosis of PE. The sensitivity of FMD is 87.5% and 95.5% for the prediction of early and late PE, respectively (Brandao et al., 2014).

Measurements of plasma VEGF, PlGF, sFlt-1, and sEng may help early detection in asymptomatic pregnant women, especially those at high risk for PE (Ali & Khalil, 2015). Angiogenic/antiangiogenic imbalance is an important feature in PE. In PE, circulating levels of sFlt-1 are increased more than one month before the onset of clinical symptoms, and PlGF is decreased in women who subsequently develop PE from the end of the first trimester (Herraiz et al., 2015). The sFlt-1/PlGF ratio is increased in PE vs Norm-Preg women and in both early and late PE (March et al., 2015). A meta-analysis of 20 different studies revealed that the overall diagnostic accuracy of sFlt-1/PlGF ratio for PE is relatively high, but suggested that the sFlt-1/PlGF has higher diagnostic efficiency in early than in late PE, which might be due to differences in pathogenesis of different forms of PE (Liu et al., 2015).

Assessment of the immune system activity could be useful in predicting PE in early pregnancy. An increase in inflammatory factors may represent an abnormal maternal immunological response, with a change in the role of monocytes and NK cells, altered release of cytokines, and increased AT_1-AA and activation of proinflammatory AT_1R (Ali & Khalil, 2015). TNF-α levels could be an early predictor of PE. Plasma obtained at gestational weeks 11–13 had higher TNF-α levels in women who developed PE in the future (Hamai et al., 1997). However, other reports suggest that plasma TNF-α levels may not represent a PE predictive marker in the first and second trimesters, but might be useful in the early third trimester (Serin et al., 2002). A more recent study suggested that an association of elevated plasma TNF-α levels and changes in uterine artery Doppler at 11–13th gestational weeks had a 100% sensitivity in the prediction of PE (Gomaa, Naguib, Swedan, & Abdellatif, 2015).

Other markers could help in the early detection of PE. Vasopressin, a hormone with vasoconstrictor activities, was found to be elevated around week 6 of pregnancy in women who then developed PE (Santillan et al., 2014). Also, miRNAs have been proposed as early biomarkers for PE. At 28 weeks of pregnancy, miRNA-206, which interacts with several genes known to be directly involved in the pathology of PE, is elevated in plasma and placenta from women who subsequently develop PE (Akehurst et al., 2015), although other studies show little predictive value of miRNAs for PE (Luque et al., 2014). Thus, despite the existence of several potential markers, their reliability in predicting PE has not been consistent and their predictive value needs to be further assessed in order to identify the best marker combinations for use in clinical settings (Ali & Khalil, 2015).

Currently, inducing labor remains as the only effective treatment for PE and few pharmacological options are available. International guidelines recommend the use of one of the following antihypertensive agents in PE: methyldopa, labetalol, other beta-blockers (acebutolol, metoprolol, pindolol, and propranolol), and Ca^{2+} channel blockers (nifedipine; Magee, Pels, Helewa, Rey, & von Dadelszen, 2014). ACE inhibitors should not be used due to their teratogenic effects and atenolol and prazosin are not recommended prior to delivery. If PE evolves to eclampsia, magnesium sulfate is often used to prevent eclamptic seizures (McDonald, Lutsiv, Dzaja, & Duley, 2012).

Sildenafil therapy has been proposed for women with severe early-onset IUGR and was associated with increased fetal growth and no maternal side effects (von Dadelszen et al., 2011). Sildenafil citrate vasodilates myometrial arteries isolated from women with IUGR-complicated pregnancies. Also, treatment of an L-NAME mouse model of HTN-Preg with sildenafil citrate restored EC integrity in the placental vessels (Motta et al., 2015).

Eculizumab, an anti-C5 antibody, was found to normalize lab values and prolong pregnancy by 17 days in a woman with PE/HELLP syndrome, suggesting the benefits of manipulating the complement system during PE. However, complement system inhibitors have the potential of increasing susceptibility to infection and their use for long-term therapy would require close monitoring (Burwick & Feinberg, 2013).

MMPs could be an important target in the management of PE. Doxycycline, a nonspecific MMP inhibitor, was thought to alleviate HTN and vascular dysfunction in PE, but was found to decrease placenta weight and cause IUGR in both Norm-Preg and HTN-Preg rats, and to reduce trophoblast invasion and placental perfusion in HTN-Preg rats (Palei, Granger, et al., 2013).

Studies suggest potential benefits of correcting the proangiogenic/antiangiogenic imbalance in PE. Edaravone is a free radical scavenger that inhibits sFlt-1 expression and reduces placental vessel abnormalities in hypoxic human TCs (Zhao et al., 2014). H_2S donors attenuate HTN, reduce sFlt-1-mediated damage, and upregulate VEGF in animal models of HTN-Preg (Holwerda et al., 2014; Wang et al., 2013). VEGF could improve the proangiogenic/antiangiogenic imbalance, but may impair bradykinin-induced vascular relaxation, and enhance basal tone and vascular permeability in PE (VanWijk et al., 2000). Modulators of PlGF could be more promising in PE, and low-molecular-weight heparin therapy was associated with increased circulating PlGF levels during third-trimester pregnancy (Yinon et al., 2015).

TNF-α antagonists such as Etanercept decrease BP, increase eNOS expression, decrease ET-1 levels, and prevent cardiac changes in RUPP rats (Gutkowska et al., 2011; LaMarca, Speed, et al., 2008). Also, IL-17 soluble receptor C could inhibit IL-17 and prevent the recruitment of host defense cells, suppress the inflammatory response, decrease AT_1-AA and ROS, and ameliorate HTN and improve pup and placental weight in RUPP rats (Cornelius et al., 2013).

9. CONCLUSION

Clinical and experimental studies have helped in understanding the mechanisms of PE. Endothelial dysfunction is a major hallmark of PE and identifying the factors that lead to endothelial damage would be key to further understand the pathogenesis of PE and help in developing appropriate therapies. Genetic, environmental, and dietary factors may cause localized abnormal placentation and altered maternal immune response, and abnormal expression of integrins, inflammatory cytokines, and MMPs leading to increased apoptosis of TCs, shallow trophoblastic invasion, and poor spiral artery remodeling, causing RUPP and placental ischemia/hypoxia. Ischemic/hypoxic placenta causes the release of bioactive factors such as sFlt-1, sEng, TNF-α, IL-6, HIF, ROS, and AT_1-AA that target ECs, cause endothelial dysfunction, decrease vasodilators, and increase vasoconstrictors and lead to HTN. However, whether the changes in vascular mediators and bioactive factors are the cause or the consequence of PE remains unclear. Further understanding of the vascular mediators, cellular mechanisms, and bioactive factors should help design more specific and efficient measures for prevention, early detection, and management of PE.

CONFLICT OF INTEREST

None.

ACKNOWLEDGMENT

This work was supported by grants from National Heart, Lung, and Blood Institute (HL-65998 and HL-111775).

REFERENCES

Abdalvand, A., Morton, J. S., Bourque, S. L., Quon, A. L., & Davidge, S. T. (2013). Matrix metalloproteinase enhances big-endothelin-1 constriction in mesenteric vessels of pregnant rats with reduced uterine blood flow. *Hypertension, 61*, 488–493.

Adekola, H., Romero, R., Chaemsaithong, P., Korzeniewski, S. J., Dong, Z., Yeo, L., et al. (2015). Endocan, a putative endothelial cell marker, is elevated in preeclampsia, decreased in acute pyelonephritis, and unchanged in other obstetrical syndromes. *The Journal of Maternal-Fetal & Neonatal Medicine, 28*, 1621–1632.

Agarwal, R., Loganath, A., Roy, A. C., Wong, Y. C., & Ng, S. C. (2001). Expression profiles of interleukin-15 in early and late gestational human placenta and in pre-eclamptic placenta. *Molecular Human Reproduction, 7*, 97–101.

Agunanne, E. E., Uddin, M. N., Horvat, D., & Puschett, J. B. (2010). Contribution of angiogenic factors in a rat model of pre-eclampsia. *American Journal of Nephrology, 32*, 332–339.

Ahmad, S., & Ahmed, A. (2004). Elevated placental soluble vascular endothelial growth factor receptor-1 inhibits angiogenesis in preeclampsia. *Circulation Research, 95*, 884–891.

Akehurst, C., Small, H. Y., Sharafetdinova, L., Forrest, R., Beattie, W., Brown, C. E., et al. (2015). Differential expression of microRNA-206 and its target genes in preeclampsia. *Journal of Hypertension, 33*, 2068–2074.

Akhilesh, M., Mahalingam, V., Nalliah, S., Ali, R. M., Ganesalingam, M., & Haleagrahara, N. (2013). Hypoxia-inducible factor-1 alpha as a predictive marker in pre-eclampsia. *Biomedical Reports, 1*, 257–258.

Alexander, B. T., Cockrell, K. L., Massey, M. B., Bennett, W. A., & Granger, J. P. (2002). Tumor necrosis factor-alpha-induced hypertension in pregnant rats results in decreased renal neuronal nitric oxide synthase expression. *American Journal of Hypertension, 15*, 170–175.

Alexander, B. T., Kassab, S. E., Miller, M. T., Abram, S. R., Reckelhoff, J. F., Bennett, W. A., et al. (2001). Reduced uterine perfusion pressure during pregnancy in the rat is associated with increases in arterial pressure and changes in renal nitric oxide. *Hypertension, 37*, 1191–1195.

Alexander, B. T., Llinas, M. T., Kruckeberg, W. C., & Granger, J. P. (2004). L-Arginine attenuates hypertension in pregnant rats with reduced uterine perfusion pressure. *Hypertension, 43*, 832–836.

Alexander, B. T., Miller, M. T., Kassab, S., Novak, J., Reckelhoff, J. F., Kruckeberg, W. C., et al. (1999). Differential expression of renal nitric oxide synthase isoforms during pregnancy in rats. *Hypertension*, *33*, 435–439.

Alexander, B. T., Rinewalt, A. N., Cockrell, K. L., Massey, M. B., Bennett, W. A., & Granger, J. P. (2001). Endothelin type a receptor blockade attenuates the hypertension in response to chronic reductions in uterine perfusion pressure. *Hypertension*, *37*, 485–489.

Ali, S. M., & Khalil, R. A. (2015). Genetic, immune and vasoactive factors in the vascular dysfunction associated with hypertension in pregnancy. *Expert Opinion on Therapeutic Targets*, *19*, 1495–1515.

Alpoim, P. N., Gomes, K. B., Pinheiro Mde, B., Godoi, L. C., Jardim, L. L., Muniz, L. G., et al. (2014). Polymorphisms in endothelial nitric oxide synthase gene in early and late severe preeclampsia. *Nitric Oxide*, *42*, 19–23.

Amaral, L. M., Cornelius, D. C., Harmon, A., Moseley, J., Martin, J. N., Jr., & LaMarca, B. (2015). 17-Hydroxyprogesterone caproate significantly improves clinical characteristics of preeclampsia in the reduced uterine perfusion pressure rat model. *Hypertension*, *65*, 225–231.

Amaral, L. M., Pinheiro, L. C., Guimaraes, D. A., Palei, A. C., Sertorio, J. T., Portella, R. L., et al. (2013). Antihypertensive effects of inducible nitric oxide synthase inhibition in experimental pre-eclampsia. *Journal of Cellular and Molecular Medicine*, *17*, 1300–1307.

Amraoui, F., Spijkers, L., Hassani Lahsinoui, H., Vogt, L., van der Post, J., Peters, S., et al. (2014). SFlt-1 elevates blood pressure by augmenting endothelin-1-mediated vasoconstriction in mice. *PLoS One*, *9*, e91897.

Anton, L., Merrill, D. C., Neves, L. A., Stovall, K., Gallagher, P. E., Diz, D. I., et al. (2008). Activation of local chorionic villi angiotensin II levels but not angiotensin (1-7) in preeclampsia. *Hypertension*, *51*, 1066–1072.

Anton, L., Olarerin-George, A. O., Hogenesch, J. B., & Elovitz, M. A. (2015). Placental expression of miR-517a/b and miR-517c contributes to trophoblast dysfunction and preeclampsia. *PLoS One*, *10*, e0122707.

Bai, K., Wang, K., Li, X., Wang, J., Zhang, J., Song, L., et al. (2013). Autoantibody against angiotensin AT1 receptor from preeclamptic patients enhances collagen-induced human platelet aggregation. *Acta Biochimica et Biophysica Sinica Shanghai*, *45*, 749–755.

Baig, S., Kothandaraman, N., Manikandan, J., Rong, L., Ee, K. H., Hill, J., et al. (2014). Proteomic analysis of human placental syncytiotrophoblast microvesicles in preeclampsia. *Clinical Proteomics*, *11*, 40.

Bakker, W. W., Henning, R. H., van Son, W. J., van Pampus, M. G., Aarnoudse, J. G., Niezen-Koning, K. E., et al. (2009). Vascular contraction and preeclampsia: Downregulation of the angiotensin receptor 1 by hemopexin in vitro. *Hypertension*, *53*, 959–964.

Bates, D. O. (2011). An unexpected tail of VEGF and PlGF in pre-eclampsia. *Biochemical Society Transactions*, *39*, 1576–1582.

Bdolah, Y., Lam, C., Rajakumar, A., Shivalingappa, V., Mutter, W., Sachs, B. P., et al. (2008). Twin pregnancy and the risk of preeclampsia: Bigger placenta or relative ischemia? *American Journal of Obstetrics and Gynecology*, *198*(428), e421–e426.

Ben Ali Gannoun, M., Zitouni, H., Raguema, N., Maleh, W., Gris, J. C., Almawi, W., et al. (2015). Association of common eNOS/NOS3 polymorphisms with preeclampsia in Tunisian Arabs. *Gene*, *569*, 303–307.

Bhavina, K., Radhika, J., & Pandian, S. S. (2014). VEGF and eNOS expression in umbilical cord from pregnancy complicated by hypertensive disorder with different severity. *BioMed Research International*, *2014*, 982159.

Bian, Z., Shixia, C., & Duan, T. (2015). First-trimester maternal serum levels of sFLT1, PGF and ADMA predict preeclampsia. *PLoS One*, *10*, e0124684.

Bobek, G., Surmon, L., Mirabito, K. M., Makris, A., & Hennessy, A. (2015). Placental regulation of inflammation and hypoxia after TNF-alpha infusion in mice. *American Journal of Reproductive Immunology*, *74*, 407–418.

Boyd, H. A., Tahir, H., Wohlfahrt, J., & Melbye, M. (2013). Associations of personal and family preeclampsia history with the risk of early-, intermediate- and late-onset preeclampsia. *American Journal of Epidemiology*, *178*, 1611–1619.

Brandao, A. H., Felix, L. R., Patricio Edo, C., Leite, H. V., & Cabral, A. C. (2014). Difference of endothelial function during pregnancies as a method to predict preeclampsia. *Archives of Gynecology and Obstetrics*, *290*, 471–477.

Brennan, L. J., Morton, J. S., & Davidge, S. T. (2014). Vascular dysfunction in preeclampsia. *Microcirculation*, *21*, 4–14.

Burwick, R. M., & Feinberg, B. B. (2013). Eculizumab for the treatment of preeclampsia/HELLP syndrome. *Placenta*, *34*, 201–203.

Cakmak, M., Yilmaz, H., Baglar, E., Darcin, T., Inan, O., Aktas, A., et al. (2016). Serum levels of endocan correlate with the presence and severity of pre-eclampsia. *Clinical and Experimental Hypertension*, *38*(2), 137–142.

Canbakan, B., Keven, K., Tutkak, H., Danisman, N., Ergun, I., & Nergizoglu, G. (2007). Circulating endothelial cells in preeclampsia. *Journal of Human Hypertension*, *21*, 558–563.

Celik, H., Avci, B., & Isik, Y. (2013). Vascular endothelial growth factor and endothelin-1 levels in normal pregnant women and pregnant women with pre-eclampsia. *Journal of Obstetrics and Gynaecology*, *33*, 355–358.

Chambers, J. C., Fusi, L., Malik, I. S., Haskard, D. O., De Swiet, M., & Kooner, J. S. (2001). Association of maternal endothelial dysfunction with preeclampsia. *JAMA*, *285*, 1607–1612.

Charnock-Jones, D. S. (2016). Placental hypoxia, endoplasmic reticulum stress and maternal endothelial sensitisation by sFLT1 in pre-eclampsia. *Journal of Reproductive Immunology*, *114*, 81–85.

Chen, Q., Lau, S., Tong, M., Wei, J., Shen, F., Zhao, J., et al. (2016). Serum uric acid may not be involved in the development of preeclampsia. *Journal of Human Hypertension*, *30*, 136–140.

Chinnathambi, V., Blesson, C. S., Vincent, K. L., Saade, G. R., Hankins, G. D., Yallampalli, C., et al. (2014). Elevated testosterone levels during rat pregnancy cause hypersensitivity to angiotensin II and attenuation of endothelium-dependent vasodilation in uterine arteries. *Hypertension*, *64*, 405–414.

Choi, S. Y., Yun, J., Lee, O. J., Han, H. S., Yeo, M. K., Lee, M. A., et al. (2013). MicroRNA expression profiles in placenta with severe preeclampsia using a PNA-based microarray. *Placenta*, *34*, 799–804.

Cindrova-Davies, T., Sanders, D. A., Burton, G. J., & Charnock-Jones, D. S. (2011). Soluble FLT1 sensitizes endothelial cells to inflammatory cytokines by antagonizing VEGF receptor-mediated signalling. *Cardiovascular Research*, *89*, 671–679.

Cohen, J. M., Kramer, M. S., Platt, R. W., Basso, O., Evans, R. W., & Kahn, S. R. (2015). The association between maternal antioxidant levels in midpregnancy and preeclampsia. *American Journal of Obstetrics and Gynecology*, *213*, 695.e1–695.e13.

Coleman, H. A., Tare, M., & Parkington, H. C. (2004). Endothelial potassium channels, endothelium-dependent hyperpolarization and the regulation of vascular tone in health and disease. *Clinical and Experimental Pharmacology & Physiology*, *31*, 641–649.

Conrad, K. P., Gandley, R. E., Ogawa, T., Nakanishi, S., & Danielson, L. A. (1999). Endothelin mediates renal vasodilation and hyperfiltration during pregnancy in chronically instrumented conscious rats. *The American Journal of Physiology*, *276*, F767–F776.

Conrad, K. P., Kerchner, L. J., & Mosher, M. D. (1999). Plasma and 24-h NO(x) and cGMP during normal pregnancy and preeclampsia in women on a reduced NO(x) diet. *The American Journal of Physiology*, *277*, F48–F57.

Cornelius, D. C., Amaral, L. M., Harmon, A., Wallace, K., Thomas, A. J., Campbell, N., et al. (2015). An increased population of regulatory T cells improves the pathophysiology of placental ischemia in a rat model of preeclampsia. *American Journal of Physiology. Regulatory, Integrative and Comparative Physiology*, *309*, R884–R891.

Cornelius, D. C., Hogg, J. P., Scott, J., Wallace, K., Herse, F., Moseley, J., et al. (2013). Administration of interleukin-17 soluble receptor C suppresses TH17 cells, oxidative stress, and hypertension in response to placental ischemia during pregnancy. *Hypertension*, *62*, 1068–1073.

Crews, J. K., Herrington, J. N., Granger, J. P., & Khalil, R. A. (2000). Decreased endothelium-dependent vascular relaxation during reduction of uterine perfusion pressure in pregnant rat. *Hypertension*, *35*, 367–372.

Cross, C. E., Tolba, M. F., Rondelli, C. M., Xu, M., & Abdel-Rahman, S. Z. (2015). Oxidative stress alters miRNA and gene expression profiles in villous first trimester trophoblasts. *BioMed Research International*, *2015*, 257090.

Cui, X. L., Brockman, D., Campos, B., & Myatt, L. (2006). Expression of NADPH oxidase isoform 1 (Nox1) in human placenta: Involvement in preeclampsia. *Placenta*, *27*, 422–431.

Cui, Y., Wang, W., Dong, N., Lou, J., Srinivasan, D. K., Cheng, W., et al. (2012). Role of corin in trophoblast invasion and uterine spiral artery remodelling in pregnancy. *Nature*, *484*, 246–250.

Davis, J. R., Giardina, J. B., Green, G. M., Alexander, B. T., Granger, J. P., & Khalil, R. A. (2002). Reduced endothelial NO-cGMP vascular relaxation pathway during TNF-alpha-induced hypertension in pregnant rats. *American Journal of Physiology. Regulatory, Integrative and Comparative Physiology*, *282*, R390–R399.

Deng, C. L., Ling, S. T., Liu, X. Q., Zhao, Y. J., & Lv, Y. F. (2015). Decreased expression of matrix metalloproteinase-1 in the maternal umbilical serum, trophoblasts and decidua leads to preeclampsia. *Experimental and Therapeutic Medicine*, *9*, 992–998.

Doridot, L., Passet, B., Mehats, C., Rigourd, V., Barbaux, S., Ducat, A., et al. (2013). Preeclampsia-like symptoms induced in mice by fetoplacental expression of STOX1 are reversed by aspirin treatment. *Hypertension*, *61*, 662–668.

Dotsch, J., Hogen, N., Nyul, Z., Hanze, J., Knerr, I., Kirschbaum, M., et al. (2001). Increase of endothelial nitric oxide synthase and endothelin-1 mRNA expression in human placenta during gestation. *European Journal of Obstetrics, Gynecology, and Reproductive Biology*, *97*, 163–167.

Echeverri, I., Ortega-Avila, J. G., Mosquera, M., Castillo, A., Jimenez, E., Suarez-Ortegon, M. F., et al. (2015). Relationship between maternal and newborn endothelial function and oxidative stress. *American Journal of Human Biology*, *27*, 822–831.

Ehsanipoor, R. M., Fortson, W., Fitzmaurice, L. E., Liao, W. X., Wing, D. A., Chen, D. B., et al. (2013). Nitric oxide and carbon monoxide production and metabolism in preeclampsia. *Reproductive Sciences*, *20*, 542–548.

Eleuterio, N. M., Palei, A. C., Rangel Machado, J. S., Tanus-Santos, J. E., Cavalli, R. C., & Sandrim, V. C. (2013). Relationship between adiponectin and nitrite in healthy and preeclampsia pregnancies. *Clinica Chimica Acta*, *423*, 112–115.

Eleuterio, N. M., Palei, A. C., Rangel Machado, J. S., Tanus-Santos, J. E., Cavalli, R. C., & Sandrim, V. C. (2015). Positive correlations between circulating adiponectin and MMP2 in preeclampsia pregnant. *Pregnancy Hypertension*, *5*, 205–208.

Eremina, V., Jefferson, J. A., Kowalewska, J., Hochster, H., Haas, M., Weisstuch, J., et al. (2008). VEGF inhibition and renal thrombotic microangiopathy. *The New England Journal of Medicine*, *358*, 1129–1136.

Esplin, M. S., Fausett, M. B., Fraser, A., Kerber, R., Mineau, G., Carrillo, J., et al. (2001). Paternal and maternal components of the predisposition to preeclampsia. *The New England Journal of Medicine*, *344*, 867–872.

Fan, X., Rai, A., Kambham, N., Sung, J. F., Singh, N., Petitt, M., et al. (2014). Endometrial VEGF induces placental sFLT1 and leads to pregnancy complications. *The Journal of Clinical Investigation*, *124*, 4941–4952.

Faupel-Badger, J. M., McElrath, T. F., Lauria, M., Houghton, L. C., Lim, K. H., Parry, S., et al. (2015). Maternal circulating angiogenic factors in twin and singleton pregnancies. *American Journal of Obstetrics and Gynecology*, *212*(636), e631–e638.

Fei, X., Hongxiang, Z., Qi, C., & Daozhen, C. (2012). Maternal plasma levels of endothelial dysfunction mediators including AM, CGRP, sICAM-1 and tHcy in pre-eclampsia. *Advances in Clinical and Experimental Medicine*, *21*, 573–579.

Feletou, M., & Vanhoutte, P. M. (2009). EDHF: An update. *Clinical Science (London, England)*, *117*, 139–155.

Fernandez-Patron, C., Zouki, C., Whittal, R., Chan, J. S., Davidge, S. T., & Filep, J. G. (2001). Matrix metalloproteinases regulate neutrophil-endothelial cell adhesion through generation of endothelin-1[1-32]. *The FASEB Journal*, *15*, 2230–2240.

Fiore, G., Florio, P., Micheli, L., Nencini, C., Rossi, M., Cerretani, D., et al. (2005). Endothelin-1 triggers placental oxidative stress pathways: Putative role in preeclampsia. *The Journal of Clinical Endocrinology and Metabolism*, *90*, 4205–4210.

Founds, S. A., Conley, Y. P., Lyons-Weiler, J. F., Jeyabalan, A., Hogge, W. A., & Conrad, K. P. (2009). Altered global gene expression in first trimester placentas of women destined to develop preeclampsia. *Placenta*, *30*, 15–24.

Fulep, E. E., Vedernikov, Y. P., Saade, G. R., & Garfield, R. E. (2001). The role of endothelium-derived hyperpolarizing factor in the regulation of the uterine circulation in pregnant rats. *American Journal of Obstetrics and Gynecology*, *185*, 638–642.

Furchgott, R. F., & Zawadzki, J. V. (1980). The obligatory role of endothelial cells in the relaxation of arterial smooth muscle by acetylcholine. *Nature*, *288*, 373–376.

Gant, N. F., Daley, G. L., Chand, S., Whalley, P. J., & MacDonald, P. C. (1973). A study of angiotensin II pressor response throughout primigravid pregnancy. *The Journal of Clinical Investigation*, *52*, 2682–2689.

Genc, H., Uzun, H., Benian, A., Simsek, G., Gelisgen, R., Madazli, R., et al. (2011). Evaluation of oxidative stress markers in first trimester for assessment of preeclampsia risk. *Archives of Gynecology and Obstetrics*, *284*, 1367–1373.

George, E. M., & Granger, J. P. (2011). Endothelin: Key mediator of hypertension in preeclampsia. *American Journal of Hypertension*, *24*, 964–969.

George, E. M., Palei, A. C., Dent, E. A., & Granger, J. P. (2013). Sildenafil attenuates placental ischemia-induced hypertension. *American Journal of Physiology. Regulatory, Integrative and Comparative Physiology*, *305*, R397–R403.

Gilbert, J. S., Babcock, S. A., & Granger, J. P. (2007). Hypertension produced by reduced uterine perfusion in pregnant rats is associated with increased soluble fms-like tyrosine kinase-1 expression. *Hypertension*, *50*, 1142–1147.

Gilbert, J. S., Gilbert, S. A., Arany, M., & Granger, J. P. (2009). Hypertension produced by placental ischemia in pregnant rats is associated with increased soluble endoglin expression. *Hypertension*, *53*, 399–403.

Gillis, E. E., Williams, J. M., Garrett, M. R., Mooney, J. N., & Sasser, J. M. (2015). The Dahl salt-sensitive rat is a spontaneous model of superimposed preeclampsia. *American Journal of Physiology. Regulatory, Integrative and Comparative Physiology*, *309*, R62–R70.

Gokina, N. I., Kuzina, O. Y., & Vance, A. M. (2010). Augmented EDHF signaling in rat uteroplacental vasculature during late pregnancy. *American Journal of Physiology. Heart and Circulatory Physiology*, *299*, H1642–H1652.

Gomaa, M. F., Naguib, A. H., Swedan, K. H., & Abdellatif, S. S. (2015). Serum tumor necrosis factor-alpha level and uterine artery Doppler indices at 11-13 weeks' gestation for preeclampsia screening in low-risk pregnancies: A prospective observational study. *Journal of Reproductive Immunology*, *109*, 31–35.

Goulopoulou, S., Hannan, J. L., Matsumoto, T., & Webb, R. C. (2012). Pregnancy reduces RhoA/Rho kinase and protein kinase C signaling pathways downstream of thromboxane receptor activation in the rat uterine artery. *American Journal of Physiology. Heart and Circulatory Physiology*, *302*, H2477–H2488.

Guimaraes, M. F., Brandao, A. H., Rezende, C. A., Cabral, A. C., Brum, A. P., Leite, H. V., et al. (2014). Assessment of endothelial function in pregnant women with preeclampsia and gestational diabetes mellitus by flow-mediated dilation of brachial artery. *Archives of Gynecology and Obstetrics*, *290*, 441–447.

Gutkowska, J., Granger, J. P., Lamarca, B. B., Danalache, B. A., Wang, D., & Jankowski, M. (2011). Changes in cardiac structure in hypertension produced by placental ischemia in pregnant rats: Effect of tumor necrosis factor blockade. *Journal of Hypertension*, *29*, 1203–1212.

Hagedorn, K. A., Cooke, C. L., Falck, J. R., Mitchell, B. F., & Davidge, S. T. (2007). Regulation of vascular tone during pregnancy: A novel role for the pregnane X receptor. *Hypertension*, *49*, 328–333.

Hall, D., Gebhardt, S., Theron, G., & Grove, D. (2014). Pre-eclampsia and gestational hypertension are less common in HIV infected women. *Pregnancy Hypertension*, *4*, 91–96.

Hamai, Y., Fujii, T., Yamashita, T., Nishina, H., Kozuma, S., Mikami, Y., et al. (1997). Evidence for an elevation in serum interleukin-2 and tumor necrosis factor-alpha levels before the clinical manifestations of preeclampsia. *American Journal of Reproductive Immunology*, *38*, 89–93.

Hammond, S., Mathewson, A. M., Baker, P. N., Mayhew, T. M., & Dunn, W. R. (2011). Gap junctions and hydrogen peroxide are involved in endothelium-derived hyperpolarising responses to bradykinin in omental arteries and veins isolated from pregnant women. *European Journal of Pharmacology*, *668*, 225–232.

He, H., Venema, V. J., Gu, X., Venema, R. C., Marrero, M. B., & Caldwell, R. B. (1999). Vascular endothelial growth factor signals endothelial cell production of nitric oxide and prostacyclin through flk-1/KDR activation of c-Src. *The Journal of Biological Chemistry*, *274*, 25130–25135.

Hefler, L. A., Reyes, C. A., O'Brien, W. E., & Gregg, A. R. (2001). Perinatal development of endothelial nitric oxide synthase-deficient mice. *Biology of Reproduction*, *64*, 666–673.

Herraiz, I., Simon, E., Gomez-Arriaga, P. I., Martinez-Moratalla, J. M., Garcia-Burguillo, A., Jimenez, E. A., et al. (2015). Angiogenesis-related biomarkers (sFlt-1/PLGF) in the prediction and diagnosis of placental dysfunction: An approach for clinical integration. *International Journal of Molecular Sciences*, *16*, 19009–19026.

Hiby, S. E., Walker, J. J., O'Shaughnessy, K. M., Redman, C. W., Carrington, M., Trowsdale, J., et al. (2004). Combinations of maternal KIR and fetal HLA-C genes influence the risk of preeclampsia and reproductive success. *The Journal of Experimental Medicine*, *200*, 957–965.

Holwerda, K. M., Burke, S. D., Faas, M. M., Zsengeller, Z., Stillman, I. E., Kang, P. M., et al. (2014). Hydrogen sulfide attenuates sFlt1-induced hypertension and renal damage by upregulating vascular endothelial growth factor. *Journal of the American Society of Nephrology*, *25*, 717–725.

Hu, X. Q., Xiao, D., Zhu, R., Huang, X., Yang, S., Wilson, S., et al. (2011). Pregnancy upregulates large-conductance Ca(2+)-activated K(+) channel activity and attenuates myogenic tone in uterine arteries. *Hypertension*, *58*, 1132–1139.

Hunter, A., Aitkenhead, M., Caldwell, C., McCracken, G., Wilson, D., & McClure, N. (2000). Serum levels of vascular endothelial growth factor in preeclamptic and normotensive pregnancy. *Hypertension*, *36*, 965–969.

Iriyama, T., Wang, W., Parchim, N. F., Song, A., Blackwell, S. C., Sibai, B. M., et al. (2015). Hypoxia-independent upregulation of placental hypoxia inducible factor-1alpha gene expression contributes to the pathogenesis of preeclampsia. *Hypertension*, *65*, 1307–1315.

Iwaki, T., Yamamoto, K., Matsuura, T., Sugimura, M., Kobayashi, T., & Kanayama, N. (2004). Alteration of integrins under hypoxic stress in early placenta and choriocarcinoma cell line BeWo. *Gynecologic and Obstetric Investigation*, *57*, 196–203.

Jain, A., Schneider, H., Aliyev, E., Soydemir, F., Baumann, M., Surbek, D., et al. (2014). Hypoxic treatment of human dual placental perfusion induces a preeclampsia-like inflammatory response. *Laboratory Investigation*, *94*, 873–880.

Jardim, L. L., Rios, D. R., Perucci, L. O., de Sousa, L. P., Gomes, K. B., & Dusse, L. M. (2015). Is the imbalance between pro-angiogenic and anti-angiogenic factors associated with preeclampsia? *Clinica Chimica Acta*, *447*, 34–38.

Javadian, P., Salmanian, B., Javadi-Paydar, M., Shamshirsaz, A. A., Ejtemaei Mehr, S., Gharedaghi, M. H., et al. (2013). Effect of morphine on the reduced uteroplacental perfusion model of pre-eclampsia in rats. *European Journal of Obstetrics, Gynecology, and Reproductive Biology*, *168*, 161–166.

Jiang, Z., Zou, Y., Ge, Z., Zuo, Q., Huang, S. Y., & Sun, L. (2015). A role of sFlt-1 in oxidative stress and apoptosis in human and mouse pre-eclamptic trophoblasts. *Biology of Reproduction*, *93*, 73.

Johnson, A. C., Tremble, S. M., Chan, S. L., Moseley, J., LaMarca, B., Nagle, K. J., et al. (2014). Magnesium sulfate treatment reverses seizure susceptibility and decreases neuroinflammation in a rat model of severe preeclampsia. *PLoS One*, *9*, e113670.

Kakigano, A., Mimura, K., Kanagawa, T., Nakayama, M., Kanayama, T., Fujita, S., et al. (2013). Imbalance of angiogenic factors and avascular edematous cystic villi in a trisomy 13 pregnancy: A case report. *Placenta*, *34*, 628–630.

Kanagal, D. V., Rajesh, A., Rao, K., Devi, U. H., Shetty, H., Kumari, S., et al. (2014). Levels of serum calcium and magnesium in pre-eclamptic and normal pregnancy: A study from coastal India. *Journal of Clinical and Diagnostic Research*, *8*, OC01–OC04.

Karakus, S., Bozoklu Akkar, O., Yildiz, C., Sancakdar, E., Cetin, M., & Cetin, A. (2016). Serum levels of ET-1, M30, and angiopoietins-1 and -2 in HELLP syndrome and preeclampsia compared to controls. *Archives of Gynecology and Obstetrics*, *293*(2), 351–359.

Kemse, N. G., Kale, A. A., & Joshi, S. R. (2014). A combined supplementation of omega-3 fatty acids and micronutrients (folic acid, vitamin B12) reduces oxidative stress markers in a rat model of pregnancy induced hypertension. *PLoS One*, *9*, e111902.

Kenny, L. C., Baker, P. N., Kendall, D. A., Randall, M. D., & Dunn, W. R. (2002). Differential mechanisms of endothelium-dependent vasodilator responses in human myometrial small arteries in normal pregnancy and pre-eclampsia. *Clinical Science (London, England)*, *103*, 67–73.

Khalil, R. A., Crews, J. K., Novak, J., Kassab, S., & Granger, J. P. (1998). Enhanced vascular reactivity during inhibition of nitric oxide synthesis in pregnant rats. *Hypertension*, *31*, 1065–1069.

Khalil, R. A., & Granger, J. P. (2002). Vascular mechanisms of increased arterial pressure in preeclampsia: Lessons from animal models. *American Journal of Physiology. Regulatory, Integrative and Comparative Physiology*, *283*, R29–R45.

Kim, Y. G., Suga, S. I., Kang, D. H., Jefferson, J. A., Mazzali, M., Gordon, K. L., et al. (2000). Vascular endothelial growth factor accelerates renal recovery in experimental thrombotic microangiopathy. *Kidney International*, *58*, 2390–2399.

Knock, G. A., & Poston, L. (1996). Bradykinin-mediated relaxation of isolated maternal resistance arteries in normal pregnancy and preeclampsia. *American Journal of Obstetrics and Gynecology*, *175*, 1668–1674.

Kourembanas, S., McQuillan, L. P., Leung, G. K., & Faller, D. V. (1993). Nitric oxide regulates the expression of vasoconstrictors and growth factors by vascular endothelium under both normoxia and hypoxia. *The Journal of Clinical Investigation*, *92*, 99–104.

Krauss, T., Pauer, H. U., & Augustin, H. G. (2004). Prospective analysis of placenta growth factor (PlGF) concentrations in the plasma of women with normal pregnancy and pregnancies complicated by preeclampsia. *Hypertension in Pregnancy*, *23*, 101–111.

Kriston, T., Venuto, R. C., Baylis, C., & Losonczy, G. (1999). Hemodynamic and renal effects of U-46619, a TXA2/PGH2 analog, in late-pregnant rats. *The American Journal of Physiology*, *276*, R831–R837.

Lain, K. Y., & Roberts, J. M. (2002). Contemporary concepts of the pathogenesis and management of preeclampsia. *JAMA*, *287*, 3183–3186.

Lamarca, B., Brewer, J., & Wallace, K. (2011). IL-6-induced pathophysiology during preeclampsia: Potential therapeutic role for magnesium sulfate? *International Journal of Interferon, Cytokine and Mediator Research*, *2011*, 59–64.

LaMarca, B. B., Cockrell, K., Sullivan, E., Bennett, W., & Granger, J. P. (2005). Role of endothelin in mediating tumor necrosis factor-induced hypertension in pregnant rats. *Hypertension*, *46*, 82–86.

LaMarca, B., Parrish, M., Ray, L. F., Murphy, S. R., Roberts, L., Glover, P., et al. (2009). Hypertension in response to autoantibodies to the angiotensin II type I receptor (AT1-AA) in pregnant rats: Role of endothelin-1. *Hypertension*, *54*, 905–909.

LaMarca, B., Speed, J., Fournier, L., Babcock, S. A., Berry, H., Cockrell, K., et al. (2008). Hypertension in response to chronic reductions in uterine perfusion in pregnant rats: Effect of tumor necrosis factor-alpha blockade. *Hypertension*, *52*, 1161–1167.

LaMarca, B., Wallukat, G., Llinas, M., Herse, F., Dechend, R., & Granger, J. P. (2008). Autoantibodies to the angiotensin type I receptor in response to placental ischemia and tumor necrosis factor alpha in pregnant rats. *Hypertension*, *52*, 1168–1172.

Lamminpaa, R., Vehvilainen-Julkunen, K., Gissler, M., & Heinonen, S. (2012). Preeclampsia complicated by advanced maternal age: A registry-based study on primiparous women in Finland 1997-2008. *BMC Pregnancy and Childbirth*, *12*, 47.

Leik, C. E., & Walsh, S. W. (2004). Neutrophils infiltrate resistance-sized vessels of subcutaneous fat in women with preeclampsia. *Hypertension*, *44*, 72–77.

Leonardo, D. P., Albuquerque, D. M., Lanaro, C., Baptista, L. C., Cecatti, J. G., Surita, F. G., et al. (2015). Association of nitric oxide synthase and matrix metalloprotease single nucleotide polymorphisms with preeclampsia and its complications. *PLoS One*, *10*, e0136693.

Levine, R. J., Lam, C., Qian, C., Yu, K. F., Maynard, S. E., Sachs, B. P., et al. (2006). Soluble endoglin and other circulating antiangiogenic factors in preeclampsia. *The New England Journal of Medicine*, *355*, 992–1005.

Lewis, D. F., Canzoneri, B. J., Gu, Y., Zhao, S., & Wang, Y. (2010). Maternal levels of prostacyclin, thromboxane, ICAM, and VCAM in normal and preeclamptic pregnancies. *American Journal of Reproductive Immunology*, *64*, 376–383.

Li, H., Ge, Q., Guo, L., & Lu, Z. (2013). Maternal plasma miRNAs expression in preeclamptic pregnancies. *BioMed Research International*, *2013*, 970265.

Li, W., Mata, K. M., Mazzuca, M. Q., & Khalil, R. A. (2014). Altered matrix metalloproteinase-2 and -9 expression/activity links placental ischemia and anti-angiogenic sFlt-1 to uteroplacental and vascular remodeling and collagen deposition in hypertensive pregnancy. *Biochemical Pharmacology*, *89*, 370–385.

Li, Q., Pan, Z., Wang, X., Gao, Z., Ren, C., & Yang, W. (2014). miR-125b-1-3p inhibits trophoblast cell invasion by targeting sphingosine-1-phosphate receptor 1 in preeclampsia. *Biochemical and Biophysical Research Communications*, *453*, 57–63.

Lillegard, K. E., Johnson, A. C., Lojovich, S. J., Bauer, A. J., Marsh, H. C., Gilbert, J. S., et al. (2013). Complement activation is critical for placental ischemia-induced hypertension in the rat. *Molecular Immunology*, *56*, 91–97.

Liu, Y., Zhao, Y., Yu, A., Zhao, B., Gao, Y., & Niu, H. (2015). Diagnostic accuracy of the soluble Fms-like tyrosine kinase-1/placental growth factor ratio for preeclampsia: A meta-analysis based on 20 studies. *Archives of Gynecology and Obstetrics*, *292*, 507–518.

Llinas, M. T., Alexander, B. T., Seedek, M., Abram, S. R., Crell, A., & Granger, J. P. (2002). Enhanced thromboxane synthesis during chronic reductions in uterine perfusion pressure in pregnant rats. *American Journal of Hypertension, 15*, 793–797.

Lockwood, C. J., Yen, C. F., Basar, M., Kayisli, U. A., Martel, M., Buhimschi, I., et al. (2008). Preeclampsia-related inflammatory cytokines regulate interleukin-6 expression in human decidual cells. *The American Journal of Pathology, 172*, 1571–1579.

Losonczy, G., Singh, J. P., Schoenl, M., Mucha, I., & Venuto, R. C. (1995). Pregnancy enhances the pressor response to thromboxane analogues in rabbits. *The American Journal of Physiology, 269*, R720–R725.

Luft, F. C. (2006). Soluble endoglin (sEng) joins the soluble fms-like tyrosine kinase (sFlt) receptor as a pre-eclampsia molecule. *Nephrology, Dialysis, Transplantation, 21*, 3052–3054.

Luksha, L., Nisell, H., & Kublickiene, K. (2004). The mechanism of EDHF-mediated-responses in subcutaneous small arteries from healthy pregnant women. *American Journal of Physiology. Regulatory, Integrative and Comparative Physiology, 286*, R1102–R1109.

Luksha, L., Nisell, H., Luksha, N., Kublickas, M., Hultenby, K., & Kublickiene, K. (2008). Endothelium-derived hyperpolarizing factor in preeclampsia: Heterogeneous contribution, mechanisms, and morphological prerequisites. *American Journal of Physiology Regulatory, Integrative and Comparative Physiology, 294*, R510–R519.

Luque, A., Farwati, A., Crovetto, F., Crispi, F., Figueras, F., Gratacos, E., et al. (2014). Usefulness of circulating microRNAs for the prediction of early preeclampsia at first-trimester of pregnancy. *Scientific Reports, 4*, 4882.

Magee, L. A., Pels, A., Helewa, M., Rey, E., & von Dadelszen, P. (2014). Diagnosis, evaluation, and management of the hypertensive disorders of pregnancy. *Pregnancy Hypertension, 4*, 105–145.

Majed, B. H., & Khalil, R. A. (2012). Molecular mechanisms regulating the vascular prostacyclin pathways and their adaptation during pregnancy and in the newborn. *Pharmacological Reviews, 64*, 540–582.

Mandala, M., Gokina, N., Barron, C., & Osol, G. (2012). Endothelial-derived hyperpolarization factor (EDHF) contributes to PlGF-induced dilation of mesenteric resistance arteries from pregnant rats. *Journal of Vascular Research, 49*, 43–49.

March, M. I., Geahchan, C., Wenger, J., Raghuraman, N., Berg, A., Haddow, H., et al. (2015). Circulating angiogenic factors and the risk of adverse outcomes among Haitian women with preeclampsia. *PLoS One, 10*, e0126815.

Masoura, S., Kalogiannidis, I., Makedou, K., Theodoridis, T., Koiou, K., Gerou, S., et al. (2014). Biomarkers of endothelial dysfunction in preeclampsia and neonatal morbidity: A case-control study. *European Journal of Obstetrics, Gynecology, and Reproductive Biology, 175*, 119–123.

Masuda, Y., Shimizu, A., Mori, T., Ishiwata, T., Kitamura, H., Ohashi, R., et al. (2001). Vascular endothelial growth factor enhances glomerular capillary repair and accelerates resolution of experimentally induced glomerulonephritis. *The American Journal of Pathology, 159*, 599–608.

Matias, M. L., Romao, M., Weel, I. C., Ribeiro, V. R., Nunes, P. R., Borges, V. T., et al. (2015). Endogenous and uric acid-induced activation of NLRP3 inflammasome in pregnant women with preeclampsia. *PLoS One, 10*, e0129095.

Matsubara, K., Matsubara, Y., Hyodo, S., Katayama, T., & Ito, M. (2010). Role of nitric oxide and reactive oxygen species in the pathogenesis of preeclampsia. *The Journal of Obstetrics and Gynaecology Research, 36*, 239–247.

Maynard, S., Epstein, F. H., & Karumanchi, S. A. (2008). Preeclampsia and angiogenic imbalance. *Annual Review of Medicine, 59*, 61–78.

Maynard, S. E., Min, J. Y., Merchan, J., Lim, K. H., Li, J., Mondal, S., et al. (2003). Excess placental soluble fms-like tyrosine kinase 1 (sFlt1) may contribute to endothelial dysfunction, hypertension, and proteinuria in preeclampsia. *The Journal of Clinical Investigation, 111*, 649–658.

Mazzuca, M. Q., Dang, Y., & Khalil, R. A. (2013). Enhanced endothelin receptor type B-mediated vasodilation and underlying [Ca(2)(+)]i in mesenteric microvessels of pregnant rats. *British Journal of Pharmacology, 169*, 1335–1351.
Mazzuca, M. Q., & Khalil, R. A. (2012). Vascular endothelin receptor type B: Structure, function and dysregulation in vascular disease. *Biochemical Pharmacology, 84*, 147–162.
Mazzuca, M. Q., Li, W., Reslan, O. M., Yu, P., Mata, K. M., & Khalil, R. A. (2014). Downregulation of microvascular endothelial type B endothelin receptor is a central vascular mechanism in hypertensive pregnancy. *Hypertension, 64*, 632–643.
McBride, C. A., Bernstein, I. M., Badger, G. J., Horbar, J. D., & Soll, R. F. (2015). The effect of maternal hypertension on mortality in infants 22, 29 weeks gestation. *Pregnancy Hypertension, 5*, 362–366.
McDonald, S. D., Lutsiv, O., Dzaja, N., & Duley, L. (2012). A systematic review of maternal and infant outcomes following magnesium sulfate for pre-eclampsia/eclampsia in real-world use. *International Journal of Gynaecology and Obstetrics, 118*, 90–96.
McMaster, M. T., Zhou, Y., & Fisher, S. J. (2004). Abnormal placentation and the syndrome of preeclampsia. *Seminars in Nephrology, 24*, 540–547.
Mills, J. L., DerSimonian, R., Raymond, E., Morrow, J. D., Roberts, L. J., 2nd, Clemens, J. D., et al. (1999). Prostacyclin and thromboxane changes predating clinical onset of preeclampsia: A multicenter prospective study. *JAMA, 282*, 356–362.
Mistry, H. D., Kurlak, L. O., & Broughton Pipkin, F. (2013). The placental renin-angiotensin system and oxidative stress in pre-eclampsia. *Placenta, 34*, 182–186.
Mitchell, B. M., Cook, L. G., Danchuk, S., & Puschett, J. B. (2007). Uncoupled endothelial nitric oxide synthase and oxidative stress in a rat model of pregnancy-induced hypertension. *American Journal of Hypertension, 20*, 1297–1304.
Molvarec, A., Czegle, I., Szijarto, J., & Rigo, J., Jr. (2015). Increased circulating interleukin-17 levels in preeclampsia. *Journal of Reproductive Immunology, 112*, 53–57.
Montagnana, M., Lippi, G., Albiero, A., Scevarolli, S., Salvagno, G. L., Franchi, M., et al. (2009). Evaluation of metalloproteinases 2 and 9 and their inhibitors in physiologic and pre-eclamptic pregnancy. *Journal of Clinical Laboratory Analysis, 23*, 88–92.
Morawska, E., Kaczmarek, M. M., & Blitek, A. (2012). Regulation of prostacyclin synthase expression and prostacyclin content in the pig endometrium. *Theriogenology, 78*, 2071–2086.
Moreno-Eutimio, M. A., Tovar-Rodriguez, J. M., Vargas-Avila, K., Nieto-Velazquez, N. G., Frias-De-Leon, M. G., Sierra-Martinez, M., et al. (2014). Increased serum levels of inflammatory mediators and low frequency of regulatory T cells in the peripheral blood of preeclamptic Mexican women. *BioMed Research International, 2014*, 413249.
Morris, R., Spencer, S. K., Kyle, P. B., Williams, J. M., Harris, A., Owens, M. Y., et al. (2015). Hypertension in an animal model of HELLP syndrome is associated with activation of endothelin 1. *Reproductive Sciences, 23*, 42–50.
Morton, J. S., & Davidge, S. T. (2013). Arterial endothelium-derived hyperpolarization: Potential role in pregnancy adaptations and complications. *Journal of Cardiovascular Pharmacology, 61*, 197–203.
Motta, C., Grosso, C., Zanuzzi, C., Molinero, D., Picco, N., Bellingeri, R., et al. (2015). Effect of sildenafil on pre-eclampsia-like mouse model induced by L-NAME. *Reproduction in Domestic Animals, 50*, 611–616.
Murphy, S. R., & Cockrell, K. (2015). Regulation of soluble fms-like tyrosine kinase-1 production in response to placental ischemia/hypoxia: Role of angiotensin II. *Physiological Reports, 3*(2), pii: e12310.
Murphy, S. R., LaMarca, B. B., Cockrell, K., & Granger, J. P. (2010). Role of endothelin in mediating soluble fms-like tyrosine kinase 1-induced hypertension in pregnant rats. *Hypertension, 55*, 394–398.

Myatt, L., & Webster, R. P. (2009). Vascular biology of preeclampsia. *Journal of Thrombosis and Haemostasis, 7*, 375–384.

Naiker, S., Naidoo, S., Naicker, T., Gathiram, P., Nadar, A., & Moodley, J. (2001). Immunolocalisation and endothelin-1 values in pre-eclampsia: An immunocytochemical study. *Journal of Obstetrics and Gynaecology, 21*, 39–45.

Nakamura, M., Sekizawa, A., Purwosunu, Y., Okazaki, S., Farina, A., Wibowo, N., et al. (2009). Cellular mRNA expressions of anti-oxidant factors in the blood of preeclamptic women. *Prenatal Diagnosis, 29*, 691–696.

Nascimento, R. A., Mendes, G., Possomato-Vieira, J. S., Goncalves-Rizzi, V. H., Kushima, H., Delella, F. K., et al. (2015). Metalloproteinase inhibition protects against reductions in circulating adrenomedullin during lead-induced acute hypertension. *Basic & Clinical Pharmacology & Toxicology, 116*, 508–515.

Nelson, S. H., Steinsland, O. S., Wang, Y., Yallampalli, C., Dong, Y. L., & Sanchez, J. M. (2000). Increased nitric oxide synthase activity and expression in the human uterine artery during pregnancy. *Circulation Research, 87*, 406–411.

Noorbakhsh, M., Kianpour, M., & Nematbakhsh, M. (2013). Serum levels of asymmetric dimethylarginine, vascular endothelial growth factor, and nitric oxide metabolite levels in preeclampsia patients. *ISRN Obstetrics and Gynecology, 2013*, 104213.

Noori, M., Donald, A. E., Angelakopoulou, A., Hingorani, A. D., & Williams, D. J. (2010). Prospective study of placental angiogenic factors and maternal vascular function before and after preeclampsia and gestational hypertension. *Circulation, 122*, 478–487.

Novotny, S. R., Wallace, K., Heath, J., Moseley, J., Dhillon, P., Weimer, A., et al. (2012). Activating autoantibodies to the angiotensin II type I receptor play an important role in mediating hypertension in response to adoptive transfer of CD4+ T lymphocytes from placental ischemic rats. *American Journal of Physiology. Regulatory, Integrative and Comparative Physiology, 302*, R1197–R1201.

Nugent, W. H., Mishra, N., Strauss, J. F., 3rd, & Walsh, S. W. (2016). Matrix metalloproteinase 1 causes vasoconstriction and enhances vessel reactivity to angiotensin II via protease-activated receptor 1. *Reproductive Sciences, 23*(4), 542–548.

Orie, N. N., Fry, C. H., & Clapp, L. H. (2006). Evidence that inward rectifier K+ channels mediate relaxation by the PGI2 receptor agonist cicaprost via a cyclic AMP-independent mechanism. *Cardiovascular Research, 69*, 107–115.

Orshal, J. M., & Khalil, R. A. (2004). Reduced endothelial NO-cGMP-mediated vascular relaxation and hypertension in IL-6-infused pregnant rats. *Hypertension, 43*, 434–444.

Ou, M., Dang, Y., Mazzuca, M. Q., Basile, R., & Khalil, R. A. (2014). Adaptive regulation of endothelin receptor type-A and type-B in vascular smooth muscle cells during pregnancy in rats. *Journal of Cellular Physiology, 229*, 489–501.

Palei, A. C., Granger, J. P., & Tanus-Santos, J. E. (2013). Matrix metalloproteinases as drug targets in preeclampsia. *Current Drug Targets, 14*, 325–334.

Palei, A. C., Spradley, F. T., Warrington, J. P., George, E. M., & Granger, J. P. (2013). Pathophysiology of hypertension in pre-eclampsia: A lesson in integrative physiology. *Acta Physiologica (Oxford, England), 208*, 224–233.

Palmer, K. R., Kaitu'u-Lino, T. J., Hastie, R., Hannan, N. J., Ye, L., Binder, N., et al. (2015). Placental-specific sFLT-1 e15a protein is increased in preeclampsia, antagonizes vascular endothelial growth factor signaling, and has antiangiogenic activity. *Hypertension, 66*, 1251–1259.

Papazoglou, D., Galazios, G., Koukourakis, M. I., Kontomanolis, E. N., & Maltezos, E. (2004). Association of -634G/C and 936C/T polymorphisms of the vascular endothelial growth factor with spontaneous preterm delivery. *Acta Obstetricia et Gynecologica Scandinavica, 83*, 461–465.

Parra, M. C., Lees, C., Mann, G. E., Pearson, J. D., & Nicolaides, K. H. (2001). Vasoactive mediator release by fetal endothelial cells in intrauterine growth restriction and pre-eclampsia. *American Journal of Obstetrics and Gynecology, 184*, 497–502.
Parrish, M. R., Ryan, M. J., Glover, P., Brewer, J., Ray, L., Dechend, R., et al. (2011). Angiotensin II type 1 autoantibody induced hypertension during pregnancy is associated with renal endothelial dysfunction. *Gender Medicine, 8*, 184–188.
Peixoto, A. B., Araujo Junior, E., Ribeiro, J. U., Rodrigues, D. B., Castro, E. C., Caldas, T. M., et al. (2016). Evaluation of inflammatory mediators in the deciduas of pregnant women with pre-eclampsia/eclampsia. *The Journal of Maternal-Fetal & Neonatal Medicine, 29*, 75–79.
Peracoli, M. T., Bannwart, C. F., Cristofalo, R., Borges, V. T., Costa, R. A., Witkin, S. S., et al. (2011). Increased reactive oxygen species and tumor necrosis factor-alpha production by monocytes are associated with elevated levels of uric acid in pre-eclamptic women. *American Journal of Reproductive Immunology, 66*, 460–467.
Pereira, R. D., De Long, N. E., Wang, R. C., Yazdi, F. T., Holloway, A. C., & Raha, S. (2015). Angiogenesis in the placenta: The role of reactive oxygen species signaling. *BioMed Research International, 2015*, 814543.
Perez-Sepulveda, A., Torres, M. J., Khoury, M., & Illanes, S. E. (2014). Innate immune system and preeclampsia. *Frontiers in Immunology, 5*, 244.
Pimentel, A. M., Pereira, N. R., Costa, C. A., Mann, G. E., Cordeiro, V. S., de Moura, R. S., et al. (2013). L-Arginine-nitric oxide pathway and oxidative stress in plasma and platelets of patients with pre-eclampsia. *Hypertension Research, 36*, 783–788.
Pinheiro, M. B., Gomes, K. B., Ronda, C. R., Guimaraes, G. G., Freitas, L. G., Teixeira-Carvalho, A., et al. (2015). Severe preeclampsia: Association of genes polymorphisms and maternal cytokines production in Brazilian population. *Cytokine, 71*, 232–237.
Poston, L., McCarthy, A. L., & Ritter, J. M. (1995). Control of vascular resistance in the maternal and feto-placental arterial beds. *Pharmacology & Therapeutics, 65*, 215–239.
Raffetto, J. D., Ross, R. L., & Khalil, R. A. (2007). Matrix metalloproteinase 2-induced venous dilation via hyperpolarization and activation of K+ channels: Relevance to varicose vein formation. *Journal of Vascular Surgery, 45*, 373–380.
Rahardjo, B., Widjajanto, E., Sujuti, H., & Keman, K. (2014). Different levels of IL-1alpha, IL-6, TNF-alpha, NF-kappaB and PPAR-gamma in monocyte cultures exposed by plasma preeclampsia and normotensive pregnancy. *Pregnancy Hypertension, 4*, 187–193.
Rahgozar, S., Amirian, T., Qi, M., Shahshahan, Z., Entezar, E. G. M., Ghasemi Tehrani, H., et al. (2015). Improved assay for quantifying a redox form of angiotensinogen as a biomarker for pre-eclampsia: A case-control study. *PLoS One, 10*, e0135905.
Rahimzadeh, M., Norouzian, M., Arabpour, F., & Naderi, N. (2016). Regulatory T-cells and preeclampsia: An overview of literature. *Expert Review of Clinical Immunology, 12*(2), 209–227.
Ramesar, S. V., Mackraj, I., Gathiram, P., & Moodley, J. (2011). Sildenafil citrate decreases sFlt-1 and sEng in pregnant l-NAME treated Sprague-Dawley rats. *European Journal of Obstetrics, Gynecology, and Reproductive Biology, 157*, 136–140.
Ramma, W., Buhimschi, I. A., Zhao, G., Dulay, A. T., Nayeri, U. A., Buhimschi, C. S., et al. (2012). The elevation in circulating anti-angiogenic factors is independent of markers of neutrophil activation in preeclampsia. *Angiogenesis, 15*, 333–340.
Regal, J. F., Lillegard, K. E., Bauer, A. J., Elmquist, B. J., Loeks-Johnson, A. C., & Gilbert, J. S. (2015). Neutrophil depletion attenuates placental ischemia-induced hypertension in the rat. *PLoS One, 10*, e0132063.
Reslan, O. M., & Khalil, R. A. (2010). Molecular and vascular targets in the pathogenesis and management of the hypertension associated with preeclampsia. *Cardiovascular & Hematological Agents in Medicinal Chemistry, 8*, 204–226.

Rios, D. R., Alpoim, P. N., Godoi, L. C., Perucci, L. O., de Sousa, L. P., Gomes, K. B., et al. (2015). Increased levels of sENG and sVCAM-1 and decreased levels of VEGF in severe preeclampsia. *American Journal of Hypertension,* pii: hpv170. [Epub ahead of print].

Roberts, J. M. (1998). Endothelial dysfunction in preeclampsia. *Seminars in Reproductive Endocrinology, 16*, 5–15.

Roberts, J. M., & Escudero, C. (2012). The placenta in preeclampsia. *Pregnancy Hypertension, 2*, 72–83.

Romero, R., Nien, J. K., Espinoza, J., Todem, D., Fu, W., Chung, H., et al. (2008). A longitudinal study of angiogenic (placental growth factor) and anti-angiogenic (soluble endoglin and soluble vascular endothelial growth factor receptor-1) factors in normal pregnancy and patients destined to develop preeclampsia and deliver a small for gestational age neonate. *The Journal of Maternal-Fetal & Neonatal Medicine, 21*, 9–23.

Rosenberg, T. J., Garbers, S., Lipkind, H., & Chiasson, M. A. (2005). Maternal obesity and diabetes as risk factors for adverse pregnancy outcomes: Differences among 4 racial/ethnic groups. *American Journal of Public Health, 95*, 1545–1551.

Royle, C., Lim, S., Xu, B., Tooher, J., Ogle, R., & Hennessy, A. (2009). Effect of hypoxia and exogenous IL-10 on the pro-inflammatory cytokine TNF-alpha and the anti-angiogenic molecule soluble Flt-1 in placental villous explants. *Cytokine, 47*, 56–60.

Sakashita, T., Higashi, Y., Soga, J., Miyoshi, H., & Kudo, Y. (2014). Circulating endothelial progenitor cells and placental abruption in women with preeclampsia. *Pregnancy Hypertension, 4*, 203–208.

Sanchez-Aranguren, L. C., Prada, C. E., Riano-Medina, C. E., & Lopez, M. (2014). Endothelial dysfunction and preeclampsia: Role of oxidative stress. *Frontiers in Physiology, 5*, 372.

Sandrim, V. C., Palei, A. C., Sertorio, J. T., Cavalli, R. C., Duarte, G., & Tanus-Santos, J. E. (2010). Effects of eNOS polymorphisms on nitric oxide formation in healthy pregnancy and in pre-eclampsia. *Molecular Human Reproduction, 16*, 506–510.

Santillan, M. K., Santillan, D. A., Scroggins, S. M., Min, J. Y., Sandgren, J. A., Pearson, N. A., et al. (2014). Vasopressin in preeclampsia: A novel very early human pregnancy biomarker and clinically relevant mouse model. *Hypertension, 64*, 852–859.

Sasaki, Y., Darmochwal-Kolarz, D., Suzuki, D., Sakai, M., Ito, M., Shima, T., et al. (2007). Proportion of peripheral blood and decidual CD4(+) CD25(bright) regulatory T cells in pre-eclampsia. *Clinical and Experimental Immunology, 149*, 139–145.

Say, L., Chou, D., Gemmill, A., Tuncalp, O., Moller, A. B., Daniels, J., et al. (2014). Global causes of maternal death: A WHO systematic analysis. *Lancet Global Health, 2*, e323–e333.

Scalera, F., Schlembach, D., & Beinder, E. (2001). Production of vasoactive substances by human umbilical vein endothelial cells after incubation with serum from preeclamptic patients. *European Journal of Obstetrics, Gynecology, and Reproductive Biology, 99*, 172–178.

Schiessl, B., Strasburger, C., Bidlingmaier, M., Mylonas, I., Jeschke, U., Kainer, F., et al. (2006). Plasma- and urine concentrations of nitrite/nitrate and cyclic Guanosinemonophosphate in intrauterine growth restricted and preeclamptic pregnancies. *Archives of Gynecology and Obstetrics, 274*, 150–154.

Schramm, A. M., & Clowse, M. E. (2014). Aspirin for prevention of preeclampsia in lupus pregnancy. *Autoimmune Diseases, 2014*, 920467.

Serin, I. S., Ozcelik, B., Basbug, M., Kilic, H., Okur, D., & Erez, R. (2002). Predictive value of tumor necrosis factor alpha (TNF-alpha) in preeclampsia. *European Journal of Obstetrics, Gynecology, and Reproductive Biology, 100*, 143–145.

Shaamash, A. H., Elsnosy, E. D., Makhlouf, A. M., Zakhari, M. M., Ibrahim, O. A., & EL-dien, H. M. (2000). Maternal and fetal serum nitric oxide (NO) concentrations in normal pregnancy, pre-eclampsia and eclampsia. *International Journal of Gynaecology and Obstetrics, 68*, 207–214.

Shah, D. A., & Khalil, R. A. (2015). Bioactive factors in uteroplacental and systemic circulation link placental ischemia to generalized vascular dysfunction in hypertensive pregnancy and preeclampsia. *Biochemical Pharmacology, 95*, 211–226.

Shen, B. Q., Lee, D. Y., & Zioncheck, T. F. (1999). Vascular endothelial growth factor governs endothelial nitric-oxide synthase expression via a KDR/Flk-1 receptor and a protein kinase C signaling pathway. *The Journal of Biological Chemistry, 274*, 33057–33063.

Shesely, E. G., Gilbert, C., Granderson, G., Carretero, C. D., Carretero, O. A., & Beierwaltes, W. H. (2001). Nitric oxide synthase gene knockout mice do not become hypertensive during pregnancy. *American Journal of Obstetrics and Gynecology, 185*, 1198–1203.

Shimokawa, H. (2010). Hydrogen peroxide as an endothelium-derived hyperpolarizing factor. *Pflügers Archiv, 459*, 915–922.

Siddiqui, A. H., Irani, R. A., Zhang, W., Wang, W., Blackwell, S. C., Kellems, R. E., et al. (2013). Angiotensin receptor agonistic autoantibody-mediated soluble fms-like tyrosine kinase-1 induction contributes to impaired adrenal vasculature and decreased aldosterone production in preeclampsia. *Hypertension, 61*, 472–479.

Sierra-Laguado, J., Garcia, R. G., & Lopez-Jaramillo, P. (2006). Flow-mediated dilatation of the brachial artery in pregnancy. *International Journal of Gynaecology and Obstetrics, 93*, 60–61.

Siljee, J. E., Wortelboer, E. J., Koster, M. P., Imholz, S., Rodenburg, W., Visser, G. H., et al. (2013). Identification of interleukin-1 beta, but no other inflammatory proteins, as an early onset pre-eclampsia biomarker in first trimester serum by bead-based multiplexed immunoassays. *Prenatal Diagnosis, 33*, 1183–1188.

Smith-Jackson, K., Hentschke, M. R., Poli-de-Figueiredo, C. E., Pinheiro da Costa, B. E., Kurlak, L. O., Broughton Pipkin, F., et al. (2015). Placental expression of eNOS, iNOS and the major protein components of caveolae in women with pre-eclampsia. *Placenta, 36*, 607–610.

Sones, J. L., Lob, H. E., Isroff, C. E., & Davisson, R. L. (2014). Role of decidual natural killer cells, interleukin-15, and interferon-gamma in placental development and preeclampsia. *American Journal of Physiology. Regulatory, Integrative and Comparative Physiology, 307*, R490–R492.

Spradley, F. T., Palei, A. C., & Granger, J. P. (2015). Increased risk for the development of preeclampsia in obese pregnancies: Weighing in on the mechanisms. *American Journal of Physiology. Regulatory, Integrative and Comparative Physiology, 309*, R1326–R1343.

Stillman, I. E., & Karumanchi, S. A. (2007). The glomerular injury of preeclampsia. *Journal of the American Society of Nephrology, 18*, 2281–2284.

Sunderland, N. S., Thomson, S. E., Heffernan, S. J., Lim, S., Thompson, J., Ogle, R., et al. (2011). Tumor necrosis factor alpha induces a model of preeclampsia in pregnant baboons (Papio hamadryas). *Cytokine, 56*, 192–199.

Tal, R. (2012). The role of hypoxia and hypoxia-inducible factor-1alpha in preeclampsia pathogenesis. *Biology of Reproduction, 87*, 134.

Tanbe, A. F., & Khalil, R. A. (2010). Circulating and vascular bioactive factors during hypertension in pregnancy. *Current Bioactive Compounds, 6*, 60–75.

Taylor, R. N., Varma, M., Teng, N. N., & Roberts, J. M. (1990). Women with preeclampsia have higher plasma endothelin levels than women with normal pregnancies. *The Journal of Clinical Endocrinology and Metabolism, 71*, 1675–1677.

Thadhani, R., Hagmann, H., Schaarschmidt, W., Roth, B., Cingoez, T., Karumanchi, S. A., et al. (2016). Removal of soluble fms-like tyrosine kinase-1 by dextran sulfate apheresis in preeclampsia. *Journal of the American Society of Nephrology: JASN, 27*(3), 903–913.

Thornburg, K. L., Jacobson, S. L., Giraud, G. D., & Morton, M. J. (2000). Hemodynamic changes in pregnancy. *Seminars in Perinatology, 24*, 11–14.

Trifonova, E. A., Gabidulina, T. V., Ershov, N. I., Serebrova, V. N., Vorozhishcheva, A. Y., & Stepanov, V. A. (2014). Analysis of the placental tissue transcriptome of normal and preeclampsia complicated pregnancies. *Acta Naturae*, *6*, 71–83.

Trowsdale, J., & Moffett, A. (2008). NK receptor interactions with MHC class I molecules in pregnancy. *Seminars in Immunology*, *20*, 317–320.

Tsatsaris, V., Goffin, F., Munaut, C., Brichant, J. F., Pignon, M. R., Noel, A., et al. (2003). Overexpression of the soluble vascular endothelial growth factor receptor in preeclamptic patients: Pathophysiological consequences. *The Journal of Clinical Endocrinology and Metabolism*, *88*, 5555–5563.

Tsukimori, K., Fukushima, K., Tsushima, A., & Nakano, H. (2005). Generation of reactive oxygen species by neutrophils and endothelial cell injury in normal and preeclamptic pregnancies. *Hypertension*, *46*, 696–700.

Turpin, C. A., Sakyi, S. A., Owiredu, W. K., Ephraim, R. K., & Anto, E. O. (2015). Association between adverse pregnancy outcome and imbalance in angiogenic regulators and oxidative stress biomarkers in gestational hypertension and preeclampsia. *BMC Pregnancy and Childbirth*, *15*, 189.

Tuzcu, Z. B., Asicioglu, E., Sunbul, M., Ozben, B., Arikan, H., & Koc, M. (2015). Circulating endothelial cell number and markers of endothelial dysfunction in previously preeclamptic women. *American Journal of Obstetrics and Gynecology*, *213*(533), e531–e537.

Valdes, G., Kaufmann, P., Corthorn, J., Erices, R., Brosnihan, K. B., & Joyner-Grantham, J. (2009). Vasodilator factors in the systemic and local adaptations to pregnancy. *Reproductive Biology and Endocrinology*, *7*, 79.

van der Heijden, O. W., Essers, Y. P., Fazzi, G., Peeters, L. L., De Mey, J. G., & van Eys, G. J. (2005). Uterine artery remodeling and reproductive performance are impaired in endothelial nitric oxide synthase-deficient mice. *Biology of Reproduction*, *72*, 1161–1168.

van Dijk, M., & Oudejans, C. B. (2011). STOX1: Key player in trophoblast dysfunction underlying early onset preeclampsia with growth retardation. *Journal of Pregnancy*, *2011*, 521826.

van Dijk, M., & Oudejans, C. (2013). (Epi)genetics of pregnancy-associated diseases. *Frontiers in Genetics*, *4*, 180.

VanWijk, M. J., Kublickiene, K., Boer, K., & VanBavel, E. (2000). Vascular function in preeclampsia. *Cardiovascular Research*, *47*, 38–48.

Velloso, E. P., Pimentel, R. L., Braga, J. F., Cabral, A. C., Reis, Z. S., Bader, M., et al. (2015). Identification of a novel agonist-like autoantibody in preeclamptic patients. *American Journal of Hypertension*, *29*(3), 405–412.

Venkatesha, S., Toporsian, M., Lam, C., Hanai, J., Mammoto, T., Kim, Y. M., et al. (2006). Soluble endoglin contributes to the pathogenesis of preeclampsia. *Nature Medicine*, *12*, 642–649.

Verdonk, K., Saleh, L., Lankhorst, S., Smilde, J. E., van Ingen, M. M., Garrelds, I. M., et al. (2015). Association studies suggest a key role for endothelin-1 in the pathogenesis of preeclampsia and the accompanying renin-angiotensin-aldosterone system suppression. *Hypertension*, *65*, 1316–1323.

Verlohren, S., Niehoff, M., Hering, L., Geusens, N., Herse, F., Tintu, A. N., et al. (2008). Uterine vascular function in a transgenic preeclampsia rat model. *Hypertension*, *51*, 547–553.

von Dadelszen, P., Dwinnell, S., Magee, L. A., Carleton, B. C., Gruslin, A., Lee, B., et al. (2011). Sildenafil citrate therapy for severe early-onset intrauterine growth restriction. *BJOG*, *118*, 624–628.

Wallace, K., Richards, S., Dhillon, P., Weimer, A., Edholm, E. S., Bengten, E., et al. (2011). CD4+ T-helper cells stimulated in response to placental ischemia mediate hypertension during pregnancy. *Hypertension*, *57*, 949–955.

Walsh, S. W. (2004). Eicosanoids in preeclampsia. *Prostaglandins, Leukotrienes, and Essential Fatty Acids, 70*, 223–232.

Wang, K., Ahmad, S., Cai, M., Rennie, J., Fujisawa, T., Crispi, F., et al. (2013). Dysregulation of hydrogen sulfide producing enzyme cystathionine gamma-lyase contributes to maternal hypertension and placental abnormalities in preeclampsia. *Circulation, 127*, 2514–2522.

Wang, W., Parchim, N. F., Iriyama, T., Luo, R., Zhao, C., Liu, C., et al. (2014). Excess LIGHT contributes to placental impairment, increased secretion of vasoactive factors, hypertension, and proteinuria in preeclampsia. *Hypertension, 63*, 595–606.

Wang, Y., Walsh, S. W., & Kay, H. H. (1992). Placental lipid peroxides and thromboxane are increased and prostacyclin is decreased in women with preeclampsia. *American Journal of Obstetrics and Gynecology, 167*, 946–949.

Wang, L., Yang, T., Ding, Y., Zhong, Y., Yu, L., & Peng, M. (2015). Chemerin plays a protective role by regulating human umbilical vein endothelial cell-induced nitric oxide signaling in preeclampsia. *Endocrine, 48*, 299–308.

Wei, Y. M., Yang, H. X., Zhu, W. W., Liu, X. Y., Meng, W. Y., Wang, Y. Q., et al. (2015). Risk of adverse pregnancy outcomes stratified for pre-pregnancy body mass index. *The Journal of Maternal-Fetal & Neonatal Medicine*, 1–5. [Epub ahead of print].

Weigand, L., Sylvester, J. T., & Shimoda, L. A. (2006). Mechanisms of endothelin-1-induced contraction in pulmonary arteries from chronically hypoxic rats. *American Journal of Physiology. Lung Cellular and Molecular Physiology, 290*, L284–L290.

Weiner, C. P., Thompson, L. P., Liu, K. Z., & Herrig, J. E. (1992). Endothelium-derived relaxing factor and indomethacin-sensitive contracting factor alter arterial contractile responses to thromboxane during pregnancy. *American Journal of Obstetrics and Gynecology, 166*, 1171–1178. discussion 1179–1181.

Williams, D. J., Vallance, P. J., Neild, G. H., Spencer, J. A., & Imms, F. J. (1997). Nitric oxide-mediated vasodilation in human pregnancy. *The American Journal of Physiology, 272*, H748–H752.

Xia, Y., Ramin, S. M., & Kellems, R. E. (2007). Potential roles of angiotensin receptor-activating autoantibody in the pathophysiology of preeclampsia. *Hypertension, 50*, 269–275.

Yang, J., Li, L., Shang, J. Y., Cai, L., Song, L., Zhang, S. L., et al. (2015). Angiotensin II type 1 receptor autoantibody as a novel regulator of aldosterone independent of preeclampsia. *Journal of Hypertension, 33*, 1046–1056.

Yang, X., Wang, F., Lau, W. B., Zhang, S., Liu, H., & Ma, X. L. (2014). Autoantibodies isolated from preeclamptic patients induce endothelial dysfunction via interaction with the angiotensin II AT1 receptor. *Cardiovascular Toxicology, 14*, 21–29.

Yi, F. X., Boeldt, D. S., Gifford, S. M., Sullivan, J. A., Grummer, M. A., Magness, R. R., et al. (2010). Pregnancy enhances sustained Ca2+ bursts and endothelial nitric oxide synthase activation in ovine uterine artery endothelial cells through increased connexin 43 function. *Biology of Reproduction, 82*, 66–75.

Yinon, Y., Ben Meir, E., Margolis, L., Lipitz, S., Schiff, E., Mazaki-Tovi, S., et al. (2015). Low molecular weight heparin therapy during pregnancy is associated with elevated circulatory levels of placental growth factor. *Placenta, 36*, 121–124.

Yinon, Y., Nevo, O., Xu, J., Many, A., Rolfo, A., Todros, T., et al. (2008). Severe intrauterine growth restriction pregnancies have increased placental endoglin levels: Hypoxic regulation via transforming growth factor-beta 3. *The American Journal of Pathology, 172*, 77–85.

Yoshida, A., Nakao, S., Kobayashi, M., & Kobayashi, H. (2000). Flow-mediated vasodilation and plasma fibronectin levels in preeclampsia. *Hypertension, 36*, 400–404.

Yu, Y., Wang, L., Liu, T., & Guan, H. (2015). MicroRNA-204 suppresses trophoblast-like cell invasion by targeting matrix metalloproteinase-9. *Biochemical and Biophysical Research Communications, 463*, 285–291.

Yuksel, M. A., Tuten, A., Oncul, M., Acikgoz, A. S., Temel Yuksel, I., Toprak, M. S., et al. (2015). Serum endocan concentration in women with pre-eclampsia. *Archives of Gynecology and Obstetrics*, *292*, 69–73.

Zarate, A., Saucedo, R., Valencia, J., Manuel, L., & Hernandez, M. (2014). Early disturbed placental ischemia and hypoxia creates immune alteration and vascular disorder causing preeclampsia. *Archives of Medical Research*, *45*, 519–524.

Zawiejska, A., Wender-Ozegowska, E., Iciek, R., & Brazert, J. (2014). Concentrations of endothelial nitric oxide synthase, angiotensin-converting enzyme, vascular endothelial growth factor and placental growth factor in maternal blood and maternal metabolic status in pregnancy complicated by hypertensive disorders. *Journal of Human Hypertension*, *28*, 670–676.

Zhao, S., Gu, Y., Lewis, D. F., & Wang, Y. (2008). Predominant basal directional release of thromboxane, but not prostacyclin, by placental trophoblasts from normal and preeclamptic pregnancies. *Placenta*, *29*, 81–88.

Zhao, Y., Zheng, Y. F., Luo, Q. Q., Yan, T., Liu, X. X., Han, L., et al. (2014). Edaravone inhibits hypoxia-induced trophoblast-soluble Fms-like tyrosine kinase 1 expression: A possible therapeutic approach to preeclampsia. *Placenta*, *35*, 476–482.

Zhou, C. C., Zhang, Y., Irani, R. A., Zhang, H., Mi, T., Popek, E. J., et al. (2008). Angiotensin receptor agonistic autoantibodies induce pre-eclampsia in pregnant mice. *Nature Medicine*, *14*, 855–862.

Zhu, X., Wu, S., Dahut, W. L., & Parikh, C. R. (2007). Risks of proteinuria and hypertension with bevacizumab, an antibody against vascular endothelial growth factor: Systematic review and meta-analysis. *American Journal of Kidney Diseases*, *49*, 186–193.

Zhu, J., Zhong, M., Pang, Z., & Yu, Y. (2014). Dysregulated expression of matrix metalloproteinases and their inhibitors may participate in the pathogenesis of pre-eclampsia and fetal growth restriction. *Early Human Development*, *90*, 657–664.

INDEX

Note: Page numbers followed by "*f*" indicate figures, and "*t*" indicate tables.

D

FSC
www.fsc.org
MIX
Papier | Fördert gute Waldnutzung
FSC® C083411

Kolibri 360 GmbH
Ferdinand-Jühlke-Straße 7
99095 Erfurt, Deutschland
produktsicherheit@kolibri360.de